Nanoparticles in Biomedical Imaging

FUNDAMENTAL BIOMEDICAL TECHNOLOGIES

Series Editor:
Mauro Ferrari, Ph.D. Houston, TX.

Volume 1: Martin, Donald. Nanobiotechnology of Biomimetic Membranes. 978-0-387-37738-4. 2006

Volume 2: Cristini, V., M. Ferrari, and P. Decuzzi. Nanoparticulate Delivery to Cancerous Lesions. 978-0-387-29085-0. Forthcoming

Volume 3: Bulte, J.W., and Mike Modo. Nanoparticles in Biomedical Imaging. 978-0-387-72026-5. 2007

Nanoparticles in Biomedical Imaging

Emerging Technologies and Applications

Edited by

Jeff W.M. Bulte
The Johns Hopkins University School of Medicine
Baltimore, USA

and

Michel M.J. Modo
Institute of Psychiatry, King's College London
London, United Kingdom

Springer

Jeff W.M. Bulte
The Johns Hopkins University
School of Medicine
Department of Radiology
720 Rutland Avenue
Baltimore, 21205
jwmbulte@mri.jhu.edu

Michel M.J. Modo
Institute of Psychiatry
King's College London
Center for the Cellular Basis of Behaviour
125 Coldharbour Lane, London SE5 9NU
United Kingdom
Mike.Modo@iop.kcl.ac.uk

ISBN 978-1-4419-2462-9 e-ISBN 978-0-387-72027-2

Printed on acid-free paper.

9 8 7 6 5 4 3 2 1

springer.com

Contents

Part 5 Quantum Dots: Applications in Optical Imaging

Preface

Within a short time frame, diagnostic nanomedicine has emerged as a multidisciplinary field that promises to deliver a more specific and sensitive detection of early disease processes. Nanoparticles (particles measuring approximately 1–100 nm) allow incorporation of a high number of detectable tracers or molecules through which they can amplify changes in signal intensity. This book contains 23 chapters dedicated to the use of nanoparticles in biomedical imaging, covering all aspects including the chemistry of nanoparticle synthesis, the physics of contrast and signal-generating mechanisms, the biological and pharmacokinetic profiles, and their use in pre-clinical animal models as well as humans. Experts throughout the world have made contributions for five major imaging modalities, namely the use of nanoparticles in magnetic resonance imaging, nuclear medicine, ultrasound imaging, computed tomography imaging, and optical imaging. We hope that this book may form a foundation for further advances in the field of nanomedicine and diagnostics.

Jeff W.M. Bulte and Michel M.J. Modo
Baltimore and London,
January 2007

List of Contributors

Jelle O. Barentsz
Department of Radiology, UMCN, Nijmegen, The Netherlands
J.Barentsz@rad.umcn.nl

Susannah H. Bloch
Department of Biomedical Engineering, University of California, Davis, CA, USA
susie_bloch@mac.com

Otto C. Boerman
Department of Nuclear Medicine, Radboud University Nijmegen Medical Centre, Nijmegen, The Netherlands
o.boerman@nucmed.umcn.nl

Jeff W.M. Bulte
Russell H. Morgan Department of Radiology and Radiological Science, Division of MR Research and the Institute for Cell Engineering, The Johns Hopkins University School of Medicine, Baltimore, MD, USA
jwmbulte@mri.jhu.edu

Shelton D. Caruthers
Philips Medical Systems, Cleveland, OH, USA
scaruthers@cmrl.wustl.edu

K.B.J.M. Cleutjens
Cardiovascular Research Institute Maastricht (CARIM), Department of Pathology, Maastricht University Hospital/Maastricht University, Maastricht, The Netherlands
kitty.cleutjens@PATH.unimaas.nl

M.J.A.P. Daemen
Cardiovascular Research Institute Maastricht (CARIM), Department of Pathology, Maastricht University Hospital/Maastricht University, Maastricht, The Netherlands
Mat.Daemen@PATH.unimaas.nl

Maxime Dahan
Laboratoire Kastler Brossel, Physics and Biology Department, Ecole normale supérieure, Paris, France
maxime.dahan@lkb.ens.fr

Celso De Mello Donega
Debye Institute, Condensed Matter and Interfaces, Utrecht University, Utrecht, The Netherlands
C.deMelloDonega@phys.uu.nl

Debamitra Dutta
Department of Materials Science and Engineering and Particle Engineering Research Center, University of Florida, Gainesville, FL, USA

Tamer Elbayoumi
Department of Pharmaceutical Sciences and Center for Pharmaceutical Biotechnology and Nanomedicine, Northeastern University, Boston, MA, USA
timmyaziz@yahoo.com

J.M.A. van Engelshoven
Cardiovascular Research Institute Maastricht (CARIM), Department of Radiology, Maastricht University Hospital/Maastricht University, Maastricht, The Netherlands
Jve@radiology.azm.nl

Dardo E. Ferrara
Division of Cardiology, Department of Medicine, Emory University, Atlanta, GA, USA
dferra2@emory.edu

Xiaohu Gao
Department of Bioengineering, University of Washington, Seattle, WA, USA
Xgao@u.washington.edu

Charles Glaus
Department of Biomedical Engineering, Emory University and Georgia Institute of Technology, Atlanta, GA, USA
glaus@gatech.edu

Samuel C. Grant
Chemical and Biomedical Engineering, The Florida State University & The National High Magnetic Field Laboratory Center for Interdisciplinary Magnetic Resonance, FL, USA
grant@magnet.fsu.edu

Arjan W. Griffioen
Angiogenesis Laboratory, Research Institute for Growth and Development, Department of Pathology/Internal Medicine, Maastricht University & University Hospital, Maastricht, The Netherlands
aw.griffioen@PATH.unimaas.nl

Seza A. Gulec
Surgical Oncology, Goshen Center for Cancer Care, Goshen, IN, USA
sgulec@goshenhealth.com

S. Heeneman
Cardiovascular Research Institute Maastricht (CARIM), Department of Pathology, Maastricht University Hospital/Maastricht University, Maastricht, The Netherlands
s.Heeneman@PATH.unimass.nl

Thilo Hölscher
Department of Radiology, University of California, San Diego, CA, USA
thoelscher@ucsd.edu

Charalambos Kaittanis
Nanoscience Technology Center,. University of Central Florida, Orlando, FL, USA
jmperez@mail.ucf.edu

Patrick H. Kee
The University of Texas Health Science Center at Houston, Houston, TX, USA

Alexander L. Klibanov Cardiovascular Division, Department of Medicine, University of Virginia, Charlottesville, VA, USA
ALK6N@hscmail.mcc.virginia.edu

Hisataka Kobayashi
Molecular Imaging Program, Center for Cancer Research, National Cancer Institute, NIH, Bethesda, MD, USA
kobayash@mail.nih.gov

Alan P. Koretsky
Laboratory of Functional and Molecular Imaging, National Institute of Neurological Disorders and Stroke, National Institutes of Health, Bethesda, MD 20892

Yuko Kono
Department of Medicine, University of California, San Diego, CA, USA
ykono@ucsd.edu

Marianne Eline Kooi
Cardiovascular Research Institute Maastricht (CARIM), Department of Radiology, Maastricht University Hospital, Maastricht, The Netherlands
ekoo@rdia.azm.nl

Rolf Koole
Debye Institute, Condensed Matter and Interfaces, Utrecht University, Utrecht, The Netherlands
R.koole@phys.uu.nl

Gregory M. Lanza
Department of Medicine, Cardiovascular Division, Washington University School of Medicine, St. Louis, MO, USA
greg@cmrl.wustl.edu

Peter Laverman
Department of Nuclear Medicine, Radboud University Nijmegen Medical Centre, Nijmegen, The Netherlands
P.Laverman@nucmed.umcn.nl

Robert F Mattrey
Department of Radiology, University of California, San Diego, CA, USA
rmattrey@ucsd.edu

Hui Mao
Department of Radiology, Emory University, Atlanta, GA, USA
hmao@emory.edu

David D. McPherson
The University of Texas Health Science Center at Houston, Houston, TX, USA
david.d.mcpherson@uth.tmc.edu

Michel M.J. Modo
Neuroimaging Research Group and Center for the Cellular Basis of Behaviour, Institute of Psychiatry, King's College London, London, UK
mike.modo@iop.kcl.ac.uk

Frederick L. Moffat
Division of Surgical Oncology, Daughtry Family Department of Surgery, University of Miami Miller School of Medicine and Sylvester Comprehensive Cancer Center, Miami, FL, USA
fmoffat@med.miami.edu

Willem J.M. Mulder
Sinai Translational and Molecular Imaging Institute and Imaging Science Laboratories, Mount Sinai School of Medicine, New York, NY, USA
wjmmulder@gmail.com

Klaas Nicolay
Biomedical NMR, Department of Biomedical Engineering, Eindhoven University of Technology, Eindhoven, The Netherlands
K.nicolay@tue.nl

Shuming Nie
Departments of Biomedical Engineering, Chemistry, Hematology and Oncology, and the Winship Cancer Institute, Emory University and Georgia Institute of Technology, Atlanta, GA, USA
snie@emory.edu

Anne M. Neubauer
Department of Medicine, Cardiovascular Division, Washington University School of Medicine, St. Louis, MO, USA
aneubauer@cmrl.wustl.edu

J. Manuel Perez
Nanoscience Technology Center, University of Central Florida, Orlando, FL, USA
jmperez@mail.ucf.edu

Matthew N. Rhyner
Departments of Biomedical Engineering, Chemistry, Hematology and Oncology, and the Winship Cancer Institute, Emory University and Georgia Institute of Technology, Atlanta, GA, USA
matt.rhyner@gatech.edu

Andreas Sachse
Bayer Schering Pharma, Global Research & Development, Berlin, Germany
andreas.sachse@bayerhealthcare.com

Swadeshmukul Santra
Nanoscience Technology Center, and Department of Chemistry and Biomolecular Science Center, University of Central Florida, Orlando, FL, USA
ssantra@mail.ucf.edu

Eyk Schellenberger
Institut für Radiologie, Charité – Universitätsmedizin Berlin, Campus Charité Mitte, Berlin, Germany
eyk.schellenberger@charite.de

Anne H. Schmieder
Department of Medicine, Cardiovascular Division, Washington University School of Medicine, St. Louis, MO, USA
Anne@cmrl.wustl.edu

Erik M. Shapiro
Molecular and Cellular Imaging Laboratory, Department of Diagnostic Radiology, Yale University School of Medicine, New Haven, CT, USA
erik.shapiro@yale.edu

Andrew M. Smith
Departments of Biomedical Engineering, Chemistry, Hematology and Oncology, and the Winship Cancer Institute, Emory University and Georgia Institute of Technology, Atlanta, GA, USA
gte402e@prism.gatech.edu

Gert Storm
Utrecht Institute for Pharmaceutical Sciences (UIPS), Utrecht University, Utrecht, The Netherlands
g.storm@pharm.uu.nl

Gustav J. Strijkers
Biomedical NMR, Department of Biomedical Engineering, Eindhoven University of Technology, Eindhoven, The Netherlands
g.j.strijkers@tue.nl

Akihiro Tanimoto
Department of Diagnostic Radiology, School of Medicine, Keio University, Shinjuku-ku, Tokyo, Japan
t-mri@tt.rim.or.jp

W. Robert Taylor
Division of Cardiology, Department of Medicine, Emory University; Department of Biomedical Engineering, Emory University and Georgia Institute of Technology; Veterans Affairs Medical Center, Atlanta, GA, USA
wtaylor@emory.edu

Paris P. Tekkis
Department of Biosurgery and Surgical Technology, St Mary's Hospital, Imperial College, London, UK
p.tekkis@imperial.ac.uk

Bijoy D. Thattaliyath
Powell Gene Therapy Center, University of Florida, Gainesville, FL,USA
bijoydt@gtc.ufl.edu

Vladimir Torchilin
Department of Pharmaceutical Sciences and Center for Pharmaceutical Biotechnology and Nanomedicine, Northeastern University, Boston, MA, USA
v.torchilin@neu.edu

Glenn A. Walter
Physiology and Functional Genomics, Powell Gene Therapy Center; University of Florida, National High Magnetic Field Laboratory, Gainesville, FL, USA
glennw@phys.med.ufl.edu

Samuel A. Wickline
Department of Medicine, Cardiovascular Division, Washington University School of Medicine, St. Louis, MO, USA
saw@howdy.wustl.edu

Patrick M. Winter
Department of Medicine, Cardiovascular Division, Washington University School of Medicine, St. Louis, MO, USA
patrick@cmrl.wustl.edu

Erik R. Wisner
Department of Surgical and Radiological Sciences, School of Veterinary Medicine, University of California, Davis, CA, USA
erwisner@ucdavis.edu

Lily Yang
Department of Surgery and Winship Cancer Institute, Emory University, Atlanta, GA, USA
lyangoz@emory.edu

1
Introduction: The Emergence of Nanoparticles as Imaging Platform in Biomedicine

Jeff. W.M. Bulte and Michel M.J. Modo

Abstract: Nanoparticles have now emerged as widely used diagnostic agents in biomedicine. They possess unique features that can be exploited with imaging. Among these are an increased sensitivity in detection through amplification of signal changes; a manipulation of surface structures and enhancement of uptake by selected cells (macrophages); altered pharmacokinetics allowing, for instance, their use as blood pool imaging agents; and physicochemical manipulations of energy (i.e., quantum dots). Specific aspects of nanoparticles are introduced to illustrate their use in biomedical imaging.

1.1. Use of Nanoparticles in Biomedical Imaging: A Historical Perspective

Along with the development of new imaging modalities that need contrast agents or tracers for (improved) visualization, it was realized that nanoparticles can form an important class of materials with unique features suitable for biomedical imaging. They can dramatically either amplify signal changes (e.g., (super)paramagnetic particles in magnetic resonance imaging), exhibit properties that induce phagocytosis and selective uptake by macrophages (e.g., liposomes in nuclear medicine and iron oxides in MR imaging) or posses unique chemical properties that makes them visible (e.g., gold particles in electron microscopy and quantum dots in fluorescent imaging). Thus, different particles are being designed to serve different purposes, i.e. (1) to enhance sensitivity ("more bang for the buck"), (2) to enhance cellular uptake, or (3) to enable modulation of energy (light, sound, and electron beams).

Historically speaking, one of the first functional uses of nanoparticles in imaging was with electron microscopy. Natural horse spleen ferritin (Rifkind et al., 1960; Singer and Schick, 1961) and synthetic colloidal gold particles (Horisberger and Rosset, 1977), both containing metallic (oxide) cores that are electron-dense, were used for the specific marking of subcellular structures after their conjugation to antibodies or selected (glyco)proteins, including lectins. For non-invasive imaging, one of the first studies involved the use of radioactive-labeled particles for myocardial perfusion imaging (Ashburn et al., 1971), setting a precedent for nanoparticle use in biomedicine. Sulfur (technetium) colloids are also among these earlier studies, with applications in lymph node imaging and cancer staging, following the use of colloidal gold (Fee et al., 1978).

Although magnetic nanoparticles were reported to have the ability to modulate MR relaxation times in the late 1970s (Ohgushi et al., 1978), their applicability as MR contrast agents were not explored until the mid-1980s (Mendonca Dias and Lauterbur, 1986; Renshaw et al., 1986a,b). Also during this time, liposomes entered the arena of non-invasive imaging as paramagnetic agents for MR imaging (Unger et al., 1988), iodinated agents for computed tomography (Havron et al., 1981; Ryan et al., 1983), and radioactive agents for nuclear medicine (Morgan et al., 1981). In addition, ultrasound bubbles were introduced for sonographic purposes (Widder and Simeone, 1986). In the 1990s, quantum dots were introduced for fluorescent imaging (Alivisatos, 1996; Chan and Nie, 1998) (Figure 1.1) including non-invasive imaging in rodents.

1.2. The Current Status: Promises and Pitfalls

Nanoparticles are now widely used, with many preparations commercially available. Several of these are being used clinically and have undergone FDA approval. An example of this is the superparamagnetic iron oxide (SPIO) liver agent Feridex® that gained approval in 1996. For many particles, off-label uses have been implemented spurring interest in their use from the wider biomedical community. The production of nanoparticles, however, is not straightforward. Nanoparticles are most commonly composed of multiple agents and polymers that convey their specific features. Proper synthesis of these is not trivial with many proprietary procedures that are difficult to access.

It is one thing to make a proper formulation of a specific nanoparticle that is suitable for a particular application, but to make this stable with a prolonged shelf life is another story. Particle aggregation and loss of colloidal properties is a major problem that is often encountered. For SPIO particles, for instance, it is key to add excess citrate and dextran during storage to "catch" iron atoms that are leaching out of the particle in addition to preventing a loss of the dextran coating. Storage procedures and other details that appear of minor importance are actually critical, but often cannot be found in published procedures.

The stability of nanoparticles is a feature that regulatory agencies scrutinize before their use in the clinic can be realized. At the one hand, even release of trace amounts of potentially toxic metals at certain pHs and semi-physiological conditions are of overall concern. This is also true for iron which although biologically compatible can be potentially harmful depending on its biochemical microenvironment and oxidation state. At the other hand, the particles ideally should be biodegradable or have proper elimination routes. As examples, SPIO particles are biodegradable, whereas ^{19}F nanoemulsions are not (they are biologically inert), although they can be eliminated by a simple exhalation and passage of gas (Bulte, 2005). It is almost certain that particles that are inert and not eliminated are unlikely to obtain clinical approval. Bangs particles (Shapiro et al., 2005) and quantum dots fall into this category.

1.3. Future Challenges for Further Advancement

Future advancement in biomedical imaging will likely involve a more specific approach for targeting nanoparticles to the molecules of interest. The challenge here is to minimize non- specific uptake (i.e., uptake by macrophages and other cells of the reticulo-endothelial system), unless, of course, the goal is to visualize the presence or absence of macrophage-type cells in certain disease processes. Particles should be sufficiently small to penetrate endothelial barriers, but should have long- circulating blood half lives in order to facilitate an accumulation at the target. This could be achieved by a manipulation of the surface charge or an

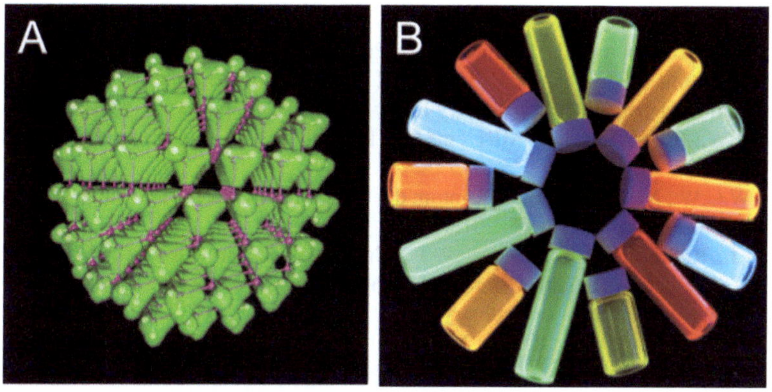

FIGURE 1.1. Quantum dot particles come in a variety of fluorescent colors depending on the size of the semiconductor crystal. The color is determined by the size of the particle and is thus tunable by synthesis. The size of the quantum dot dictates the band gap, with the band gap being the energy required to lift an electron from its valence band, filled with electrons, to its conduction band, which is empty. When an electron falls from the conduction band back down to the valence band, eliminating a hole, the lost energy is emitted as light whose color then corresponds to the band gap. The quantum dot in (**A**) is composed of GaAs and those in (**B**) are composed of CdSe and CdTe. In (**A**), green represents the total electron charge density for a quantum dot of 465 atoms. Image (**A**) is taken from http://www.lbl.gov/Science-Articles/Archive/Quantum-Dot-Electronics.html and image (**B**) from http://www.rsc.org/chemistryworld/News/2005/September/19090501.asp.

incorporation of stealth coatings. Yet, nanoparticles should be large enough to enable amplification of signal changes upon target binding.

Ideally, nanoparticle formulations should be stable for prolonged times and composed of materials that are biocompatible or biodegradable, unless the concentrations are so low that the tracer principle applies. A relatively new area is the use of diagnostic "smart" agents that can act as sensors for probing the enzyme/protein environment. These beacons may shed more light on how certain tumor processes take place. New types of diagnostic nanoparticles are constantly being developed. Some involve new agents for older imaging techniques (e.g., radiopaque bismuth particles for CT imaging) (Rabin et al., 2006), and older formulations for new imaging techniques (e.g., liposomes for CEST MR imaging) (Aime et al., 2005). With more variations and integrations of imaging techniques on the horizon, the future of nanoparticles as an imaging platform is bright and mighty. The following chapters each cover current applications of nanoparticle imaging agents, which may pave the way toward further refinements and innovations in biomedical imaging.

References

Aime, S., Castelli, D.D., Terreno, E., 2005. Highly sensitive MRI chemical exchange saturation transfer agents using liposomes. Angew Chem Int Ed Engl 44, 5513–5515.

Alivisatos, A.P., 1996. Semiconductor clusters, nanocrystals, and quantum dots. Science 271, 933–937.

Ashburn, W.L., Braunwald, E., Simon, A.L., Peterson, K.L., Gault, J.H., 1971. Myocardial perfusion imaging with radioactive-labeled particles injected directly into the coronary circulation of patients with coronary artery disease. Circulation 44, 851–865.

Bulte, J.W., 2005. Hot spot MRI emerges from the background. Nat Biotechnol 23, 945–946.

Chan, W.C.W., Nie, S.M., 1998. Quantum dot bioconjugates for ultrasensitive nonisotopic detection. Science 281, 2016–2018.

Fee, H.J., Robinson, D.S., Sample, W.F., Graham, L.S., Holmes, E.C., Morton, D.L., 1978. The determination of lymph shed by colloidal gold scanning in patients with malignant melanoma: a preliminary study. Surgery 84, 626–632.

Havron, A., Seltzer, S.E., Davis, M.A., Shulkin, P., 1981. Radiopaque liposomes: a promising new contrast material for computed tomography of the spleen. Radiology 140, 507–511.

Horisberger, M., Rosset, J., 1977. Colloidal gold, a useful marker for transmission and scanning electron microscopy. J Histochem Cytochem 25, 295–305.

Mendonca Dias, M.H., Lauterbur, P.C., 1986. Ferromagnetic particles as contrast agents for magnetic resonance imaging of liver and spleen. Magn Reson Med 3, 328–330.

Morgan, J.R., Williams, K.E., Davies, R.L., Leach, K., Thomson, M., Williams, L.A., 1981. Localisation of experimental staphylococcal abscesses by 99MTC-technetium-labelled liposomes. J Med Microbiol 14, 213–217.

Ohgushi, M., Nagayama, K., Wada, A., 1978. Dextran-magnetite: new relaxation agent and its application to T2 measurements in gel. J Magn Reson 29, 599–601.

Rabin, O., Manuel Perez, J., Grimm, J., Wojtkiewicz, G., Weissleder, R., 2006. An X-ray computed tomography imaging agent based on long-circulating bismuth sulphide nanoparticles. Nat Mater 5, 118–122.

Renshaw, P.F., Owen, C.S., Evans, A.E., Leigh, J.S., Jr., 1986a. Immunospecific NMR contrast agents. Magn Reson Imag 4, 351–357.

Renshaw, P.F., Owen, C.S., McLaughlin, A.C., Frey, T.G., Leigh, J.S., Jr., 1986b. Ferromagnetic contrast agents: a new approach. Magn Reson Med 3, 217-225.

Rifkind, D.A., Hsu, K.C., Morgan, Seegal, B.C., Knox, A.W., Rose, H.M., 1960. Use of ferritin-conjugated antibody to localize antigen by electron microscopy. Nature 187, 1094-1095.

Ryan, P.J., Davis, M.A., Melchior, D.L., 1983. The preparation and characterization of liposomes containing X-ray contrast agents. Biochim Biophys Acta 756, 106–110.

Shapiro, E.M., Skrtic, S., Koretsky, A.P., 2005. Sizing it up: cellular MRI using micron-sized iron oxide particles. Magn Reson Med 53, 329–338.

Singer, S.J., Schick, A.F., 1961. Properties of specific stains for electron microscopy prepared by conjugation of antibody molecules with ferritin. J Biophys Biochem Cytol 9, 519–&.

Unger, E., Needleman, P., Cullis, P., Tilcock, C., 1988. Gadolinium-DTPA liposomes as a potential MRI contrast agent. Work in progress. Invest Radiol 23, 928–932.

Widder, D.J., Simeone, J.F., 1986. Microbubbles as a contrast agent for neurosonography and ultrasound-guided catheter manipulation: in vitro studies. AJR Am J Roentgenol 147, 347–352.

Part 1
(Para)magnetic Nanoparticles: Applications in Magnetic Resonance Imaging

2
MR Lymphangiography Using Nano-Sized Paramagnetic Contrast Agents with Dendrimer Cores

Hisataka Kobayashi

Abstract: Imaging of the lymphatic system is difficult because its channels and lymph nodes are small and not directly accessible. Currently, two clinical methods, the direct lymphangiography and the lymphoscintigraphy, are used to visualize parts of the human lymphatic system, but have significant limitations. Nano-sized contrast agents have recently been evaluated to be advantageous for visualization of the lymphatic system because of their appropriate physical sizes and the potential of signal amplification. In this review, the magnetic resonance lymphangiography using a series of nano-sized paramagnetic contrast agents, gadolinium labeled dendrimers, is discussed focusing on the synthetic method, their pharmacokinetic characteristics for selection of appropriate agents, and applications for visualization of various lymphatic disorders in mouse models.

2.1. Introduction

The lymphatic system is difficult to evaluate because its channels and lymph nodes are small and not directly accessible. Small animal imaging of the lymphatics would greatly further research in this field; however, the available methods are suboptimal. Two clinical methods used to visualize parts of the human lymphatic system have also been explored in mouse studies, but have significant limitations. X-ray lymphangiography uses an iodine oil agent, which must be injected directly into lymph vessels in the extremity. This requires a high degree of technical skill since lymphatic channels are difficult to cannulate in humans and it is even more challenging to perform in animals. Direct X-ray lymphangiography sometimes causes life-threatening complications including lung embolization, pulmonary edema, and adult respiratory distress syndrome if the oily iodinated contrast media intravasates into the vascular system and it has, thus, been nearly abandoned clinically (Silvestri et al., 1980). Lymphoscintigraphy is another clinical method that has been attempted in animals. In lymphoscintigraphy, radiolabeled human serum albumin or aggregated albumin is injected intracutaneously or subcutaneously and the radiolabeled agent migrates to draining deep lymphatic vessels and nodes (Nawaz et al., 1990; Perrymore et al., 1996). While effective clinically, this method is not adequate in small animals due to the poor image resolution of radionuclide imaging.

Another approach to lymphatic imaging in animals is to employ magnetic resonance lymphangiography (MRL). Magnetic resonance imaging has the advantage of high resolution, without ionizing radiation, and has been readily adapted to small animal imaging with the use of dedicated surface coils. Individual lymphatic channels and nodes can be distinguished on MRL. A variety of MR contrast agents have been used for MRL including Gd-DTPA, liposomes, and Gd(III) macro-molecular chelates and iron oxides. Such agents have been tested in a variety of animal species including pig, rabbit, and rat models (Fujimoto et al., 2000; Harika et al., 1995; Misselwitz et al., 1999; Staatz et al., 2002).

While the study of the lymphatics in small animal models is important for research, there are few applications for direct clinical translation. Sentinel node imaging, on the other hand, has become extremely important in the surgical management of cancers, such as breast cancer and melanoma. Sentinel lymph node (SLN) biopsy has recently become a standard of care in breast cancer surgery because the disease status of the SLN accurately reflects the status of more distant axillary lymph nodes (Bass et al., 1999; Giuliano et al., 1997; Veronesi et al., 2001). The theory of sentinel node imaging is that, among all the nodes draining an organ (e.g., a breast), there is only one or at most just a few that first receive the lymphatic drainage. If the tumor has spread to the lymph nodes, it is these "sentinel" nodes that would be positive. The presence of positive sentinel nodes has implications for prognosis and management (Dowlatshahi et al., 1997; Fisher et al., 1983; Goldhirsch et al., 2001). The adoption of the sentinel node strategy has meant dramatically less invasive surgery which has gained widespread acceptance from surgeons and patients. Therefore, the testing of techniques for sentinel node imaging in small animals has recently gained attention as ways are sought to improve sentinel node imaging. Thus, by removing the SLN, the status of the entire lymphatic bed can be determined. Clinically, the two most commonly used methods employ peritumoral injections of either isosulfan blue dye or a radionuclide labeled sulfur colloid or aggregated albumin. In the "blue dye" technique, the SLN is detected by direct visualization of blue staining of the lymph node. The dye method requires dissection of tissue until the "dyed" SLN is detected. Blue dye can stain the skin for many months after injection. In lymphoscintigraphy, accumulation of the isotope, typically technetium-99 m labeled sulfur colloid, can be detected with a hand held gamma counter that directs the surgeon to begin dissection in a specific region (Alazraki et al., 2000; Alazraki et al., 2001). Although lymphoscintigraphy has a limited spatial resolution, it is clinically useful. It is important to note that lymphoscintigraphy detects normal draining nodes and does not necessarily detect metastatic disease in the nodes. An alternative method that detects metastatic nodes is to inject 2-fluoro[^{18}F]-deoxy-D-glucose (FDG) intravenously and look for abnormal uptake in the neck and axillary regions (Guller et al., 2003). Because it is a PET emitter, FDG is harder to detect using a hand held device. Moreover, it may not detect only the SLN, since it is systemically administrated.

Experimental methods to detect sentinel nodes are being developed. When conventional low molecular weight contrast agents for MRI or CT are injected, it is not always easy to detect sentinel nodes in pig and dog models (Suga et al., 2003). However, MR lymphography performed in the pig using macromolecular contrast agents, such as Gadomer-17 or dendrimers, have been more successful (Torchia and Misselwitz, 2002; Torchia et al., 2001). A family of polymers known as dendrimers appear particularly well suited to imaging the lymphatics with MRI. Dendrimers can be synthesized to different sizes, but will remain monodispersed. Thus, a library of dendrimer-based nano-sized macromolecular paramagnetic MRI contrast agents was synthesized to investigate the best agents for MRL (Kobayashi et al., 2006; Kobayashi et al., 2003a). Herein, we describe the synthesis of dendrimer-based contrast agents and the results of preclinical studies in mouse models using these agents.

2.2. Preparation and Quality Control of Nano-Sized Dendrimer-Based Paramagnetic MRI Contrast Agents

The details of preparation methods for each agent have been previously published in Kobayashi et al. (2003b, 2001a,b,d). In order to load maximum gadolinium ions on a single dendrimer molecule, minor modification of reaction condition was made in each dendrimer. Other researchers used similar but different synthetic methods and chelates, and reported variable conjugation ratio of the chelate molecules and gadolinium ions to dendrimer molecules (Bryant et al., 1999; Wiener et al., 1994). All of the dendrimers investigated for MRL ranged from generation-2 (G2; 3 nm) to generation-10 (G10; 15 nm) for polyamidoamine (PAMAM) and from generation-2 to generation-4 for polypropylenimine diaminobutane (DAB). The dendrimer substrate was obtained from commercial sources (Dendritech, Inc., Midland, MI or Aldrich Chemical Co., Milwaukee, WI). These compounds are highly soluble in aqueous solution and both possess a spherical surface topology composed of primary amino groups that increase exponentially in number with each generation (Tomalia et al., 1990; Wu et al., 1994) (Figure 2.1). The well-defined structure, monodispersity of size, and large number of available reactive surface amino groups have led to the use of dendrimers as substrates for the attachment of large numbers of chelating agents for creating macromolecular MR contrast agents (Bryant et al., 1999; Kobayashi and Brechbiel, 2003, 2004, 2005; Wiener et al., 1994). During our synthesis, dendrimers were initially concentrated in phosphate buffer at mild basic pH, and thereafter reacted with molar equivalent amounts of DTPA-derivatives as chelating moieties (Kobayashi et al., 2001d,e; Wu et al., 1994), equal to the number of surface amine residues on the dendrimer molecule and

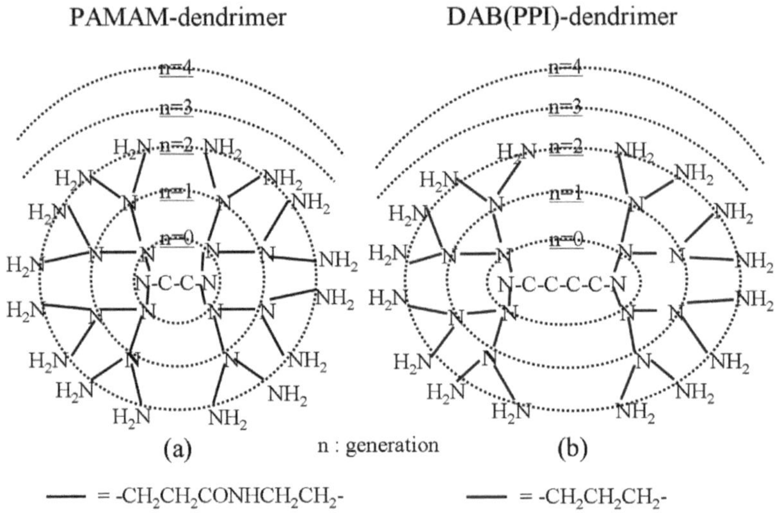

FIGURE 2.1. Schema of dendrimer cores used for contrast agents.

repeated this reaction 2 to 4 times to conjugate sufficient numbers of chelates with each dendrimer molecules. As a quality control study, a radiolabeling assay that employed either ^{111}In or ^{153}Gd as a tracer demonstrated the number of 1B4M-DTPA molecules conjugated to each dendrimer (Kobayashi et al., 2001e). When the reaction was stopped, more than 90% of the exterior primary amine groups on these dendrimers theoretically reacted with the 1B4M-DTPA thereby providing consistent end products suitable for animal studies (Kobayashi et al., 2001e). However, this analysis and the calculation were based on the weight of totally dehydrated and desalted chelate molecules. As a result, this HPLC-based assay may overestimate the number of chelate molecules on a dendrimer because of the water and salt content in the lyophilized chelate molecules (Kobayashi and Brechbiel, 2005).

Dendrimer-1B4M-DTPA conjugates were mixed with Gd(III) citrate at pH 4.5 and, after formation of the complex, the excess Gd(III) in each preparation was removed by diafiltration with an appropriate molecular weight filtration membrane. As a quality control, a replacement assay that employed ^{153}Gd as a tracer demonstrated that the number of 1B4M-DTPA molecules of the dendrimer-1B4M-DTPA conjugates chelating Gd(III) atoms ranged from 75% to 90% (Kobayashi et al., 2001e). The larger agents generally were able to hold a smaller proportion of Gd ions. The purified contrast agents were separated by size-exclusion high performance liquid chromatography (SE-HPLC) using Sephadex series of size-exclusion columns and analyzed by the 280 nm UV absorption and ^{153}Gd radioactivity. The number of Gd(III) ions per a dendrimer-based contrast agent was determined by the replacement assay and independently validated by the inductively coupled plasma mass spectrometry (ICP-MS) assay. The number of Gd(III) ions per a contrast agent was consistent on both assays (Kobayashi and Brechbiel, 2003).

The size of the dendrimer-based contrast agents were analyzed by SE-HPLC, mass spectroscopy, and light scattering methods. Although the size of the agents varied with technique, variations in size were generally small (~15%) (Kobayashi et al., 2001a; Yordanov et al., 2003).

2.3. MRI Methods for MR Lymphangiography

In order to test dendrimer-based MRL contrast agents in vivo, a variety of mouse models ranging from normal to diseased were evaluated. In order to perform MRL, a 1.5 T superconductive magnet unit was employed (Signa Horizon or Signa LX, Milwaukee, WI) and either a commercially available high-resolution wrist coil or dual 3 in. phased array surface coil with a custom mouse holder or a custom 1 in. bird-cage surface coil were utilized for all MRI studies to permit high-resolution images because of advantageous T1 relaxivity of dendrimer-based agents at 1.5 T and easy clinical translation. For MR lymphangiography, the T1-weighted 3D-fast spoiled gradient echo technique with fat-suppression technique and serial 3D data acquisition or 3D-fast imaging employing steady-state acquisition was used for all of the mice studied. A single MR study lasted 2–5 min and 5–10 serial studies were performed for dynamic MRL. Coronal images were usually reconstructed with 0.6–0.8 mm thick sections and 0.15–0.4 mm overlapping sections. The field of view was 6–8 × 3–4 cm and the size of each voxel was therefore $0.1\text{–}0.3 \times 0.1\text{–}0.3 \times 0.6\text{–}0.8\ mm^3$. Serial dynamic data were analyzed using commercially available workstations. Whole body 3D-MRIs were reconstructed with the maximum intensity projection (MIP) method to show up the enhanced lymphatic system. Details of individual experiments can be found in Kobayashi et al. (2006, 2005, 2003a,c, 2004, 2001c).

2.4. MR Lymphangiography with Nano-Sized Paramagnetic MRI Contrast Agent

2.4.1. *Optimal Size of the Agents*

In order to determine the optimal size of Gd-dendrimer for lymphatic vessel and lymph node imaging, each agent was tested in vivo. Intracutaneous injection at the phalanges revealed that larger nano-sized agents worked better for visualizing the lymphatic vessels than smaller agents. MR lymphangiograms with the largest agent studied (G8; ~13 nm) showed the best depiction of the lymphatic system among all the tested agents (Figure 2.2). The relatively large size of these molecules meant that once the agent was absorbed by the lymphatic system it tended to remain within it. In comparison, low molecular weight Gd-DTPA only faintly demonstrated the distal large lymph nodes such as the axillary and external iliac lymph nodes early after injection. In order to detect sentinel lymph nodes, various size dendrimers were tested after mammary pad injection. Among the agents tested, a G6 agent of ~10 nm in diameter proved to be the most effective (Figure 2.3). In comparison, the larger G8 dendrimer showed slower

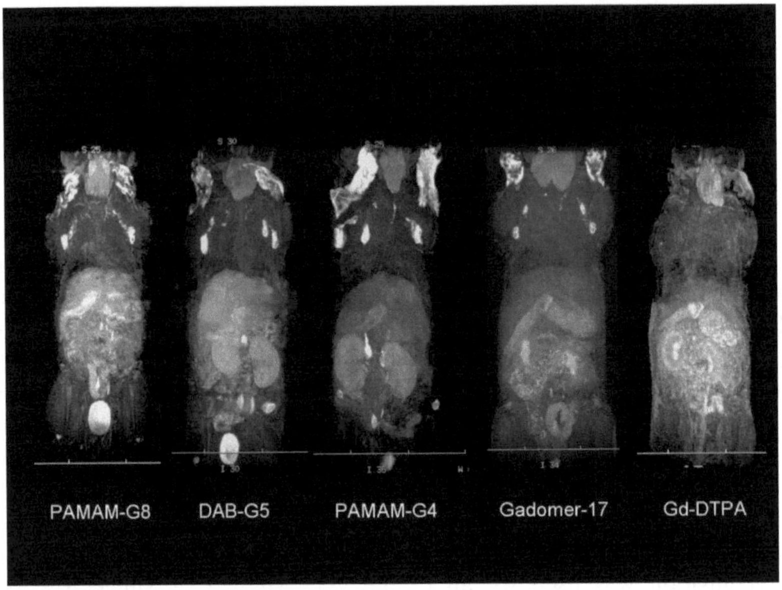

FIGURE 2.2. Whole body dynamic 3D-micro-MR lymphangiography of mice injected intracutaneously into all four middle phalanges with 0.005 mmol Gd/kg of PAMAM-G8, DAB-G5, PAMAM-G4, Gadomer-17, or Gd-DTPA are shown. The images obtained at 45 min post-injection are shown for all contrast agents tested. Maximum intensity projections are illustrated.

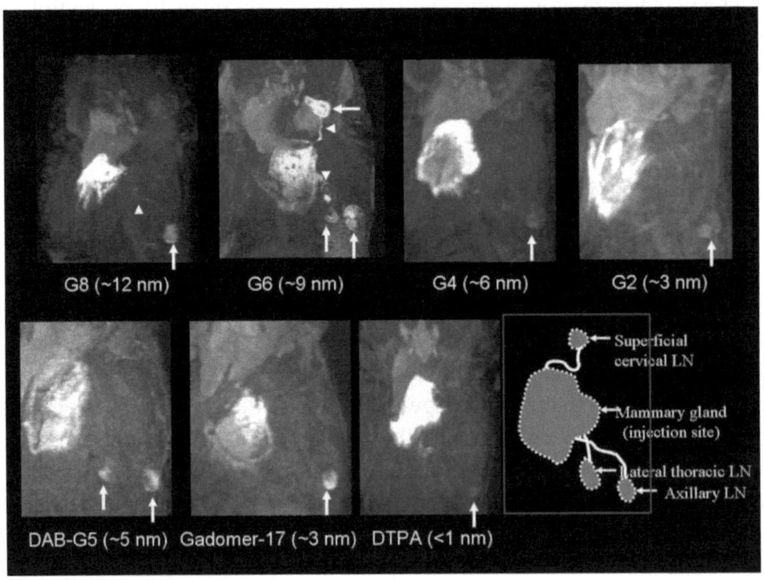

FIGURE 2.3. A series of 3D dynamic MR lymphangiograms obtained 12 min after the injection of the five contrast agents including three Gd-dendrimers. The G6 contrast agent of ~9 nm in diameter was taken up by both lymph nodes and lymphatic vessels and achieved the largest enhancement ratio among all agents examined.

enhancement of the axillary lymph node (Kobayashi et al., 2006). Smaller nano-sized agents were not as good as G6 or G8 in depicting the lymphatics and SLN. Among the smaller agents, DAB-G5 and Gadomer-17 were acceptable to visualize lymph nodes probably because of their less hydrophilic properties than PAMAM series of contrast agents (Kobayashi et al., 2006).

The ideal lymphatic imaging agent needs to be small enough to enter the lymphatic vessels, yet large enough to be retained within the lymphatics and not leak from the capillary vessels. Lymphographic contrast agents must be at least 4 nm in diameter to enable efficient retention within the lymphatics (Alazraki et al., 2001; Kobayashi et al., 2003a). Molecules smaller than 4 nm in diameter tend to diffuse into the surrounding tissue, resulting in poor signal-to-background ratios. Larger molecules, on the other hand, diffuse more slowly from the interstitial space and, thus, accumulate more slowly in the sentinel node, requiring a longer imaging window for visualizing nodes. The larger diameter G8 agent (~12 nm), used in our previous report for deep lymphatic imaging studies (Kobayashi et al., 2003c), is too large for rapid uptake by lymphatic vessels arising from the breast, although it is suitable for imaging lymphatics of the extremities. In contrast, the G6 contrast agent is retained in the lymphatic vessels, resulting in efficient enhancement of lymphatic vessels and lymph nodes. Both G2 and G4 agents were too small to stay within the lymphatic vessels, resulting in convection away from the injection site and only minimal enhancement in the lymph nodes (Figure 2.4).

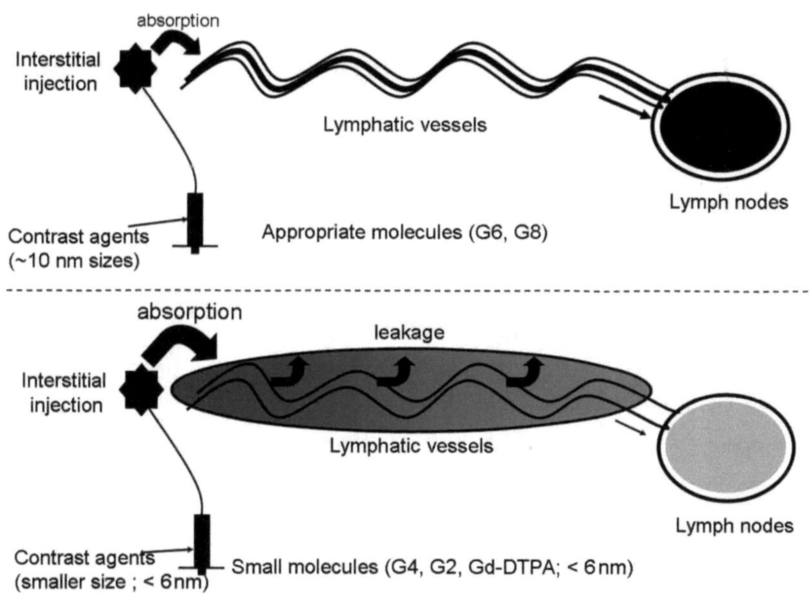

FIGURE 2.4. *Top*: G6 dendrimers are taken up by lymphatics and retained within the lymphatics resulting in opacification of the lymph nodes (black). Lower molecular weight contrast agents (*bottom*) are absorbed by the lymphatics, but leak out from them resulted in lower lymph node concentrations (gray).

The use of agents with appropriate sizes enables us to clearly depict systemic or local lymphatic vessels and lymph nodes in normal mice.

2.4.2. Injection Routes: Intradermal vs. Subcutaneous

Interestingly, we observed that when we injected the nano-sized contrast agents at different depths from the skin surface, the visualization of lymphatic system was affected. Intradermal injections of contrast agents visualized the lymphatics more quickly and with more signal intensity than deeper subcutaneous injections (Figure 2.5) (Kobayashi et al., 2003a). This may partially explain why sentinel lymph node imaging of the breast cancer using the intramammary gland (peri-tumoral; subcutaneous) injection required smaller agents and larger doses than systemic lymphangiography using intracutaneous injection into the skin (Kobayashi et al., 2006). As an alternative way of injecting lymphatic imaging agents for sentinel node detection, the intradermal areolar or subareolar injection has recently been proposed and used for the breast cancer growing in any location (Kern, 2001). The intradermal areola injection might decrease the required dose of contrast agent, time of imaging, and allow the use of larger agents. Still, areola injection is technically difficult in the mouse, because of the small size of the nipple compared to the needle.

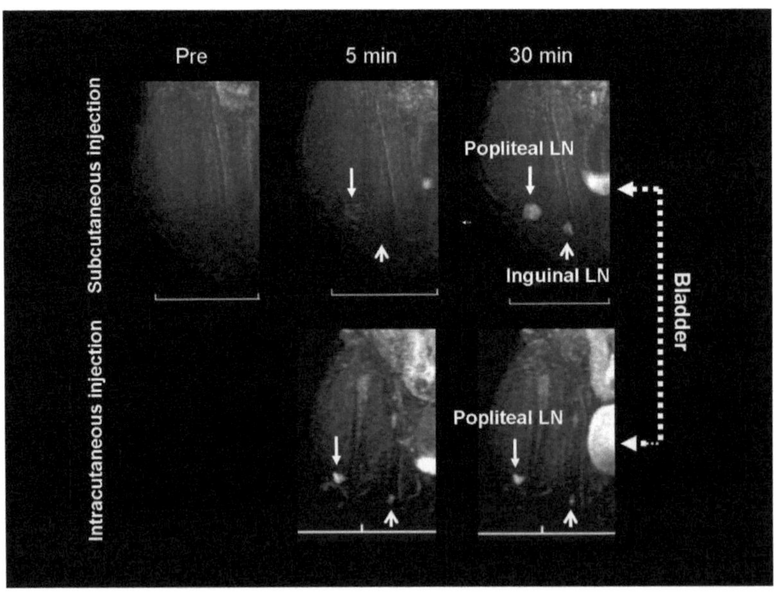

FIGURE 2.5. Dynamic MR lymphangiograms with G6 contrast agent injected into the foot either subcutaneously or intracutaneously. Intracutaneous injection showed more rapid and more intense uptake in both popliteal and inguinal LNs compared to subcutaneous injection.

2.5. Application of MR Lymphangiography with Nano-Sized Contrast Agents for Disease Models

2.5.1. Anatomic Abnormalities

MR lymphangiography has been tested in a variety of animal models of diseases. For instance, MR lymphangiography with Gd-dendrimer contrast agents was able to depict the location of intranodal lymphomatous infiltration similarly to direct X-ray lymphangiography (Figure 2.6). Although involvement of lymphoma rarely blocked the lymphatic flow, lymph node metastasis of solid cancers blocked the lymphatic flow. When this occurred, MR lymphangiography could detect the collateral lymphatic flow bypassing the blocked lymph node (Figure 2.7) (Kobayashi et al., 2003a). In addition, this method was also able to distinguish between the appearance of infection-induced hyperplasia of lymphocytes within lymph nodes from either chronic lymphoproliferative or neoplastic conditions (Kobayashi et al., 2003c). The enhanced resolution of this method might have wide applicability to the study of immunology and cancer in both experimental animal models and clinical medicine. Additionally, since a very small dose of the agent allowed visualization of the entire deep lymphatic system, this result suggests that this application might be feasible for human use.

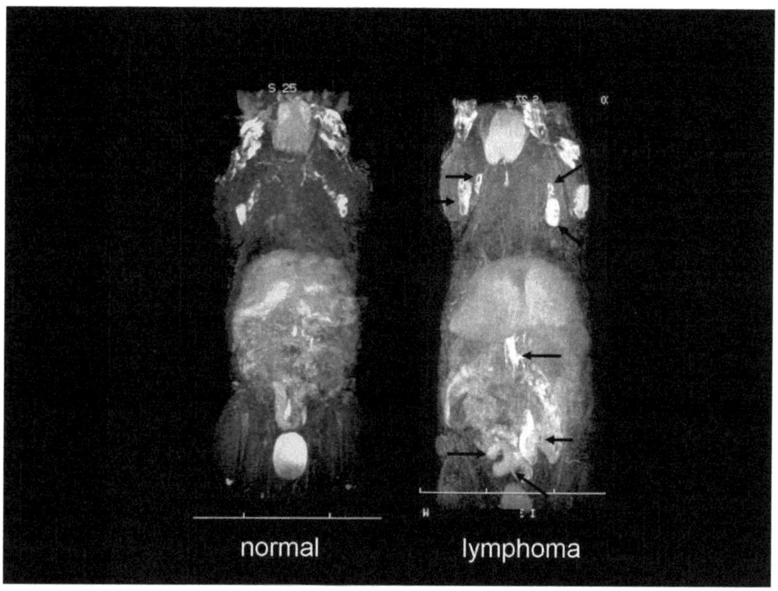

FIGURE 2.6. Whole body 3D-MR lymphangiography (MIP) of a normal or an IL-15 transgenic, lymphoproliferative/lymphoma model mouse obtained with G8 contrast agents (arrows indicate abnormal lymph nodes swelling).

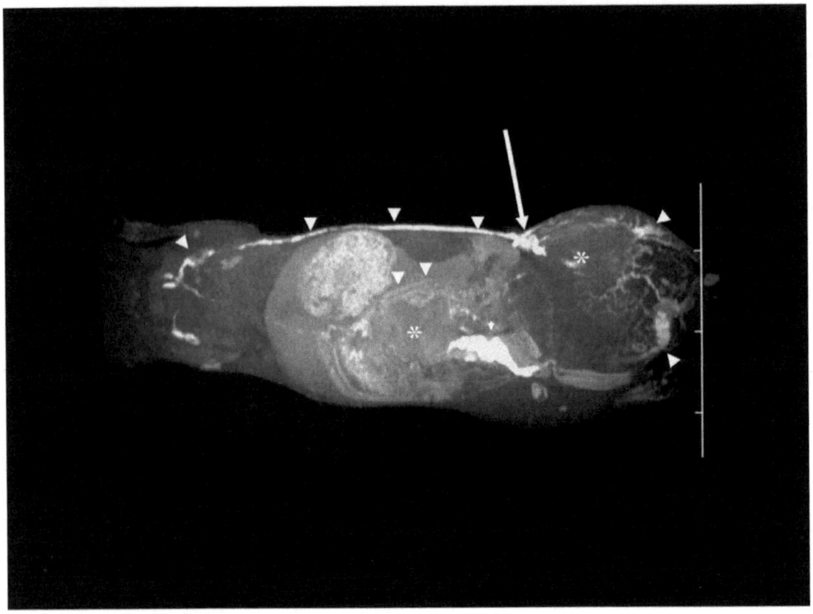

FIGURE 2.7. Whole body 3D-micro-MR lymphangiography of mice with lymph node metastasis of MC-38 cells injected with G8 agent. Large inguinal and abdominal tumors (asterisks) accompanied by a normal left inguinal lymph node (long arrow) are seen. In addition, dilated lymphatic vessels surrounding the tumor and collateral lymphatic vessels, which communicate with the thoracic duct via the axillary lymph node, are depicted (arrowheads).

2.5.2. Functional Abnormalities

In the case of lymphedema where lymphatic malformations alter lymphatic drainage patterns, the lymphatic flow may be slowed down. Currently, lymphoscintigraphy is the only clinically available method to examine lymphatic function. However, the poor spatial resolution of lymphoscintigraphy does not allow it to be applied to small animals. Even in humans, spatial resolution is inadequate to identify specific lymphatic vessels. In k-cycline transgenic mice, a lymphedema model, the dynamic MR lymphangiogram with a Gd-dendrimer contrast agent was able to diagnose the slowed lymphatic flow characteristics of lymphedema (Figure 2.8) (Sugaya et al., 2005). Another potential clinical application of MR lymphangiography is to investigate the lymphatic function of the sentinel node, which may provide insight into lymphatic flow from cancer tissue to the regional lymph nodes. For this application, the dose might be further minimized for specific local use to detect sentinel nodes metastasis of breast cancer (Figure 2.9a) (Kobayashi et al., 2004). Studies in animals were able to demonstrate detection of multiple lymph nodes, normal flow, and the lymphatic vessels as well. Nodal metastases were detectable as filling defects within the nodes. Thus, MRL using G6 dendrimer could be useful for staging, monitoring,

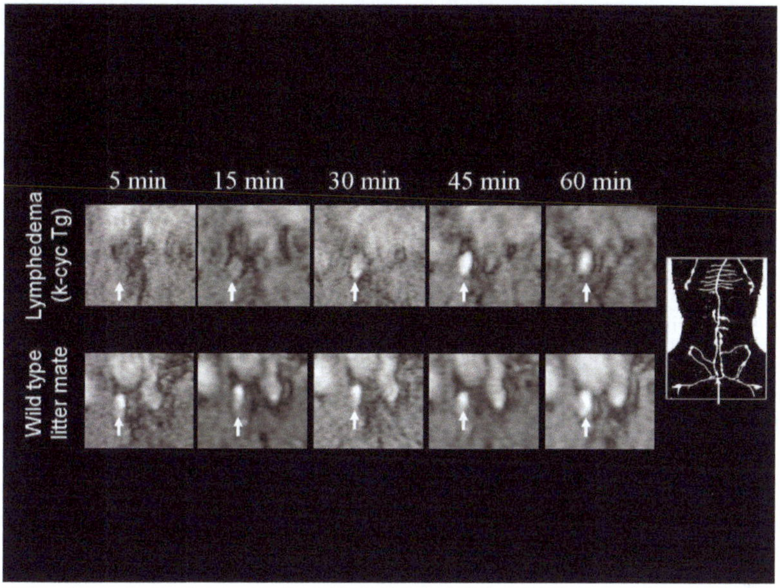

FIGURE 2.8. Dynamic MR lymphangiograms of the G8 agent showed delayed detection of the right external iliac node (arrows) in a lymphedema model mouse (k-cycline transgenic; *upper*), compared to a wild-type littermate (*lower*).

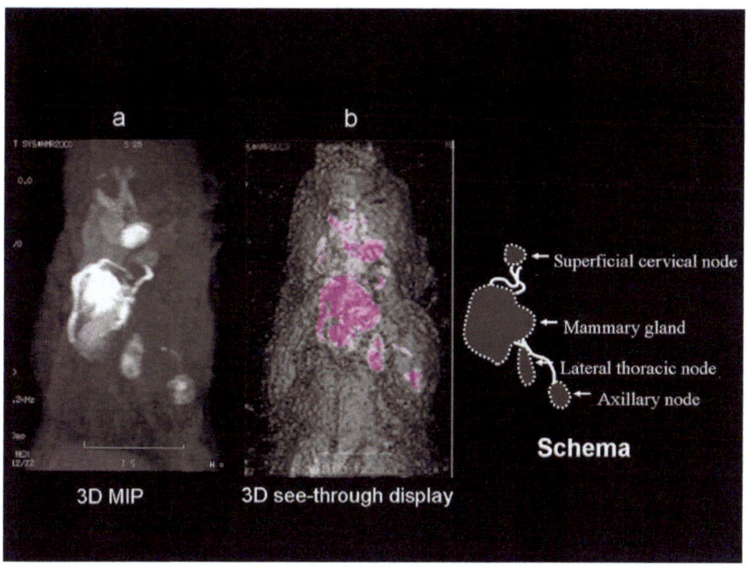

FIGURE 2.9. A chest 3D-MR mammo-lymphangiogram (maximum intensity projection; left anterior oblique view 30°) of a normal mouse obtained with G6 contrast agent at 30 min post-intramammary glands (*) injection clearly depicts draining (sentinel) lymph nodes. 3D see-through display demonstrates the location of lymphatic vessels and sentinel lymph nodes through the skin (3D volume rendering image).

and planning surgical interventions in breast cancer patients to achieve medical and cosmetic objectives. Additionally, serial 3D images of the lymph system can be processed and presented using surface rendering to create a see-through surface (skin) in which the relation of the underlying SLN to the skin surface could be determined (Figure 2.9b).

2.6. Conclusion

A brief description of the preparation of Gd-dendrimer nano-size paramagnetic contrast agents has been presented, with its optimization based on the size, the possible influence of injection depth, and an overview of their applications. Interestingly, the lymphatic system is exquisitely sensitive to the physical size of the contrast agent employed and the correct match of each Gd-dendrimer to the specific MRL application is important. Therefore, lymphatic imaging appears to be a good example of the successful use of nanotechnology to solve pressing issue in biology and the medicine.

Acknowledgement. This research was supported by the Intramural Research Program of the NIH, National Cancer Institute, Center for Cancer Research. I would like to thank to Dr. Peter L. Choyke, Molecular Imaging Program, NCI/NIH, for his editorial assistance.

References

Alazraki, N.P., Styblo, T., Grant, S.F., Cohen, C., Larsen, T., Aarsvold, J.N. 2000. Sentinel node staging of early breast cancer using lymphoscintigraphy and the intraoperative gamma-detecting probe. Semin Nucl Med 30, 56–64.

Alazraki, N.P., Styblo, T., Grant, S.F., Cohen, C., Larsen, T., Waldrop, S., Aarsvold, J.N. 2001. Sentinel node staging of early breast cancer using lymphoscintigraphy and the intraoperative gamma detecting probe. Radiol Clin North Am 39, 947–956.

Bass, S.S., Cox, C.E., Ku, N.N., Berman, C., Reintgen, D.S. 1999. The role of sentinel lymph node biopsy in breast cancer. J Am Coll Surg 189, 183–194.

Bryant, L.H., Jr., Brechbiel, M.W., Wu, C., Bulte, J.W., Herynek, V., Frank, J.A. 1999. Synthesis and relaxometry of high-generation (G = 5, 7, 9, and 10) PAMAM dendrimer-DOTA-gadolinium chelates. J Magn Reson Imaging 9, 348–352.

Dowlatshahi, K., Fan, M., Snider, H.C., Habib, F.A. 1997. Lymph node micrometastases from breast carcinoma: reviewing the dilemma. Cancer 80, 1188–1197.

Fisher, B., Bauer, M., Wickerham, D.L., Redmond, C.K., Fisher, E.R. 1983. Relation of number of positive axillary nodes to the prognosis of patients with primary breast cancer. An NSABP update. Cancer 52, 1551–1557.

Fujimoto, Y., Okuhata, Y., Tyngi, S., Namba, Y., Oku, N. 2000. Magnetic resonance lymphography of profundus lymph nodes with liposomal gadolinium-diethylene-triamine pentaacetic acid. Biol Pharm Bull 23, 97–100.

Giuliano, A.E., Jones, R.C., Brennan, M., Statman, R. 1997. Sentinel lymphadenectomy in breast cancer. J Clin Oncol 15, 2345–2350.

Goldhirsch, A., Glick, J.H., Gelber, R.D., Coates, A.S., Senn, H.J. 2001. Meeting highlights: International Consensus Panel on the Treatment of Primary Breast Cancer. Seventh International Conference on Adjuvant Therapy of Primary Breast Cancer. J Clin Oncol 19, 3817–3827.

Guller, U., Nitzsche, E., Moch, H., Zuber, M. 2003. Is Positron Emission Tomography an Accurate Non-invasive Alternative to Sentinel Lymph Node Biopsy in Breast Cancer Patients? J Natl Cancer Inst 2003; 95: 95, 1040–1043.

Harika, L., Weissleder, R., Poss, K., Zimmer, C., Papisov, M.I., Brady, T.J. 1995. MR lymphography with a lymphotropic T1-type MR contrast agent: Gd-DTPA- PGM. Magn Reson Med 33, 88–92.

Kern, K.A. 2001. Breast lymphatic mapping using subareolar injections of blue dye and radiocolloid: illustrated technique. J Am Coll Surg 192, 545–550.

Kobayashi, H., Brechbiel, M.W. 2003. Gadolinium-based macromolecular MRI contrast agents. Mol Imag 2, 1–10.

Kobayashi, H., Brechbiel, M.W. 2004. Dendrimer-based nanosized MRI contrast agents. Curr Pharm Biotechnol 5, 539–549.

Kobayashi, H., Brechbiel, M.W. 2005. Nano-sized MRI contrast agents with dendrimer cores. Adv Drug Deliv Rev 57, 2271–2286.

Kobayashi, H., Kawamoto, S., Bernardo, M., Brechbiel, M.W., Knopp, M.V., Choyke, P.L. 2006. Delivery of gadolinium-labeled nanoparticles to the sentinel lymph node: comparison of the sentinel node visualization and estimations of intra-nodal gadolinium concentration by the magnetic resonance imaging. J Control Release 111, 343–351.

Kobayashi, H., Kawamoto, S., Brechbiel, M.W., Bernardo, M., Sato, N., Waldmann, T.A., Tagaya, Y., Choyke, P.L. 2005. Detection of lymph node involvement in hematologic malignancies using micromagnetic resonance lymphangiography with a gadolinum-labeled dendrimer nanoparticle. Neoplasia 7, 984–991.

Kobayashi, H., Kawamoto, S., Choyke, P.L., Sato, N., Knopp, M.V., Star, R.A., Waldmann, T.A., Tagaya, Y., Brechbiel, M.W. 2003a. Comparison of dendrimer-based macromolecular contrast agents for dynamic micro magnetic resonance lymphangiography. Magn Reson Med 50, 758–766.

Kobayashi, H., Kawamoto, S., Jo, S., Bryant, L.H., Jr., Brechbiel, M.W., Star, R.A. 2003b. Macromolecular MRI contrast agents with small dendrimers: pharmacokinetic differences between sizes and cores. Bioconjug Chem 14, 388–394.

Kobayashi, H., Kawamoto, S., Saga, T., Sato, N., Hiraga, A., Ishimori, T., Akita, Y., Mamede, M.H., Konishi, J., Togashi, K., Brechbiel, M.W. 2001a. Novel liver macromolecular MR contrast agent with a polypropylenimine diaminobutyl dendrimer core: comparison to the vascular MR contrast agent with the polyamidoamine dendrimer core. Magn Reson Med 46, 795–802.

Kobayashi, H., Kawamoto, S., Saga, T., Sato, N., Hiraga, A., Konishi, J., Togashi, K., Brechbiel, M.W. 2001b. Micro-MR angiography of normal and intratumoral vessels in mice using dedicated intravascular MR contrast agents with high generation of polyamidoamine dendrimer core: reference to pharmacokinetic properties of dendrimer-based MR contrast agents. J Magn Reson Imag 14, 705–713.

Kobayashi, H., Kawamoto, S., Sakai, Y., Choyke, P.L., Star, R.A., Brechbiel, M.W., Sato, N., Tagaya, Y., Morris, J.C., Waldmann, T.A. 2004. Lymphatic drainage imaging of breast cancer in mice by micro-magnetic resonance lymphangiography using a nano-size paramagnetic contrast agent. J Natl Cancer Inst 96, 703–708.

Kobayashi, H., Kawamoto, S., Star, R.A., Waldmann, T.A., Tagaya, Y., Brechbiel, M.W. 2003c. Micro-magnetic resonance lymphangiography in mice using a novel dendrimer-based magnetic resonance imaging contrast agent. Cancer Res 63, 271–276.

Kobayashi, H., Saga, T., Kawamoto, S., Sato, N., Hiraga, A., Ishimori, T., Konishi, J., Togashi, K., Brechbiel, M.W. 2001c. Dynamic micro-magnetic resonance imaging of liver micrometastasis in mice with a novel liver macromolecular magnetic resonance contrast agent DAB-Am64-(1B4M-Gd)(64). Cancer Res 61, 4966–4970.

Kobayashi, H., Sato, N., Hiraga, A., Saga, T., Nakamoto, Y., Ueda, H., Konishi, J., Togashi, K., Brechbiel, M.W. 2001d. 3D-micro-MR angiography of mice using macromolecular MR contrast agents with polyamidoamine dendrimer core with references to their pharmacokinetic properties. Magn Reson Med 45, 454–460.

Kobayashi, H., Sato, N., Kawamoto, S., Saga, T., Hiraga, A., Haque, T.L., Ishimori, T., Konishi, J., Togashi, K., Brechbiel, M.W. 2001e. Comparison of the macromolecular MR contrast agents with ethylenediamine-core versus ammonia-core generation-6 polyamidoamine dendrimer. Bioconjug Chem 12, 100–107.

Misselwitz, B., Platzek, J., Raduchel, B., Oellinger, J.J., Weinmann, H.J. 1999. Gadofluorine 8: initial experience with a new contrast medium for interstitial MR lymphography. Magma 8, 190–195.

Nawaz, M.K., Hamad, M.M., Abdel-Dayem, H.M., Sadek, S., Eklof, B.G. 1990. Tc-99m human serum albumin lymphoscintigraphy in lymphedema of the lower extremities. Clin Nucl Med 15, 794–799.

Perrymore, W.D., Harolds, J.A. 1996. Technetium-99m-albumin colloid lymphoscintigraphy in postoperative lymphocele. J Nucl Med 37, 1517–1518.

Silvestri, R.C., Huseby, J.S., Rughani, I., Thorning, D., Culver, B.H. 1980. Respiratory distress syndrome from lymphangiography contrast medium. Am Rev Respir Dis 122, 543–549.

Staatz, G., Spuntrup, E., Buecker, A., Misselwitz, B., Gunther, R.W. 2002. T1-weighted MR-lymphography after intramammary administration of Gadomer-17 in pigs. Rofo Fortschr Geb Rontgenstr Neuen Bildgeb Verfahr 174, 29–32.

Suga, K., Yuan, Y., Ogasawara, N., Okada, M., Matsunaga, N. 2003. Localization of breast sentinel lymph nodes by MR lymphography with a conventional gadolinium contrast agent. Preliminary observations in dogs and humans. Acta Radiol 44, 35–42.

Sugaya, M., Watanabe, T., Yang, A., Starost, M.F., Kobayashi, H., Atkins, A.M., Borris, D.L., Hanan, E.A., Schimel, D., Bryant, M.A., et al. 2005. Lymphatic dysfunction in transgenic mice expressing KSHV k-cyclin under the control of the VEGFR-3 promoter. Blood 105, 2356–2363.

Tomalia, D.A., Naylor, A.M., Goddard III, W.A. 1990. Starburst dendrimers: Molecular-level control of size, shape, surface chemistry, topology, and flexibility from atoms to macroscopic matter. Angew Chem-Int Ed 29, 138–175.

Torchia, M.G., Misselwitz, B. 2002. Combined MR lymphangiography and MR imaging-guided needle localization of sentinel lymph nodes using Gadomer-17. Am J Roentgenol 179, 1561–1565.

Torchia, M.G., Nason, R., Danzinger, R., Lewis, J.M., Thliveris, J.A. 2001. Interstitial MR lymphangiography for the detection of sentinel lymph nodes. J Surg Oncol 78, 151–156.

Veronesi, U., Galimberti, V., Zurrida, S., Pigatto, F., Veronesi, P., Robertson, C., Paganelli, G., Sciascia, V., Viale, G. 2001. Sentinel lymph node biopsy as an indicator for axillary dissection in early breast cancer. Eur J Cancer 37, 454–458.

Wiener, E.C., Brechbiel, M.W., Brothers, H., Magin, R.L., Gansow, O.A., Tomalia, D.A., Lauterbur, P.C. 1994. Dendrimer-based metal chelates: a new class of magnetic resonance imaging contrast agents. Magn Reson Med 31, 1–8.
Wu, C., Brechbiel, M.W., Kozak, R.W., Gansow, O.A. 1994. Metal-chelate-dendrimer-antibody constructs for use in radioimmunotherapy and imaging. Bioorg Med Chem Lett 4, 449–454.
Yordanov, A.T., Kobayashi, H., English, S.J., Reijnders, K., Milenic, D., Krishna, M.C., Mitchell, J.B., Brechbiel, M.W. 2003. Gadolinium-labeled dendrimers as biometric nanoprobes to detect vascular permeability. J Mater Chem 13, 1523–1525.

3
Use of USPIOs for Clinical Lymph Node Imaging

Jelle O. Barentsz and Paris P. Tekkis

Abstract: The detection of lymph node metastases is critical for the choice of pre-operative chemo-radiotherapy, surgical treatment, and patient prognosis. Ultrasmall superparamagnetic iron oxide (USPIO) enhanced magnetic resonance imaging (MRI) is now being used as a potential biomarker for the detection of lymph node metastases. USPIO is transported into the interstitial space and reaches the lymph nodes via the lymphatic circulation acting as a 'negative contrast' agent, which can potentially identify metastases independent of lymph node size. We assessed here the diagnostic precision and application of magnetic resonance imaging (MRI) in conjunction with USPIO contrast as a biomarker for detecting LNMs pre-operatively, compared with the gold-standard post-operative histopathology. A total of 19 prospective studies have been published between 1994 and 2005, comprising 3,004 lymph nodes in 631 patients who underwent comparable MR imaging with and without USPIO. The overall sensitivity and specificity for MRI with USPIO were 88.1% and 96.2% respectively, with an area under the ROC curve of 84.2% and a diagnostic odds ratio of 123.1. When unenhanced MRI was evaluated, there was a significant reduction in the overall sensitivity and specificity (63% and 92.7% respectively) and a diagnostic odds ratio of 26.7. USPIO-enhanced MRI had a higher sensitivity and specificity for lymph node status in the abdomen and pelvis compared with the chest or head and neck. Similarly, the diagnostic precision was better when a 3T MRI scanner was used at 2.6 mg/kg contrast dose, than when using a 1.5T field strength. Thus, USPIO is a promising nanomarker, which may potentially increase the precision for the pre-operative diagnosis of lymph node metastases. It may be used for providing guidelines for selecting patients for targeted neo-adjuvant chemotherapy and extended lymphadenectomy thus possibly reducing the incidence of local recurrence.

3.1. Introduction

Pelvic lymph node metastases have a significant impact on the prognosis of patients with malignancies. In prostate cancer, for example, even micro metastases in a single node rule out surgical cure by the available treatment protocols.

For bladder cancer, lymph node metastasis are also significant. More than five lymph node metastasis or extra-capsular growth precludes curative surgical treatment. Similarly, in the case of rectal cancer, involvement of the mesorectal lymph nodes would be a relative indication for pre-operative radiotherapy or chemotherapy in order to downstage the tumour and minimize the likelihood of circumferential margin involvement and local recurrence. Thus, the status of the lymph nodes largely dictates the management of the primary tumour.

Surgical open pelvic lymph node dissection (PLND) with histopathological examination is currently the most reliable method of assessing lymph node status. Pelvic lymphadenectomy and in particular extended pelvic lymphadenectomy is an invasive procedure associated with potential complications and side effects. A non-invasive, reliable method for detecting and staging nodal metastasis in the pre-operative phase may redirect clinicians towards less invasive treatment strategies, such as chemotherapy or radiotherapy either as neoadjuvant or as definitive treatment. Routine cross-sectional imaging modalities, such as CT and MRI, lack the desired sensitivity in identifying metastases as they largely rely on size criteria only, and small metastases in normal-sized nodes can be missed. Moreover, differences in signal intensity on MR images between normal and cancerous nodes, as well as gadolinium enhancement, have also proven to be unreliable.

Although very promising in metastatic lung cancer, the role of ^{18}FDG PET-scanning is limited in the urinary tract region, as ^{18}F-fluorodeoxyglucose accumulates in the urinary bladder and kidneys. This makes an evaluation of metastases at these sites difficult. Also, in various tumours, like prostate and bladder cancer, this method is further limited by its low uptake in metastatic nodes. Although the sensitivity of ^{18}FDG PET is slightly better (67%) compared to those of CT and unenhanced MRI, this value is, however, not high enough to replace pelvic lymph node dissection.

Radiologists, surgeons, and oncologists all recognize the limitations of current size and morphological imaging criteria evaluating lymph nodes non-invasively. MR lymphangiography (MRL) using the intravenously administered contrast agent Ferumoxtran-10 with a long plasma circulation time is a novel cellular imaging tool for the evaluation of nodal involvement non-invasively. Ferumoxtran-10 belongs to a class of nanoparticle-based contrast agents that are referred to as USPIO (ultrasmall superparamagnetic iron oxide). It is commercially known as Sinerem® in Europe (Laboratoire Guerbet, Aulnay sous Bois, France), and as Combidex® in the US (Advanced Magnetics, Cambridge, MA), but are not yet approved by FDA. According to early studies, MRL is an accurate tool to differentiate benign from malignant lymph nodes (Vassallo et al., 1994; Weissleder et al. 1990). Currently, the available contrast agent is a freeze-dried iron oxide agent reconstituted in normal saline and administered intravenously by a slow drip infusion and can easily be performed on outpatient basis. The first MRI scan is done to anatomically identify and visualize the nodes. The second scan is completed 24 hours after the injection of the contrast agent and serves to evaluate the contrast enhancement of the identified lymph nodes.

3.2. Mechanism of Action

After IV injection, the ferumoxtran-10 particles are taken up by macrophages, and are transported to the interstitial space and from there through the lymph vessels to the lymph nodes (Figure 3.1). Thus, this contrast agent is cell-specific for macrophages. Once within normally functioning nodes, the intracellular

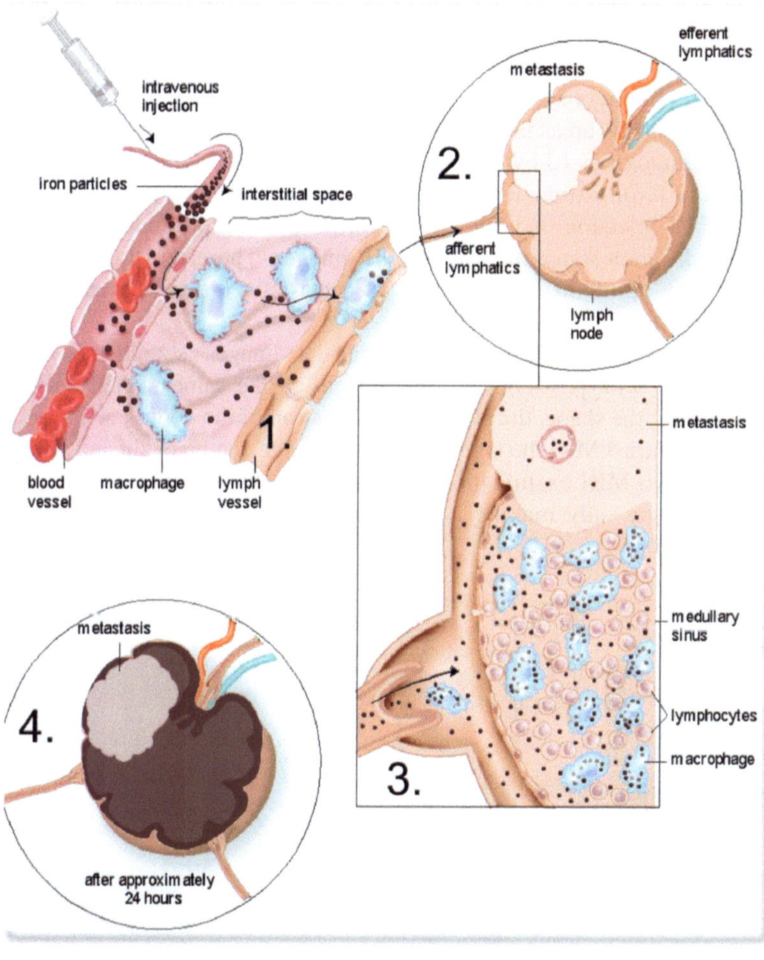

FIGURE 3.1. Uptake mechanism of ferumoxtran-10. IV injected particles slowly extravasate from vascular to interstitial space (**1**) and are then transported to lymph nodes via lymphatic vessels (**2**). In lymph nodes, particles are internalized by macrophages (**3**), and these intracellular iron-containing particles cause normal nodal tissue to have low signal intensity. Disturbances of lymph flow or nodal architecture by metastases lead to abnormal accumulation patterns depicted by the lack of decreased signal intensity (**4**). Reprinted with permission from Deserno et al. (2004).

ferumoxtran-10 within the macrophages reduces the signal intensity of normal node tissue, because of the T2*-susceptibility effects induced by the iron oxide, producing a signal drop or *negative* contrast enhancement. In areas of lymph nodes that are involved with malignant cells, macrophages are replaced by cancer cells. Therefore, in these areas there is no uptake of the ferumoxtran-10 particles. In addition, due to increased vascular permeability and increased diffusion in cancer tissue, there is minimal leakage of ferumoxtran-10 particles into the extra-cellular space of malignant metastatic areas, which produces a low local concentration and non-clustering of ferumoxtran-10 particles at these sites (Gerlowski and Jain 1986). Through their T1-relaxivity this can induce an increase in signal intensity on T1-weighted images, producing *positive* enhancement (Bellin et al., 1998). Thus, the ability of post-ferumoxtran-10 MRI to identify metastatic areas in the lymph nodes depends primarily on the degree of uptake of ferumoxtran-10 by the macrophages in normal lymph node tissue and the leakage of ferumoxtran-10 particles in the metastatic area itself. Twenty-four hours after IV injection of ferumoxtran-10, normal lymph node and malignant tissue have different signal intensity on MR images. Therefore, this non-invasive technique may result in the detection of metastatic deposits in normal-sized nodes (Figure 3.2).

Optimal evaluation of post-ferumoxtran-10 images should be done by comparing pre- with post-contrast MR images in the same plane. On the pre-contrast images, the shape, the size, and the location of the nodes can be assessed. On the post-contrast MRI the signal intensity change can be evaluated. However, this requires two MRI examinations, which limits this technique. This problem can be solved by only making a post-ferumoxtran-10 MR exam, using both

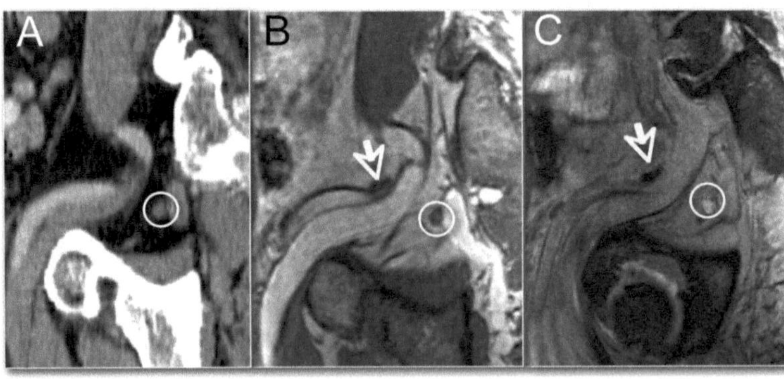

FIGURE 3.2. Normal node and small positive node in a 60-year-old male with prostate cancer. **A** CT scan in semi-sagittal plane shows normal size (6 mm) node (circle). **B** Post-ferumoxtran-10 T_1-weighted TSE MR image (which is insensitive to iron) shows two gray normal size nodes (circle, arrow). **C** On post-ferumoxtran-10 T_2^*-weighted MEDIC MR image (which is iron sensitive) one node is black (arrow) and the other is white (circle). On histopathology the black node was normal and the white node completely metastatic.

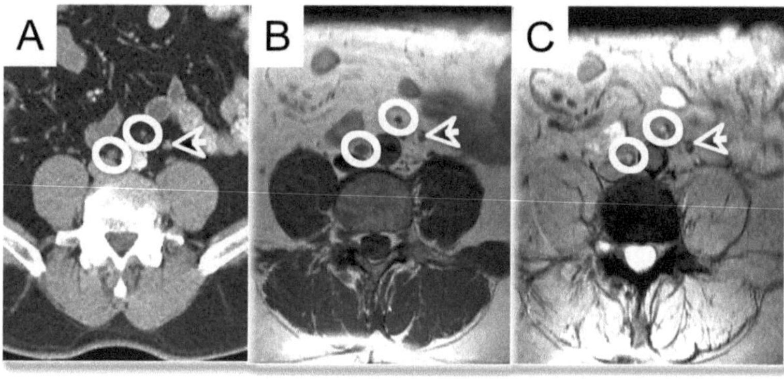

FIGURE 3.3. **A** 58-year-old patient treated for prostate cancer with nodal recurrence. CT and MRI obtained 2 years after lymphadenectomy, prostatectomy and hormonal therapy, with the PSA value increasing from 0 to 1.8. **A** CT scan in axial plane shows three normal-sized nodes (circle right 5 mm; circle left 2 mm). **B** Post-ferumoxtran-10 T_1-weighted TSE MR image (which is insensitive to iron) shows the same three gray nodes (circles, arrow). **C** On post-ferumoxtran-10 T_2^*-weighted MEDIC MR image (which is iron sensitive) one node is black (arrow) and the other two are white (circles). On histopathology, the black node was normal and the white ones were metastatic.

a sequence which is insensitive for iron using a T1- or proton-weighted TSE sequences and a sequence which is sensitive for iron. For the latter purpose, a good sequence is a high resolution T2*-weighted MEDIC (TE~18ms). The T1/PW TSE sequences yield high-resolution images without (susceptibility) artefacts (i.e., insensitive to iron), whereas the T2*-sequences provides information about the iron content of the nodes. It is, therefore, important to apply both sequences with the same resolution and slice-positioning parameters. In pelvic tumours (prostate cancer), this can be best done in a plane parallel to the psoas muscles (obturator, or semi-sagittal plane, Figure 3.2), and the axial plane, covering the para-aortic until the femoral region (Figure 3.3). In the 'reading' of the images, a one-by-one comparison gives the best results (Figure 3.2). Incidentally in this way, even a 2 mm metastatic node can be found (Figure 3.3).

3.3. Clinical Value

When using high-resolution MR techniques, small metastases can be prospectively recognized in small-sized lymph nodes (5–10 mm) (Harisinghani & Barentsz et al., 2003). These small lymph nodes would be considered to be benign in plain MRI or CT examinations. In addition, hyperplastic-enlarged nodes can be correctly recognized as non-metastatic, based on their low-signal intensity. This results in improved sensitivity (~90%), with remaining equal high specificity (~95%) in various tumours (Mack et al., 2002; Michael, 2002; Harisinghani et al., 2003; Deserno et al. 2004; Will et al., 2006).

3.4. Overview of Various Cancers

A quantitative meta-analysis was performed on prospective studies, comparing MRI with and without USPIO, against histological diagnosis after surgery or biopsy (Will et al., 2006). The primary endpoints were sensitivity, specificity, area under the receiver operating characteristic (ROC) curve, and diagnostic odds ratio of the imaging technique used. Sensitivity analysis was performed to evaluate the diagnostic precision of USPIO-enhanced MRI for different body regions imaged (head and neck, chest, abdomen and pelvis), studies using a 1.5 T magnet and studies using 1.7 or 2.6 mg/kg doses of ferumoxtran-10.

A total of 19 prospective studies were published between 1994 and 2005 and were used for data analysis. These comprised 3,004 lymph nodes in 631 patients who underwent comparable MR imaging with histological verification as shown in Table 3.1. The overall sensitivity and specificity for MRI with USPIO (17 studies) were 88.1% and 96.2% respectively, with an area of 84.2% under the ROC curve and a diagnostic odds ratio of 123.1. When unenhanced MRI was evaluated, there was a significant reduction in the overall sensitivity and specificity (63% and 92.7% respectively), with an area of 84.2% under the ROC curve and a diagnostic odds ratio of 26.7 (Table 3.2). When analyzed specifically by body region, USPIO-enhanced MRI had a high sensitivity and specificity for lymph node status in the abdomen and pelvis compared with the chest or head and neck. The highest sensitivity and specificity was noted for the two studies, reporting results only for prostate cancer. This may be explained by the fact that the lymph flow in the legs, pelvis and abdomen is more pronounced compared to thorax and head and neck. Similarly, the diagnostic precision of USPIO-enhanced MRI appeared to be better when a 3 T MRI scanner was used at 2.6 mg/kg contrast dose, than when using a 1.5 T field strength (Heeaskkers, 2006).

Analysis by nodal size (without USPIO injection) was reported by two studies, with just one specifically examining lymph nodes of less than 5 mm in diameter (Table 3.2). For lymph nodes between 5 and 10 mm, the sensitivity and specificity were 96.4% and 99.3%, respectively, and for lymph nodes under 5 mm the sensitivity and specificity was reduced to 41.1% and 98.1%, respectively. These results, although less than the sensitivity reached for nodes of a larger diameter, were superior to that obtained without USPIO. Even at 3T MRI has a limited resolution with 1 mm being at the limit of the highest definition. As sagittal cut section thickness is commonly around 3 mm, it is clear that micro-metastases with partial node destruction can be missed.

3.4.1. Prostate Cancer

Specifically, in prostate cancer contrast-enhanced CT and conventional MRI have a low sensitivity (40%), which is improved to 100% on a patient level and to 90% on a nodal level (Harisinghani et al., 2003). When using ferumoxtran-10-enhanced MRI, patients may be reliably selected for prostatectomy or radiotherapy without the need for invasive and costly procedures, such as open and

Table 3.1. Study, patient and technical characteristics of reported studies. Adapted from Will et al. (2006).

Study	No. of subjects	No. of lymph nodes	Inclusion criteria	Body region	Tumour type	Dose (mg Fe/kg)	Type of MRI	MRI timing
Anzai et al. (1994)	11	91	13	Head and neck	a	1.7	1.5 T	10–48 h
Anzai et al. (2003)	134	276	1, 2,3,4	Neck and body	a,b,c,e,f	2.6	0.2 T, 1.0 T, 1.5 T	24–36 h
Bellin et al. (1998)	28	60	1,2,3,7,14,15	Pelvis	d,g,h,i,l,n,o,p	1.7	0.5 T, 1.0 T, 1.5 T	22–26 h
Deserno et al. (2004)	58	172	2,7	Pelvis	o	2.6	1.5 T	24–36 h
Harisinghani et al. (1998)	11	27	2,7	Abdomen and Pelvis	d,g,h,m	1.7 or 2.6	1.5 T	24–36 h
Harisinghani et al. (1999)	19	49	2,14	Abdomen and Pelvis	D,g,h,j,k,l	1.7 or 2.6	1.5 T	24–36 h
Harisinghani et al. (2002)	50	168	2,7	Pelvis	g	2.6	1.5 T	24–36 h
Harisinghani et al. (2003)	80	334	2,7	Pelvis	g	2.6	1.5 T	24 h
Hoffman et al. (2000)	9	101	13	Head and neck	a	1.7	1.5 T	24–36 h
Keller et al. (2004)	9	91	5,6,7	Pelvis	h,i,p,q	2.6	1.5 T	18–28 h
Kernstine et al. (1999)	9	18	1,8,9	Thoracic	b	2.6	–	24–36 h
Koh et al. (2004)	12	74	2,7,12	Pelvis	d	2.6	1.5 T	24–36 h
Mack et al. (2002)	27	1029	1.7,11	Head and neck	a	2.6	1.5 T	24–36 h
Michel et al. (2002)	18	–	7.10	Axilla	c	2.6	1.5 T	24–36 h
Nguyen et al. (1999)	16	27	2	Thoracic	b	2.6	1.5 T	24–36 h
Pannu et al. (2000)	6	–	1,7	Mediastinum	b	2.6	1.5 T	–

(Continued)

TABLE 3.1. (Continued)

Study	No. of subjects	No. of lymph nodes	Inclusion criteria	Body region	Tumour type	Dose (mg Fe/kg)	Type of MRI	MRI timing
Rockall et al. (2005)	44	335	1,7	Pelvis and Para-aortic	h	2.6	1.5 T	24–36 h
Sigal et al. (2002)	81	100	7,11,12	Head and neck	a	2.6	1.0 T, 1.5 T	24–36 h
Stets et al. (2002)	9	52	2,7	Axilla	c	2.6	1.0 T	24–36 h
Total	631	3,004						

S/N = number of subjects/number of nodes; – = data unavailable

Inclusion criteria: 1 = age $>$ 18 years; 2 = confirmed primary cancer; 3 = possibility of nodal metastases; 4 = at least 1 lymph node visible on precontrast MRI; 5 = biopsy proven carcinoma of the cervix, uterine corpus, or vulva; 6 = uterine carcinoma suspected on clinical, ultrasound, or CT grounds; 7 = all scheduled for surgery or other histological node positivity confirmatory procedure at the time of inclusion; 8 = patients with lung lesions suspicious of malignancy; 9 = candidate for mediastinoscopy or thoracotomy; 10 = patients with cytological or histological proof of breast cancer; 11 = proven squamous cell carcinoma of the head and neck; 12 = absence of previous chemo/radio therapy of the area concerned; 13 = suspected extracranial head and neck cancer; 14 = abdominal or pelvic lymph node visible at pre-operative MR or CT imaging; 15 = age $<$ 85 years.

Tumour type: a = unspecified head and neck; b = lung and mediastinum; c = breast; d = colon or rectum; e = unspecified abdominal; f = unspecified pelvic; g = prostate; h = uterus or endometrial; i = ovarian; j = Merkel cell tumour; k = lymphoma; l = testicles or seminoma; m = retroperitoneal; n = kidney; o = bladder; p = cervical; q = vulval.

TABLE 3.2. Sensitivity, specificity, diagnostic odds ratio (DOR) and area under the receiver operating characteristic curve (AUROC) for studies evaluating the diagnostic precision of USPIO-enhanced MRI for the detection of lymph mode metastases. Adapted from Wills et al. (2006).

Pooled analysis	No. of pts/LN	No. of studies	Sensitivity (95% CI)	Specificity (95% CI)	DOR (95% CI)	AUROC (SE)
Un-enhanced MRI	334/1270	6	63.0% (57.2%–68.6%)	92.7% (90.9%–94.2%)	26.7 (8.4–84.4)	84.2% (11.0%)
USPIO enhanced MRI	607/3004	17	88.1% (85.3%–90.6%)	96.2% (95.1%–96.8%)	123.0 (5.9–256.9)	96.5% (1.0%)
Analysis by body region and cancer type						
Head and neck	128/1321	4	88.8% (82.7%–93.3%)	97.9% (97.0%–98.7%)	325.8 (22.5–4713.5)	97.0% (2.4%)
Chest	34/97	3	70.2% (55.1%–82.7 %)	88.0% (75.7%–95.5%)	32.9 (8.4–128.9)	92.0% (1.0%)
Abdomen and pelvis	311/1310	9	91.8% (88.6%–94.6%)	96.0% (94.6%–97.1%)	211.1 (118.2–376.9)	97.7% (0.6%)
Prostatic cancer only	130/502	2	91.5% (85.6%–95.5%)	96.7% (94.3%–98.3%)	266.1 (107.8–656.7)	–
Analysis by nodal size						
5–10 mm	98/190	2	96.3% (81.3%–100.0%)	98.8% (95.6%–99.9%)	178.5 (0.3–123332.0)	–
<5 mm	80/185	1	41.2% (18.4%–67.1%)	98.1% (93.0%–98.8%)	–	–
Analysis by contrast dose and 1.5T MRI						
1.5 Tesla MRI	469/2683	12	89.1% (85.9%–91.7%)	97.1% (96.4%–97.8%)	189.8 (75.6–476.5)	97.7% (0.8%)
1.7 mg/kg contrast dose	48/252	3	89.1% (83.1%–93.5%)	88.3% (83.9%–91.8%)	162.1 (13.8–1905.0)	97.4% (2.5%)
2.6 mg/kg contrast dose	529/2676	12	93.3% (91.2%–94.9%)	93.9% (92.8%–94.8%)	32.0 (50.9–283.3)	96.8% (1.1%)

CI = confidence interval; DOR = diagnostic odds ratio; Sensitivity = true positive/(true positive + false negative); specificity = true negative/(true negative + false positive); DOR = [(true positive rate/false positive rate)] / [(1-true positive rate)/(1-false positive rate)].

laparoscopic PLND. Furthermore, if the node is greater than 5 mm, the presence of malignancy can be confirmed by image-guided biopsy and, thus, PLND in these patients must be avoided. This was the case in 5 of 80 (6%) patients in the study of Harisinghani et al. (2003). All five nodes were confirmed positive. Finally, with ferumoxtran-10-enhanced MRI, all pelvic nodes are visualized. Harisinghani and Barentsz (2003) showed that in 11% of their patients, as a result of ferumoxtran-10 injection, metastatic nodes were detected by MRI, which were outside the classical field of lymph node resection. In patients with a suspicion for a recurrence, for example in patients with a PSA-relapse after treatment, this technique may show metastatic nodes even when they are small (Figure 3.3), thus allowing earlier adequate therapy. Finally, identifying small pathologic nodes will facilitate more appropriate use of sophisticated radiation therapy. For example, when positive nodes are accurately identified, precise intensity modulated radiotherapy can be performed. This results in an increased dose on the malignant nodes and a decreased dose, with reduced side effects, on normal tissues.

3.4.2. Urinary Bladder Cancer

In 58 patients with proven bladder cancer, MRI performed before and after injection of ferumoxtran-10 was compared with histopathologic results in surgically removed lymph nodes (Deserno et al., 2004). One hundred and seventy two nodes imaged after injection of ferumoxtran-10 were matched and correlated with the results of node dissection. Of these, 122 were benign and 50 were malignant. With nodal size and shape criteria, the accuracy, sensitivity, specificity and positive and negative predictive values on pre-contrast images were 92%, 76%, 99%, 97%, and 91%, respectively; the corresponding values on post-contrast images were 95%, 96%, 95%, 89%, and 98%. In the depiction of pelvic metastases, sensitivity and negative predictive value significantly improved on post-contrast images compared with those obtained pre-contrast, from 76% to 96% ($P < 0.001$) and from 91% to 98% ($P < 0.01$), respectively. At post-contrast imaging, metastases (4–9 mm) were prospectively found in 10 of 12 normal-sized nodes (<10 mm); these metastases were not detected on pre-contrast images. Post-contrast images also showed lymph nodes that were missed at pelvic node dissection in two patients. Thus ferumoxtran-10–enhanced MRI significantly improves nodal staging in patients with bladder cancer by depicting metastases even in normal-sized lymph nodes.

3.4.3. Head and Neck Cancer

In head and neck cancers, 25% lymph nodes are positive despite negative preoperative imaging (contrast CT and US biopsy) as metastatic nodes are small (5–10 mm). In addition, PET is non-specific and does not provide anatomic location. Therefore, extensive surgery (i.e., radical neck dissection) is performed in virtually all patients. This results in cosmetic deformity and a complication rate of 36%–54%. Mack et al. (2002) reported that ferumoxtran-10 MRI was accurate in 26/27 (96%) patients, which resulted in a reduced extent of surgery in 26% of patients.

3.4.4. Breast Cancer

Early results with ferumoxtran-10 MRI in breast cancer show a sensitivity of 78%, a specificity of 96% and a negative predictive value of 97% (Figure 3.4) (Michel et al. 2002). The sentinel lymph node procedure in breast cancer has a 3% to 10% false negative rate; furthermore, positive internal mammary lymph nodes are missed in 17%. Finally, the sentinel lymph node is the only positive node in 61% lymph node positive patients. These patients all undergo axillary dissection, with subsequent high rate of clinically significant lymph edema. As a result from its high negative predictive value, in patients with a negative ferumoxtran-10 MRI, axillary dissection may potentially be avoided. Further studies are under its way to validate this statement.

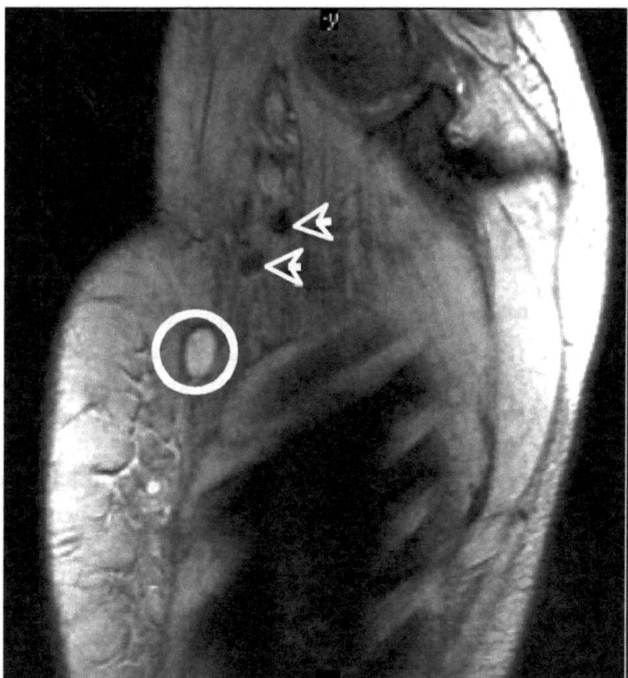

FIGURE 3.4. **A** 55-year-old patient with breast cancer and positive lymph node. Sagittal post-ferumoxtran-10 T_2^*-weighted MEDIC MR image shows one white (circle) and two dark nodes (arrows). The white node showed metastases on histopathology, the dark ones were normal.

3.4.5. Upper Gastrointestinal Cancer

One published study has documented the application of USPIO-MRI for the pre-operative staging and assessment of lymph node metastases in 17 patients with advanced gastric cancer (Tatsumi et al., 2006). Of 781 lymph nodes dissected during surgery, the imaging results of 194 nodes were correlated with their histopathological findings. Fifty-nine lymph nodes from 11 patients had histopathological metastases. In non-affected normal lymph nodes, a dark signal intensity on MRI was evidently caused by the diffuse uptake of the contrast medium by macrophages resident in the lymph nodes, which phagocytosed the iron oxide particles. The number of phagocytic macrophages was decreased in metastatic lymph nodes and they demonstrated three enhancement patterns: (i) lymph nodes with overall dark signal intensity due to the diffuse uptake of USPIO; (ii) lymph nodes with partial high signal intensity due to partial uptake; and (iii) no blackening of lymph nodes due to no uptake of USPIO. Patterns (ii) and (iii) were defined as metastatic giving a sensitivity and specificity of USPIO-enhanced MRI of 100% (59/59) and 92.6% (125/135), respectively. Lymph nodes in the retroperitoneal and para-aortic regions were more readily identified and diagnosed on the MR images than those in the perigastric region. With regard to oesophageal cancer in a recent study of 16 patients (Nishimura et al., 2006) undergoing surgical resection, of the 408 resected lymph nodes, imaging results of 133 nodes were compared with histopathologic findings. Twenty-four lymph nodes had histopathologic metastases, reporting a sensitivity of 100% and specificity of 95.4% for the diagnosis of metastatic nodes using USPIO-enhanced MRI.

3.4.6. Rectal Cancer

Similarly, one study of 12 patients with rectal cancer has been published (Koh et al., 2004). Appearances of 74 nodes on in vivo MRI were compared with histopathologic findings. Six lymph nodes contained metastases which demonstrated eccentric and uniform high signal intensity patterns of USPIO uptake, reporting a sensitivity and specificity of 67% and 96% respectively. An abstract presentation by Rasheed et al. (2006) of 13 patients with rectal cancer who underwent MRI before and 24 hours after intravenous infusion of USPIO contrast reported a good correlation between the histopathogical lymph nodes and the ones seen on USPIO-unenhanced MRI ($r = 0.73$; $P = 0.005$). Taking into account the recent study on the use of USPIO in gastric cancer (Tatsumi et al., 2006), in addition to published meta-analytical work, the pooled sensitivity and specificity of USPIO-enhanced MRI for the detection of lymph node metastases in the abdomen and pelvis would be 93% and 96% respectively (Figure 3.5).

Finally, we believe that novel 3D reconstruction techniques would be of particular help in displaying and analyzing the massive amount of high-resolution data. In this sense it should be feasible to display both normal and abnormal lymph nodes and their location with respect to important surgical landmarks

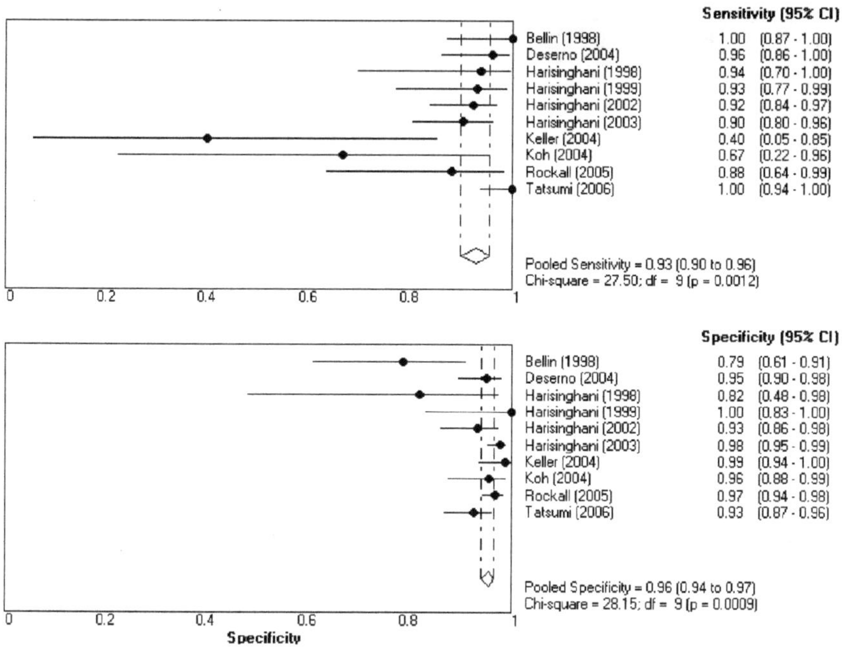

FIGURE 3.5. Updated sensitivity and specificity of USPIO-enhanced MRI for the detection of lymph node metastases in the abdomen and pelvis.

like vessels, obturator nerves, and ureters in 3D. In this respect, a Lymph Node Task Card is a works-in-progress post-processing package (Siemens Princeton, USA) allowing simultaneous viewing of 3D sequences in four segments, with in line MIP and VRT capabilities for rendering datasets. In addition, automatic fusion of CTA and ferumoxtran-MRI data is possible with this tool (Figure 3.6). This allows a better detection by clinicians as to where the positive nodes are located and increases the precision in the planning of automated radiation oncology.

3.5. Conclusion

A macrophage-specific MR contrast agent allows the detection of small and otherwise undetectable lymph node metastases in patients with cancer. This has an important clinical impact, as the diagnosis will be more precise and less invasive to obtain. Subsequently, this will reduce morbidity and health care costs. However, thorough knowledge of sequence parameters and planes, lymph node anatomy, appearance of normal and abnormal nodes and pitfalls is essential when using this technique. This implies a very important role for education by expert radiologists, MR-manufacturers and contrast agent companies.

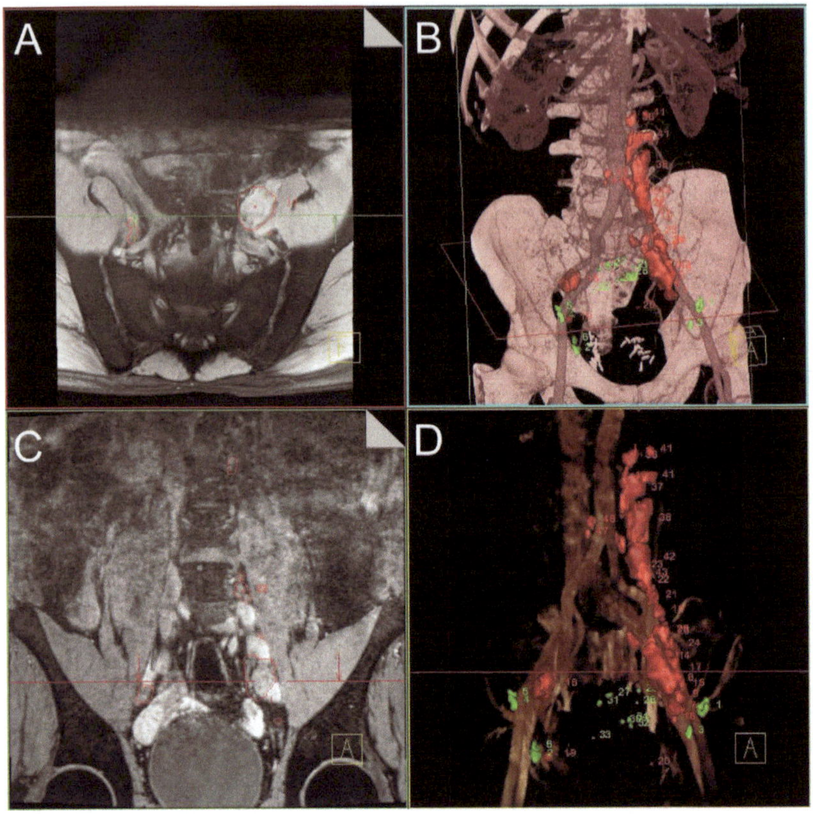

FIGURE 3.6. **A** 51-year-old patient with prostate cancer and generalized lymph node metastases. Siemens Task Card showing MR images (**A,C,D**) 24 hours post-ferumoxtran-10 and (**B**) CTA-MRL fusion. **A** High resolution T_2^*-weighted sequence. **B** MRL-CTA merged image showing pathologic nodes in red and normal nodes in green. With the help of a segmentation technique pathologic nodes ("white" on **A** and **C**) are indicated by a red color and normal node as green, and thus relation between vessels and nodes can be seen. **C** 3D T_2-weighted sequence. **D** T_1-weighted MRA sequence shows vessels and segmented nodes.

References

Anzai, Y., Blackwell, K.E., Hirschowitz, S.L., Rogers, J.W., Sato, Y., Yuh, W.T., et al. 1994. Initial clinical experience with dextran-coated superparamagnetic iron oxide for detection of lymph node metastases in patients with head and neck cancer. Radiology 192:709–715.

Anzai, Y., Piccoli, C.W., Outwater, E.K., Stanford, W., Bluemke, D.A., Nurenberg, P., et al. 2003. Evaluation of neck and body metastases to nodes with ferumoxtran 10-enhanced MRI: phase III safety and efficacy study. Radiology 228:777–788.

Bellin, M.F., Roy, C., Kinkel, K., Thoumas, D., Zaim, D. et al. 1998. Lymph node metastases: safety and effectiveness of MRI with Ultrasmall Superparamanetic Iron Oxide Particles. Initial clinical experience. Radiology 207:799–808.

Deserno, W.M., Harisinghani, M.G., Taupitz, M., Jager, G.J., Witjes, J.A., Mulders, P.F., Hulsbergen van de Kaa, C.A., Kaufmann, D., and Barentsz, J.O. 2004. Urinary bladder cancer: preoperative nodal staging with ferumoxtran-10-enhanced MRI. Radiology 233:449–456.

Gerlowski, L.E., and Jain, R.K. 1986. Microvascular permeability of normal and neoplastic tissues. Microvasc Res 31:288–305.

Harisinghani, M.G., Saini, S., Hahn, P.F., Weissleder, R., and Mueller, P.R. 1998. MR imaging of lymph nodes in patients with primary abdominal and pelvic malignancies using ultrasmall superparamagnetic iron oxide (Combidex). Acad Radiol 5(Suppl 1):S167–S169.

Harisinghani, M.G., Saini, S., Weissleder, R., Hahn, P.F., Yantiss, R.K., Tempany, C., et al. 1999. MR lymphangiography using ultrasmall superparamagnetic iron oxide in patients with primary abdominal and pelvic malignancies: radiographic-pathologic correlation. AJR Am J Roentgenol 172:1347–1351.

Harisinghani, M.G. and Barentsz, J.O. (co-first authors), Hahn, P.F., Deserno, W., de la Rosette, J., Saini, S., et al. 2002. MR lymphangiography for detection of minimal nodal disease in patients with prostate cancer. Acad Radiol 9(Suppl 2):S312–S313.

Harisinghani, M., Barentsz, J., Hahn, P., Deserno, W., et al. 2003. Noninvasive detection of clinically occult lymph-node metastases in prostate cancer. NEJM 348: 2491–2499.

Hoffman, H.T., Quets, J., Toshiaki, T., Funk, G.F., McCulloch, T.M., Graham, S.M., et al. 2000. Functional magnetic resonance imaging using iron oxide particles in characterizing head and neck adenopathy. Laryngoscope 110:1425–1430.

Keller, T.M., Michel, S.C., Frohlich, J., Fink, D., Caduff, R., Marincek, B., et al. 2004. USPIO-enhanced MRI for preoperative staging of gynecological pelvic tumors: preliminary results. Eur Radiol 14:937–944.

Kernstine, K.H., Stanford, W., Mullan, B.F., Rossi, N.P., Thompson, B.H., Bushnell, D.L., et al. 1999. PET, CT, and MRI with Combidex for mediastinal staging in non-small cell lung carcinoma. Ann Thorac Surg 68:1022–1028.

Koh, D.M., Brown, G., Temple, L., Raja, A., Toomey, P., Bett, N., et al. 2004. Rectal cancer: mesorectal lymph nodes at MR imaging with USPIO versus histopathologic findings–initial observations. Radiology 31:91–99.

Mack, M.G., Balzer, J.O., Straub, R., Eichler, K., and Vogl, T.J. 2002. Superparamagnetic Iron Oxide–enhanced MR Imaging of Head and Neck Lymph Nodes. Radiology 222:239–244.

Michel, S.C.A., Keller, T.M., Fröhlich, J.M., Fink, D., Caduff, R., Seifert, B., Marincek, B., and Kubik-Huch, R. 2002. Preoperative Breast Cancer Staging: MR Imaging of the Axilla with Ultrasmall Superparamagnetic Iron Oxide Enhancement. Radiology 225:527–536.

Nguyen, B.C., Stanford, W., Thompson, B.H., Rossi, N.P., Kernstine, K.H., Kern, J.A., et al. 1999. Multicenter clinical trial of ultrasmall superparamagnetic iron oxide in the evaluation of mediastinal lymph nodes in patients with primary lung carcinoma. J Magn Reson Imag 10:468–473.

Nishimura, H., Tanigawa, N., Hiramatsu, M., Tatsumi, Y., Matsuki, M., and Narabayashi, I. 2006. Preoperative esophageal cancer staging: magnetic resonance imaging of lymph node with ferumoxtran-10, an ultrasmall superparamagnetic iron oxide. J Am Coll Surg. 202:604–611.

Pannu, H.K., Wang, K.P., Borman, T.L., and Bluemke, D.A. 2000. MR imaging of mediastinal lymph nodes: evaluation using a superparamagnetic contrast agent. J Magn Reson Imag 12:899–904.

Rasheed, S., Guenther, T., Talbot, I., McDonald, P., Northover, J., Stirling, J., Culver, L., Glynne-Jones, R., and Padhani, A.R. 2006. USPIO - enhanced rectal cancer specimen MRI: how well does it correlate with node identification at histopathology? Colorectal Dis. 8:721–721.

Rockall, A.G., Sohaib, S.A., Harisinghani, M.G., Babar, S.A., Singh, N., Jeyarajah, A.R., et al. 2005. Diagnostic performance of nanoparticle-enhanced magnetic resonance imaging in the diagnosis of lymph node metastases in patients with endometrial and cervical cancer. J Clin Oncol 23:2813–2821.

Sigal, R., Vogl, T., Casselman, J., Moulin, G., Veillon, F., Hermans, R., et al. 2002. Lymph node metastases from head and neck squamous cell carcinoma: MR imaging with ultrasmall superparamagnetic iron oxide particles (Sinerem MR) – results of a phase-III multicenter clinical trial. Eur Radiol 12:1104–1113.

Stets, C., Brandt, S., Wallis, F., Buchmann, J., Gilbert, F.J., and Heywang-Kobrunner, S.H. 2002. Axillary lymph node metastases: a statistical analysis of various parameters in MRI with USPIO. J Magn Reson Imag 16:60–68.

Tatsumi, Y., Tanigawa, N., Nishimura, H., et al. 2006. Preoperative diagnosis of lymph node metastases in gastric cancer by magnetic resonance imaging with ferumoxtran-10. Gastric Cancer 9:120–128.

Vassallo, P., Matei, C., Heston, W.D.W., et al. 1994. AMI-227-enhanced MR Lymphography: usefulness for differentiating reactive from tumor bearing lymph nodes. Radiology 193:501–506.

Weissleder, R., Elizondo, G., Wittenberg, J. et al. 1990. Ultrasmall paramagnetic iron oxide: an intravenous contrast agent for assessing lymph nodes with MR imaging. Radiology 175:494–498.

Will, O., Purkayastha, S., Chan, C., Athanasiou, T., Darzi, A.W., Gedroyc, W., and Tekkis, P.P. 2006. Diagnostic precision of nanoparticle-enhanced MRI for lymph node metastases: a meta-analysis. Lancet Oncol 7(1):52–60.

4
Use of SPIOs for Clinical Liver Imaging

Akihiro Tanimoto

Abstract: Superparamagnetic iron oxide (SPIO) particles are as MR contrast media composed of iron oxide crystals coated with dextran or carboxydextran. These particles are sequestered by phagocytic Kupffer cells in the normal reticuloendothelial system (RES), but are not retained in tumor tissue. Consequently, there are significant differences in T_2/T_2^* relaxation between normal RES tissue and tumors. This results in an increased lesion conspicuity and detectability. The introduction of SPIO has been expected to substantially increase the detectability of hepatic metastases. For focal hepatocellular lesions, it has been documented that SPIO-enhanced MR imaging exhibits slightly better diagnostic performance than dynamic helical CT in the detection of hypervascular hepatocellular carcinoma (HCC). A combination of dynamic and static MR-imaging technique using T_1- and T_2-imaging criteria appears to provide clinically more useful patterns of enhancement. SPIO-enhanced MR imaging also provides information useful for differential diagnosis, via enhancement of hemangiomas and RES-containing tumors. With the exploitation of rapid T_2^*-sensitive sequences, SPIO-enhanced dynamic MR imaging may become comparable to gadolinium-enhanced dynamic MR imaging and dynamic studies with multi-detector-row CT. SPIO-enhanced MR imaging plays an important role in therapeutic decision-making for patients with malignant hepatic tumors.

Keywords: Liver, contrast media, SPIO, MRI, neoplasm, metastasis, hepatocellular carcinoma.

4.1. Introduction

Because of the unique physiologic properties of the liver, improved depiction and characterization of focal hepatic lesions is possible with liver-specific MR contrast agents. Superparamagnetic iron oxide (SPIO) particles are opsonized and sequestered by phagocytic Kupffer cells of the normal reticuloendothelial system (RES). Phagocytosed SPIO particles in Kupffer cells produce strong T_1, T_2, and T_2^* relaxation effects in the liver parenchyma. Malignant tumors have no Kupffer cells and therefore do not retain no SPIO and exhibit no change in signal. This leads to increased tumor-liver contrast, which can be exploited to decrease threshold size for lesion detection.

SPIO has been widely used for liver MR imaging in the last several years (Ward et al., 2003; Reimer et al., 2000; Poeckler-Schoeniger et al., 1999). Appropriate use of tissue-specific agents would permit accurate detection and characterization of focal hepatic lesions. Ferumoxides and ferucarbotran are commercially available as SPIO particles, and the latter is a bolus-injectable agent, which allows shortened examination time and dynamic perfusion MR imaging. In this chapter, current clinical evaluation of SPIO for focal hepatic lesions is reviewed with the literature and our experience with it.

4.2. SPIO Particles

A SPIO particle is a conglomerate of numerous iron oxide crystals coated with dextran or carboxydextran. The mean size of SPIOs ranges from approximately 60 to 250 nm, making it subject to phagocytosis by Kupffer cells. Sequestered SPIO particles are metabolically biodegradable and bioavailable. They therefore exhibit a rapid turnover into the body iron stores and incorporation into erythrocyte hemoglobin. Two SPIO particle formulations are now clinically available, ferumoxides and ferucarbotran (Table 4.1). Superparamagnetism is a property intermediate between those of paramagnetic and ferromagnetic materials. Superparamagnetic materials are comprised of crystals of certain materials, such as magnetite (Fe_3O_4) and maghemite (Fe_2O_3), large enough to form a solid-phase microscopic volume or "domain" in which atomic unpaired electron spins are aligned by positive exchange forces. When a domain is a volume of material (5–35 nm diameter for Fe_3O_4) possessing uniform magnetization and a specific direction, superparamagnetism is displayed. Magnetic field gradients induced by superparamagnetic particles contribute to the dephasing of protons that move by diffusion in the vicinity of a particle, resulting in significant T_2/T_2^* relaxation.

TABLE 4.1. Features of commercially available SPIO particles.

Commercial name	Feridex, Endorem	Resovist
Generic name	Ferumoxides	Ferucarbotran
Particle size (nm)	100–250	57
r1 ($mM^{-1}sec^{-1}$)	23	24
r2 ($mM^{-1}sec^{-1}$)	100	168
Core material	(Fe_2O_3) m·(FeO) n	$Fe_2O_3 + Fe_3O_4$
Method of administration	Drip infusion	Bolus injection

4.3. Clinical Applications

4.3.1. MR-Imaging Parameters for SPIO-MRI

Enhancement with SPIO depends on susceptibility effects, which vary with different pulse sequences, so pulse sequence choice is critically important in determining the diagnostic effectiveness of SPIO, and optimized parameters are essential to maximize lesion detection. Pulse sequence optimization for SPIO-enhanced MR imaging has recently been discussed (Ward et al., 2000; Kim et al., 2002a,b). Basically, pre- and post-contrast MR images consist of three imaging parameters: T_2-weighted (fast) spin echo (SE), T_1-weighted gradient echo (GRE), and T_2^*-weighted GRE. We compared signal-to-noise ratio (SNR) and lesion detectability among T_2-weighted fast spin echo (FSE), echo planar imaging (EPI), short TE fast spoiled GRASS (SPGR), and long TE fast SPGR. Long TE fast SPGR exhibited the best SNR and detectability. This sequence provides little lesion-liver contrast before contrast enhancement, but a large decrease in signal occurs after SPIO administration (Figure 4.1). Therefore, lesion-liver contrast after SPIO is independent of inherent T_1 and/or T_2 difference between lesions and liver tissue, but highly dependent on Kupffer cell distribution and phagocytic function. The flip angle on long TE fast SPGR was set at 45–60 degrees, such that solid masses retain higher signal intensity than the surrounding liver and are distinguishable from liver cysts, which become almost as low in intensity as the surrounding liver tissue. SPIO-enhanced MR imaging is basically a modality for delineating phagocytic activity and cannot be used to assess the lesion vascularity and/or viability. This weakness could be circumvented with the use of perfusion MR imaging.

Perfusion study using echo planar imaging (EPI) yields negative enhancement of hypervascular tumors (Ichikawa et al., 1999), and one-stop shop diagnosis (involving both dynamic and RES-targeted MR imaging) for hypervascular hepatocellular carcinoma (HCC) may be feasible in the future. Parallel imaging will improve the image quality of single-shot EPI by correcting magnetic field inhomogeneity. On dynamic MR imaging using T_1-weighted GRE, enhancement was positive in the liver for at least 30 seconds after bolus injection of SPIO (Reimer, 1998). However, positive enhancement of hypervascular HCC in early phase of T1W-GRE is too weak to assess the tumor viability. Recent advances in parallel imaging techniques have allowed for further improvements in image quality by reducing blurring and minimizing susceptibility-induced artifacts. One trial study of whole-body cancer screening using positron emission tomography (PET)-like projection images based on diffusion-weighted imaging (DWI) data has been reported (Takahara et al., 2004). Recently, combined use of diffusion-weighted imaging and SPIO was attempted for a small number of patients, and post-SPIO DWI exhibited improved contrast-to-noise ratio between malignant lesions and liver (Naganawa et al., 2005). Further study is needed to determine whether the combination of DWI and SPIO can improve the detection of malignant liver lesions.

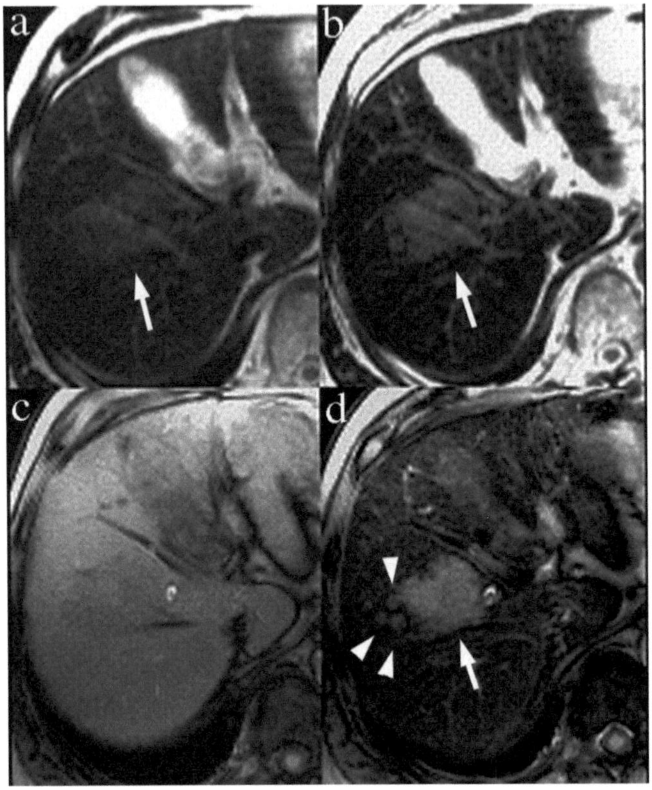

FIGURE 4.1. Lesion-liver contrast on FSPGR with long echo time (TE). On T_2-weighted FSE, a hyperintense mass is noted before SPIO (**a**, arrow). The lesion is more clearly visualized after SPIO (**b**, arrow). However, T_2-weighted FSE images mask the effect of SPIO because of its inherent T_2 contrast. On FSPGR with long TE, liver SNR is excellent but lesion-liver contrast is poor before SPIO (**c**). After SPIO administration, liver signal is markedly decreased and HCC (**d**, arrow) with satellite nodules (**d**, arrowheads) is clearly demonstrated. Because the inherent lesion-liver contrast is weak, the contrast after SPIO is due to Kupffer cell function alone.

4.3.2. Comparative Studies

Despite the advent of newer treatment modalities, surgical resection is still considered the principal treatment for focal malignant hepatic lesions. In the pre-operative evaluation of focal hepatic lesions, imaging methods that can reliably depict all malignant lesions are necessary, since imaging findings affect the choice of surgical intervention. To clarify the clinical role of SPIO-enhanced MR imaging in multi-modality decision-making, numerous comparative studies have been conducted, particularly in Asia and Europe, because of the high prevalence of HCC in these continents. However, conclusions drawn from such comparative

TABLE 4.2. Rates of detection of hypervascular HCC for the same patients: A comparison.

	Dynamic CT	Dynamic MRI	SPIO-MRI
Ø < 1 cm ($n = 29$)	12 (41%)*	23 (79%)	20 (69%)
Ø >= 1 cm ($n = 43$)	38 (88%)	41 (95%)	42 (98%)
Total ($n = 72$)	50 (69%)*	64 (89%)	62 (86%)

studies should be carefully weighed since imaging equipment and parameters were not uniform among institutions (Table 4.2).

4.3.2.1. SPIO-MRI vs Non-enhanced MRI

In the early years of SPIO-enhanced MR imaging, Stark et al. reported that SPIO-enhanced images of the liver obtained with standard pulse sequence techniques significantly increased the number of hepatic lesions detected and reduced the threshold size for detection to 3 mm (Stark et al., 1988). SPIO-enhanced MR imaging is more accurate than non-enhanced MR imaging for the detection of focal hepatic lesions (Figure 4.2), and combined analysis of non-enhanced and SPIO-enhanced images is more accurate in the characterization of focal hepatic lesions on SPIO-enhanced images alone (Reimer et al., 2000; Stark et al., 1988).

4.3.2.2. SPIO-MRI vs Dynamic CT

SPIO-enhanced MRI is particularly advantageous for detecting hepatic metastases, because the surrounding liver sustains normal phagocytic activity, and metastatic liver tumors have no Kupffer cells. Therefore, the diagnosis is more straightforward than HCC derived from liver cirrhosis. Ward et al. compared SPIO-enhanced MR imaging with dual phase dynamic CT for 51 hepatic resection candidates with known colorectal metastases (Ward et al., 1999). The mean sensitivity of MR was significantly higher than that of CT ($p < 0.02$): 79.8% for MR and 75.3% for CT for all lesions, and 80.6% for MR and 73.5% for CT for malignant lesions. SPIO-enhanced MR imaging was more sensitive than dual-phase CT in the depiction of colorectal metastases (Ward et al., 1999). The development of multi-detector row helical CT (MDCT) allows multiple images to be obtained simultaneously during spiral acquisition. This technological advance has allowed effective scanning of the liver at collimations less than 5 mm in a single breath hold. Haider et al. (2002) investigated the value of collimations less than 5 mm in detecting hepatic metastases 1.5 cm or smaller by using MDCT. They found that no significant difference was noted in the pooled sensitivity for metastatic lesions (80% [20 of 25 lesions]) and in the conspicuity of lesions at different collimations (5.00, 3.75, and 2.50 mm) (Haider et al., 2002).

Recently, Onishi et al. (2006) compared SPIO-enhanced MR imaging with MDCT. There was no significant difference in the sensitivity of detection of hepatic metastases between contrast-enhanced MDCT alone and SPIO-enhanced

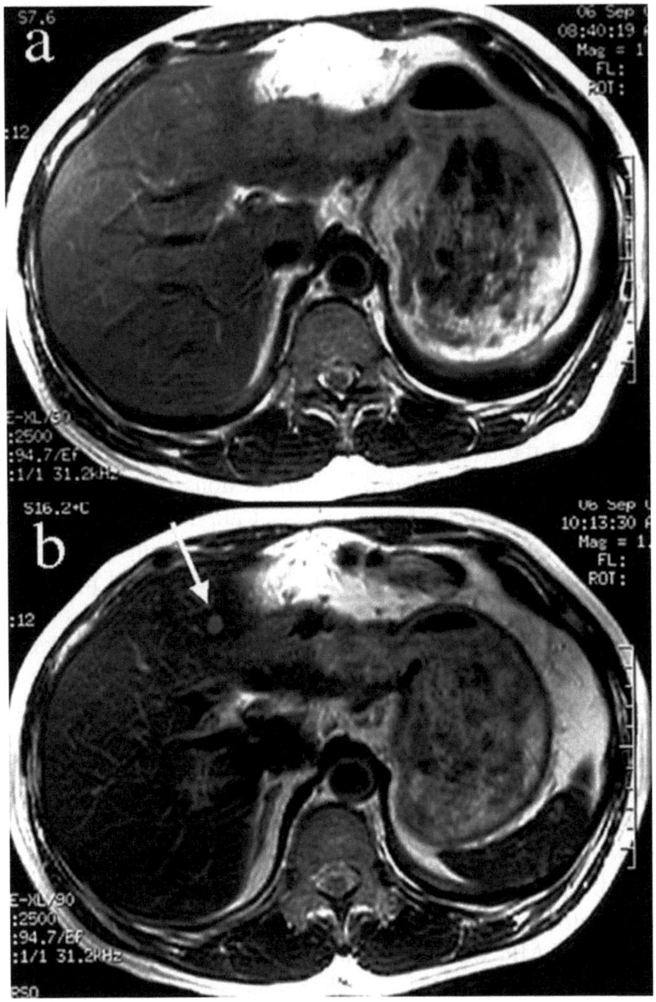

FIGURE 4.2. Liver metastasis from gastric cancer in a 52-year-old male: Unenhanced vs SPIO-enhanced MR imaging. No lesion is found on unenhanced T_2-weighted FSE (**a**). SPIO-enhanced MR imaging reveals a small mass exhibiting decreased phagocytic activity in the lateral segment (**b**; arrow).

MR images alone. However, the addition of SPIO-enhanced MR imaging to contrast-enhanced MDCT can improve sensitivity in the detection of hepatic metastases. Another study using MDCT, Gd-enhanced MR imaging and SPIO-enhanced MR imaging showed that accuracy for gadolinium-enhanced MR imaging and SPIO-enhanced MR imaging was similar, but both techniques were significantly more accurate than CT (Ward et al., 2005). Because they used a four-detector row CT scanner, further studies will be needed to assess diagnostic capability of recent 64-detector row CT scanner.

It has been documented that SPIO-enhanced MR imaging is more sensitive than dual-phase spiral CT in depicting hypervascular HCC (Reimer et al., 2000). Lee et al. (2003) showed that the mean sensitivity of SPIO-enhanced MR imaging was significantly higher (70.6%, $P < 0.05$) than that of dual-phase spiral CT (58.1%). Kim et al. (2005) compared SPIO (ferucarbotran)-enhanced MR imaging with triple-phase MDCT for pre-operative detection of HCC. In their study, the mean sensitivities of MR imaging and triple-phase MDCT were 90.2% and 91.3%, respectively, their mean specificities were 97.0% and 95.3%, respectively. SPIO-enhanced MR imaging was as accurate as triple-phase MDCT in pre-operative detection of HCC (Kim et al., 2005). SPIO-enhanced MR imaging provides information supplementary to that obtained with dynamic CT, particularly by excluding pseudolesions. SPIO-enhanced MR imaging may be preferable because it features no exposure to irradiation.

4.3.2.3. SPIO-MRI vs Dynamic MRI

Vogl et al. (1996) assessed the efficacy of static and dynamic MR imaging using SPIO versus Gd-DTPA in patients with focal liver lesions. Detection rate was improved for metastatic lesions revealing 36 lesions unenhanced versus 53 focal lesions using SPIO-enhanced MR imaging, whereas Gd-DTPA-enhanced scans showed no additional lesion versus unenhanced and Resovist-enhanced MR imaging (Vogl et al., 1996). Also, SPIO decreased the signal intensity of benign tumors and helped differentiate benign from malignant tumors (specificity: SPIO-MRI 93%, gadolinium-MRI 81.5%) (Vogl et al., 1996). In view of sensitivity and specificity for hepatic metastases, SPIO-enhanced MR imaging has priority to dynamic CT and Gd-enhanced MR imaging.

Several studies have shown that Gd-based dynamic MR imaging is slightly better than SPIO-enhanced MR imaging in the detection of small HCCs (Tang et al., 1999, Pauleit et al., 2002). In lesion conspicuity, Gd-enhanced MR imaging is superior to SPIO-enhanced MR imaging (Tang et al., 1999). Paulait et al., (2002) reported similar findings, but they used only respiratory-triggered T2-weighted turbo spin echo as a pulse sequence for SPIO enhancement. However, SPIO yields additional information when imaging findings on dynamic MR imaging are questionable because of intrahepatic arterioportal shunt (AP shunt) and/or post-therapeutic liver damage (Oudkerk et al., 1997) (Figure 4.3). Pseudolesions caused by AP shunt can be circumvented by the use of SPIO, since Kupffer cell function is usually maintained in liver parenchyma exhibiting AP shunt (Oudkerk et al., 1997). Ward et al. (2000) reported the usefulness of double-contrast MR imaging, i.e. combined SPIO- and Gd-dynamic MR imaging on the same day, for diagnosis of HCC. SPIO-enhanced MR imaging (mean accuracy = 0.76) was more accurate than unenhanced MR imaging (mean accuracy = 0.64, $P < 0.04$), and double-contrast MR imaging (mean accuracy = 0.86) was more accurate than SPIO-enhanced imaging ($P < 0.05$). Both HCCs and dysplastic nodules were correctly characterized with all three techniques, although observer confidence in lesion characterization was greatest

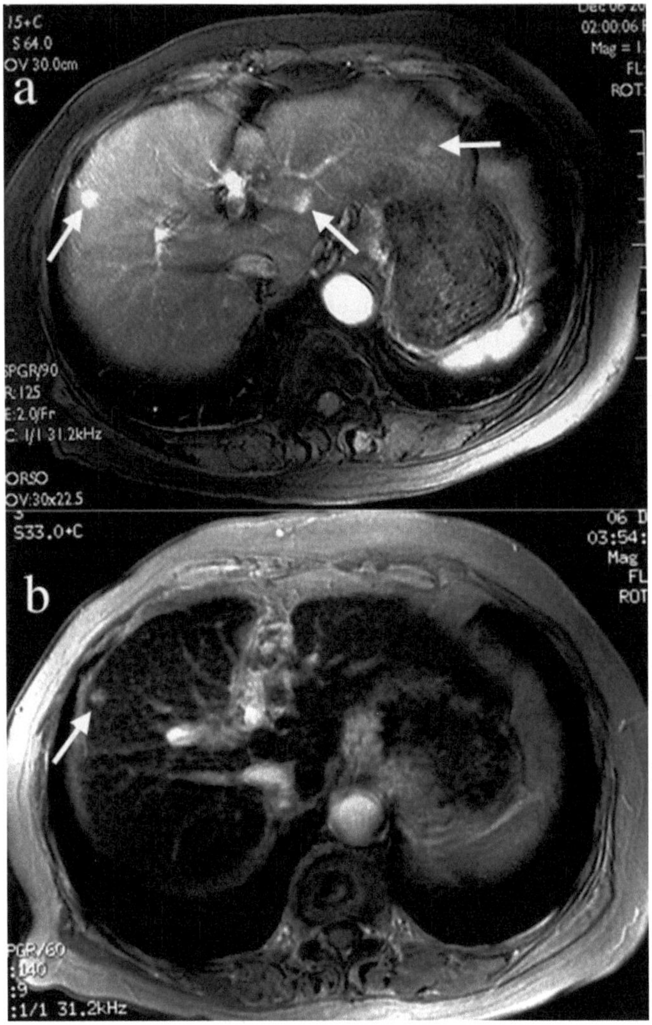

FIGURE 4.3. HCC in a 71-year-old male: dynamic MR imaging vs SPIO-enhanced MR imaging. On dynamic MR imaging (**a**), multiple enhancing lesions are noted in the liver (arrows). SPIO-MR imaging demonstrated only one lesion in S8 (**b**; arrow). At surgery, the lesion in S8 was proven to be HCC, and the other lesions were not detected and were considered pseudolesions due to AP-shunt.

with double-contrast MR imaging (Figure 4.4). Double-contrast MR imaging significantly improves the diagnosis of HCC compared with SPIO-enhanced and SPIO-non–enhanced imaging ($P < 0.01$).

4.3.2.4. SPIO-MRI vs Paramagnetic Hepatobiliary Agents-Enhanced MRI

Paramagnetic hepatobiliary compounds are referred to as "value-added versions" of extracellular Gd compounds because they increase tumor-liver contrast in

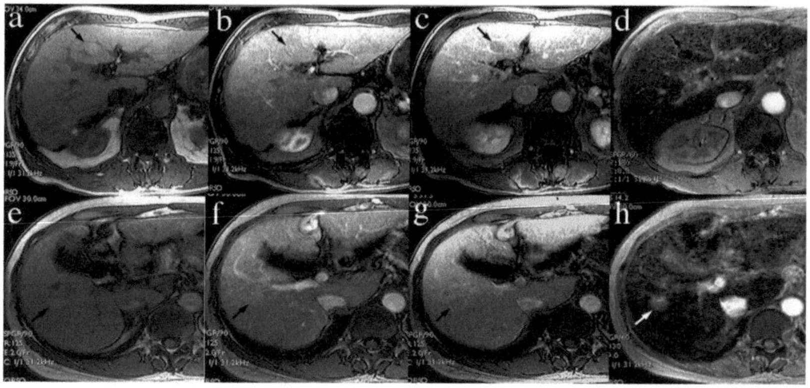

FIGURE 4.4. Comparison between dynamic MR imaging and SPIO-enhanced MR imaging in differentiating borderline lesions. Dysplastic nodule in a 75-year-old male (**a–d**) and, well-differentiated HCC in a 65-year-old male (**e–h**). On pre-contrast T1W-GRE, both lesions (**a** and **e**, arrows) exhibit high intensity and are hypovascular on dynamic MR imaging (**b, c, f**, and **g**; arrows), and they cannot be distinguished from each other. On SPIO-enhanced MR imaging, DN takes up SPIO (**d**, arrow) but HCC exhibits decreased phagocytic activity (**h**, arrow).

both perfusion phase and hepatobiliary phase. These agents are partially taken up by hepatocytes, yielding positive and sustained enhancement of the liver parenchyma on T_1-weighted images. There are few reports regarding comparison of efficacy between SPIO and paramagnetic hepatobiliary agents in the diagnosis of focal hepatic lesions. Kim et al. (2004, 2005) have recently reported comparative studies between SPIO and gadobenate dimeglumine (Gd-BOPTA) for the detection of hepatic metastases or HCC. In the detection of liver metastases, the mean accuracy (Az values) and sensitivity of Gd-BOPTA-enhanced delayed phase imaging (0.982, 95.5%) were comparable to those of SPIO-enhanced imaging (0.984, 97.2%) (Kim et al., 2005). They concluded that Gd-BOPTA-enhanced delayed phase imaging showed comparable diagnostic performance to SPIO-enhanced imaging for the detection of liver metastases. In another study, the authors concluded that Gd-BOPTA-enhanced 3D dynamic imaging exhibited better diagnostic performance than SPIO-enhanced imaging in the detection of HCC (Kim et al., 2004). The mean sensitivity and positive predictive value of SPIO-enhanced imaging were 81.0% and 85.0%, respectively, while those of Gd-BOPTA-enhanced MRI were 91.4% and 88.1%, respectively (Kim et al., 2004). For gadoxetic acid (Gd-EOB-DTPA), one animal study showed that the detectability of tumors with SPIO-enhanced MR imaging was comparable to that of Gd-EOB-enhanced "static" MR imaging (Tanimoto et al., 1997). Gadoxetic acid might have been better than SPIO for use with dynamic MR imaging.

4.3.2.5. SPIO-MRI vs CTAP/CTHA

CT during arterial portography (CTAP) plus CT hepatic arteriography (CTHA) is a relatively invasive combination of modalities, but has been considered the most sensitive method for detection of focal hepatic lesions (Li et al., 1999). SPIO-enhanced MR imaging has been found at least as accurate as CT during arterial portography (CTAP) in the detection of liver metastases (Figure 4.5) (Senéterre et al., 1996). Additional merit has also been documented for SPIOs in the differentiation between true lesions and pseudo-lesions (perfusion defect on CTAP caused by localized AP shunt) (Oudkerk et al., 1997). Therefore, SPIO-enhanced MR imaging is a good alternative to CTAP, which is both invasive and expensive. The combination of CTAP and CTHA is superior to CTAP alone

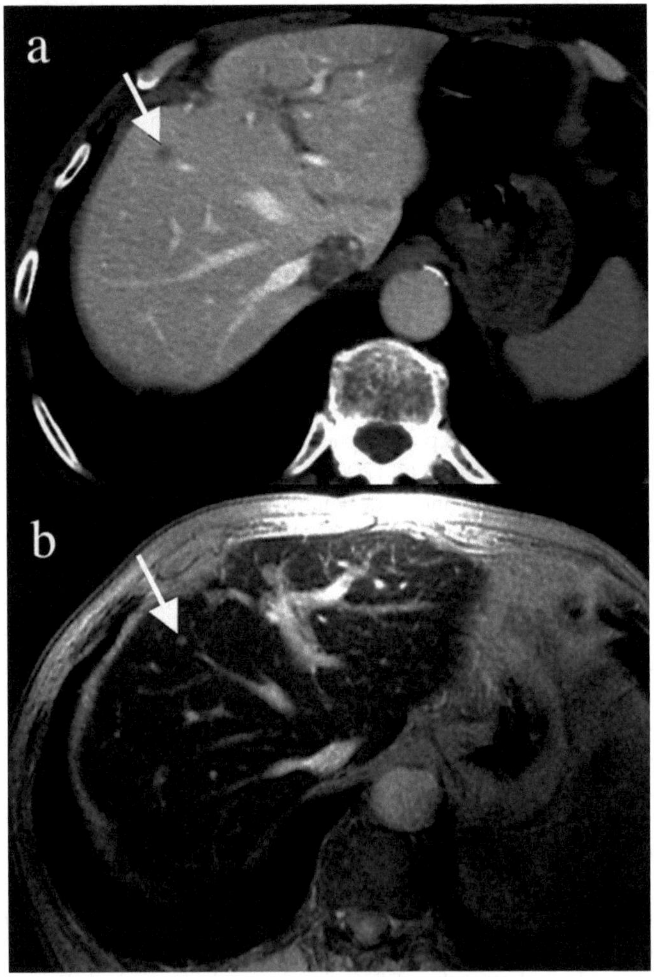

FIGURE 4.5. Liver metastasis from colonic cancer in a 74-year-old male. On CTAP, a small perfusion defect is noted in S4 (**a**, arrow). On SPIO-enhanced T_2^*-weighted GRE, a small mass exhibiting decreased phagocytic activity is noted in the same region (**b**, arrow).

for the detection of hypervascular HCCs (Murakami et al., 1997; Makita et al., 2000). However, the use of CTAP plus CTHA is limited by pseudolesions due to intrahepatic AP shunts particularly in the setting of chronic liver damage, and its specificity is relatively low (Makita et al., 2000). Some authors do not recommend CTAP plus CTHA for pre-operative evaluation of HCC, because of its invasiveness, cost, and its unacceptably high false-positive rate without substantially increasing sensitivity compared with triple-phase helical CT (Jang et al., 2000).

Kwak et al. (2004) compared Gd-enhanced dynamic MR imaging, SPIO-enhanced MR imaging, and combined Gd-enhanced dynamic and SPIO-enhanced MR imaging with combined CTAP and CTHA in the detection of HCC using receiver operating characteristic (ROC) analysis for 24 patients with 38 HCCs. The diagnostic accuracies of combined CTAP/CTHA and combined Gd-enhanced dynamic/SPIO-enhanced MR imaging were significantly higher than those of Gd-enhanced dynamic MR imaging or SPIO-enhanced MR imaging by themselves ($p < 0.005$). The mean specificities of combined CTAP and CTHA (93%) and combined Gd-enhanced dynamic and SPIO-enhanced MR imaging (95%) were significantly higher than those of Gd-enhanced dynamic MR imaging (87%) and SPIO-enhanced MR imaging (88%; $p < 0.05$). Combined Gd-enhanced dynamic and SPIO-enhanced MR imaging may obviate the need for more invasive combined CTAP and CTHA for pre-operative evaluation of patients with HCC. In our study, SPIO-enhanced MR imaging exhibited a diagnostic efficacy equivalent to that of CTAP plus CTHA as a pre-operative test for HCCs (Tanimoto et al., 2005). SPIO-enhanced MR imaging has the potential to replace CTAP plus CTHA in certain clinical settings, and is recommended for the pre-operative work-up of candidates for surgical resection of malignant hepatic tumors. Because CTAP plus CTHA is approximately three times more expensive than SPIO-enhanced MR imaging, the use of SPIOs is also economically favorable. However, it should be noted that inhomogeneity of SPIO uptake can occur because of reduction in Kupffer cell density caused by inflammation, scarring, and regeneration in cirrhosis (Elizondo et al., 1990).

4.3.3. Differential Diagnosis

The advantages of SPIO-enhanced MR imaging have also been reported in tissue characterization. Some types of hepatic lesions such as focal nodular hyperplasia (FNH), hepatocellular adenoma, dysplastic nodules (DN), and well-differentiated HCC sustain phagocytic activity and may demonstrate iron oxide uptake (Grandin et al., 1995; Vogl et al., 1996). Sustained phagocytic activity is a feature of FNH, and is helpful in differential diagnosis with SPIO-enhanced MR imaging (Grandin et al., 1995; Vogl et al., 1996) (Figure 4.6). One criterion, of a threshold signal loss of 10% on SPIO-enhanced MR images, has been established to distinguish benign from malignant lesions (sensitivity 88%, specificity 89%) by receiver-operating characteristic analysis (Vogl et al., 1996). Successful differentiation between HCC and DN is clearly of great importance for the early

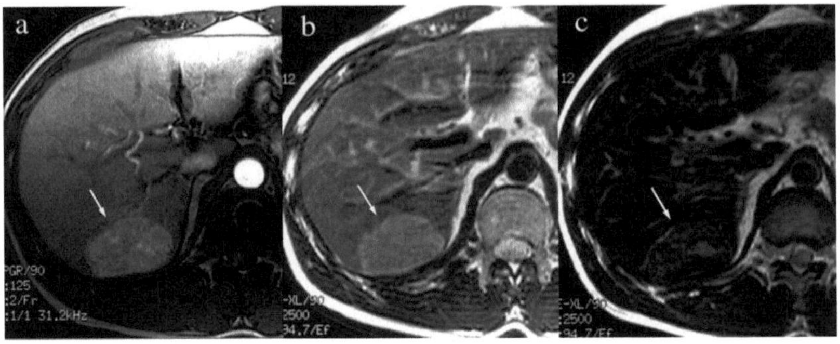

FIGURE 4.6. Focal nodular hyperplasia in a 34-year-old male. In the arterial phase of dynamic MR imaging, a well-enhanced tumor is noted in S7 (**a**, arrow). Pre-contrast T_2W-FSE reveals a high-intensity mass (**b**, arrow). On SPIO-enhanced T_2W-FSE, the tumor takes up SPIO and its signal intensity is decreased (**c**).

and precise treatment of HCC in cirrhotic liver. One study found no significant difference in the number of Kupffer cells between well-differentiated HCC and surrounding liver tissue (Tanaka et al., 1996). It should thus be noted that phagocytic activity might overlap among borderline lesions. Another study reported that the ratio of the intensity of tumorous lesion to that of non-tumorous areas on SPIO-enhanced MR images (SPIO intensity ratio) correlated inversely with the Kupffer-cell-count ratio in HCCs and dysplastic nodules and increased as the degree of differentiation of HCCs decreased (Imai et al., 2000). This indicates that the uptake of SPIO in HCCs decreased as the degree of differentiation of HCCs declined.

We found that some well-differentiated HCCs exhibited signal decrease similar to the surrounding liver on T_2W-FSE images, but less signal decrease than surrounding liver on T_2^*W-GRE images (Table 4.3) (Tanimoto, 2006). Conversely, DNs exhibited strong decrease in signal on both T_2W and T_2^*W images (Figure 4.7). This finding could be explained by the fact that magnetic susceptibility effects in T_2^*W-GRE imaging depends on the intracellular SPIO cluster size (Tanimoto et al., 2001). In well-differentiated HCCs, Kupffer cell density would be maintained but Kupffer cell function reduced compared to

TABLE 4.3. Contrast patterns on T2W and T2*W imaging of borderline lesions.

Rank	T2W-FSE		T2*W-GRE	
	w-HCC	DN	w-HCC	DN
5 (homogeneously lower signal than the liver)	6	3	0	4*
4 (inhomogeneously lower signal than the liver)	0	3	1	2
3 (same signal as the liver)	5	0	4	0
2 (inhomogeneously higher signal than the liver)	7	0	14	0
1 (homogenously higher signal than the liver)	7	0	5	0

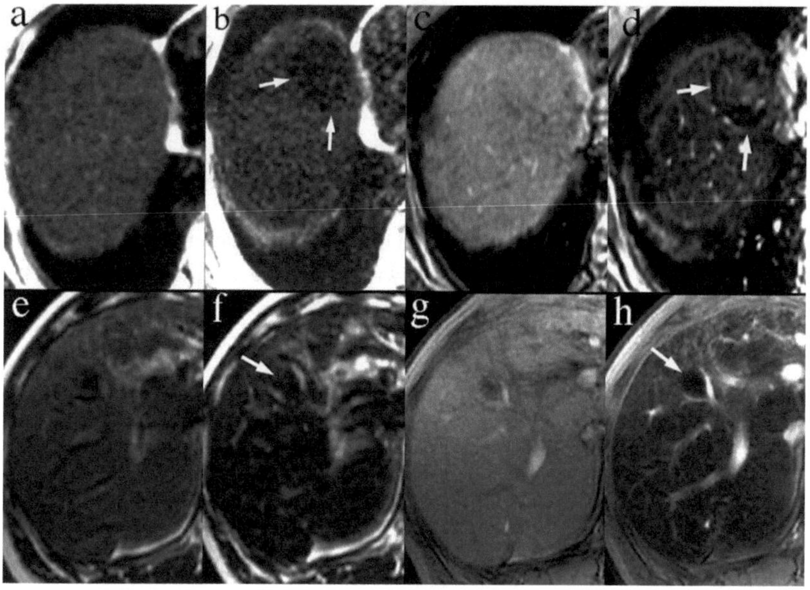

FIGURE 4.7. Differentiation between DN and well-differentiated HCC by SPIO-enhanced MR imaging. Both well-differentiated HCC in a 64-year-old female (**a**) and DN in a 55-year-old male (**e**) exhibit homogeneous signal decrease on T_2W-FSE after SPIO administration (**b, f**; arrows). Differentiation between HCC and DN is therefore difficult with this sequence. On T_2^*W-GRE, well-differentiated HCC exhibits heterogeneous signal decrease (**d**, arrow), but DN exhibits homogeneous signal decrease (**h**, arrow). T_2^*W-GRE thus aids differentiation between HCC and DN.

surrounding liver. Small intracellular SPIO clusters in well-differentiated HCCs produce little SNR decrease on T_2^*W-GRE images, but cause significant SNR loss on T_2W-FSE images. Conversely, large intracellular SPIO clusters in DNs cause prominent SNR loss on T_2^*W-GRE imaging, likely because of magnetic susceptibility effects related to fixed field inhomogeneity (Tanimoto et al., 2001). This is likely due to mechanisms of T_2 shortening that are independent of magnetic susceptibility effects, but related to the greater free water interaction made possible by the relatively larger surface area of the small clusters (Tanimoto et al., 1994, 2001).

SPIOs are known to be a negative enhancer of MR images due to its strong T_2 and T_2^* relaxation effects, but can be used as a positive enhancer on T_1-weighted images due to its strong T_1 relaxation effect (Chambon et al., 1993). The diagnostic usefulness of SPIO-enhanced T_1W-GRE imaging has been previously documented in tissue characterization (Takahama et al., 2003; Lee et al., 2003). Differentiation between hemangiomas and malignant tumors is sometimes difficult on the basis of their signal intensity on unenhanced or gadolinium-enhanced MR imaging. Differentiation between hemangiomas and solid masses relies on the combination of the T_1 blood pool effect, which positively enhances hemangiomas, and the T_2 effect, which negatively enhances hemangiomas and the surrounding liver (Montet et al., 2004) (Figure 4.8). Positive ring enhancement of focal hepatic lesions on SPIO-enhanced T_1W-GRE images is known to be a sign of malignancy (Reimer and Tombach, 1998; Kanematsu et al., 2003). This feature is most typical of hepatic metastases (Figure 4.9), but is frequently seen in HCC as well (Kim et al., 2002). In our study, positive ring enhancement was noted in 13 of 18 HCCs (72%) (Tanimoto et al., 2005). There have been a few reports on the mechanism of ring enhancement (Kanematsu et al., 2003). Peritumoral Kupffer cell density was increased in the ring enhancement (+) group compared with the ring enhancement (−) group (Tanimoto et al., 2005). In the ring enhancement (+) group, the tumor size measured on T_2W was

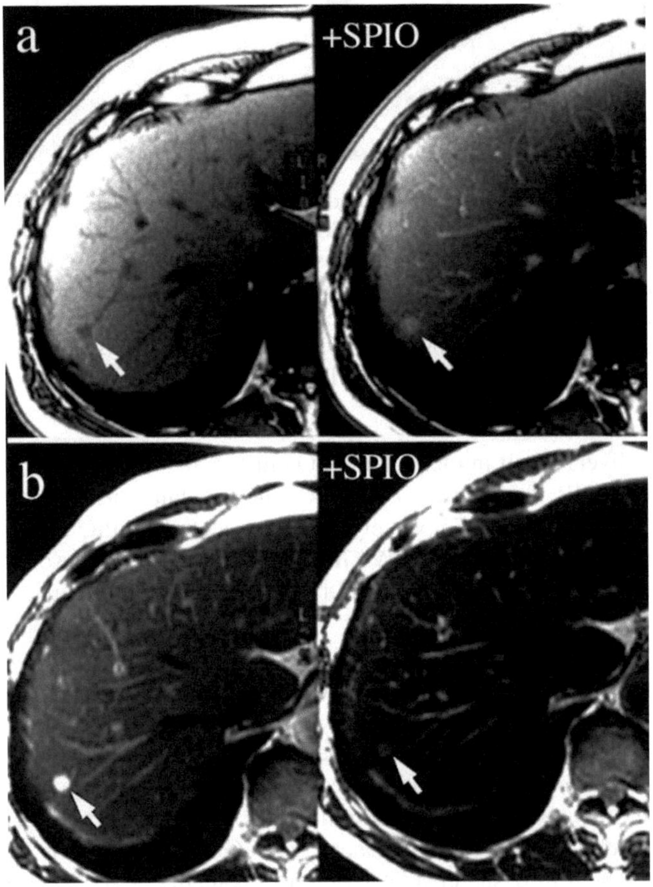

FIGURE 4.8. Liver hemangioma in a 60-year-old male. A hemangioma in S7 (arrow) exhibits low intensity on T_1W-GRE (**a**, left) and high intensity on T_2W-FSE (**b**, left). After SPIO administration, the tumor is positively enhanced on T_1W-GRE (**a**, right) and negatively enhanced on T_2W-FSE (**b**, right).

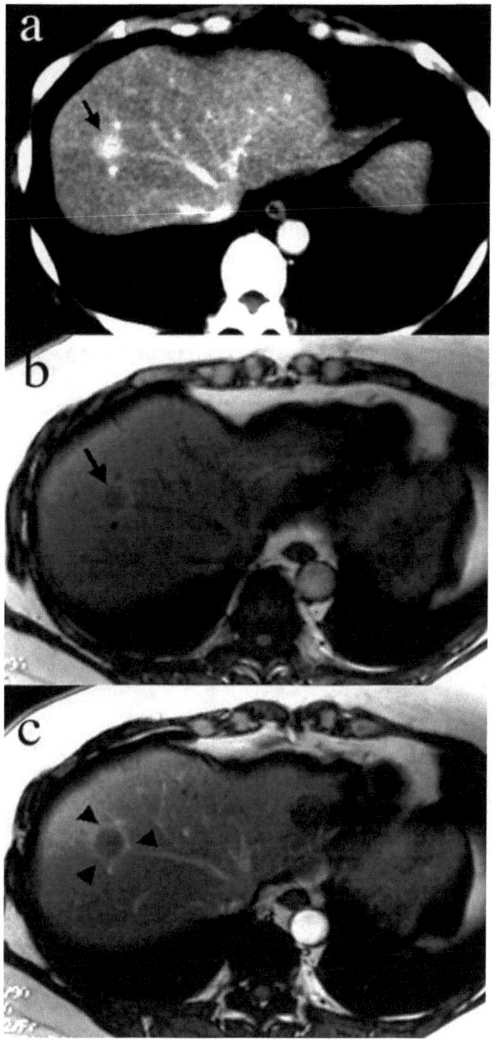

FIGURE 4.9. Liver metastasis from colonic cancer in a 52-year-old female: Ring enhancement. Contrast-enhanced CT exhibits a well-enhancing mass in S8, mimicking liver hemangioma (**a**, arrow). The T1W-GRE image exhibits a hypointense mass (**b**, arrow). After SPIO administration, positive ring enhancement is noted around the tumor (**c**, arrowheads).

smaller than that measured on either T_1W or T_2^*W, suggesting a sustained T_1 relaxation effect and a decreased T_2^* relaxation effect on the peritumoral region. Ring enhancement on SPIO-enhanced T_1W may correlate with increased Kupffer cell density and decreased SPIO clustering in Kupffer cells in the peritumoral region.

4.3.4. *Shortcomings of SPIO in Cirrhotic Liver*

There are some shortcomings of SPIO-enhanced MR imaging in the diagnosis of HCC. First, SPIO-enhanced MR imaging is basically unable to assess lesion vascularity. Perfusion MR imaging using SPIO is still a work-in-progress (Figure 4.10). Second, the decrease in signal intensity of cirrhotic liver with SPIO is limited compared to that in normal liver (Elizondo et al., 1990; Kuwatsuru et al., 1997). The percentage of signal-intensity loss and liver-lesion contrast-to-noise ratio on SPIO-enhanced images was significantly higher in patients with mild liver cirrhosis (Child's class A) than in patients with severe liver

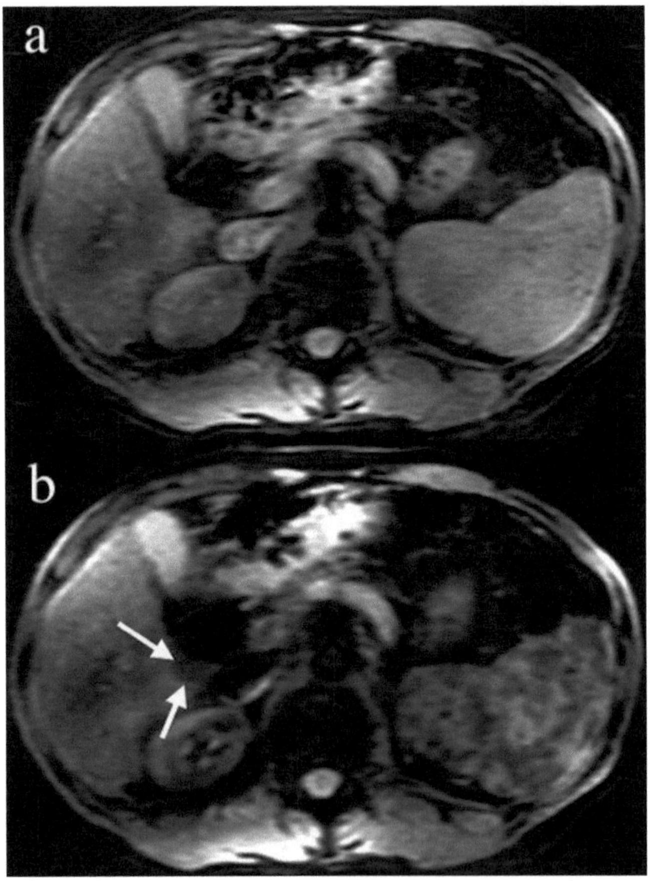

FIGURE 4.10. HCC in a 58-year-old male. Perfusion MR imaging. With use of the parallel imaging technique, the image quality of single-shot echo planar imaging is improved. The tumor in S6 becomes low in signal intensity following bolus injection of SPIO (arrows). Imaging parameters are as follows: spin echo type echo planar imaging; Repetition time (TR) = 4000 msec (millisecond), TE = 58 msec (millisecond), b-value = 500 s/mm^2, number of excitation = 8, matrix size = 128 × 128, slice thickness = 7 mm, parallel imaging factor = 2.

cirrhosis (Child's class C) (Tang et al., 1999). Lesion-liver contrast enhancement is weaker for HCCs arising in cirrhotic liver than for hepatic metastases occurring in normal liver. Prolongation of the imaging window after SPIO administration might improve lesion-liver contrast, but structural and functional inhomogeneity in cirrhosis could cause false-positive lesions after SPIO administration. Third, HCCs may contain various numbers of Kupffer cells, and some well-differentiated HCCs exhibit signal decrease after SPIO administration. These factors all worsen lesion-liver contrast and the rate of detection of HCC. In cirrhotic liver, T_2^*-weighted GRE imaging is more sensitive to decreased phagocytic activity than T_2W-FSE imaging (Tanimoto et al., 2002).

4.4. Conclusions

SPIO-enhanced MR imaging is an effective means of pre-therapeutic evaluation and follow-up diagnosis of liver tumors, featuring improved detection and capacity to differentiate lesions. It also improves the selection of patients who are candidates for curative liver surgery, since with it invasive surgery can be avoided if multiple lesions are present. With the exploitation of perfusion MR imaging, the sensitivity of this technique will substantially increase. From the perspective of comprehensive medical care, one-stop diagnosis with SPIO will be an attractive alternative to existing multi-modality-based decision-making.

References

Chambon, C., Clement, O., Le Blanche, A., Schouman-Claeys, E., Frija, G., 1993. Superparamagnetic iron oxides as positive MR contrast agents: in vitro and in vivo evidence. Magn Reson Imag 11, 509–519.

Elizondo, G., Weissleder, R., Stark, D.D., Guerra, J., Garza, J., Fretz, C.J., Todd, L.E., Ferrucci, J.T., 1990. Hepatic cirrhosis and hepatitis: MR imaging enhanced with superparamagnetic iron oxide. Radiology 174, 797–801

Grandin, C., Van Beers, B.E., Robert, A., Gigot, J.F., Geubel, A., Pringot, J., 1995. Benign hepatocellular tumors: MRI after superparamagnetic iron oxide administration. J Comput Assist Tomogr 19, 412–418.

Haider, M.A., Amitai, M.M., Rappaport, D.C., O'Malley, M.E., Hanbidge, A.E., Redston, M., Lockwood, G.A., Gallinger, S., 2002. Multi-detector row helical CT in preoperative assessment of small (< or = 1.5 cm) liver metastases: is thinner collimation better? Radiology 225, 137–142.

Ichikawa, T., Arbab, A.S., Araki, T., Touyama, K., Haradome, H., Hachiya, J., Yamaguchi, M., Kumagai, H., Aoki, S., 1999. Perfusion MR imaging with a superparamagnetic iron oxide using T2-weighted and susceptibility-sensitive echoplanar sequences: evaluation of tumor vascularity in hepatocellular carcinoma. AJR Am J Roentgenol 173, 207–213.

Imai, Y., Murakami, T., Yoshida, S., Nishikawa, M., Ohsawa, M., Tokunaga, K., Murata, M., Shibata, K., Zushi, S., Kurokawa, M., Yonezawa, T., Kawata, S., Takamura, M., Nagano, H., Sakon, M., Monden, M., Wakasa, K., Nakamura, H., 2000. Superparamagnetic iron oxide-enhanced magnetic resonance images of hepatocellular carcinoma: Correlation with histological grading. Hepatology 32, 205–212.

Jang, H.J., Lim, J.H., Lee, S.J., Park, C.K., Park, H.S., Do, Y.S., 2000. Hepatocellular carcinoma: are combined CT during arterial portography and CT hepatic arteriography in addition to triple-phase helical CT all necessary for preoperative evaluation? Radiology 215, 373–380.

Kanematsu, M., Matsuo, M., Yamada, Y., Semelka, R.C., Kondo, H., Goshima, S., Hoshi, H., Moriyama, N., 2003. Perilesional hyperintense rim of malignant hepatic tumors on ferumoxide-enhanced T1-weighted gradient-echo MR images: correlation between MR imaging and histopathologic findings. J Magn Reson Imaging 18, 40–48.

Kim, S.H., Choi, D., Kim, S.H., Lim, J.H., Lee, W.J., Kim, M.J., Lim, H.K., Lee, S.J., 2005. Ferucarbotran-enhanced MRI versus triple-phase MDCT for the preoperative detection of hepatocellular carcinoma. AJR Am J Roentgenol 184, 1069–1076.

Kim, S.H., Choi, D., Lim, J.H., Lee, W.J., Jang, H.J., Lim, H.K., Lee, S.J., Cho, J.M., Kim, S.K., Kim, G.C., 2002a. Optimal pulse sequence for ferumoxides-enhanced MR imaging used in the detection of hepatocellular carcinoma: a comparative study using seven pulse sequences. Korean J Radiol 3, 87–97.

Kim, J.H., Kim, M.J., Suh, S.H., Chung, J.J., Yoo, H.S., Lee, J.T., 2002b. Characterization of focal hepatic lesions with ferumoxides-enhanced MR imaging: utility of T1-weighted spoiled gradient recalled echo images using different echo times. J Magn Reson Imaging 15, 573–583.

Kim, Y.K., Kim, C.S., Lee, Y.H., Kwak, H.S., Lee, J.M., 2004. Comparison of superparamagnetic iron oxide-enhanced and gadobenate dimeglumine-enhanced dynamic MRI for detection of small hepatocellular carcinomas. AJR Am J Roentgenol 182, 1217–1223.

Kim, Y.K., Lee, J.M., Kim, C.S., Chung, G.H., Kim, C.Y., Kim, I.H., 2005. Detection of liver metastases: gadobenate dimeglumine-enhanced three-dimensional dynamic phases and one-hour delayed phase MR imaging versus superparamagnetic iron oxide-enhanced MR imaging. Eur Radiol 15, 220–228.

Kuwatsuru, R., Brasch, R.C., Muhler, A., Mathur, A., Vexler, V.S., Rosenau, W., Mintorovitch, J., Berthezene, Y., Cohen, F., Shames, D.M., 1997. Definition of liver tumors in the presence of diffuse liver disease: comparison of findings at MR imaging with positive and negative contrast agents. Radiology 202, 131–138.

Kwak, H.S., Lee, J.M., Kim, C.S., 2004. Preoperative detection of hepatocellular carcinoma: comparison of combined contrast-enhanced MR imaging and combined CT during arterial portography and CT hepatic arteriography. Eur Radiol 14, 447–457.

Lee, J.M., Kim, I.H., Kwak, H.S., Youk, J.H., Han, Y.M., Kim, C.S., 2003. Detection of small hypervascular hepatocellular carcinomas in cirrhotic patients: comparison of superparamagnetic iron oxide-enhanced MR imaging with dual-phase spiral CT. Korean J Radiol 4, 1–8.

Lee, J.M., Kim, C.S., Youk, J.H., Lee, M.S., 2003. Characterization of focal liver lesions with superparamagnetic iron oxide-enhanced MR imaging: value of distributional phase T1-weighted imaging. Korean J Radiol 4, 9–18.

Li, L., Wu, P.H., Mo, Y.X., Lin, H.G., Zheng, L., Li, J.Q., Lu, L.X., Ruan, C.M., Chen, L., 1999. CT arterial portography and CT hepatic arteriography in detection of micro liver cancer. World J Gastroenterol 5, 225–227.

Makita, O., Yamashita, Y., Arakawa, A., Nakayama, Y., Mitsuzaki, K., Ando, M., Namimoto, T., Takahashi, M., 2000. Diagnostic accuracy of helical CT arterial portography and CT hepatic arteriography for hypervascular hepatocellular carcinoma in chronic liver damage. An ROC analysis. Acta Radiol 41, 464–469.

Murakami, T., Oi, H., Hori, M., Kim, T., Takahashi, S., Tomoda, K., Narumi, Y., Nakamura, H., 1997. Helical CT during arterial portography and hepatic arteriography for detecting hypervascular hepatocellular carcinoma. AJR Am J Roentgenol 169, 131–135.

Montet, X., Lazeyras, F., Howarth, N., Mentha, G., Rubbia-Brandt, L., Becker, C.D., Vallee, J.P., Terrier, F., 2004. Specificity of SPIO particles for characterization of liver hemangiomas using MRI. Abdom Imag 29, 60–70.

Naganawa, S., Sato, C., Nakamura, T., Kumada, H., Ishigaki, T., Miura, S., Maruyama, K., Takizawa, O., 2005. Diffusion-weighted images of the liver: comparison of tumor detection before and after contrast enhancement with superparamagnetic iron oxide. J Magn Reson Imaging 21, 836–840.

Onishi, H., Murakami, T., Kim, T., Hori, M., Iannaccone, R., Kuwabara, M., Abe, H., Nakata, S., Osuga, K., Tomoda, K., Passariello, R., Nakamura, H., 2006. Hepatic metastases: detection with multi-detector row CT, SPIO-enhanced MR imaging, and both techniques combined. Radiology 239, 131–138.

Oudkerk, M., van den Heuvel, A.G., Wielopolski, P.A., Schmitz, P.I., Borel Rinkes, I.H., Wiggers, T., 1997. Hepatic lesions: Detection with ferumoxide-enhanced T1-weighted MR imaging. Radiology 203, 449–456.

Pauleit, D., Textor, J., Bachmann, R., Conrad, R., Flacke, S., Layer, G., Kreft, B., Schild, H., 2002. Hepatocellular carcinoma: detection with gadolinium- and ferumoxides-enhanced MR imaging of the liver. Radiology 222, 73–80.

Poeckler-Schoeniger, C., Koepke, J., Gueckel, F., Sturm, J., Georgi, M., 1999. MRI with superparamagnetic iron oxide: efficacy in the detection and characterization of focal hepatic lesions. Magn Reson Imag 17, 383–392.

Reimer, P., Jahnke, N., Fiebich, M., Schima, W., Deckers, F., Marx, C., Holzknecht, N., Saini, S., 2000. Hepatic lesion detection and characterization: value of non-enhanced MR imaging, superparamagnetic iron oxide-enhanced MR imaging, and spiral CT-ROC analysis. Radiology 217, 152–158.

Reimer, P., Muller, M., Marx, C., Wiedermann, D., Muller, R., Rummeny, E.J., Ebert, W., Shamsi, K., Peters, P.E., 1998. T1 effects of a bolus-injectable superparamagnetic iron oxide, SH U 555 A: dependence on field strength and plasma concentration—preliminary clinical experience with dynamic T1-weighted MR imaging. Radiology 209, 831–836.

Reimer, P., Tombach, B., 1998. Hepatic MRI with SPIO: detection and characterization of focal liver lesions. Eur Radiol 8, 1198–1204.

Senéterre, E., Taourel, P., Bouvier, Y., Pradel, J., Van Beers, B., Daures, J.P., Pringot, J., Mathieu, D., Bruel, J.M., 1996. Detection of hepatic metastases: ferumoxides-enhanced MR imaging versus unenhanced MR imaging and CT during arterial portography. Radiology 200, 785–792.

Stark, D.D., Weissleder, R., Elizondo, G., Hahn, P.F., Saini, S., Todd, L.E., Wittenberg, J., Ferrucci, J.T., 1988. Superparamagnetic iron oxide: clinical application as a contrast agent for MR imaging of the liver. Radiology 168, 297–301.

Takahama, K., Amano, Y., Hayashi, H., Ishihara, M., Kumazaki, T., 2003. Detection and characterization of focal liver lesions using superparamagnetic iron oxide-enhanced magnetic resonance imaging: comparison between ferumoxides-enhanced T1-weighted imaging and delayed-phase gadolinium-enhanced T1-weighted imaging. Abdom Imag 28, 525–530.

Takahara, T., Imai, Y., Yamashita, T., Yasuda, S., Nasu, S., Van Cauteren, M., 2004. Diffusion weighted whole body imaging with background body signal suppression (DWIBS): technical improvement using free breathing, STIR and high resolution 3D display. Radiat Med 22, 275–282.

Tanaka, M., Nakashima, O., Wada, Y., Kage, M., Kojiro, M., 1996. Pathomorphological study of Kupffer cells in hepatocellular carcinoma and hyperplastic lesions in the liver. Hepatology 24, 807–812.

Tang, Y., Yamashita, Y., Arakawa, A., Namimoto, T., Mitsuzaki, K., Abe, Y., Katahira, K., Takahashi, M., 1999. Detection of hepatocellular carcinoma arising in cirrhotic livers: comparison of gadolinium- and ferumoxides-enhanced MR imaging. AJR Am J Roentgenol 172, 1547–1554.

Tanimoto, A., Kuribayashi, S., 2006. Application of superparamagnetic iron oxide to imaging of hepatocellular carcinoma. Eur J Radiol 58, 200–216.

Tanimoto, A., Oshio, K., Suematsu, M., Pouliquen, D., Stark, D.D., 2001. Relaxation effects of clustered particles. J Magn Reson Imag 14, 72–77.

Tanimoto, A., Pouliquen, D., Kreft, B.P., Stark, D.D., 1994. Effects of spatial distribution on proton relaxation enhancement by particulate iron oxide. J Magn Reson Imag 4, 653–657.

Tanimoto, A., Satoh, Y., Yuasa, Y., Jinzaki, M., Hiramatsu, K., 1997. Performance of Gd-EOB-DTPA and superparamagnetic iron oxide particles in the detection of primary liver cancer: a comparative study by alternative free-response receiver operating characteristic analysis. J Magn Reson Imaging 7, 120–124.

Tanimoto, A., Wakabayashi, G., Shinmoto, H., Okuda, S., Kuribayashi, S., Mukai, M., 2005. The mechanism of ring enhancement in hepatocellular carcinoma on superparamagnetic iron oxide-enhanced T1-weighted images: an investigation into peritumoral Kupffer cells. J Magn Reson Imag 21, 230–236.

Tanimoto, A., Wakabayashi, G., Shinmoto, H., Nakatsuka, S., Okuda, S., Kuribayashi, S., 2005. Superparamagnetic iron oxide-enhanced MR imaging for focal hepatic lesions: a comparison with CT during arterioportography plus CT during hepatic arteriography. J Gastroenterol 40, 371–380.

Tanimoto, A., Yuasa, Y., Shinmoto, H., Jinzaki, M., Imai, Y., Okuda, S., Kuribayashi, S., 2002. Superparamagnetic iron oxide mediated hepatic signal change in patients with and without cirrhosis: Pulse sequence effects and Kupffer cell function. Radiology 222, 661–666.

Vogl, T.J., Hammerstingl, R., Schwarz, W., Kummel, S., Muller, P.K., Balzer, T., Lauten, M.J., Balzer, J.O., Mack, M.G., Schimpfky, C., Schrem, H., Bechstein, W.O., Neuhaus, P., Felix, R., 1996. Magnetic resonance imaging of focal liver lesions. Comparison of the superparamagnetic iron oxide resovist versus gadolinium-DTPA in the same patient. Invest Radiol 31, 696–708.

Vogl, T.J., Hammerstingl, R., Schwarz, W., Mack, M.G., Muller, P.K., Pegios, W., Keck, H., Eibl-Eibesfeldt, A., Hoelzl, J., Woessmer, B., Bergman, C., Felix, R., 1996. Superparamagnetic iron oxide—enhanced versus gadolinium-enhanced MR imaging for differential diagnosis of focal liver lesions. Radiology 198, 881–887.

Ward, J., Chen, F., Guthrie, J.A., Wilson, D., Lodge, JP., Wyatt, J.I., Robinson, P.J., 2000. Hepatic lesion detection after superparamagnetic iron oxide enhancement: comparison of five T2-weighted sequences at 1.0 T by using alternative-free response receiver operating characteristic analysis. Radiology 214, 159–166.

Ward, J., Guthrie, J.A., Scott, D.J., Atchley, J., Wilson, D., Davies, M.H., Wyatt, J.I., Robinson, P.J., 2000. Hepatocellular carcinoma in the cirrhotic liver: Double-contrast MR imaging for diagnosis. Radiology 216, 154–162.

Ward, J., Guthrie, J.A., Wilson, D., Arnold, P., Lodge, J.P., Toogood, G.J., Wyatt, J.I., Robinson, P.J., 2003. Colorectal hepatic metastases: detection with SPIO-enhanced breath-hold MR imaging—comparison of optimized sequences. Radiology 228, 709–718.

Ward, J., Naik, K.S., Guthrie, J.A., Wilson, D., Robinson, P.J., 1999. Hepatic lesion detection: comparison of MR imaging after the administration of superparamagnetic iron oxide with dual-phase CT by using alternative-free response receiver operating characteristic analysis. Radiology 210, 459–466.

Ward, J., Robinson, P.J., Guthrie, J.A., Downing, S., Wilson, D., Lodge, J.P., Prasad, K.R., Toogood, G.J., Wyatt, J.I., 2005. Liver metastases in candidates for hepatic resection: comparison of helical CT and gadolinium- and SPIO-enhanced MR imaging. Radiology 237, 170–180.

5
The Emerging Role of USPIOs for MR Imaging of Atherosclerosis

M.E. Kooi, S. Heeneman, M.J.A.P. Daemen, J.M.A. van Engelshoven, and K.B.J.M. Cleutjens

Abstract: Atherosclerosis is an inflammatory disease of the arteries, which causes more than 19 million deaths worldwide every year. Most complications of atherosclerosis are caused by rupture of a vulnerable plaque. Currently, there are no imaging techniques available in clinical practice that can accurately identify a vulnerable plaque. There is an emerging role for Ultrasmall Superparamagnetic Iron Oxide Particles (USPIOs) in MR imaging of atherosclerosis. This review will first give an update on the biology of atherosclerosis. Next, we will focus on the physicochemical properties of USPIOs and describe animal and human studies that showed USPIO uptake mainly in macrophages of atherosclerotic plaque, which could be detected as areas of focal signal loss in MR images. Novel superparamagnetic particles are currently being developed which enable visualization of other cell types and functions than macrophage uptake that are relevant for atherosclerosis. We conclude that USPIOs are a highly promising imaging tool for atherosclerosis and larger prospective studies are warranted to prove the value in daily clinical practice.

5.1. Atherosclerosis

Atherosclerotic cardiovascular disease is an inflammatory disease of the arteries, which causes more than 19 million deaths worldwide every year (Naghavi et al., 2003). In atherosclerosis, atherosclerotic plaques develop over decades, ultimately leading to advanced lesions with a lipid-rich, necrotic core separated from the lumen by a fibrous cap. A thick fibrous cap ($> 65\,\mu$m) indicates stability of the plaque since it prevents the thrombogenic lipid-rich necrotic core from contacting the blood in the lumen and promoting thrombus formation, which can lead to local or distal arterial occlusion with subsequent ischemia. About 70 % of fatal acute myocardial infarctions and/or sudden coronary deaths is caused by rupture of the fibrous cap (Figure 5.1) while in only 2 out of 7 cases the ruptured plaque is stenotic before the acute event (Naghavi et al., 2003).

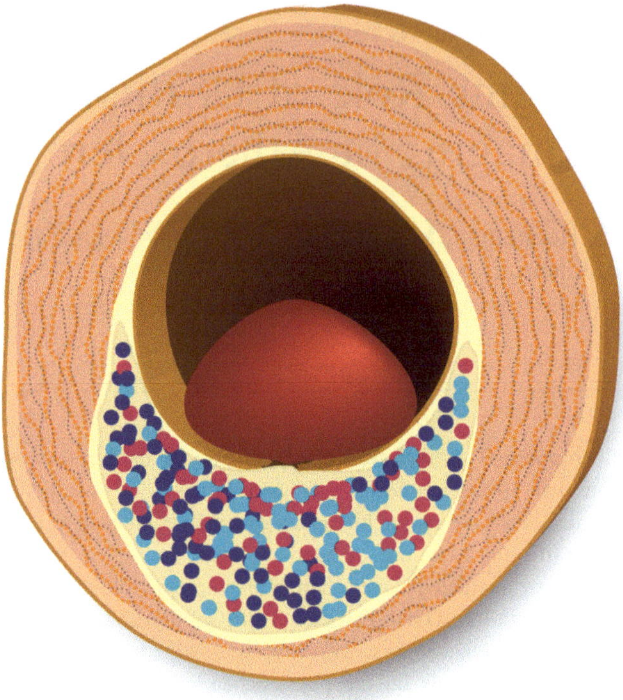

FIGURE 5.1. Rupture of the fibrous cap of a vulnerable plaque can lead to thrombus formation. Disruption usually occurs at sites with a thin fibrous cap in macrophage-rich regions. Modified from Figure 1 of Leiner et al. (2005).

Thus, the risk of disruption depends on plaque vulnerability rather than on the degree of vascular stenosis.

During the past decade it has become evident that atherosclerosis is a chronic inflammatory disease (Hansson et al., 2002; Hansson, 2005; Tedgui and Mallat et al., 2006). In particular, the macrophage has emerged as the key cellular mediator of inflammation in atherosclerotic lesions, since macrophages participate in all phases of atherogenesis. After subtle endothelial injury, vascular endothelial cells become activated and start to express several types of leukocyte adhesions molecules (including VCAM-1 and E-selectin), which causes monocytes and lymphocytes to adhere to the site of endothelial activation. Once these monocytes and lymphocytes have attached, chemokines (including monocyte chemoattractant protein-1, MCP-1), produced by the underlying intima, stimulate them to migrate into the subendothelial space. Once resident in the arterial intima, monocytes differentiate into macrophages expressing both toll-like receptors and scavenger receptors. These scavenger receptors (including CD36 and SR-BI) internalize a broad range of molecules and particles including bacterial endotoxins, apoptotic cell fragments, and oxidized LDL, ultimately leading to foam cell formation. Toll-like receptors also bind molecules with pathogen-like molecular patterns, but in contrast to scavenger receptors they

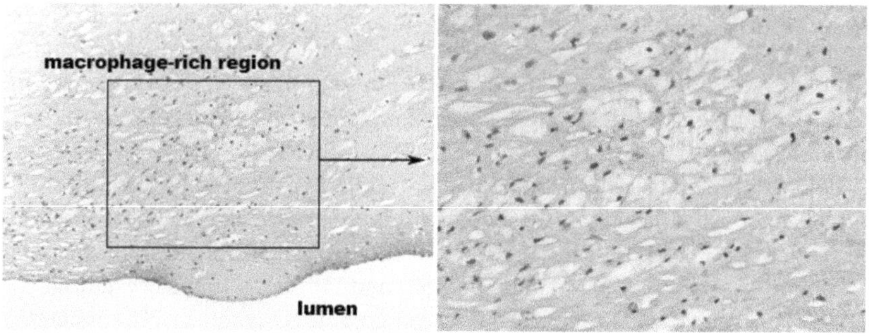

FIGURE 5.2. Histological section of a macrophage-rich advanced human carotid atherosclerotic plaque.

can initiate a signal cascade that leads to cell activation. In addition to MCP-1, macrophage colony-stimulating factor (M-CSF) also appears to play an important role in the activation of various macrophage functions implicated in atherogenesis. Besides monocytes/macrophages, also T and B cells, natural killer T cells, antigen presenting dendritic cells, and mast cells contribute to plaque inflammation.

Plaque rupture preferentially occurs where the fibrous cap is thin. At these sites, activated immune cells are abundant. Especially, activated macrophages, T cells, and mast cells produce pro-inflammatory cytokines (including IL-1, TNF-α), proteolytic enzymes [e.g., matrix metalloproteinases (MMPs) and cysteine proteases (Cathespins)], coagulation factors, radicals, and vasoactive molecules transforming a stable plaque into a vulnerable plaque that can rupture and thereby induce thrombus formation. Furthermore, postmortem studies showed prominent macrophage accumulation in ruptured lesions, while stable plaques of patients with stable angina definitely contain less macrophages (Falk et al., 1995). Figure 5.2 shows an advanced macrophage-rich atherosclerotic plaque. Given the key role of macrophages in the development of atherosclerosis, imaging of inflammation is expected to be extremely valuable for the evaluation of therapy, e.g. using novel intermediate imaging as endpoint instead of major adverse clinical events as endpoint. Moreover, since an abundance of macrophages is found especially in ruptured lesions, such new imaging techniques might enable identification of patients at risk for future ischemic events, such as stroke. This is of major importance since decision-making in current clinical practice is still mainly based on degree of stenosis, which is a rather poor predictor for future clinical ischemic events.

5.2. Physicochemical Properties of Superparamagnetic Iron Oxide Particles

Recently, several authors explored the possible application of Superparamagnetic Iron Oxide Particles (SPIO) as MR imaging tool for atherosclerosis. These particles are internalized by macrophages and the biological amplification through intracellular accumulation results in a strong signal reduction in T_2/T_2^*-weighted MR images (Corot et al., 2004). These nanoparticles consist of a small iron oxide core with a diameter of 4–10 nm stabilized with a biodegradable coating to prevent particle aggregation and to optimize the pharmacokinetic behavior (Wang et al., 2001).

The crystalline superparamagnetic core usually consists of magnetite (Fe_3O_4), which has the advantage that its biocompatibility has already been proven (Gupta and Gupta et al., 2005). In the presence of an external magnetic field, magnetic ordering of thousands of paramagnetic Fe^{2+} and Fe^{3+} ions in the crystalline core occurs. This phenomenon is referred to as superparamagnetism. The advantage of superparamagnetic substances is that they have a greater resultant magnetic field in the presence of an external field than paramagnetic substances and a substantial larger susceptibility than soluble paramagnetic materials. Similar to paramagnetic materials, the magnetization disappears when the external magnetic field is removed (Wang et al., 2001).

Both T_1 and T_2/T_2^* relaxation times are considerably shortened by superparamagnetic particles of iron oxide. The r_2 relaxivity of 100 $l \cdot mmol^{-1} \cdot s^{-1}$ and the r_1 relaxivity of 20 $l \cdot mmol^{-1} \cdot s^{-1}$ at 39°C and 0.47 T of a typical superparamagnetic iron oxide agent are substantially larger than those of paramagnetic gadolinium (Gd)-chelates (r_2 relaxivity of 5.7 $l \cdot mmol^{-1} \cdot s^{-1}$ and r_1 relaxivity of 4.5 $l \cdot mmol^{-1} \cdot s^{-1}$ for Gd-DTPA at 39°C and 0.47 T) (Jung and Jacobs et al., 1995). These high relaxivities enable visualization of low superparamagnetic iron oxide concentrations in MR images. It has been stated that superparamagnetic iron oxide particles allow in vivo detection with MR imaging at concentrations as low as 1 μg Fe/kg tissue (Weissleder et al., 1992). Internalization of nanoparticles in macrophages, and subsequent aggregation, dramatically modifies their magnetic properties. Generally, effective T_1-relaxivity is reduced whereas T_2/T_2^* relaxivity is increased after compartmentalization (Bjornerud and Johansson et al., 2004). Therefore, internalized USPIOs are visualized as areas of signal loss on T_2/T_2^* weighted images.

The total particle size (including iron oxide core and coating material) in solution can be larger than the dry size, due to the absorbed, hydrated coating layer. Therefore, these particles are characterized by their hydrodynamic diameter (Wang et al., 2001). Depending on the hydrodynamic diameter, three classes of superparamagnetic iron oxide particles can be distinguished (Bjornerud and Johansson et al., 2004).

1) Monocrystalline Iron Oxide Nanoparticles (MIONs); mean hydrodynamic diameter < 20 nm, particles contain only one superparamagnetic crystalline core.
2) Ultrasmall Superparamagnetic Iron Oxide Particles (USPIOs); 20 nm $<$ mean hydrodynamic diameter < 50 nm, particles contain only one superparamagnetic crystalline core.
3) Superparamagnetic Particles of Iron Oxide (SPIOs); mean hydrodynamic diameter > 50 nm, particles contain several superparamagnetic cores.

TABLE 5.1. Physicochemical characteristics of various superparamagnetic iron oxide agents which have been tested for uptake in macrophages/atherosclerotic plaques.

Generic name/ Laboratory code	Classification	Trade name	Developing company	Coating	Particle size	Rabbit plasma half-life	Human plasma half-life	r1 at 0.47 T ($l \cdot mmol^{-1} \cdot s^{-1}$)	r2 at 0.47 T ($l \cdot mmol^{-1} \cdot s^{-1}$)
DDM 43/34/106	USPIO	–	IDF	Carboxydextran (Schmitz et al., 2000)	Mean hydrodynamic diameter 25 nm (Schmitz et al., 2000)	≈ 6 hours (Schmitz et al., 2000)	–	39.0 (Schmitz et al., 2000)	65.6 (Schmitz et al., 2000)
Ferumoxtran-10	USPIO	Sinerem® Combidex®	Guerbet Advanced Magnetics	Dextran	Mean diameter 35 nm	≤ 48 hours*	24–36 hours	22–25	80–85
Ferumoxytol	USPIO	–	Advanced Magnetics	Carbohydrate (Yancy et al., 2005)	Volume weighted diameter 30 nm (Yancy et al., 2005)	≤ 6 hours*	–	38 (Yancy et al., 2005)	83 (Yancy et al., 2005)
Ferumoxide/ Ami-25	SPIO	Feridex® Endorem®	Advanced Magnetics Berlex Laboratories Guerbet	Dextran (Wang et al., 2001)	80–150 mm (Wang et al., 2001)	–	≈ 6 min (Wang et al., 2001)	98.3 (Wang et al., 2001)	23.9 (Wang et al., 2001)

(Continued)

Table 5.1. *Continued*

Generic name/ Laboratory code	Classification	Trade name	Developing company	Coating	Particle size	Rabbit plasma half-life	Human plasma half-life	r1 at 0.47 T ($l \cdot mmol^{-1} \cdot s^{-1}$)	r2 at 0.47 T ($l \cdot mmol^{-1} \cdot s^{-1}$)
Vacuum filtered Feridex through 25 nm filter	USPIO	–	–	Dextran	12 nm (Briley-Saebo et al., 2006)	16.7 hours (Briley-Saebo et al., 2006)	–	10 @ 60 MHz (Briley-Saebo et al., 2006)	–
VCAM-1-targeted fluorescent cross-linked iron oxide particles, similar to MIONs	MION	–	–	Dextran	39 nm (Kelly et al., 2005)	–	–	21.1 (Kelly et al., 2005)	62.6 (Kelly et al., 2005)
Fluorescent cross-linked iron oxide particles, similar to moins	MION	–	–	Dextran	32 nm (Pande et al., 2006)	–	–	–	–

*The plasma half-life was determined from signal intensity in blood in daily MR angiograms (Yancy et al., 2005) which is not a very accurate method to determine plasma half-life.

Unlike the low-molecular weight gadolinium chelates, which are commonly used as MR contrast agents in clinical practice, superparamagnetic iron oxide particles do not easily leak into the interstitium and therefore act as blood pool agents as long as the endothelium is not damaged by pathological processes. They are eliminated from blood by uptake into the reticuloendothelial system (RES) cells in the liver (Kuppfer cells), spleen, and bone marrow (Bjornerud and Johansson et al., 2004). The larger SPIOs are more efficiently taken up by the RES cells and therefore generally have a shorter blood half-life time (<10 minutes in humans) (Reimer and Tombach et al., 1998) than USPIOs (3–36 hours). Different coating materials can explain the large range of blood half-life times of different USPIOs. The dextran-coated USPIO ferumoxtran-10 (Table 5.1) has an increased plasma half-life (24–36 hours in humans) (Corot et al., 2004) compared the other approved USPIOs for human applications with plasma half-life times of 3–6 hours (Simon et al., 2006; Wang et al., 2001). This prolonged plasma half-life results in a wider biodistribution, enabling internalization by macrophages in deep pathological tissues such as the atherosclerotic plaque (Corot et al., 2004).

The amount of iron that is used for clinical imaging with superparamagnetic iron oxide particles is small compared to total body iron stores (Reimer et al., 1995). There are no indications that USPIOs lead to cell activation, since a very low level of interleukin-1 release during in vitro endocytosis of USPIOs by macrophages was observed at high USPIO concentrations (Raynal et al., 2004). As iron oxide particles degrade, the iron enters the normal plasma iron pool and eventually it is secreted from the body as the body iron stores turn over (Wang et al., 2001).

In a clinical investigation the safety of the USPIO ferumoxtran-10 was evaluated in 152 patients (Anzai et al., 2003). Several minor adverse events (28 %), mostly headache, back pain, urticaria, and vasodilatation, were associated with ferumoxtran-10 administration, but usually these adverse events resolved within 24 hours. No serious adverse event occurred in this study and therefore these authors concluded that ferumoxtran-10 is a safe MR contrast agent.

5.3. USPIO-Enhanced MRI of Macrophages in Atherosclerotic Plaques

5.3.1. Animal Studies

MR imaging of the small pulsating arterial vessel wall itself is already quite a challenge and therefore it remained questionable whether imaging of plaque inflammation with USPIOs would be achievable. During the last few years, several investigators have demonstrated the feasibility of in vivo MR imaging of rabbit atherosclerotic plaques with USPIOs (Table 5.2) (Hyafil et al., 2006; Ruehm et al., 2001; Schmitz et al., 2000; Yancy et al., 2005). Uptake of USPIOs in the vessel wall of atherosclerotic rabbits lead to strong focal signal loss

TABLE 5.2. Summary of animal studies investigating uptake of superparamagnetic particles of iron oxide in atherosclerotic plaque.

Type of study	iron oxide nanoparticle	Dose	Findings	Reference
WHHL* rabbits	DDM 43/34/106	50–200 μmol Fe/kg	• Focal signal loss confined to aortic wall on 31 % ± 11 % of the MR images in 200 μmol Fe/kg group • Number of images with focal signal loss in 50 μmol Fe/kg groups is not significantly different from controls • Histology showed uptake of USPIOs in macrophages and occasionally in endothelial cells • Some macrophage-rich regions showed no USPIO uptake	(Schmitz et al., 2000, 2002)
WHHL* rabbits vs NZW† rabbits	Ferumoxtran-10	1000 μmol Fe/kg	• Marked signal loss in aortic wall of WHHL rabbits 4–5 days post-USPIO infusion • Histology and electron microscopy showed marked Fe uptake in macrophages of plaque in WHHL rabbits • No such changes were observed in control rabbits or hyperlipidemic rabbits that did not receive ferumoxtran-10	(Ruehm et al., 2001)
Balloon-injured hypercholes-terolemic NZW† rabbits	Ferumoxtran-10 vs Ferumoxytol	500 μmol Fe/kg	• Plasma half-life of ferymoxytol much shorter than that of ferumoxtran-10 • Decreased signal on in vivo MR images 2, 4, and 8 weeks following injury with ferumoxtran-10, but not with ferumoxytol and not in control animals • In contrast, in vitro macrophage phagocytosis 4–6 fold higher with ferumoxytol than ferumoxtran-10 • Absolute iron content correlated with ex vivo MRI signal intensity ($r = -0.82$, $P < 0.0001$) • Large differences in USPIO distribution in plaques 2, 4, and 8 weeks post-injury	(Yancy et al., 2005)

Apo E $^{-/-‡}$ mice on atherogenic diet vs wild-type mice	Ferumoxide/ AMI-25	1000 μmol Fe/kg	• Cytokine-treated mice showed larger USPIO uptake vs. non-treated mice • No USPIO uptake in wild-type mice	(Litovsky et al., 2003)
Balloon-injured hypercholes-terolemic NZW[†] rabbits	Ferumoxtran-10	1000 μmol Fe/kg	• Marked signal loss was detected only in injured aortas by in vivo MRI • Indirect measure of in vivo signal loss correlated with macrophage infiltration ($r = 0.82$, $P < 0.05$) and USPIO uptake ($r = 0.78$, $P < 0.05$) on corresponding histological sections	(Hyafil et al., 2006)
Balloon-injured hypercholes-terolemic NZW[†] rabbits	vacuum filtrated Ferumoxide/AMI-25	84 μmol Fe/kg	• Modified GRE sequence with variable dephasing gradients can be used to generate positive USPIO contrast in in vivo MR images of atherosclerotic plaque	(Briley-Saebo et al., 2006)
Apo E $^{-/-‡}$ mice on atherogenic diet vs wild-type mice	Fluorescent cross-linked iron oxide particles, similar to MIONs	0.2 nmol fluorochrome/gram weight[§]	• Direct visualization of fluorescent superparamagnetic iron oxide nanoparticles with fluorescence microscopy • Majority of nanoparticles were taken up by macrophages in plaque, but occasionally endothelial cells and smooth muscle cells were noted to contain nanoparticles	(Pande et al., 2006)

(Continued)

TABLE 5.2. *Continued*

Type of study	iron oxide nanoparticle	Dose	Findings	Reference
Apo E $^{-/-‡}$ mice on atherogenic diet vs wild-type mice	VCAM-1-targeted fluorescent cross-linked iron oxide particles, similar to MIONs	89 μmol Fe/kg	• Extensive MR signal loss in post-contrast images of cholesterol-fed Apo E$^{-/-‡}$ mice while the control nanoparticle showed no uptake • No significant uptake of contrast agent at same dose in wild type mice • Contrast agent co-localized with VCAM-1-expressing cells in atherosclerotic lesions	(Kelly et al., 2005)

*Watanabe heriditable hyperlipidemic rabbits

†New Zealand white rabbits

‡Apolipoprotein E deficient mice

§These near-infrared fluorescent magnetonanoparticles were used for detection of macrophage activity in atherosclerosis with laser scanning fluorescence microscopy.

in the gradient echo MR images of atherosclerotic rabbits, while no effect on control animals was observed (Hyafil et al., 2006; Ruehm et al., 2001; Yancy et al., 2005). Histology showed that macrophages were the dominant cell type for USPIO uptake; however, some endothelial cells and smooth muscle cells were occasionally also noted to contain iron oxide particles (Pande et al., 2006; Schmitz et al., 2000). Electron microscopy confirmed the uptake of USPIO in cytoplasm of macrophages (Ruehm et al., 2001).

Hyafil et al. (2006) showed that an indirect in vivo measure for USPIO-induced susceptibility artifacts in MR images of balloon-injured hypercholesterolemic rabbit aortas correlated well with RAM-11 positive areas (macrophages; $r = 0.82$, $P < 0.05$) and Perl's-positive areas (USPIOs; $r = 0.78$, $P < 0.05$), demonstrating that MRI can quantify uptake of USPIOs in macrophages. The indirect measure that was used in this study was the percentage of reduction in luminal areas, which is attributed to the extension of susceptibility artifacts outside anatomic borders of the vessel wall. Magnetic susceptibility artifacts could not be quantified directly in this study, because dark semicircular artifacts were observed at the water–fat interface in the frequency-encoding direction, also in the pre-contrast images. A high ex vivo correlation between signal intensity in MR images and iron content as quantified with inductively coupled plasma-mass spectrometry (ICP-MS) ($r = -0.92$, $P < 0.001$) was found by Yancy et al. (2005). In vivo quantification in the latter study was hampered by heterogeneity of plaque deposition, thickness of the MR image slice, and spatial in-plane resolution (Yancy et al., 2005). Schmitz et al. (2002) determined the accuracy for ex vivo MRI detection of USPIO accumulation in rabbit aorta vessel wall, using histology as standard of reference. They found a high accuracy, expressed as area under the receiver–operator characteristic (ROC) curve, of 0.85 and 0.88 for reader 1 and 2, respectively, with an interobserver agreement of 0.67. A drawback was the rather large number of false positive findings in 19 % of the segments, partly caused by calcifications and mural thrombi. False-positive findings might be lower in the in vivo situation, when only those areas of new focal signal loss in post-contrast images are attributed to USPIO accumulation. Shrinkage of the histological specimens and subsequent matching problems of in vivo MR images with histology made in vivo analysis of accuracy impossible.

Different types and doses of USPIOs were used by the various groups (Table 5.2). The largest post-contrast signal decrease seemed to be observed by Ruehm et al. (2001) (Figure 5.3) and Hyafil et al. (2006) both using a very high dose of 1 mmol Fe/kg ($\approx$ 20 times the normal clinical dose) and an USPIO agent with a high plasma half-life time (ferumoxtran-10, Sinerem®, Guerbet), allowing a larger time-window for the uptake of iron oxide particles by macrophages of plaque. Interestingly, in the first study conventional extracellular Gd-DOTA (Dotarem®) failed to reveal any abnormality in Watanabe hereditable hyperlipidemic (WHHL) rabbits, while abnormalities were clearly visible in the post-USPIO images (Ruehm et al., 2001).

The effect of USPIO dose was demonstrated by Schmitz et al. (2000), who observed a significant larger amount of areas of focal signal loss clearly confined

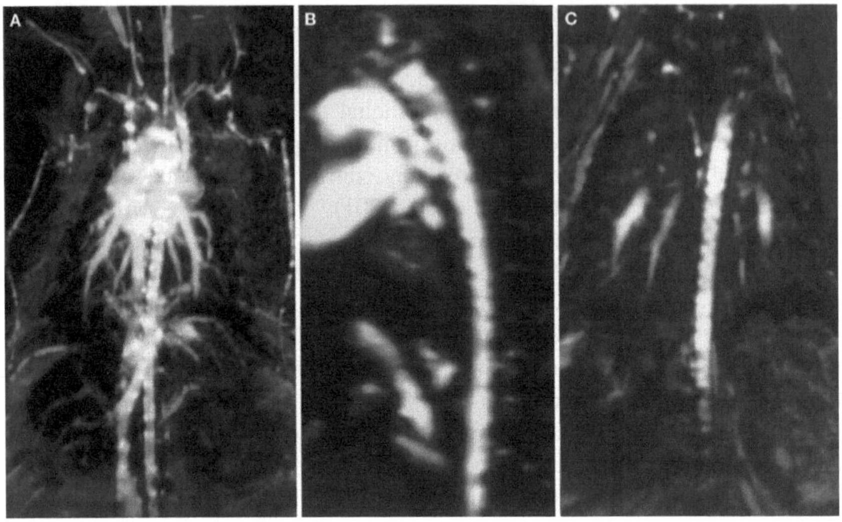

FIGURE 5.3. **A** Coronal MIP. **B** Sagittal oblique. **C** Coronal oblique reformatted images of contrast-enhanced 3D MRA data sets of a Watanabe hereditable hyperlipidemic (WHHL) rabbit obtained 5 days after intravenous injection of ferumoxtran-10. Note susceptibility effects originating within vessel wall and representing Fe uptake in macrophages embedded in plaque. Reprinted from Ruehm et al.(2001), with permission of Lippincott Williams & Wilkins.

to the aortic wall on $31 \pm 11\,\%$ of the MR images in the 200 μmol Fe/kg group, while the number of images with this finding in the 50 μmol Fe/kg groups were not significantly different from the control group without USPIOs. Histology confirmed this observation since a significant larger number of aortic segments of the high dose group were iron positive on a Prussian blue staining, while this was not the case for the low-dose groups. These findings may also partly be explained by a difference in post-contrast delay, which were 48 hours for the high-dose and 24 and 8 hours for the low-dose groups. A slight difference in histology findings was indeed observed between the groups with 24 and 8 hours delay, since in $21 \pm 24\,\%$ of the aortic sections of the first group a thin subendothelial rim of iron-positive cells was observed, while this was never observed in the latter group.

The distribution of USPIO in atherosclerotic plaque depends on the animal model. Yancy et al. (2005) demonstrated large differences in USPIO distribution in the aorta vessel wall of balloon-injured hypercholesterolemic New Zealand white rabbits, 2, 4, and 8 weeks post-injury. Ex vivo MRI (Figure 5.4) and histology (Figure 5.5) showed USPIO uptake in a rim in the intima along the internal elastic lamina 2 weeks post-injury, while a focal deposition of USPIO was observed in distinct areas throughout the intima and occasionally in media 4 and 8 weeks post-injury. Although the iron distribution was very different in these groups, ex vivo MRI signal decreases (↓ 41 %–45 %) were comparable.

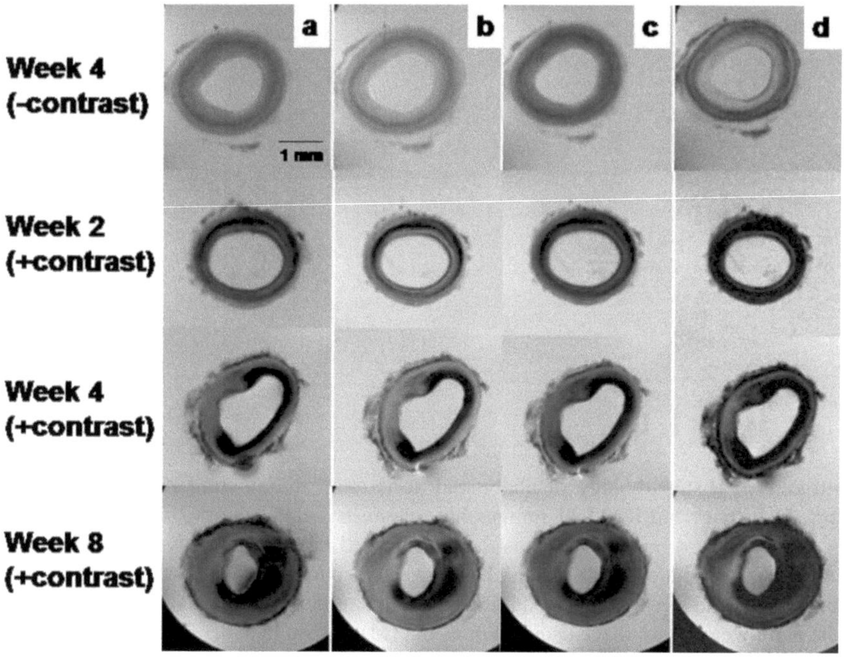

FIGURE 5.4. Ex vivo MRI of the abdominal aorta. Representative single slice transverse gradient echo (**a**), proton density (**b**), T_1 (**c**), and T_2 (**d**) weighted spin echo images of the abdominal aorta from a 4 week post-injury control (−contrast) and 2, 4, and 8 week post-injury ferumoxtran-10 treated (+contrast) animals are shown. Images provided by courtesy of Jucker. Reprinted from Yancy et al. (2005), with permission of Wiley-Liss, Inc., a subsidiary of John Wiley & Sons, Inc.

In another animal model of 12–28-month-old WHHL rabbits, iron accumulation was predominantly observed in a rim on the luminal side of the aortic wall (Schmitz et al., 2000).

Based on cell morphology and RAM-11 staining, it appeared that USPIO accumulation was associated with smaller macrophages rather than larger foam cells (Yancy et al., 2005). Electron microscopy supported this observation, since macrophages containing cytoplasmic Fe particles were surrounded by foam cells without cytoplasmic iron (Ruehm et al., 2001). A disadvantage of electron microscopy is that only a very small sample is investigated which might not be representative.

A long plasma half-life time, allowing prolonged exposure of the vessel wall to USPIOs, is crucial for the capability of a certain type of USPIO to label atherosclerotic plaque. Yancy et al. (2005) compared the uptake of two different types of USPIOs with a large difference in plasma half-life time, ferumoxtran-10 and ferumoxytol, in rabbit atherosclerotic plaque. Rabbit plasma half-life as estimated from daily MR angiograms was approximately 8-fold higher for ferumoxtran-10. Although in vitro macrophage phagocytosis of ferumoxytol

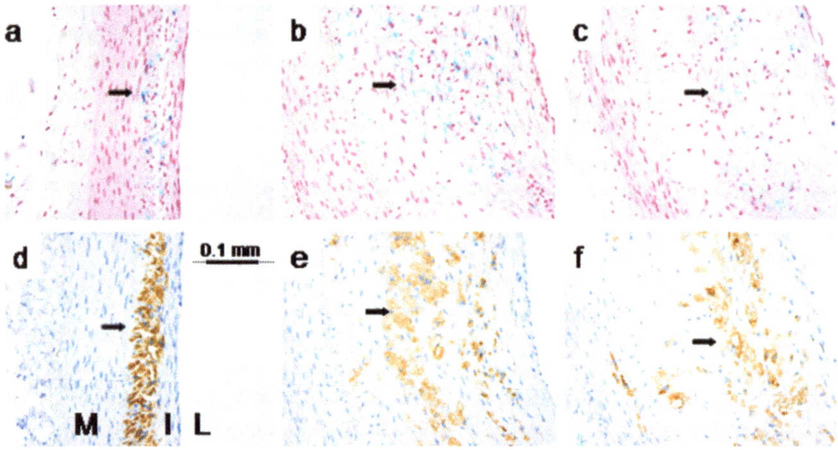

FIGURE 5.5. Histopathology of abdominal aorta from ferumoxtran-10 treated animals. Perl's iron (**a–c**) and RAM-11 immunohistochemistry (**d–f**) staining of abdominal aorta sections from 2, 4, and 8 week post-balloon injured rabbits, respectively. Iron accumulation was associated with macrophage in all sections (see arrows in each slide). Macrophage and iron were located primarily in the intimal (I) layer and only at the 8 week time point were macrophage detected in media (M) as seen in panel f. ×20 magnification. L = lumen. Images provided by courtesy of Jucker. Reprinted from Yancy et al.(2005), with permission of Wiley-Liss, Inc., a subsidiary of John Wiley & Sons, Inc.

was 4–6 fold greater than with ferumoxtran-10, ferumoxtran-10 was detected in vivo with MRI in all vessels while in vivo detection of ferumoxytol was negligible. With ex vivo MRI the uptake of ferumoxytol could also be visualized although the signal decrease was less than that in the ferumoxtran-10 groups. Iron quantification of in vitro plaque specimens with ICP-MS and histology (Perl's stain) confirmed the lower iron uptake in the ferumoxytol groups.

MRI reading is generally hampered by high USPIO concentrations in the lumen at low post-contrast imaging times. Schmitz et al. (2000) observed a circumferential area of low signal intensity which could not be attributed precisely to either the vessel wall or lumen at a post-contrast imaging time of 8 hours. This phenomenon is probably a susceptibility artifact due to the high USPIO concentration in the lumen. Indeed, no iron uptake was observed in corresponding histological sections (Schmitz et al., 2000). Similarly, Ruehm et al.. (2001) were unable to delineate the aortic wall on the first 2 days after contrast injection due to extensive susceptibility artifacts, and only after 4 days the wall could be delineated in all animals. Therefore, a sufficiently large post-contrast imaging time needs to be chosen.

In histology, the uptake of iron oxide nanoparticles is usually visualized indirectly, using a Perl's or Prussian blue staining. An interesting development is the use of near-infrared fluorescent iron oxide particles, which enables direct visualization of these nanoparticles in plaque with fluorescence miscroscopy (Pande et al., 2006).

Uptake of USPIOs causes a negative contrast (signal loss) in gradient echo images, which complicates image interpretation, since differentiation between USPIO uptake and dark-appearing areas like calcified tissue becomes laborious. Briley-Saebo et al. (2006) have shown in an initial study that a Gradient echo Acquisition for Superparamagnetic Particles (GRASP) sequence can generate positive USPIO-contrast in rabbits with balloon-injured hypercholesterolemic aortic wall. Signal enhancement observed using GRASP corresponded well with signal loss in the T_2^* weighted images. The value of this promising new pulse sequence for atherosclerotic plaque imaging remains to be validated in larger studies.

5.3.2. Patient Studies

Incidental signal loss in the arterial wall of the aorta and pelvic arteries in patients who had originally received USPIOs (ferumoxtran-10; Sinerem®, Guerbet) for staging lymph node metastases lead to the exciting hypothesis that USPIOs might be used to image macrophage infiltration also in the human vessel wall (Schmitz et al., 2001). Kooi et al. (2003) showed that indeed in patients with severe symptomatic carotid stenosis, ferumoxtran-10 (Sinerem®, Guerbet) uptake predominantly in macrophages of atherosclerotic plaque can be detected as pronounced post-contrast signal loss in gradient echo MR images, even though the dose in patients (2.6 mg Fe/kg) is about 20 times lower than the typical dose used in rabbit studies (Table 5.3). These findings were later confirmed by Trivedi et al. (2004, 2006). Histology showed that USPIO is taken up predominantly by rupture-prone and ruptured lesions (lesions with a thin or ruptured fibrous cap) since 27 of 36 (75 %) of these lesions showed USPIO uptake while this was only the case in 1 of 14 (7 %) of the stable lesions with a thick fibrous cap (Kooi et al., 2003). This indicates that USPIO-enhanced MRI is a highly promising non-invasive method to identify high-risk plaques.

Both Kooi et al. (2003) and Trivedi et al. (2004, 2006) showed that ferumoxtran-10 was primarily taken up by macrophages, confirming earlier findings in animal studies. Double staining for macrophages and iron showed extensive intracellular, granular localization of iron-containing particles in the cytoplasm of macrophages, especially in the vicinity of small capillaries (Figure 5.6) (Kooi et al., 2003). Electron microscopy showed multiple USPIOs per cell in an intracellular granular iron distribution in phagosomes of macrophages (Kooi et al., 2003; Trivedi et al., 2004) (Figure 5.7). Trivedi et al. (2006) reported that focal Perl's staining was usually in the subendothelial fibrous cap region, while diffuse Perl's staining was visualized at every depth in the plaque. The same authors also found that Perl's staining appeared to be intracellular as well as extracellular in location (Trivedi et al., 2006). In accordance with earlier animal studies, quite a few histological sections with an abundance of macrophages did not show USPIO uptake, suggesting that functional state of endothelium and macrophages and other factors might determine USPIO uptake (Kooi et al., 2003; Trivedi et al., 2004, 2006). Kooi et al. (2003) also showed

TABLE 5.3. Summary of human studies investigating uptake of superparamagnetic particles of iron oxide in atherosclerotic plaque.

Type of study	iron oxide nanoparticle	Dose	Findings	Reference
Patients who had originally received ferumoxtran-10 for lymph node metastases	Ferumoxtran-10	2.6 mg Fe/kg = 46 μmol Fe/kg	• Pronounced signal loss in wall of aortic and pelvic arteries suggests that ferumoxtran-10 accumulates in human atherosclerotic plaques	(Schmitz et al., 2001)
Symptomatic patients with severe carotid stenosis	Ferumoxtran-10	2.6 mg Fe/kg = 46 μmol Fe/kg	• Uptake of USPIOs in macrophages of atherosclerotic plaque in 10 of 11 patients • Uptake of USPIO in 75 % of the ruptured/rupture prone lesions and 7 % of the stable lesions • Significant signal loss in MR images 24 hours post-infusion but not 72 hours post-infusion • Histology and electron microscopy showed USPIO uptake in macrophages • Incidental uptake of USPIO by smooth muscle cells, myofibroblasts, and endothelial cells • Quite a few histological sections with an abundance of macrophages did not show USPIO uptake	(Kooi et al., 2003)
Symptomatic patients with severe carotid stenosis	Ferumoxtran-10	2.6 mg Fe/kg = 46 μmol Fe/kg	• Uptake of USPIO in macrophages of atherosclerotic plaque in 7 of 8 patients • Areas of signal reduction in MR images, corresponding to USPIO/macrophage-positive histological section in all these 7 patients • Histology and electron microscopy showed uptake of USPIOs in macrophages • Largest signal loss 24–36 hours post-USPIO infusion • Quite a few histological sections with an abundance of macrophages did not show USPIO uptake	(Trivedi et al., 2004)

Symptomatic patients with severe carotid stenosis	Ferumoxtran-10	2.6 mg Fe/kg = 46 μmol Fe/kg	• Uptake of USPIO in macrophages in 24 of 30 plaques • Areas of signal loss on post-USPIO MR images, corresponding to USPIO/macrophages-positive histological sections, in 24 of 27 patients • Qualitative MRI was highly sensitive (92.5 %) and moderately specific (64 %). • Agreement in location of MR signal loss and iron staining was good (Cohen kappa = 0.60) • Significant correlation between magnitude of relative MR signal changes in regions with focal signal loss and iron-positive cell count ($r = -0.63$, $P < 0.001$) • Quite a few histological sections with an abundance of macrophages did not show USPIO uptake	(Trivedi et al., 2006)
Symptomatic patients with severe carotid stenosis	Ferumoxtran-10	2.6 mg Fe/kg = 46 μmol Fe/kg	• 19 of 20 patients showed signal loss on the symptomatic as well as on the contralateral asymptomatic side	(Tang et al., 2006)

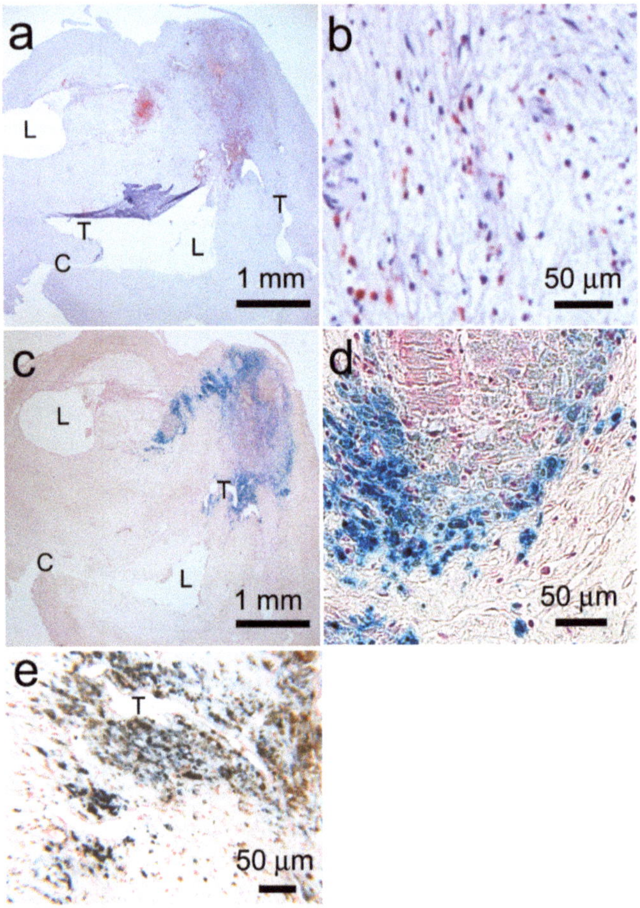

FIGURE 5.6. CD68 (**a** and **b**), Perl's (**c** and **d**), and CD68/Perl's (**e**) double staining of the endarterectomy specimen of a patient who received USPIOs. Double staining showed colocalization of macrophages and Perl's iron staining. The red coloring (**a** and **b**) and brown coloring (**e**) are indicative of macrophages, and the blue coloring (**c**, **d**, and **e**) is indicative of the accumulation of USPIOs. The double lumen is indicated with L, the surgical cut with C, and tears with T. Reprinted from Kooi et al. (2003), with permission of Lippincott Williams & Wilkins.

incidental ASMA/Perl's iron and CD31/Perl's iron double staining, which is indicative for limited USPIO uptake by smooth muscle cells, myofibroblasts, and endothelial cells. Trivedi et al. (2006) found no colocalization with smooth muscle and/or endothelial cells, but this can easily be overlooked because the number of these cell types that are Perl's positive is very limited indeed.

A large number of images of atherosclerotic plaque showed pronounced signal loss 24–36 hours post-contrast infusion (Kooi et al., 2003; Trivedi et al., 2006) (Figure 5.8). Since calcified regions in plaque also result in focal signal loss

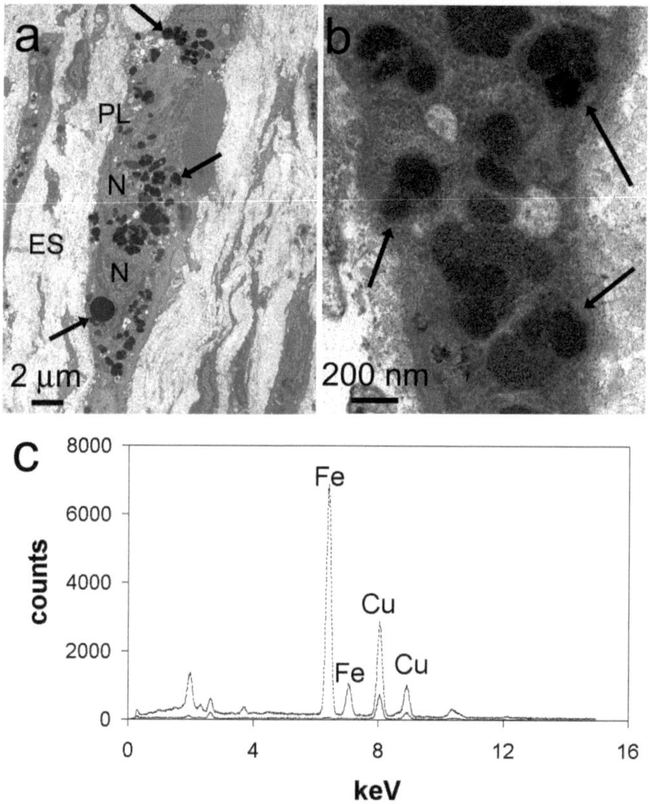

FIGURE 5.7. Electron microscopy and Energy-Dispersive X-ray (EDX) analysis of Perl's positive regions of an endarterectomy specimen of a patient who received USPIOs. **a** Presence of numerous particles within a single macrophage (arrows). The plasma membrane is indicated with PL, the cell nucleus with N, and the extracellular space with ES. **b** Detail of phagosomes in macrophages containing clusters of iron particles (arrows). **c** EDX spectrum of particles in macrophages (dotted line). It is shown that these particles contain a large amount of iron. The solid line corresponds to an EDX spectrum of a reference area. Reprinted from Kooi et al.(2003), with permission of Lippincott Williams & Wilkins.

in MR images, only those areas with new focal signal loss in post-contrast images are considered to be a positive MR finding for USPIO uptake. The magnitude of the measured signal loss depends on the used MR pulse sequence as well as whether regions-of-interest contain entire vessel wall or only area of focal or diffuse signal loss. Qualitative MRI was highly sensitive (92.5 %) and moderately specific (64 %) for detection of USPIOs, using histology as standard of reference. The agreement in location of MR signal loss and iron staining was good (Cohen kappa = 0.60) (Trivedi et al., 2006) and there was a significant correlation between the magnitude of relative MR signal changes in regions with

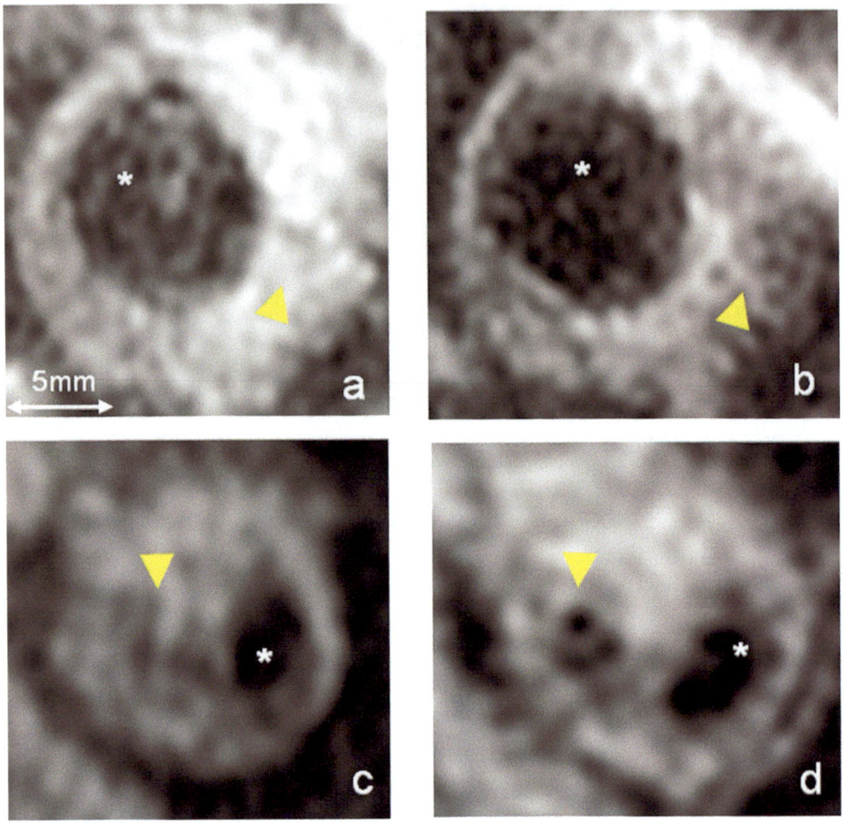

FIGURE 5.8. Pattern of USPIO-induced signal loss. Axial MR images (T_2*W [TE=5.6 ms]) through the common carotid artery. **a, b** Diffuse USPIO effect. **c, d** Focal USPIO signal effect. **a, c** Pre-USPIO images. **b, d** Thirty-six hours after infusion (*lumen, arrow heads: USPIO signal effect). Images provided by courtesy of Gillard. Reprinted from Trivedi et al. (2006), with permission of Lippincott Williams & Wilkins.

focal signal loss and Perl's positive cell count ($\rho = -0.63$, $P < 0.001$) (Trivedi et al., 2006). These analysis might have been hampered by the varying delay between infusion and surgery of 6.9 ± 4.8 days with a range of 2–18 days so that with a fixed delay these correlations might be higher.

The temporal effect on signal loss in MR images was investigated by Trivedi et al. (2004). Largest signal loss was found 24 and 36 hours post-USPIO infusion, while the magnitude of signal loss was lower after 48 hours and there was no significant signal loss 72 hours post-contrast. This confirmed earlier findings by Kooi et al. (2003) who found significant signal loss after 24 hours, while no significant decrease occurred 72 hours post-infusion. In histology, USPIO uptake could still be visualized, even though the time between infusion and surgery was on average 5–7 days (Kooi et al., 2003; Trivedi et al., 2006).

The amount of USPIOs in plaque at these time points however seems to be too low to visualize with MRI.

All studies so far have been performed on symptomatic patients. An initial study determined signal loss in the carotid vessel wall on the symptomatic as well as the contralateral asymptomatic side (Tang et al., 2006). It was shown that in 19 of 20 patients bilateral signal loss was observed (Tang et al., 2006), highlighting the possible systemic nature of vulnerable plaque (Buffon et al., 2002; Madjid et al., 2004; Mauriello et al., 2005). Future studies are warranted to investigate differences in USPIO uptake between symptomatic and asymptomatic patients, to determine the effect of drugs, such as statin therapy on USPIO-enhanced images and to study the relation between USPIO uptake and recurrent strokes.

5.3.3. Factors Regulating Uptake of USPIOs

The route of USPIO uptake in atherosclerotic plaque is still unknown. Preliminary data of Yancy et al. (2005) showed less ferumoxtran-10 uptake in normocholesterolemic vs hypercholesterolemic rabbits suggesting that a pro-inflammatory stimulus enhances uptake. Incidental USPIO accumulation in endothelial cells in rabbit as well as human plaque suggests that USPIOs may enter the plaque via transcytosis. Alternatively, macrophages may be newly recruited USPIO-loaded monocytes. This is supported by the study of Litovsky et al. (2003) who showed more iron-positive macrophages in atherosclerotic plaque of cytokine-treated apo E knockout mice 6 days after SPIO infusion, since cytokines enhance the recruitment of monocytes from the blood. The effect of cytokine treatment for the smaller USPIOs, which are generally thought to be more suitable for plaque imaging, remains to be investigated.

Competition experiments indicated that the cellular uptake of SPIOs (ferumoxides) involves scavenger receptor SR-A-mediated endocytosis (Raynal et al., 2004). Improvements of the in vitro techniques are needed to verify the SR-A endocytotic pathway for the USPIO ferumoxtran-10, which is currently used for atherosclerotic plaque imaging.

An in vitro study with cultured murine macrophage-like cells showed that endocytosis of dextran-coated SPIOs by activated macrophages is enhanced by the cytokines IL-4 and IFN-γ, while it is down-regulated in the presence of the HMG-CoA reductase inhibitor lovastatin (Rogers and Basu et al., 2005). Thus, the number of macrophages as well as their phagocytotic capacity may influence SPIO uptake in plaque. The effect of the above-mentioned factors on the smaller USPIOs with prolonged plasma half-life still has to be determined.

Several investigators reported, in both animal and human studies, that although macrophages were the dominant cell type of USPIO uptake, some regions with an abundance of macrophages showed no iron uptake while morphology was similar as iron-positive areas (Kooi et al., 2003; Schmitz et al., 2000; Trivedi et al., 2004, 2006; Yancy et al., 2005), suggesting that functional state of the endothelium and macrophages, local inflammation, and other factors might determine USPIO uptake. In line with these findings, Trivedi et al. (2006) found a poor, but significant, correlation between total number of Perl's positive cells (USPIOs) and CD68-stained macrophages ($\rho = 0.29$, $P < 0.001$) in human plaques. In contrast, Hyafil et al. (2006) observed a linear relationship between Perl's (USPIOs) and RAM-11 positive areas (macrophages) in rabbits with injured aortas ($r = 0.98$, $P < 0.05$), and Pande et al. (2006) found a strong correlation between average near-infrared fluorescent (NIRF) signal of NIRF-MIONs and the number of macrophages in carotid plaques of $ApoE^{-/-}$ mice. The findings in the patient study may have been hampered by clearance of USPIOs from the plaque since the delay between infusion and surgery was 6.9 ± 4.8 days with a range of 2–18 days while the largest signal loss in MR images due to USPIO uptake occurred 24–36 hours post-infusion (Trivedi et al., 2004).

5.3.4. MR Pulse Sequences

Trivedi et al. (2004) showed that signal loss was maximal on T_2*-weighted images with a large echo time; however, these images have an inherently low signal-to-noise ratio which makes image interpretation laborious. Therefore, one needs to compromise between enough T_2* weighting without losing too much signal-to-noise ratio and thus image quality. With the current state of techniques, a pulse sequence with an intermediate echo time is probably preferred. Trivedi et al. (2006) demonstrated acceptable image quality using a T_2*-weighted pulse sequence with an echo time/repetition time of 5.6 ms/heart beat (Figure 5.8).

In future studies, image interpretation might be facilitated by using new pulse sequences which generate positive USPIO contrast (Cunningham et al., 2005; Mani et al., 2006) or alternatively, generating a positive contrast image by calculating the susceptibility gradient map from the 3D anatomical gradient echo images (Dahnke et al., 2006). An initial animal study with positive USPIO contrast showed promising results (Briley-Saebo et al., 2006).

Image interpretation can be complicated by artifacts. Several authors reported the occasional appearance of a dark band adjacent to the hyperintense signal of the lumen mainly in T_1-weighted gradient echo images that was not associated with USPIO uptake (Kooi et al., 2003; Schmitz et al., 2000). This phenomenon is probably caused by a susceptibility artifact of USPIO in the lumen and might be prevented using double inversion recovery black blood pre-pulses since Trivedi et al. (2004) did not report this phenomenon.

5.4. Novel Iron Oxide Nanoparticles for Imaging of Atherosclerosis

A highly promising development is the coupling of cell-specific or molecule-specific affinity ligands, such as antibodies, peptides, or small molecules, to the dextran coating of iron oxide nanoparticles, enabling visualization of not only macrophages, but also other cell types which are important in atherosclerosis.

Recently, a high throughput approach for the selection of promising novel nanoparticles was introduced by Weissleder et al. (2005) who screened a large library of 146 of such novel nanoparticles against different cell lines to identify agents with high specificity for, e.g., endothelial cells and activated human macrophages. Fluorescent cross-linked iron oxide particles, which are in physicochemical and biological properties similar to monocrystalline iron oxide

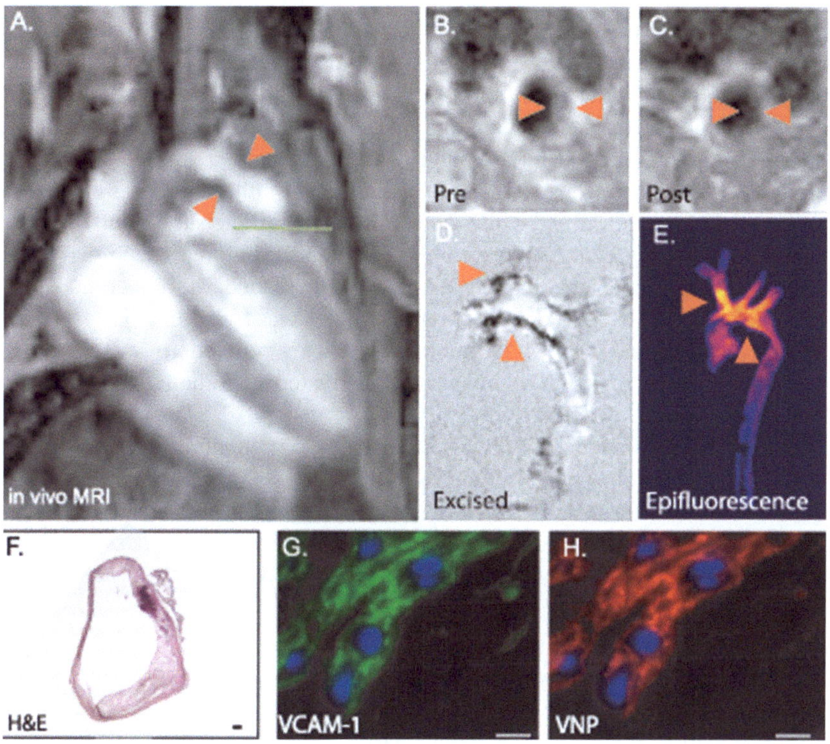

FIGURE 5.9. MRI of atherosclerotic lesions using novel iron oxide nanoparticles. **A** Cholesterol-fed apoE$^{-/-}$ mice were imaged via MRI using gadolinium-protected graph copolymer to delineate vascular lumen and structural aortic abnormalities such as narrowing (arrows). Axial images (at the level of the green line) were then obtained before (**B**) and 24 hours after (**C**) administration of the magnetic VCAM-1-Targeted Nanoparticle (VNP). Note the decrease in signal intensity of the eccentrically thickened aortic wall between the two arrows. **D** Ex vivo MRI confirms extensive low signal changes in the aortic wall induced by VNP accumulation (arrows), further corroborated by macroscopic epifluorescence imaging of Cy 5.5 in VNP (**E**). **F–H** Histological validation of MR findings. **F** Hematoxylin/eosin section of the aorta image in **B** and **C** confirms eccentric wall thickening (×2). Comparative immunofluorescence of VCAM-1 expression (**G**, green) and VNP accumulation (**H**, red). Nuclei in **G** and **H** are counterstained with DAPI (blue). Note that there is extensive colocalization of VCAM-1 expression and VNP. Images provided by courtesy of Weissleder. Reprinted from Kelly et al. (2005), with permission of Lippincott Williams & Wilkins.

nanoparticles (MIONs), were labeled with different synthetic small molecules. The screening identified nanoparticles that could discriminate among different cell types as well as among different physiological states of the same cell type, e.g. macrophage activation status.

The feasibility of using such a novel nanoparticle for imaging of atherosclerosis has been explored by Kelly et al. (2005). An MR contrast agent consisting of ≈ 6 peptides with high affinity for VCAM-1 as identified with iterative phage display, coupled to a Cross-Linked Iron Oxide (CLIO) nanoparticle similar to MION, was internalized by VCAM-1 expressing cells while the control particle without the specific peptides failed to accumulate in those cells (Kelly et al., 2005). The cellular specificity of the VCAM-targeted nanoparticle (VNP) was clearly modified, since accumulation in macrophages was 11 times lower than in endothelial cells. Initial imaging findings with a limited number of animals ($n = 3$) showed extensive MR signal loss in atherosclerotic plaques of cholesterol-fed apoE$^{-/-}$ mice while the control nanoparticle showed no uptake at the same dose (5 mg Fe/kg body weight) (Figure 5.9). Fluorescence microscopy confirmed uptake in VCAM-expressing cells of the plaque. Moreover, no significant uptake of the VCAM-1-targeted nanoparticles was found in wild-type mice. Alternatively, an anti-VCAM-1 antibody has been coupled to a cross-linked iron oxide particle for in vivo imaging of activated endothelium in a murine inflammatory model (Tsourkas et al., 2005).

Another novel nanoparticle was created by conjugating an anti-human E-selectin antibody fragment to a cross-linked iron oxide particle. E-selectin expression in an Matrigel model with functional blood vessels was detectable with in vivo MRI (Kang et al., 2006). This contrast agent could possibly detect early changes in endothelial phenotype in atherosclerosis.

5.5. Future Perspectives

Animal as well as patient studies have clearly demonstrated the feasibility of in vivo imaging of accumulated macrophages in atherosclerotic plaque. Given the key role of macrophages in all stages of plaque development and the abundance of macrophages especially in vulnerable plaques, USPIO-enhanced MRI could be an extremely important clinical imaging tool.

Only a limited number of patient studies have been performed so far. Larger multi-center trials are warranted with symptomatic and asymptomatic patients with severe as well as moderate stenosis to confirm these results. The predictive value of USPIO-enhanced MR findings for future ischemic events needs to be determined in follow-up studies and the relation between USPIO uptake and drug use, such as statin therapy, needs to be investigated. New animal and patient studies could elucidate the mechanism of USPIO uptake in plaque.

The generation of negative contrast by internalized USPIOs has been a drawback for application of USPIOs. Pre-contrast, as well as post-contrast, imaging has been required to distinguish USPIO uptake from other dark-appearing areas, such as calcified tissue. New pulse sequences can generate positive USPIO contrast or alternatively a positive contrast image can be generated by calculating the susceptibility gradient map from the 3D anatomical gradient echo images. Generation of positive as well as negative contrast images might take away the need to acquire pre-contrast images. The value of these positive contrast images needs to be validated in large studies.

The in vitro uptake of nanoparticles by macrophages for the larger SPIOs is much higher than for USPIOs, but due to fast plasma clearance these nanoparticles are hardly taken up by macrophages in deep pathological tissue, such as atherosclerotic plaque, in vivo. Extraction of blood monocytes, ex vivo labeling with SPIOs, and reinjection of these labeled monocytes might significantly improve the MR signal, but this still needs to be tested in animal studies (Metz et al., 2004).

The signal loss is largest in T_2^*-weighted images with a long echo time; however, these images have an inherently low signal-to-noise ratio (SNR). The SNR could be improved by using higher field strengths, improved radio-frequency coils, and novel pulse sequences. A great challenge will be imaging of USPIO uptake in coronary plaques, possibly aided by the use of intravascular coils.

Novel iron oxide superparamagnetic nanoparticles enable imaging of not only macrophages, but also other cell types and their functional state, such as activation status of macrophages and VCAM-1-expressing endothelial cells. The value of these novel contrast agents needs to be proven in larger animal studies, followed by patient studies.

Possible future applications for USPIOs in atherosclerosis could be the evaluation of therapy, imaging-aided selection of patients at risk for future ischemic events, and selecting individualized treatment strategies based on the molecular profile of atherosclerotic plaque.

5.6. Conclusions

During the last decade it has become established that atherosclerosis is an inflammatory disease. Macrophages play a central role in all stages of plaque development, and they are abundant especially in the vulnerable plaque. The use of USPIOs enables in vivo non-invasive visualization of macrophage-rich areas in atherosclerotic plaque. Larger patient studies are warranted to investigate the value of USPIOs in daily clinical practice. Novel iron oxide nanoparticles with molecule- or cell-specific affinity ligands coupled to the dextran coating enable visualization of not only macrophages but also functional state and other cell types that are relevant to atherosclerosis. The value of these novel contrast agents needs to be validated in larger animal studies.

References

Anzai, Y., Piccoli, C.W., Outwater, E.K., et al., 2003. Evaluation of neck and body metastases to nodes with ferumoxtran 10-enhanced MR imaging: phase III safety and efficacy study. Radiology 228, 777–788.

Bjornerud, A., Johansson, L., 2004. The utility of superparamagnetic contrast agents in MRI: theoretical consideration and applications in the cardiovascular system. NMR Biomed 17, 465–477.

Briley-Saebo, K.C., Mani, V., Hyafil, F., et al., 2006. Positive contrast MR imaging of in vivo atherosclerosis in a rabbit model using ultrasmall iron oxide particles. Proc. Intl. Soc. Mag. Reson. Med. 14, 563.

Buffon, A., Biasucci, L.M., Liuzzo, G., et al., 2002. Widespread coronary inflammation in unstable angina. N Engl J Med 347, 5–12.

Corot, C., Petry, K.G., Trivedi, R., et al., 2004. Macrophage imaging in central nervous system and in carotid atherosclerotic plaque using ultrasmall superparamagnetic iron oxide in magnetic resonance imaging. Invest Radiol 39, 619–625.

Cunningham, C.H., Arai, T., Yang, P.C., et al., 2005. Positive contrast magnetic resonance imaging of cells labeled with magnetic nanoparticles. Magn Reson Med 53, 999–1005.

Dahnke, H., Liu, W., Frank, J., et al., 2006. Optimal positive contrast of labeled cells via conventional 3D imaging. Proc. Intl. Soc. Mag. Reson. Med. 14, 361.

Falk, E., Shah, P.K., Fuster, V., 1995. Coronary plaque disruption. Circulation 92, 657–71.

Gupta, A.K., Gupta, M., 2005. Synthesis and surface engineering of iron oxide nanoparticles for biomedical applications. Biomaterials 26, 3995–4021.

Hansson, G.K., Libby, P., Schonbeck, U., et al., 2002. Innate and adaptive immunity in the pathogenesis of atherosclerosis. Circ Res 91, 281–291.

Hansson, G.K., 2005. Inflammation, atherosclerosis, and coronary artery disease. N Engl J Med 352, 1685–1695.

Hyafil, F., Laissy, J.P., Mazighi, M., et al., 2006. Ferumoxtran-10-enhanced MRI of the hypercholesterolemic rabbit aorta: relationship between signal loss and macrophage infiltration. Arterioscler Thromb Vasc Biol 26, 176–181.

Jung, C.W., Jacobs, P., 1995. Physical and chemical properties of superparamagnetic iron oxide MR contrast agents: ferumoxides, ferumoxtran, ferumoxsil. Magn Reson Imaging 13, 661–674.

Kang, H.W., Torres, D., Wald, L., et al., 2006. Targeted imaging of human endothelial-specific marker in a model of adoptive cell transfer. Lab Invest 86, 599–609.

Kelly, K.A., Allport, J.R., Tsourkas, A., et al., 2005. Detection of vascular adhesion molecule-1 expression using a novel multimodal nanoparticle. Circ Res 96, 327–336.

Kooi, M.E., Cappendijk, V.C., Cleutjens, K.B., et al., 2003. Accumulation of Ultrasmall Superparamagnetic Particles of Iron Oxide in Human Atherosclerotic Plaques Can Be Detected by In Vivo Magnetic Resonance Imaging. Circulation 107, 2453–2458.

Leiner, T., Gerretsen, S., Botnar, R., et al., 2005. Magnetic resonance imaging of atherosclerosis. Eur Radiol 15, 1087–1099.

Litovsky, S., Madjid, M., Zarrabi, A., et al., 2003. Superparamagnetic iron oxide-based method for quantifying recruitment of monocytes to mouse atherosclerotic lesions in vivo: enhancement by tissue necrosis factor-alpha, interleukin-1beta, and interferon-gamma. Circulation 107, 1545–1549.

Madjid, M., Zarrabi, A., Litovsky, S., et al., 2004. Finding vulnerable atherosclerotic plaques: is it worth the effort? Arterioscler Thromb Vasc Biol 24, 1775–1782.

Mani, V., Briley-Saebo, K.C., Itskovich, V.V., et al., 2006. Gradient echo acquisition for superparamagnetic particles with positive contrast (GRASP): sequence characterization in membrane and glass superparamagnetic iron oxide phantoms at 1.5T and 3T. Magn Reson Med 55, 126–135.

Mauriello, A., Sangiorgi, G., Fratoni, S., et al., 2005. Diffuse and active inflammation occurs in both vulnerable and stable plaques of the entire coronary tree: a histopathologic study of patients dying of acute myocardial infarction. J Am Coll Cardiol 45, 1585–1593.

Metz, S., Bonaterra, G., Rudelius, M., et al., 2004. Capacity of human monocytes to phagocytose approved iron oxide MR contrast agents in vitro. Eur Radiol 14, 1851–1858.

Naghavi, M., Libby, P., Falk, E., et al., 2003. From vulnerable plaque to vulnerable patient: a call for new definitions and risk assessment strategies: Part I. Circulation 108, 1664–1672.

Pande, A.N., Kohler, R.H., Aikawa, E., et al., 2006. Detection of macrophage activity in atherosclerosis in vivo using multichannel, high-resolution laser scanning fluorescence microscopy. J Biomed Opt 11, 21009.

Raynal, I., Prigent, P., Peyramaure, S., et al., 2004. Macrophage endocytosis of superparamagnetic iron oxide nanoparticles: mechanisms and comparison of ferumoxides and ferumoxtran-10. Invest Radiol 39, 56–63.

Reimer, P., Rummeny, E.J., Daldrup, H.E., et al., 1995. Clinical results with Resovist: a phase 2 clinical trial. Radiology 195, 489–496.

Reimer, P., Tombach, B., 1998. Hepatic MRI with SPIO: detection and characterization of focal liver lesions. Eur Radiol 8, 1198–1204.

Rogers, W.J., Basu, P., 2005. Factors regulating macrophage endocytosis of nanoparticles: implications for targeted magnetic resonance plaque imaging. Atherosclerosis 178, 67–73.

Ruehm, S.G., Corot, C., Vogt, P., et al., 2001. Magnetic Resonance Imaging of Atherosclerotic Plaque With Ultrasmall Superparamagnetic Particles of Iron Oxide in Hyperlipidemic Rabbits. Circulation 103, 415–422.

Schmitz, S.A., Coupland, S.E., Gust, R., et al., 2000. Superparamagnetic iron oxide-enhanced MRI of atherosclerotic plaques in Watanabe hereditable hyperlipidemic rabbits. Invest Radiol 35, 460–471.

Schmitz, S.A., Taupitz, M., Wagner, S., et al., 2001. Magnetic resonance imaging of atherosclerotic plaques using superparamagnetic iron oxide particles. J Magn Reson Imaging 14, 355–361.

Schmitz, S.A., Taupitz, M., Wagner, S., et al., 2002. Iron-oxide-enhanced magnetic resonance imaging of atherosclerotic plaques: postmortem analysis of accuracy, inter-observer agreement, and pitfalls. Invest Radiol 37, 405–411.

Simon, G.H., von Vopelius-Feldt, J., Wendland, M.F., et al., 2006. MRI of arthritis: Comparison of ultrasmall superparamagnetic iron oxide vs. Gd-DTPA. J Magn Reson Imag 23, 720–727.

Tang, T., Howarth, S.P., Miller, S.R., et al., 2006. Assessment of inflammatory burden contralateral to the symptomatic carotid stenosis using high-resolution ultrasmall, superparamagnetic iron oxide-enhanced MRI. Stroke 37, 2266–2270.

Tedgui, A., Mallat, Z., 2006. Cytokines in atherosclerosis: pathogenic and regulatory pathways. Physiol Rev 86, 515–581.

Trivedi, R.A., U-King-Im, J.M., Graves, M.J., et al., 2004. In vivo detection of macrophages in human carotid atheroma: temporal dependence of ultrasmall superparamagnetic particles of iron oxide-enhanced MRI. Stroke 35, 1631–1635.

Trivedi, R.A., Mallawarachi, C., U-King-Im, J.M., et al., 2006. Identifying Inflamed Carotid Plaques Using In Vivo USPIO-Enhanced MR Imaging to Label Plaque Macrophages. Arterioscler Thromb Vasc Biol 26, 1601–1606.

Tsourkas, A., Shinde-Patil, V.R., Kelly, K.A., et al., 2005. In vivo imaging of activated endothelium using an anti-VCAM-1 magnetooptical probe. Bioconjug Chem 16, 576–581.

Wang, Y.X., Hussain, S.M., Krestin, G.P., 2001. Superparamagnetic iron oxide contrast agents: physicochemical characteristics and applications in MR imaging. Eur Radiol 11, 2319–2331.

Weissleder, R., Lee, A.S., Khaw, B.A., et al., 1992. Antimyosin-labeled monocrystalline iron oxide allows detection of myocardial infarct: MR antibody imaging. Radiology 182, 381–385.

Weissleder, R., Kelly, K., Sun, E.Y., et al., 2005. Cell-specific targeting of nanoparticles by multivalent attachment of small molecules. Nat Biotechnol 23, 1418–1423.

Yancy, A.D., Olzinski, A.R., Hu, T.C., et al., 2005. Differential uptake of ferumoxtran-10 and ferumoxytol, ultrasmall superparamagnetic iron oxide contrast agents in rabbit: critical determinants of atherosclerotic plaque labeling. J Magn Reson Imag 21, 432–442.

6
(Super)paramagnetic Nanoparticles: Applications in Noninvasive MR Imaging of Stem Cell Transfer

Glenn A. Walter, Swadeshmukul Santra, Bijoy Thattaliyath, and Samuel C. Grant.

Abstract: Common techniques to monitor muscle stem cell transplants typically rely on ex vivo genetic modification to allow expression of reporter genes. Specific reporter genes allow for graft identification during post-mortem analysis. The use of these invasive techniques makes even simple and practical questions difficult and labor intensive to answer. The design and improvement of the stem cell-based therapies will be greatly facilitated by the development of sensitive, noninvasive, and nondestructive techniques for tracking stem cells following implantation or infusion. MR offers such an opportunity, due to its ability to obtain high-resolution 3D images repeatedly and noninasively. Reliable tracking by MR imaging can be achieved by labeling cells with (super)paramagnetic nanoparticles. In this chapter, we discuss recent advances in the development and applications of (super)paramagnetic nanoparticles for in vivo cell tracking and targeting.

6.1. Introduction

6.1.1. The Need for Noninvasive Imaging of Cellular Therapies for Regenerative Medicine

Stem cells can be defined as undifferentiated cells that are capable of self-renewal, proliferation, and differentiation (Figure 6.1) (Blau et al., 2001; Leri et al., 2005). These unique cells can be further divided into three broad types of cells based on their ability to differentiate into any cell type. They are totipotent, pluripotent, and multipotent cells. Totipotent stem cells have the ability to form any cell type and form an entire organism. Pluripotent and mulitpotent stem cells are limited in their ability to form a new organism. Pluripotent stem cells can give rise to all three germ layers and multipotent stem cells have limited cell lineages that they can form (Blau et al., 2001). Perhaps the most far-reaching potential application of stem cells is the generation of new and genetically corrected

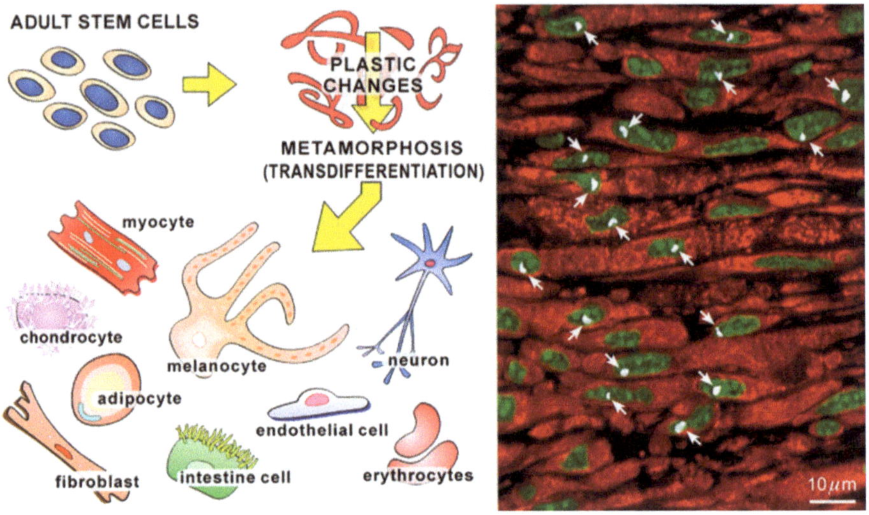

FIGURE 6.1. Plasticity of adult stem cells. Stem cells acquire cell phenotypes different from their organ of origin. The confocal image illustrates the transdifferentiation of male c-kit positive bone marrow–derived cells (BMCs) into cardiomyocytes after infarction in a female mouse. Myocytes are labeled by cardiac myosin heavy chain (red) and nuclei by propidium iodide (PI; green). The Y chromosome in nuclei is shown by white dots. Reprinted from Leri et al., (2005). Reproduced with permission from the American Physiological Society.

cells and tissues that could be used for "cell therapies" forming the basis of a innovative and growing field of regenerative medicine. A commonality of many diseases is the disruption of cellular function and/or destruction of tissues of the body. Engineered tissues could be used to replace or repair these defective or destroyed tissues, or to provide a cellular platform to continuously secrete therapeutic proteins. Pluripotent stem cells could be stimulated in a controlled manner to develop into specialized cells, offering the possibility of a renewable source of replacement cells and tissues to treat a multitude of diseases, such as Parkinson's, Alzheimer's (Cogle et al., 2004; Laywell et al., 1999, 2000; Laywell et al., 2002; Lindvall and Kokaia, 2006; Muller et al., 2006; Scheffler et al., 1999, 2005; Steindler, 2002; Steindler and Pincus, 2002), various heart diseases (Chien, 2006; Deasy and Huard, 2002; Payne et al., 2005), osteoarthritis, rheumatoid arthritis (Palmer et al., 2005), muscular dystrophy (Chakkalakal et al., 2005), and senescence (2006, 2006). The overall potential of these cells seems to be limitless (see the official **National Institutes of Health** resource for stem cell research: http://stemcells.nih.gov/).

Embryonic stem cells (ESCs), which are derived from the inner mass of the blastocyst, possess unique properties (Blau et al., 2001; Leri et al.,, 2005; Thomson et al., 1998. These cells can be grown in vitro and propagated indefinitely in their undifferentiated state while retaining a normal karyotype. It has been shown that they can be differentiated in vitro into all cell types. Similarly, following the in vivo delivery of human ESC lines to skeletal muscle or to the testis, they still retain their capacity to forming all cell lineages. Unfortunately, ESC lines can also form teratomas. Teratomas are tumors that include mature endodermal, mesodermal, and ectodermal structures (Thomson et al., 1998). Likewise, ESCs injected intravenously can engraft and colonize all organs with the possible development of neoplastic lesions (Wang et al., 2002).

Adult stem cell can be derived from numerous tissues (Blau et al., 2001). For example, adult bone marrow contains, in addition to differentiated cells, a heterogeneous population of undifferentiated cells with stem cell properties, being composed of hematopoietic stem cells (HSCs) and mesenchymal stem cells (MSCs) (Blau et al., 2001; Leri et al., 2005). These two subsets of cells have been employed in the repair of damaged organs giving rise to tissues distinct from the organ of origin (Lakshmipathy and Verfaillie, 2005; Mathur and Martin, 2004; Quesenberry et al., 2004; Benedetti et al., 2000; Muller et al., 2006). Multipotent and pluripotent stem cells have also been isolated from muscle, heart, liver, and adipose tissue. HSC transplantation into bone marrow–depleted animals has been shown to result in wide spread repopulation of a depleted hematopoietic cell pool (Brazelton, 2005). This situation is ideally suited for noninvasive imaging modalities that can detect tissue regeneration and function correction on the entire organ level. Adult stem cells, on the other hand, can be limited in their ability to form certain functional cell types and in their regenerative capacity. Therefore, adult stem cell transplantation into other adult tissues raises a number of challenges for the field of regenerative medicine and for tracking its efficacy noninvasively. The first challenge is to develop imaging modalities with the sensitivity to track a limited number of cells. The second challenge is the development of imaging modalities that are capable of discrimination between the few viable and nonviable cells that persist after first week of transplantation (Mintz-Hittner et al., 2004).

Throughout the body, stem cells can be found in tissue specific niches, which are defined as microenvironments that nurture stem cells and maintain tissue homeostasis (Yin, 2006; Anversa et al., 2006; Moore and Lemischka, 2006). These niches are necessary to maintain a balance between stem cell quiescence and activity; thus, providing a safeguard against excessive stem cell proliferation that could lead to cancer. The biological signals that result in progenitor cell migration and tissue regeneration are still poorly understood. Beyond the mere isolation of stem cells, cell and molecular biologists are now changing their attention to isolating the signals involved in stem cell homing and regeneration of the target tissue (Brazelton, 2005). In part, these efforts are hampered by an inability to noninvasively follow a small number of cells as they navigate toward damaged or targeted regions. Conventionally, finding an engrafted stem cell has relied on laborious, standard histological techniques (Brazelton and Blau, 2005), which truly represent the problem of "finding a needle in a haystack." Therefore, it is a challenge of noninvasive imaging to identify stem cell niches and monitor the effects of cellular signals that influence primitive cells to leave its niche to regenerate or replace old and dying cells (Anversa et al. 2006). The delivery of both ESCs and adult stem cells have been imaged by magnetic resonance imaging (MRI) in many different target tissues (Table 6.1) and will be discussed in further detail later in this chapter. Although MRI likely will not be able to match the resolution of confocal or multiphoton microscopy, we do anticipate that it will be able to noninvasively locate the "haystacks" containing functional

TABLE 6.1. Some examples and properties of commercial SPIO agents for cell tracking

Class	Trade and common names	Tissue	Stem or progenitor cell Type	Mean particle size
VSOP	C125	Brain (Stroh et al., 2005a)	ESC (Stroh et al., 2005a)	9 nm
SPIO	Feridex, Endorem, Ferumoxide, Resovist	Cardiac (Arai et al., 2006; Bulte et al., 2005; Cahill et al., 2004b; He et al., 2007; Himes et al., 2004; Kraitchman et al., 2003, 2005; Leor et al., 2006; Sykova and Jendelova, 2005; Tallheden et al., 2006) Brain (Heyn et al., 2006; Magnitsky et al., 2005; Walczak et al., 2005; Watson et al., 2006) Kidney (Bos et al., 2004; Hauger et al., 2006) Liver (Bos et al., 2004; Hauger et al., 2006) SC (Lepore et al., 2006), Bone (Daldrup-Link et al., 2005a) Muscle (Cahill et al., 2004a)	ESC (Arai et al., 2006; Himes et al., 2004; Tallheden et al., 2006) HSC (Arbab et al., 2005) MSC (Arbab et al., 2004a, 2005; Bos et al., 2004; Hauger et al., 2006; He et al., 2007; Kraitchman et al., 2003, 2005; Walczak et al., 2005) MDSC (Cahill et al., 2004a) Mb (Cahill et al., 2004b; Garot et al., 2003) NSC (Magnitsky et al.,2005; Walczak et al., 2005; Watson et al., 2006) Mac (Heyn et al., 2006; Leor et al., 2006) HUVEC (Daldrup-Link et al., 2005a; Gimi et al., 2006) NPC (Lepore et al., 2006)	80–150 nm
USPIO	Sinerem, Combidex, Ferumoxtran, Clariscan	Cardiac (He et al., 2007; Kustermann et al., 2005),	MSC (Ittrich et al., 2005) EVCM (Kustermann et al., 2005) Mono (Metz et al., 2004) HUVEC (Mantyla et al., 2006) (Ju et al., 2006) Sm (He et al., 2007) ESC (Ref Hoehn)	20–40 nm
MPIO	Bangs Laboratories	Cardiac (Stuckey et al., 2006) Brain (Shapiro et al., 2006a) Bone (Mayer-Kuckuk et al., 2005)	NPC (Shapiro et al., 2006a) BMSC (Mayer-Kuckuk et al., 2005; Stuckeyet al., 2006)	0.96, 1.63, 2.79 μm

(Continued)

TABLE 6.1. (Continued)

Class	Trade and common names	Tissue	Stem or progenitor cell Type	Mean particle size
Cell sorting	MACS-Microbeads	Brain (Jendelova et al., 2005; Sykova and Jendelova, 2005, 2006) SC (Sykova and Jendelova, 2005, 2006) Cardiac (Weber et al., 2004a)	EPC (Weber et al., 2004a) ESC (Sykova and Jendelova, 2005, 2006) MSC (Sykova and Jendelova, 2005, 2006) BMSC (Jendelova et al., 2005)	50 nm

VSOP, very small superparamagnetic iron oxide particles; SPIO, superparamagnetic iron oxide nanoparticle; USPIO, ultrasmall SPIO; MPIO, micron-sized iron oxide particles; ESC, embryonic stem cell; HSC, hematopoietic stem cell; MSC, mesenchymal stem cell; MDSC, muscle-derived stem cell; Mb, myoblast; EPC, endothelial progenitor cells; NSC, neural stem cell; EVCM, embryonic ventricular cardiomyocyte; Mono, monocytes; HUVEC, human umbilical vein endothelial cell; BMSC, bone marrow stem cells; NPC, neuroprogenitor cells.

stem cells within the living animal. Furthermore, noninvasive imaging not only has the added advantage of tracking the migration of these cells in vivo but also has the potential to sequentially follow their fate and efficacy in tissue regeneration, correction, and maintenance.

6.2. Noninvasive Imaging of Stem Cell Transfer

Noninvasive imaging modalities have been developed that allow for extremely detailed images of living organisms. Dedicated systems have been explored that allow small laboratory animals to be imaged noninvasively (Min and Gambhir, 2004; Rudin and Weissleder, 2003; Weissleder and Ntziachristos, 2003). Magnetic resonance imaging, radionuclide imaging, computed tomography (CT), ultrasound imaging (US), and optical imaging have all been adapted for experimental studies of biological processes and disease (Allport and Weissleder, 2001). In order to perform noninvasive monitoring of stem cell migration and integration, imaging modalities have to meet the following general criteria: capable of three-dimensional (3D), single cell imaging, high sensitivity, nontoxic, high specificity, lack of label dilution, no label transfer to adjacent cells, and ability for *in situ* confirmation. Although other technologies—including radionuclide imaging (Massoud, 2003; Min and Gambhir 2004; Wu et al., 2003) and bioluminescence imaging (Contag and Ross, 2002; Massoud and Gambhir, 2003, 2004; Wu et al., 2003)—meet several of these criteria, this chapter focuses on the development of paramagnetic nanoparticles that can be used for simultaneous MR and optical imaging to monitor stem cell therapies in vivo. Multimodal imaging is an important aspect of cell tracking due to the unique strengths that each modality can offer and the potential for the corroboration between different imaging modalities as well ex vivo verification (Michalet et al., 2005; Mulder et al., 2006a; van Tilborg et al., 2006). Here we outline two imaging modalities that have made extensive use of nanoparticle contrast agents.

6.2.1. Optical Imaging

Traditional epifluorescence or reflectance fluorescence imaging has been used to produce two-dimensional images of flurophores in superficial tissues and living animals (Weissleder and Ntziachristos, 2003). Fluorescent proteins, such as green fluorescent protein (GFP) from the jellyfish *Aequorea victoria*, were some of the first proteins to be used for live cell and in vivo imaging (Hadjantonakis et al., 2003). Site-directed mutagenesis and the isolation of fluorescent proteins from different organisms have lead to multiwavelengths applications that span excitation wavelengths from 350 to 630 nm and emission wavelengths from 450 to 650 nm (Hadjantonakis et al., 2003). A drawback of the most popular form of GFP (EGFP) for in vivo imaging is its low-emission wavelength (510 nm), which overlaps with the autofluorescence of many tissues. In addition, photon scattering at these wavelengths is extensive, making it suitable for shallow depth–penetration applications (<0.08 cm; Reviewed in Weissleder and Ntziachristos (2003)) but preventing its use in tomographic imaging of thick tissue. Bioluminescence can be used to overcome some of the issues encountered with endogenous fluorophores. In bioluminescence imaging, a substrate (typically luciferin) is administered to an animal that has been designed to carry the luciferase (enzyme) such that when the substrate and enzyme meet, the luciferase is oxidized and emits light to enable detection. The application of bioluminescence for in vivo tracking is promising. However, current drawbacks include positional uncertainty of light-emitting cells due to nonhomogeneous scattering, light penetration issues, and an apparent need for a stable expression of luciferase. These endogenous probes have the advantage that they are produced by the cell and can be used to track cell lineages and cellular function.

New tomographic imaging techniques are being developed to produce 3D images of flurophores in deep tissue. These techniques are most efficient at near infrared (NIR) wavelengths. The power of NIR imaging is the ease in which a NIR photon (700–900 nm) passes through thick biological tissues, including bone, skin, and muscles in vivo. NIR spectroscopic methods are commonly used to monitor blood volume, oxygenation, and flow dynamics in humans and animals (Boushel et al., 2000; Jobsis, 1977). Not only is the tissue absorption coefficient dramatically reduced at NIR wavelengths, but also so is tissue autofluorescence. These conditions result in favorable circumstances for deep tissue imaging (Reviewed in Frangioni (2003); Sevick-Muraca et al. (2002); Weissleder and Ntziachristos (2003)). Indeed, with a high quantum efficiency NIR-charge-coupled device (CCD) and the appropriate illumination and optics, NIR images have been acquired with in-plane resolutions of 10–20 μm, providing the ability to resolve microscopic bone structures in mice (Zaheer et al., 2001). NIR chromophores have been used for optical imaging using diffuse optical tomography (DOT), fluorescence reflectance imaging

(FRI), fluorescence-mediated molecular tomography (FMT), and photoacoustic tomography (PAT). PAT is based on the measurement of laser-induced ultrasonic waves and results in images with high optical contrast and ultrasonic resolution (Ntziachristos et al., 2005). This imaging modality has been successfully applied to imaging vasculature structures and tumor angiogenesis located a few millimeters beneath the skin (Wang et al., 2004a,b; Zhang et al., 2006).

Organic fluorescent dyes are the most commonly used fluorophores. Dyes, such as fluorescein isothiocyanate (FITC) and carboxyfluorescein diacetatesuccinimidyl ester (CFSE), have been used in various biological applications, such as fluorescent-labeled antibodies and molecules that are used to stain cells or organelles (Lyons and Parish, 1994; Weston and Parish, 1990). The main limitations of using organic dyes, however, are as follows: (1) prone to rapid photobleaching; (2) not well suited for simultaneous multicolor imaging applications; (3) emission/excitation is often susceptible to changes in the local chemical environment; (4) emission from dyes can overlap with autofluorescence from tissues; and (5) have low quantum yields beyond ~850 nm (Frangioni, 2003; Kim et al., 2004).

A promising alternative replacement for organic fluorescent dyes are Quantum dots (QDs), particularly for single cell labeling and tracking (Pinaud et al., 2006; Medintz et al., 2005; Stroh et al., 2005b; Michalet et al., 2005). Quantum dots are semiconductor crystals with all three dimensions in the 1–10 nm size range that luminescence due to quantum confinement effects. QDs provide a new class of biomarkers that could overcome the limitations of organic dyes. The use of QDs as luminescence probes in cell imaging has increased significantly since the first examples from Alivisatos' (Bruchez et al., 1998) and Nie's (Chan and Nie, 1998) groups. In many respects, these luminescent nanocrystals behave like a transitional stage between bulk semiconductors and single atoms. In this size regime, the electrons exhibit quantum mechanical effects (Alivisatos, 1996; Empedocles and Bawendi, 1999; Gaponenko, 1998; Steigerwald and Brus, 1990; Weller, 1993a,b,1998; Zhang, 1997). Quantum dots have been used to image superficial tumors, transplanted cells, and real-time tracking of single HSC moving through the vasculature in vivo (Stroh et al., 2005b). The coupling of either multiphoton or tomographic optical methods with high-resolution, 3D MRI will provide a powerful tool to track stem cell migration to and from their respective niches in vivo.

The important features that should be built into the fluorescence-based part of the multimodal nanoparticle are as follows (Sharma et al., 2006):

- In vitro and in vivo stability;
- Resistance to photo bleaching;
- High quantum yield and high absorbency;
- Resistance to metabolic disintegration and nontoxicity;
- Emission in the NIR 700–900 nm window;
- Adequate dispersibility in the biological environment; and
- Ability to conjugate to carry a large payload of paramagnetic contrast agents.

6.2.2. MRI

The ability to obtain images with resolutions approaching that of a single cell (<50 μm) makes MRI particularly well suited for noninvasive studies of stem cell transplants. MRI has become a well-established clinical tool for the noninvasive imaging of anatomical changes in many organ systems. Similarly, high-resolution magnetic resonance microscopy (MRM) has also made tremendous strides forward due to advances in magnet technology. MRM images can now be achieved routinely with resolutions well below 100 microns (Benveniste and Blackband, 2006; Maronpot et al., 2004; Ciobanu and Pennington, 2004; Ciobanu et al., 2002). MRM brings not only new capabilities, but also technical challenges to the field of MRI (Tyszka et al., 2005; Pirko et al., 2005). It is rare that MR images are acquired with isotropic resolutions less than 10 μm and it has been previously estimated that the fundamental limitation to spatial resolution is on the order of 1 μm (Tyszka et al., 2005). In practice, the achievable imaging resolution will be limited by available gradient strength and sensitivity that is achievable in a given amount of time.

A number of strategies have been employed to improve MR sensitivity by increasing signal to noise. This can be achieved by increasing the magnitude of the induced nuclear magnetism or the efficiency of signal detection (Beck et al., 2002). With the advent of high-field, whole body magnets (3–9.4 T) and ultra-high field animal imagers (11–21 T) equipped with high performance gradients and phase array RF capabilities, the potential for increased sensitivity can be realized in vivo, but it is still unknown as to whether these technological improvements directly translate into an enhanced ability to perform molecular and cellular imaging. Furthermore, it is widely recognized that numerous MR contrast mechanisms are dependent upon magnetic field strength, offering opportunities for enhanced and emergent contrast mechanisms that could be exploited at high fields for cellular imaging. Figure 6.2 illustrates the great potential of MR microscopy at high fields, namely 21.1 T. In this example, the magnetic field–dependent, susceptibility-weighted contrast afforded by a gradient-recall echo acquisition provides excellent delineation of neuroanatomical structures in an excised, perfusion-fixed C57BL/6J mouse brain while also providing a high spatial resolution of $18 \times 18 \times 35$ μm^3 (Fu et al., 2005). To achieve this level of detail, it is usually necessary to image samples ex vivo using optimized MRM coils and to acquire data over a long acquisition period. On the other hand, in vivo imaging is complicated by the need to image internal structures and by limitations in the total imaging time available. For instance, the achievable resolution in an animal experiment is constrained by the need to maintain a physiologically relevant status, which limits the imaging time, methods of restraint, and exposure to anesthesia.

Working at higher fields also does not come without significant technological challenges. Proton relaxation times (T_1, T_2, and T_2^*) change in a direction that is not favorable (longer T_1, shorter T_2s) for tissue imaging. RF penetration decreases, and both macroscopic and microscopic magnetic susceptibility variations in tissue cause intrinsic T_2^* values to be much lower. These alternations in

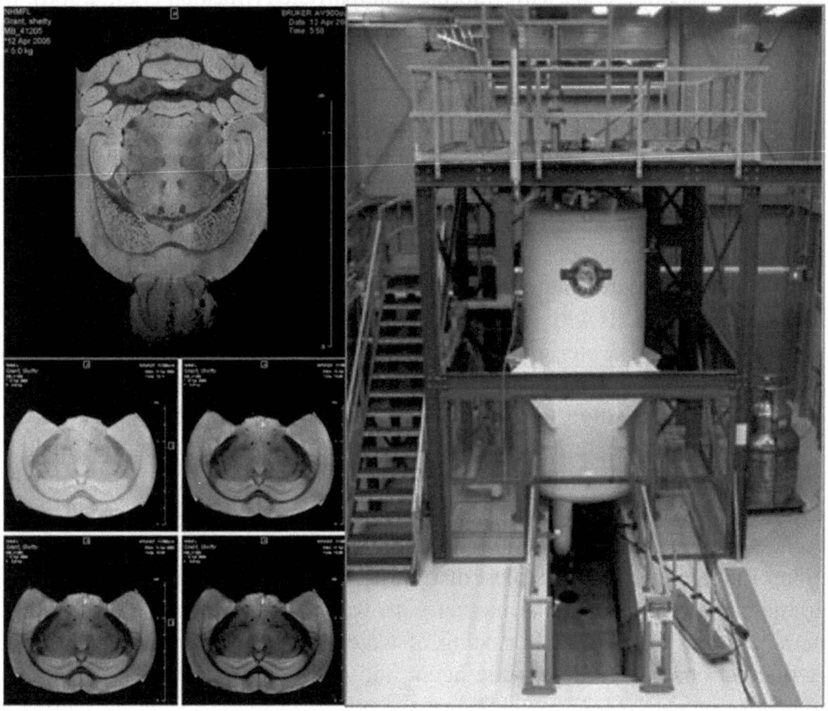

FIGURE 6.2. **A** Gradient-recall echo images of an excised, perfusion-fixed C57BL/6J mouse brain in a 10 mm linear birdcage recorded at 21.1 T. A true 3D gradient-recall echo dataset was acquired at a resolution of 18 × 18 × 35 μm in 29 h using the following imaging parameters: matrix = 1024 × 512 × 256; TE/TR = 12.5/100 ms; averages = 8; FOV = 1.9 × 0.92 × 0.8 cm; bandwidth = 100 kHz; 50° tip angle = 25 μs hard pulse. A series of 2D multi-slice gradient-recall echo images (40 × 40 × 250 μm resolution) were acquired to sample using the following parameters: matrix = 230 × 200; TE = 4.5–24.5 ms; TR = 1.5 s; averages = 2; FOV = 9.2 × 8.0 mm; slice thickness = 250 μm; number of slices = 17; 90° pulse = 2 ms three-lobe sinc pulse; and imaging time per dataset = 10 min. **B** The ultra-wide bore 900 MHz superconducting NMR magnet at the National High Magnetic Field Laboratory in Tallahassee, Florida. This cryostat stands 4.9 m tall, weighs over 13,600 kg and has a stored energy of 38 MJ containing 2,400 L liquid helium at atmospheric pressure. With the room temperature shim set in place the inner diameter is 89 mm. Reprinted from Fu et al. (2005) with permission from Elsevier.

contrast mechanisms and RF performance can be utilized to provide increased image information at high fields, but they require optimized acquisition parameters, pulse sequence design, and coil technology to take full advantage of this information while maintaining accurate anatomical representations.

Stem cells located within their niche do not have endogenous MR contrast. AQ1
For stem cells to be visualized by MR, stem cells need to be labeled by either

endogenous or exogenous contrast agents. This contrast has previously been achieved by labeling cells with fluorine nanoparticles (Masuda et al., 2006), iron binding genes (Deans et al., 2006), or paramagnetic nanoparticles (see below): The strength of a contrast is quantitatively represented as the relaxivity (Caravan, 2006; Caravan et al., 1999), r_1 or r_2, where the subscript refers to either the longitudinal ($1/T_1$) or the transverse rate ($1/T_2$). Relaxivity is simply the change in relaxation rate after the introduction of the contrast agent normalized to the concentration of contrast agent or cell number. T_1 agents usually have r_2/r_1 ratios of 1–2, whereas that value for T_2 agents, such as iron oxide particles, is as high as 10 or more (Caravan et al., 1999).

The most successful methods to date for MR-based tracking of cells have relied on labeling cells with (super)paramagnetic contrast agents that alter proton relaxation times of water protons and generate contrast on T_1-weighted, T_2-weighted, or gradient recall-echo (T_2^*-weighted) images. Paramagnetic materials have a small but positive magnetic susceptibility, meaning that such molecules tend to align with an externally applied magnetic field and act as atomic magnetic dipoles. Typically, paramagnetic materials, such as transition and lanthanide metals, possess unpaired electrons in a partially filled electron shell. In the absence of an external magnetic field, thermal motion produces a random orientation of paramagnetic molecules, disrupting any net magnetic moment. In MRI, paramagnetic contrast agents interact with water protons through dipole–dipole interactions resulting in a shortening of water MR relaxation times. Therefore, the interaction of a paramagnetic agent and the water surrounding it increases the relaxation rates ($1/T_1$ and $1/T_2$) of the water protons causing an increase in signal on T_1-weighted images and a decrease in signal on T_2-weighted images.

Positive contrast agents are commonly made up of paramagnetic materials, mainly those based on metal ions with large numbers of unpaired electrons such as Mn^{2+} (five unpaired electrons), and Gd^{3+} (seven unpaired electrons). With positive contrast agents, the T_1 and T_2 times of water protons are similarly reduced, but because T_1 is usually 1–2 orders of magnitude greater than T_2, the shortening of T_1 produces a more marked increase in signal (a "positive" effect) on T_1-weighted images. Gadolinium chelates (Gd-DTPA, Gd-DTPA-BMA)(Caravan et al., 1999) are paramagnetic contrast agents used in experimental and clinical studies. Paramagnetic agents tend to shorten water T_1 relaxation time greater than the T_2 and T_2^*.

Superparamagnetism is primarily associated with very small crystalline iron oxides (1–14 nm) (Thorek et al., 2006; Arbab et al., 2006a). The term superparamagnetic refers to the characteristic nature of the nanosized particles when placed in a magnetic field. Due to their single crystal nature, superparamagnetic iron oxides (SPIO) form a single magnetic domain as the entire crystal aligns with an applied magnetic field. This results in the induction of a large net magnetic moment with the ability to cause substantial disturbances in the surrounding magnetic field, leading to rapid dephasing of surrounding protons and changes in proton relaxation rates (Thorek et al., 2006; Arbab et al., 2006a). It should be noted that, with superparamagnetism, it is the effect of the superparamagnetic contrast agent on its surrounding which is imaged. Superparamagnetic contrast agents tend to shorten proton T_2 and T_2^* values more drastically than T_1 generating negative MR contrast (Arbab et al., 2006a; Bulte, 2004; Modo et al., 2005; Thorek et al., 2006).

6.3. Nanoparticle Contrast Agents for Cell Labeling

6.3.1. Iron Oxides

Superparamagnetic iron oxide–containing nanomaterials are commonly used as contrast agents, which can dephase neighboring protons in a region equivalent to many times the size of the particle itself (Brown et al., 1999). Typically, SPIOs consist of two components, a superparamagnetic iron oxide core and a hydrophilic coating. The SPIO core can be composed of magnetite (Fe_3O_4) and/or maghemite (γFe_2O_3). This SPIO core is commonly formed by the co-precipitation of ferric and ferrous ions (Gupta and Gupta, 2005; Renshaw et al., 1986), where the resultant size, shape, and composition of nanoparticles (NPs) can be controlled by the salts used, the ratio of Fe^{2+} to Fe^{3+}, pH, and the ionic strength of the media. In order to achieve good dispersity and to maintain size stability of the resulting magnetic particles, hydrophilic polymers, such as dextran and polyvinylalcohol, are usually added to act as coating agents or stabilizing matrixes. It is becoming more common to create stable aqueous magnetic suspensions using a variety of other nanofabrication methods based on precipitation in highly constrained domains utilizing sol–gel preparations, polymer matrix-mediated synthesis, microemulsions, and vesicles (Gupta and Gupta, 2005). The size of the SPIO core can range greatly from 2 to 3 nm, tens of nanometers for polymer-coated polycrystalline iron oxide nanoparticles (Hinds et al., 2003), to micron-sized particles (Weber et al., 2004b; Dunning et al., 2006) (Table 6.1). The most widely used SPIO is a commercially available product that is a dextran-coated superparamagnetic particle (Tallheden et al., 2006; (Hauger et al., 2006; Miyoshi et al., 2005; Walczak et al., 2005; Hoehn et al., 2002b) or ultrasmall SPIO (USPIO) (Sun et al., 2005). SPIO nanoparticles shorten the T_1, T_2, and T_2^* relaxation times of water within a labeled cell and within the surrounding voxels containing the cell (Anderson et al., 2006; Heyn et al., 2006; Kirik et al., 2005; Modo et al., 2005; Shapiro et al., 2006b; Sharma et al., 2006). MR scans that are sensitive to susceptibility and T_2^* effects have been shown to be able to detect contrast from extremely small amounts of SPIO (Bulte and Kraitchman, 2004; Shapiro et al., 2005). In these cases, true cellular resolution is not necessary because it is the effect of SPIO on the surrounding volume that is imaged. In fact, the effects of SPIO usually extend well beyond the physical extents of the cells in which they are loaded.

The in vivo detection limit of cellular MRI has been determined by a number of groups using different SPIO compounds. This minimal number of detectable cells depends on many factors including magnetic field strength, the total amount or cellular iron load, and the cellular distribution of SPIO. The effect of the

SPIO on a surrounding media can be up to 50 times the size the iron core (Dunning et al., 2006). Single cells have been imaged both in vitro using culture models and in fixed embryo samples by MRI. Using a model of direct cerebral injection of SPIO-labeled embryonic stem (ES) cells into the rat brain, Hoehn et al. (2002a) determined a in vivo detection limit of 500 cells at 7 T, whereas Stroh et al. established an in vivo detection limit between 20 and 100 SPIO-labeled ES cells following injection into the rat brain at 17.6 T. Within the cell, individual nanometer-sized SPIOs tend to aggregate into much larger clusters of particles within the cell's endosomes. On the other hand, the much larger micron-sized paramagnetic iron oxide particles (MPIO) already have diameters of microns and have allowed for single particles to be visualized by T_2^*-weighted sequences with spatial low resolutions as low as 100 μm (Hinds et al., 2003; Shapiro et al., 2006a,b). It has been demonstrated that both in vitro and in vivo SPIO-labeled, single cell detection is possible at clinical MRI field strengths (1.5–3 T). Calculations of the minimum amount of iron needed to detect a single cell at a resolution of $100 \times 100 \times 200 \mu m^3$ at 1.5 T using the FIESTA sequence indicated only 1.4–3.0 pg of iron per cell to be necessary. Recently, it has been shown that individual cells can be imaged at 9.4 T with 40 μm^3 resolution containing ~1 pg of iron per cell with gradient recall-echo imaging (Smirnov et al., 2006). These are iron-loading levels that can easily be achieved in many stem and progenitor cell types.

6.3.2. Paramagnetic NPs

SPIOs generate MRI contrast by the production of signal voids on T_2- or T_2^*-weighted images. Whereas new tailored imaging sequences can partially recover this loss in signal intensity, inducing "white marker" contrast by exploiting phase and frequency differences generated by the SPIO (Mani et al., 2006; Cunningham et al., 2005), an alternative would be to produce positive (T_1) contrast by the use of paramagnetic contrast agents. This dependence on the selective interaction with cellular water may have additional advantages for molecular and cellular imaging. For instance, it opens up the possibility of imaging the cell directly, as well as the design of smart probes that are responsive to changes in water exchange due to gene expression (Louie et al., 2000). On the other hand, this dependence raises the possibility of changes in contrast due to cellular compartmentalization (Terreno et al., 2006) and increased toxicity if the unchelated paramagnetic agent is released outside of the endosome.

Gadolinum-based contrast agents have been used successfully for both molecular and cellular imaging. Artemov et al. (2004) demonstrated successful targeting to Her-2/neu receptors and fibrin using relatively low payloads of 12.5 Gd ions per ligand. Overall, this approach has had limited success due to the necessity to increase the payload of contrast agent per ligand or cell to overcome the much weaker relaxitivity of Gd compared to the SPIO. It has been estimated that the number of Gd-chelates needed to visualize cells is around 100–1000 chelates or 10–100 μM per cell (Arbab et al., 2006a). This payload can be problematic for stem cell labeling due to the small amount of noncarrier-mediated uptake into many types of progenitor cells (Zheng et al., 2005). Therefore, approaches must be used to concentrate Gd contrast agents within the cell to a larger degree necessary than that required for SPIO. This cellular uptake has been achieved using cell translocation peptides such as HIV-tat (Bhorade et al., 2000) peptides, polyarginine peptides (Allen et al., 2004), or lipophilic complexes (Zheng et al., 2005). In addition, unchelated forms of lanthanides are toxic to the cell. Careful examination of cellular function and paramagnetic complex stability at low pH (i.e., inside a endosome/lysosome) need to be performed. However, there is little doubt that cell and molecular MR imaging will benefit from ongoing research focused on designing paramagnetic complexes endowed with improved relaxivity within the cellular milieu.

An alternative approach for increasing cellular paramagnetic contrast agent is to increase paramagnetic payload of the contrast agents through the generation of fullerenes (Anderson et al., 2006), nanoparticulates (Modo et al., 2002; Santra et al., 2005b; Sharma et al., 2006; Sitharaman et al., 2005; Tan and Zhang, 2005), or liposomes (Mulder et al., 2006a,b). Such complexes have large surface areas that can be functionalized to contain contrast agents, progenitor cell–specific ligands, and cell penetration peptides. With the capacity to carry 50,000–90,000 gadolinium ions per particle, $\sim$250 nm liquid perfluorocarbon nanoparticle emulsions by Wickline and Lanza et al. have been functionalized with paramagnetic chelates and homing ligands (fibrin, integrins) incorporated within an outer phospholipid surfactant monolayer (Lanza et al., 2004a,b; Morawski et al., 2005; Schmieder et al., 2005). Payload increases boost the molecular relaxivity of the nanoparticle to many times greater than the relaxivity of the Gd chelate alone, resulting in the detection of as few as 100 pmol/L of nanoparticles with a contrast-to-noise ratio of 5.57 even at 1.5 T (Morawski et al., 2004).

In most cases, these paramagnetic complexes can be coupled to optical dyes (i.e., quantum dots), which allow for the ex vivo corroboration of the MRI results or coupling to in vivo optical measurements capable of single cell tracking (Stroh et al., 2005b). For instance, during the last decade, the water-in-oil (W/O) microemulsion technique has become a very powerful tool to synthesize monodisperse nanoparticles (NPs) of various kinds (Arriagada and Osseo-Asare, 1999; Lin et al., 2003; Qhobosheane et al., 2001; Santra et al., 2001a,b 2004; Xiao et al., 2003), including magnetic NPs (Santra et al., 2001b), semiconductor quantum dots (Xiao et al., 2003; Yang et al., 2004), dye-doped NPs (Qhobosheane et al., 2001; Santra et al., 2001a,b), etc. at room temperature (Figure 6.3) (Santra et al., 2005a). The beauty of the W/O microemulsion technique is its versatility. It does not require drastic experimental conditions such as high temperature or high pressure. A variety of "tailor-made" NPs, such as core-shell or doped NPs, can be readily synthesized. It is also possible to synthesize monodisperse NPs as small as a couple of nanometers to several hundreds of nanometers by controlling the synthesis parameters. Previous studies report the synthesis of bifunctional contrast agents for dual (fluorescence and magnetic) imaging in vivo (Daldrup-Link et al., 2004; Huber et al., 1998;

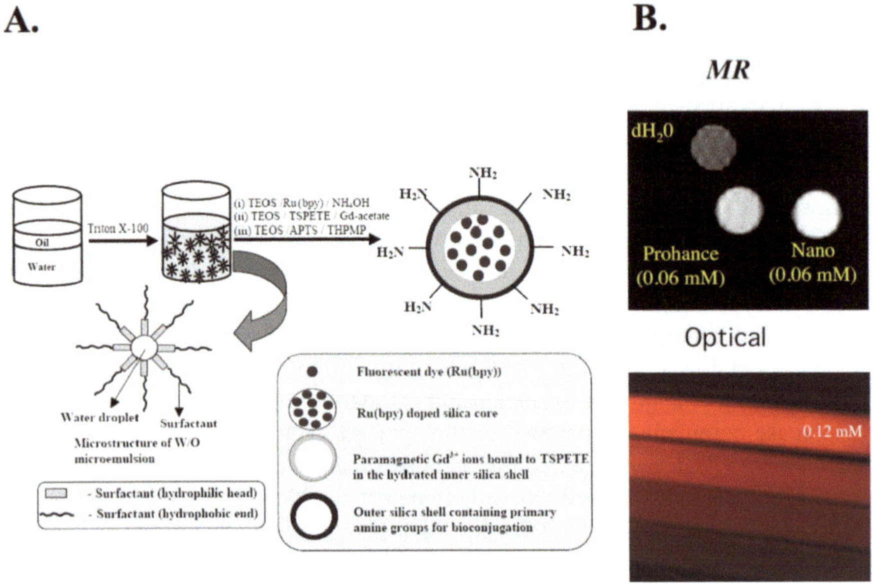

FIGURE 6.3. Diagram of a multifunctional nanoparticle (**A**) generation using W/O microemulsion technique that is both MR (**B**; *top*) and optically active (Rubpy; **B**; *bottom*). T_1 weighted MRI of PBS, and equal concentrations of a clinical available contrast agent and the Gd-Rubpy NP. **A** is reprinted from Santra et al. (2005a). Reproduced with permission from WILEY-VCH Verlag.

Kircher et al., 2003; Sosnovik et al., 2005; Veiseh et al., 2005). In this contrast agent design, organic fluorescent dyes were used, which may have limitations for real-time imaging as these dyes often undergo a rapid photobleaching process. Qdots, also known as nanocrystals, are a nontraditional type of semiconductor used for many biological-imaging techniques due to their photostability, brightness, and tuneability (Michalet et al., 2005). In vivo applications of Qdots have just begun to be explored (Gao et al., 2002, 2004, 2005; Gao, 2003; Jiang et al., 2004; Smith et al., 2004; Voura et al., 2004). John Frangioni (Kim et al., 2004; Michalet et al., 2005) and Shuming Nie's (Gao et al., 2004) research groups have successfully demonstrated in vivo bioimaging applications of Qdots for cancer detection (see also Chapter 22 in this book). Frangioni's group took the advantage of near-IR Qdots to perform deep tissue imaging (Kim et al., 2004; Lim et al., 2003; Parungo et al., 2005). Recently a bifunctional paramagnetic 7.5 nm Qdot nanoparticle was synthesized for stem cell tracking by both optical and (Figure 6.4) MR imaging (Yang et al., 2006).

Manganese (Mn), one of the first paramagnetic MR contrast agents, has also been used to label cells, resulting in a decreased intracellular T_1 time. Mn enters cells in vivo through calcium channels in the cell membrane, and has been used systemically to image both neuronal and cardiac activity (Arbab et al., 2006a). In cell phantoms, Aoki et al. (2006) have shown that $MnCl_2$ can act as a positive

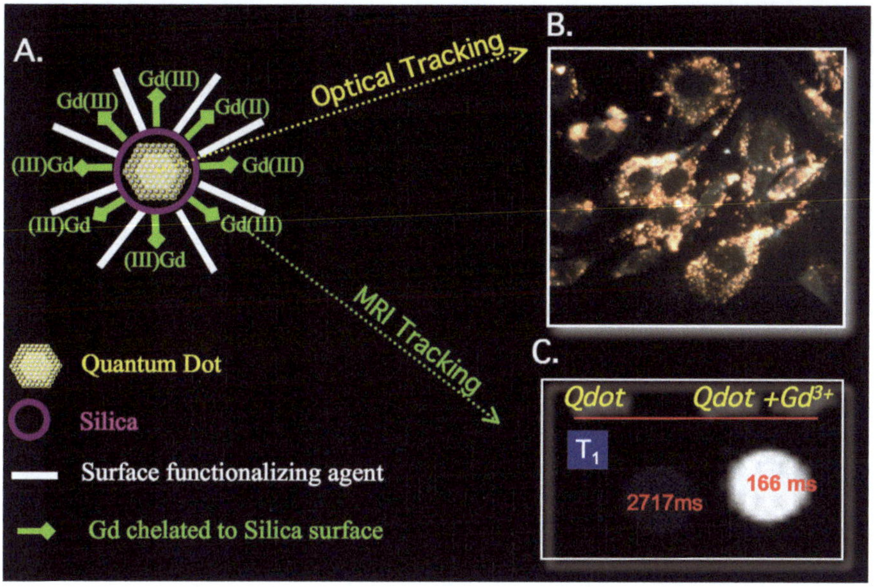

FIGURE 6.4. Multimodal quantum dots that generate dual functionalities of yellow fluorescence and paramagnetism. Water-soluble silica-coated ZnS-passivated CdS:Mn (CdS:Mn/ZnS core/shell) quantum dots (**A**) are functionalized with Gd^{III} ions to impart the paramagnetic property (**C**) and the Qdot can be used to track labeled cells. (**B**) Multimodal quantum dots were used to label muscle-derived stem cells and their uptake confirmed in vitro.

contrast agent in allogeneic natural killer cells at 11.7 T. Based on large T_1 changes measured in cell pellets, they estimated that 50–100 cells in a 100 μm^3 voxel should be detectable. Work still needs to be performed on the amount of time that Mn will be retained within the cell and whether any of the toxic effects associated with Mn are observed.

6.3.3. Paramagnetic Complexes as High-Field Contrast Agents

With the advent of high-field whole body magnets (3–9.4 T) and animal systems (up to 21 T), with the potential for increased sensitivity, it will be important to translate these technological improvements directly into enhanced molecular and cellular images. Among the lanthanide ions, Gd is known as the best low-field T_1 relaxation agent because of its long electronic relaxation time. Complexes with other paramagnetic lanthanides (III) have been commonly used as shift agents in NMR spectroscopy or as susceptibility agents in MRI (Arteaga et al., 1999; Beache et al., 1998; Dubois et al., 2005; Haraldseth et al., 1996; Johnson et al., 2000; Villringer et al., 1988). While Gd does not increase water relaxivity at magnetic field strengths above 4T, other lanthanide complexes have been shown to increase their relaxivity quadratically with field strength (Bulte et al., 1998; Caravan et al., 2001; Kellar et al., 1998; Vander Elst et al., 2002a,b). This field dependence is illustrated in Figure 6.5 for Dysprosium-DTPA-BBMA in 2% agarose imaged from 4.7 to 21 T. For select lanthanide ions [Dy(III), Pr(III), Sm(III), Ho(III), Er(III), and Yb(III)], this increase in relaxivity at high fields is due to very short electronic relation times that result in a large static magnetic moment, referred to as Curie relaxation (Bulte et al., 1998; Caravan et al., 2001; Vander Elst et al., 2002a). It had previously been suggested by Aime et al. that the predicted reduction of T_2 with field strength can be used in high-field NMR and as negative contrast agent for MRI (Bulte et al., 1998; Kellar et al., 1998; Vander Elst et al., 2002a).

Another class of contrast agents that stand to directly benefit from ultra-high magnetic fields are chemical exchange saturation transfer (CEST) contrast agents (Aime et al., 2002a,b, 2005a; Gillies et al., 2004; Snoussi et al., 2003; Terreno et al., 2004; Trokowski et al., 2004; Ward, 2000; Ward and Balaban 2000; Woessner et al., 2005; Zhang et al., 2003a,b; Zhou et al., 2004). CEST agents typically contain an exchangeable pool of protons that can be used to transfer magnetization to bulk water. If the conditions are optimal, the saturation of a small pool of exchangeable protons will result in a measurable decrease in the 40–55 molar tissue water signal typically used to image. Despite resulting in a decrease in signal intensity similar to SPIO contrast agents, CEST images have the advantage that they can be turned on and off for different lanthanide (iii) paramagnetic chelates (PARACEST agents) through the use of frequency selective pre-saturation pulses. PARACEST agents are particularly useful due to

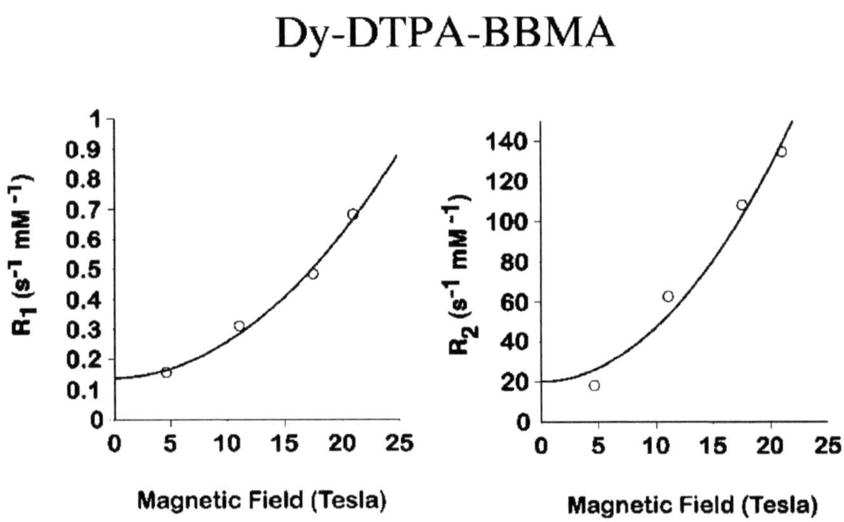

FIGURE 6.5. Experimentally determined increase in r_1 and r_2 that occurs with Dy-DTPA-BBMA with increasing magnetic field between 4.7 and 21 T in 2% agarose based on spin-echo images.

their large chemical shifts and long life times of bound water, which can result in a more efficient transfer of magnetization to bulk water increasing contrast. It has been predicted that most LnDOTA-4AmCE complexes should function as PARACEST agents at 11.75 T, but only a few (Eu^{3+}, Tb^{3+}, Dy^{3+}, and Ho^{3+}) will be useable at low fields (1.5 T) (Zhang et al., 2003a). On the other hand, some of the potentially most effective PARACEST agents, such as those based on Yb^{3+}, may not be usable below 11 T. It has also been predicted that PARACEST agents will likely become more efficient at higher B_0 fields. Block equation modeling has predicted that when the bound water pool of a PARACEST agent (500 ppm and exchange time (t_M) of 3 μs) is completely saturated, then 5 % of the bulk water signal will be eliminated by ~10 μM of the agent. This corresponds to a concentration that is well below the detection limit of a low molecular weight Gd^{3+}-based contrast agent with a r_1 of 4–5 $mM^{-1}s^{-1}$ (Zhang et al., 2003a). Moreover, similar to the substantial increase in relaxivity of Gd^{3+}-based agents by conjugation to a polymer or formation of a nanoaggregates, it may be possible to extend the lower detection limit into the submicromolar range. PARACEST contrast agents also operate at specific RF irradiation frequencies such that contrast in different cell populations can be turned on and off using different irradiation pulses. Moreover, PARACEST contrast is not generated based on T_1 and T_2^* differences and will not obscure or be complicated by underlying tissue anatomy, hopefully increasing their cell specificity. Since PARACEST is extremely sensitive to water exchange, it will provide an ideal mechanism for activatable contrast agents but again water comparmentation within the cellular milieu (endosome vs the cytoplasm) may lead to differing amount of PARACEST contrast. PARACEST agents have been used to image changes in temperature (Hekmatyar et al., 2005; Terreno et al., 2004), glucose (Trokowski et al., 2004; Zhang et al., 2003b), pH (Aime et al., 2002a,b; Gillies et al., 2004; Ward and Balaban, 2000), and liposomes (Aime et al., 2005b).

6.4. Labeling Stem Cells with Superparamagnetic Contrast Agents

We have seen that at high magnetic fields and by using special imaging sequences MR can be sensitive to very small amounts of SPIO, but to ensure that MRI can track the cells they are often heavily loaded with superparamagnetic contrast agents. Frequently, uptake efficiency is dependent on the type of cell used. For instance, a small number of progenitor cells have been shown to be phagocytotic in nature and can be efficiently labeled by simple incubation with SPIO contrast agent. However, many ferumoxides alone cannot be used to efficiently label nonphagocytic or nonrapidly dividing mammalian cells in vitro (Arbab et al., 2003). To overcome this problem, the imaging community has learned from the lessons of molecular and cell biologists, particularly from their experiences with hard to transduce cells using plasmid DNA. This has lead to the use of commonly used molecular biology transfection agents (cationic transfection agent, liposomes, electroporation, gene gun, and viruses) to improve the efficiency of labeling the cell of interest with the contrast agents for MR and optical imaging.

6.4.1. Endocytosis

In the majority of the stem cells studied, spontaneous noncarrier-mediated uptake of contrast agent into the cell has been found to be very inefficient. Unlabeled cells usually contain less, between 0.01 and 0.1, picograms of iron per cell (Arbab et al., 2003). Although most types of stem progenitor cells do not readily take up MR contrast agents, the phagocytic activity of certain types of cells, such as macrophages (Heyn et al., 2006), chord blood (Daldrup-Link et al., 2005a), endothelial progenitor cells (EPCs) (Crich et al., 2004), human monocytes (Metz et al., 2004), and immortalized rat progenitor cells (Sun et al., 2005), has shown to endocytose very high levels of SPIO. For instance, Heyn et al. (2006) have found that a macrophage cell line will consume and retain extremely high levels of iron (60.9 ± 1.6 pg Fe/cell). Anionic magnetic nanoparticles (Billotey et al., 2003; Riviere et al., 2005; Smirnov et al., 2006) also have been developed and used for intracellular labeling of macrophages (Billotey et al., 2003), HeLa (Smirnov et al., 2006), and smooth muscle cells (Riviere et al., 2005). The negative surface charge of the particles induces an uptake three orders of magnitudes higher than conventional dextran-coated SPIOs (Billotey et al., 2003). Amazingly, a number of cells can even endocytose MPIOs which range in diameter from 0.96 μm to as large as 5.80 μm resulting in iron contents between 100 and 387 pg (Shapiro et al., 2005) per cell.

6.4.2. Facilitated Cell Targeting

A number of techniques have been developed to increase the uptake of superparamagnetic contrast agents through macropinocytosis (Arbab et al., 2006a). The use of coating or attaching cell-penetrating peptides to the superparamagnetic contrast agent or nanoparticle can greatly facilitate cell loading (Allen et al.,, 2004; 2000, 2000; Bullok et al., 2006; Lewin et al., 2000; Liu et al., 2006; Santra et al., 2004, 2005c; Zhao et al., 2002; Zhao and Weissleder, 2004). Further, optimizing the number of cell-penetrating peptides on targeted nanoparticles can result in a 100-fold increase in SPIO labeling of lymphocytes as was shown by Zhao et al. (2002), by using multiple HIV-TAT-derivatized CLIOs (cross-linked superparamagnetic iron oxides).

Another approach for intracellular magnetic labeling is the use of internalizing monoclonal antibodies (MABs), which results in a receptor-mediated uptake of superparamagnetic contrast agents. For example, it has been shown that the mouse anti-transferrin receptor (Tfr) MAB OX-26 (Aghi and Chiocca, 2005) induces internalization of the transferrin upon binding. The conjugation to OX-26 improves cellular uptake of drugs by receptor-mediated endocytosis. Bulte et al.,

1999. have used an OX-26 MAB conjugated to SPIO (MION-46L-OX-26) to magnetically label oligodendrocyte progenitors (Bulte et al. 1999) and neural precursor cells (Bulte et al. 2003) leading to extensive endosomal accumulation of SPIO. The ubiquitous nature of the Tfr represents a convenient system for cell targeting, but the use of MAB is not limited to Tfr. Commercially available magnetic beads commonly used for cell sorting have also been used to internally label dendritic cells by targeting cell surface accessory molecules (anti CD-11c) (Ahrens et al., 2003). MABs share the disadvantages of any antibody in terms of the need for a specific antibody for each cell type, possible species specificity, as well as the high production cost.

Dendrimers are known to have a high affinity for cellular membranes and have been used to transport foreign DNA into cells (Dennig and Duncan, 2002; Hudde et al., 1999; Kukowska-Latallo et al., 1996; Shah et al., 2000; Yoo and Juliano, 2000). We have previously shown that nonspecific, highly efficient labeling of muscle derived stem cells can be achieved using magnetodendrimers (Walter et al., 2004). This compound consists of carboxylated poly(amidoamine) 4.5 generation starburst dendrimers, encapsulating a superparamagnetic iron-oxide core. This labeling process results in a relaxation enhancement five times greater than that achieved using the monoclonal antibodies to the transferrin receptor (Bulte et al., 2002a, 2001). Furthermore, this labeling technique has been implemented successfully to monitor the bio-distribution of oligodendroglial progenitor transplants into dysmelinated rat brain at 4.7 T and 1.5 T (Bulte et al., 2001). However, magnetodendrimers have the disadvantage that they are not commercially available.

Recently, a number of groups (Frank et al., 2002; Hoehn et al., 2002) have utilized labeling techniques using commercially available cellular transfection reagents and clinically approved MR contrast agents. The use of both cationic and liposomal based–transfection agents, which are commonly used by molecular biologists for cell transfection using plasmid DNA, has been shown to be economical and easy to institute in numerous types of stem and progenitor cells (Frank et al., 2003; Matuszewski et al. 2005; Montet-Abou et al., 2005; Hoehn et al., 2002b). Many transfection agents have been used including dextrans, phosphates, artificial lipids (e.g., FuGene™, Lipofectin™, Lipofectamine™), proteins (e.g., poly-l-lysine, protamine sulfate), and dendrimers (e.g., Superfect™, Polyfect™) (Arbab et al., 2004b; reviewed in Arbab et al. (2006a)). Commonly during nanoparticle fabrication, the zeta potential is adjusted so as to provide optimal dispersion in water (Sharma et al., 2006). Similar to dextran-coated SPIO particles, surface-functionalized silica nanoparticles have negative zeta potentials and when mixed with cationic transfection agents (i.e., poly-l-lysine) they form ionic complexes that are endocytosed by stem and progenitor cells (Figure 6.7).

Other approaches that have been used to successfully label cells are the use of electroporation (Walczak et al., 2005) and viruses (Toyoda et al., 2004). It has recently been reported that electroporation efficiently labels stem cells within a 20 ms time period (Walczak et al., 2005). This fast labeling is advantageous because the longer the cells are cultured the greater are the risks that they will be altered by the artificial environment in which they are maintained. Electroporation is also attractive since it does not rely on any transfection agent to label the cell, possibly reducing aggregate formation and the variability in amount of cell-to-cell iron labeling. However, care needs to be taken and optimized protocols used with electroporation. By its very nature, electroporation is initially damaging to the cell membrane. Post-labeling, careful viability, and differentiation assays need to be performed. Similar to gene therapy, it was also found that viruses can be used not only to transfect cells with genes, but also to deliver SPIO (Toyoda et al., 2004). This has been achieved using the capsids from hemagglutinating virus of Japan (HVJ; Sendai virus) envelope (HVJ-E) in combination with Ferumoxides to effectively label mouse astrocyte with SPIO (Toyoda et al., 2004).

6.5. Delivery of Therapeutic Cell Transplants

There are two general routes of stem cell transplant delivery. *Local injection* has been shown to be the most efficient way to straightforwardly deliver cells directly to a known region of damage or necrosis. Indeed there is great industrial interest in developing catheter-based delivery systems to transplant cells into the damaged or infarcted myocardium. It is also the most straightforward method for imaging because there is prior knowledge of cell density as well as the location where the image contrast should be generated. The majority of imaging modalities have been developed based upon this a priori knowledge in combination with initially high cell density to guide and interpret changes in image contrast. Unfortunately, massive cell death associated with stem cell transplantation can lead to nonspecific image contrast, generated in part by scavenger cells, such as macrophages, as they attempt to remove dead cells and label. From an in vivo imaging perspective, the initial low viable cell number places extremely high demands on (1) sensitivity—ranging from single cell detection to tissue volumes on the order of mm^3 and (2) cell specificity in the presence of massive cell death. Direct cell transplantation into tissues in which stem cells show large migratory behavior, such as the brain (Bulte et al., 2002b, 1999; Hoehn et al., 2002a; Zelivyanskaya et al., 2003; Lindvall and Kokaia, 2006; Muller et al., 2006) and spleen (Shapiro et al., 2006b), has the advantage of monitoring the rapid cell movement away from its initial site of cell transplantation avoiding the initial nonspecific image contrast. *Vascular* delivery has the advantage that cells can be delivered systemically to all the tissues in the body. This broad delivery pattern is especially important for the treatment of diseases that affect tissues throughout the entire body and in those tissues that are not readily accessible for direct injections. This is a major problem facing cell therapies for the muscular dystrophies in which all the muscles can be affected including the obscured diaphragm muscle. In highly vascularized tissue, migration distances from the capillary to the tissue are small. An additional advantage of systemic delivery for noninvasive imaging is that dead cells are readily removed by the circulatory system in most tissues. Unfortunately, cell density is diluted by the total body mass resulting in an incorporation rate at least an order of magnitude less than that when adult stem cells are directly injected into the tissues. This significant reduction in numbers poses a major challenge for stem cell tracking. It should also be noted that frequently stem cell recruitment following vascular delivery might necessitate injury at the target site, resulting in expression of chemokines and chemotatic signals, as well as vessel damage (Sampaolesi et al., 2003; Skuk and Tremblay, 2003; Torrente et al., 2001)

6.6. Tracking of Stem Cells and Progenitors

6.6.1. Tissue Constructs

Tissue engineering is playing an ever-increasing role in regenerative medicine. It is becoming more apparent that the ability of stem and progenitor cells to regenerate and replace fully functioning cells depends on the underlying tissue environment. This includes chemotatic (Lapidot et al., 2005), mitogentic (Dimmeler et al., 2005) as well as the overall mechanical properties (Engler et al., 2006) of the damaged tissue and extracellular matrix (Leri et al.,, 2005). The use of SPIO-labeled cells both in vitro and in vivo will help to advance the understanding of cell dependence on the above-mentioned mechanisms. To create a valid tracking method, the label must not be toxic or the cells' ability to differentiate and divide must be altered, while leaving the chemotatic response unaltered. SPIO labeling has been shown not to alter chemotaxis (Arbab et al., 2006b, 2005; Garden et al., 2006) in in vitro migration assays of adult hematopoietic stem cells (HSCs) (Arbab et al., 2005), EPCs (Arbab et al., 2006b), and T cells (Garden et al., 2006). These measurements are traditionally carried out in transwell chambers that typically have permeable membranes that restrict cell–cell or cell–matrix interactions to soluble factors and are not equipped for dynamic monitoring. The need to investigate these cell–cell and cell–matrix interactions opens up a new realm of possibilities for MRI to image cell migration in tissue-engineered constructs both in vitro (Gimi et al., 2006) and in vivo (Crich et al., 2004). This would allow for real-time visualization and determination of the extracellular matrix on chemotaxis in 3D space. This approach has recently been used by Gimi et al. (2006) to monitor the migratory behavior of SPIO-labeled HUVECs as an in vitro model of angiogenesis. In another model of angiogenesis, Crich et al. (2004) imaged Gd-HDO3-labeled EPCs forming endothelial tubes grown in Matrigel plugs both in vitro and in vivo. MRI not only offers the opportunity to visualize the cell migration within these constructs, but allows for the coupling to other MR modalities, such as diffusion and elastography, to provide simultaneous information on cell dynamics, as well as the biomechanical properties of the extracellular matrix (Othman et al., 2005; Xu et al., 2006).

6.6.2. Central Nervous System (CNS)

Visualizing transplanted stem and progenitor cells in experimental diseases and models of damage to the central nervous system have provided good examples of the use of superparamagnetic nanoparticles to track cell delivery and migration. Numerous cell types have been followed using MRI following transplantation, including neuroprogenitor cells (Jiang et al., 2005; Lepore et al., 2006; Magnitsky et al., 2005; Miyoshi et al., 2005; Modo et al., 2002, 2004; Shapiro et al., 2006a; Walczak et al., 2005; Watson et al., 2006; Zhang et al., 2003c), bone marrow cells (Jendelova et al., 2003; Sykova et al., 2006; Zhang et al., 2004), mesenchymal stem cells (Sykova et al., 2006), astrocytes (Toyoda et al., 2004), embryonic stem cells (Hoehn et al., 2002b; Stroh et al., 2005a), and macrophages (Heyn et al., 2006). MR studies of cell migration within the CNS is particularly attractive due to the amazingly large migratory rate of cells that have entered the rostral migratory stream (RMS) which is involved in the continual repopulation of forebrain structures including the olfactory bulb (Scheffler et al., 2005; Steindler et al., 1996). Migratory cells within the brain's RMS have been found to migrate at a rate of ~120 μm/h (Alvarez-Buylla et al., 2000) and stem cells can be found to migrate up to 13 mm away from the site of transplantation in rat spinal cords (Lee et al., 2004), both are certainly within the resolution of high-field MRM. MRM has been extremely successful in tracking the migration of SPIO-labeled cells away from regions of direct injection in the brain and within the spinal cord.

Severe motor dysfunction can occur in patients due to a loss of oligodendrocytes, which are responsible for myelinated axons. Transplantation of oligodendroglial progenitors leads to myelination and functional recovery of dysmyelinated rat strains, i.e. the myelin-deficient Long Evans shaker rat (Bulte and Kraitchman, 2004). SPIO-labeled oligodendrocyte progenitors cells have been visualized post-transplantation in both the spinal cord (Bulte et al., 1999) and the brain (Bulte et al., 2001; Franklin et al., 1999). An early study showed that SPIO labeling of oligodendrocyte progenitor cells by an anti-Tfr MAB could be detected ex vivo at 4.7 T, and labeled cells retained their in vivo capacity to migrate and re-myelinate axons in myelin-deficient rats (Bulte et al., 1999). In vivo results were obtained using neuro-stem cells (NSCs), derived oligodendroglial progenitors, which were SPIO-labeled with a magnetodendrimer and transplanted in the ventricles of neonatal Long Evans shaker rats (Bulte et al., 2001). MRI has also been used to monitor transplantation of other cells that are able to cause re-myelinatation such as Schwann cells (Lee et al., 2004) and olfactory ensheathing cells (OECs) (Lee et al., 2004).

Spinal cord injury (SCI) is a major disabling health problem that adults face today. In the United States, it is estimated that there are approximately 240,000 individuals that have sustained a SCI, with roughly 11,000 new injuries occurring each year. In addition, the relative number of new injuries is rising and expected to increase approximately 20% by 2010, compared to the 1994 prevalence. A promising therapy for SCI is regeneration of the damaged part of the cord by exogenous stem and progenitor cells (Cummings et al., 2005; Enzmann et al., 2006; Pearse and Bunge, 2006). MRI has provided a proven method in which SPIO-labeled cells can be monitored following targeting to damaged SCI by either direct injection (Lee et al., 2004) or vascular delivery strategies (Jendelova et al., 2004; Sykova and Jendelova, 2005, 2006; Sykova et al., 2006) or implanted tissue constructs (Sykova et al., 2006). Cell imaging specifically following local delivery strategies in SCI damage is complicated by the large changes in chord structure and MR contrast due to the SCI lesion which results in endogenous iron deposition (Berens et al., 2005; Lee et al., 2004; Velardo et al., 2004; Wirth et al., 1992). In the uninjured spinal cord, SPIO-labeled cells have been shown to migrate large distances from the site of injection (Lee et al., 2004; Lepore et al., 2006). Despite the ability of SPIO-labeled OECs to migrate up to 9 mm in the intact spinal cord, Lee et al. (2004) showed diminished migration of cells in transection model of SCI. Both SPIO and unlabeled cells failed to cross the transection gap 8 weeks post-grafting. In contrast, when differentiated ES cells are transplanted into a contusion SCI rat model, SPIO-labeled cells resulted in MR contrast up to 13 mm away from the injection site (Bulte and Kraitchman, 2004). These studies again indicate the importance of using imaging modalities to sequentially track the migration of different cell types in situations where the underlying tissue environment has been altered. Following the vascular delivery of SPIO-labeled mesenchymal stem cells (MSCs), dramatic therapeutic and imaging results have been observed 1-week post-SCI. Intravenously delivered MSCs not only resulted in MR image contrast changes, but significantly improved hindlimb motor function in rats with a spinal cord compression lesion (Jendelova et al., 2004; Sykova and Jendelova, 2005, 2006; Sykova et al., 2006). In order to overcome the physical barriers limiting cell migration, SCI researchers are exploring methods to modify the extracellular matrix within the lesion site. This includes the extensive use of tissue-engineered constructs to help bridge the gap between adjacent regions of healthy spinal cord and to provide an appropriate mechanical and chemical environment and for neuronal cell migration and regeneration. SPIO-labeled MSCs have been visualized within these tissue-engineered constructs post-implantation into SCI lesions (Sykova et al., 2006). Again, similar to MR imaging in isolated tissue constructs, MR diffusion and MR elastography measurements could be used to determine the dependence of the regenerated neurons to bridge the lesion on extracellular matrix environment (Othman et al., 2005; Xu et al., 2006).

Stroke is the third leading cause of death and a major cause of long-term disability in the United States (Warlow et al., 2003). To date, therapeutic interventions for cerebral ischemia, hypoxia, and hemorrhage are limited to protecting surviving ischemic neurons and in augmenting the recovery process (Gilman, 2006; Warlow et al., 2003). The use of stem cells for neurogenesis and circuitry restitution following such an ischemic injury creates new opportunities for therapy. These new cell-based therapies require a noninvasive method to follow cell delivery post-transplantation in preclinical and clinical trials. MRI has been used to track the delivery and migration of both iron oxide- and Gd-labeled stem and progenitor cells in uninjured brains (Heyn et al., 2006; Himmelreich et al., 2005; Magnitsky et al., 2005; Shapiro et al., 2006a; Stroh et al., 2005a; Watson et al., 2006; Zelivyanskaya et al., 2003), models of cerebral ischemia

(Hoehn et al., 2002a, 2002b; Jiang et al., 2005; Modo et al., 2002, 2004; Weber et al., 2006; Zhang et al., 2003c, 2002), and cancer (Anderson et al., 2005; Zhang et al., 2004). Using SPIO-labeled NSCs, Watson et al. (2006) found that in uninjured animals the migration and the developmental fate of transplanted cells is dependent on the region in which cells were delivered (cerebral cortex, hippocampus, thalamus) with the most widespread migration in the cortex. In contrast for brains that have been injured or contain invasive CNS tumors, both neural progenitors and reactive astrocytes readily migrate across these established anatomical boundaries in response to chemtotatic factors (Boockvar et al., 2005; Lindvall and Kokaia, 2006; Muller et al., 2006; Zhang et al., 2005). MRI has been able to track this process noninvasively. For example, MR-imaged migration of SPIO-labeled ES cells along the corpus callosum to the other hemisphere at the ischemic site of injury following middle cerebral artery occlusion (Hoehn et al., 2002b). In 2003, Zhang et al. (2003c) used MRI not only to visualize transplanted neuroprogenitor cells in rats but to quantitatively determine their in vivo migratory rate of ~65 μm/h toward an ischemic lesion. Multimodal contrast agents have been used to track stem and progenitor cells transplanted following ischemic injury (Modo et al., 2002, 2004) consisting of an optical contrast agent for either confocal or histological verification coupled with Gd for MRI monitoring. Modo et al. (2004) demonstrated that by pre-labeling neural stem cells with the bimodal contrast agent Gadolinium-RhodamIne Dextran (GRID), the transhemispheric migration of implanted neural stem cells contralateral to a stroke lesion can be followed in vivo by serial MRI and corroborated by subsequent histological analyses. Recently, a paramagnetic functionalized Qdots nanoparticle (Figure 6.5) has been developed (Yang et al., 2006) and used to monitor multipotent astrocytic stem cells transplanted following ischemic injury in a murine model of neonatal perinatal stroke.

An additional advantage of high-resolution imaging of the CNS with MRI is that the efficacy of cell therapies can also be determined noninvasively. Magnetic resonance spectroscopy can be used to assess the restoration of normal metabolite following cellular transplantation (Honma et al., 2006). Contrast- and diffusion-enhanced MR imaging can be used to image normal tissue permeability, angiogenesis, and tissue integrity (Honma et al., 2006; Zhang et al., 2002). A recent study by Jiang et al. (2006) is an outstanding example of combining MRI for both cell tracking and the characterization of changes in underlying tissue composition through the use of diffusion tensor imaging to perform noninvasive fiber tracking to assess white matter reorganization after stroke and treatment with SPIO-labeled NSCs.

6.6.3. Cardiac

Cellular therapies for primary cardiomyopathies have been proposed that aim to replace damaged cardiomyocytes and, over the last decade, cardiac cellular transplantation techniques have made significant gains. A variety of cell types have been evaluated as potential candidates for cardiac repair (Kessler and Byrne,

1999). Neonatal cardiomyocytes have been shown to survive and integrate when transplanted in healthy or injured myocardium (Leor et al., 1996; Scorsin et al., 1997, 1996). Immortalized cells derived from an atrial tumor also have been shown to survive and establish gap junctions in healthy, but not infarcted, porcine myocardium (Watanabe et al., 1998). Mesenchymal stem cells (MSCs) are an adult stem cell population that has received attention due to the suitability of MSCs for cell and ex vivo gene therapies. Initial reports indicated that human MSCs do not express MHC II markers, and express only low levels of MHC I (a characteristic that may allow the cells to escape surveillance by host immune cells) (Pittenger et al., 2000). Particular interest developed for the use of skeletal muscle myoblasts, as these cells have the potential for autologous transplantation (Ghostine et al., 2002; Menasche et al., 2001; Scorsin et al., 2000). Survival and differentiation of C2C12 myoblasts toward a slow-twitch muscle phenotype after arterial delivery into the murine heart has been demonstrated (Robinson et al., 1996). However, given the inherent differences between skeletal and cardiac myocytes, the ideal candidate for cellular cardiomyoplasty may be a less-committed progenitor cell that can be autologously isolated and undergo full cardiac differentiation. Muscle derived stem cells (MDSCs) have been transplanted into the murine myocardium and shown to integrate in SCID (Severe Combined Immunodeficiency) mice (Sakai et al., 2002) and to form gap junctions similar to true cardiomyocytes (Oshima et al., 2005). The functional impact of using bone-marrow cells for cellular cardiomyoplasty was demonstrated when Tomita et al. (1999) reported that adherent cells isolated from rat bone-marrow and treated with 5-azacytidine could improve heart function in a cryo-injury model of cardiomyopathy.

Given that noninvasive monitoring techniques will be essential for the clinical evaluation of stem cell transplants, many groups have explored the use of SPIO-labeled cells prior to transplantation into the beating myocardium. These efforts have included the use of hESC (Tallheden et al., 2006; Arai et al., 2006), MSCs, myoblasts, bone marrow (Walczak et al., 2005; 2006, 2006), macrophages (Leor et al., 2006), smooth muscle (Riviere et al., 2005), and EPCs (Weber et al., 2004b). An advantage of cardiovascular MRI is that both standard T_1- and T_2-weighted MR imaging sequences that are sensitive to SPIO-labeled cells can be simultaneously coupled with other diagnostic MR-imaging sequences used to determine viability, global, and regional cardiac function. It has been a long hope and the basis of cell therapies that transplanted cells would be able to "mend a broken heart" replacing fibrotic scar tissue with functional cardiomyocytes which are electrically connected to the rest of the heart and is the basis of many ongoing clinical trials (Chang et al., 2006; Laflamme and Murry, 2005). To this end both small and large animal preclinical models have been developed to track cell transplantation following an experimentally induced myocardial infarction. The challenge with small animals is the extremely high heart rate and small tissue mass that require optimized MR pulse sequences and hardware. Using a MR gradient recalled imaging sequence, gated to the ECG, cardiac images were acquired post-transplantation of SPIO-labeled myoblast along the entire long axis of the beating heart, with an in-plane resolution of $195 \times 195\ \mu m^2$. Hypo-intense regions were clearly distinguishable in the left anterior wall of the animals receiving labeled transplants (Figure 6.6). The hypo-intense areas were present at all time points, although they did appear to become smaller and oriented with the curvature of the myocardial wall over time. Importantly, animals receiving unlabeled myoblasts did not exhibit discreet regions of intramyocardial hypo-intensity on T_1-weighted images. Histological analyses (Prussian Blue staining) demonstrated that the regions of intramyocardial MR signal loss correspond to the location of the engrafted myoblasts that had differentiated into myotubes (Figure 6.6). The rational for moving toward large animal studies is the ability to use image-guided catheter devices, similar to those used in clinical trials, providing endocardial cell delivery to targeted regions within the heart. This demand for interventional cardiovascular MR has pushed the development of low-field MRI devices (1.5 T) which are capable of simultaneous measurement of areas at risk (delayed contrast enhancement), 3D imaging, guidance of catheter delivery in real time and the imaging of SPIO-labeled cells once delivered to the heart (Dick et al., 2003; Saeed et al., 2005). The number and the region in which the cells are deposited (i.e., in the peri-infarct, infracted region, healthy myocardium) are extremely important to the overall success of the transplant.

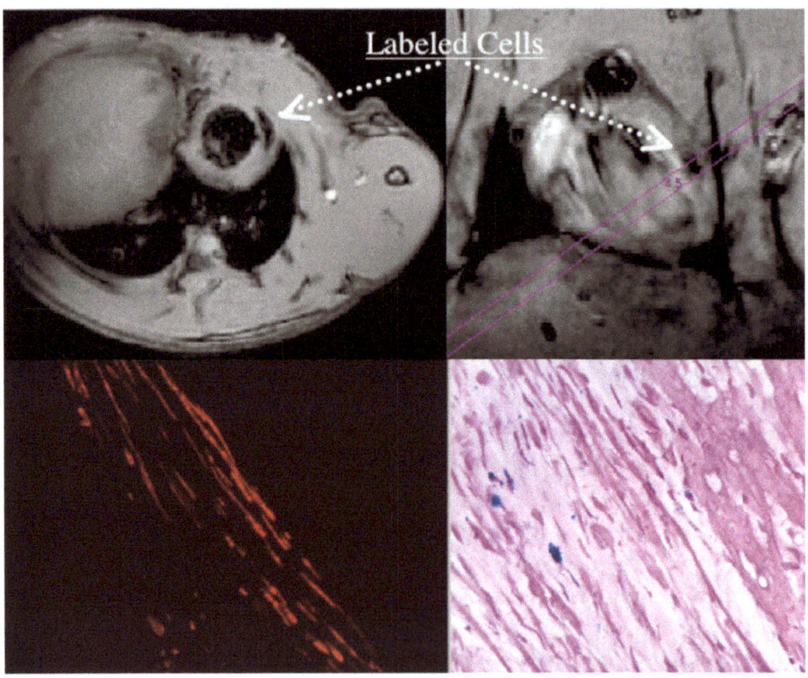

FIGURE 6.6. Short (*top-left*) and long (*top-right*) axis MRIs of myoblast transplant in viable myocardium 24 h post-injection. Expression of skeletal fast-myosin (red) by the engrafted myoblasts (*bottom-left*). Prussian blue stain of engrafted myoblasts (*bottom-right*). Modified from Cahill et al. (2004b).

Kraitchman et al. have demonstrated this dependence on location of cell delivery in a canine model of myocardial infarct. SPIO-labeled cells that were delivered to the peri-infarct and healthy myocardium were observed to have migrated to the infracted area. It is no doubt that with the numerous ongoing clinical trials using stem cells for heart failure the use of SPIO-labeled cells will play an expanding role in therapeutic monitoring and cell tracking.

6.6.4. *Musculoskeletal*

An immediate application of this technology to musculoskeletal disorders is the monitoring of cell therapies to mitigate the devastating effects of muscular dystrophy. Muscular dystrophy is an inherited disease that is known to result in skeletal muscle weakness and cardiac and respiratory failure, due to chronic bouts of muscle damage and regeneration which eventually exhausts the endogenous pool of stem cells leading to organ failure and death.

Initial transplantation strategies in muscular dystrophy focused primarily on the delivery of myoblasts to the dystrophic muscle. Unfortunately, the therapeutic efficacy of early myoblast transfer studies was limited by massive myoblast cell death observed immediately following in vivo delivery (Cossu and Mavilio, 2000). It has been hypothesized that early cell death after transplantation of myoblasts is due to cellular and humoral immune responses, and the induction of apoptosis (Smythe et al., 2000, 2001). Muscle-derived stem cells (MDSC) have shown a potential ability to repair dystrophic skeletal and cardiac muscle (Deasy, 2002; Jankowski et al., 2002; Torrente et al., 2001). This cell population can undergo in vivo differentiation to regenerate lost myofibers and restore dystrophin expression (Huard et al., 1994; Huard et al., 2003; Ikezawa et al., 2003; Jankowski et al., 2002; Lee et al., 2000; Qu-Petersen et al., 2002). They are also capable of reconstituting the hematopoietic stem cell compartment of lethally irradiated *mdx* mice (Cao et al., Cao et al.). Both characteristics combined are indicative of a unique stem cell population with a less-committed phenotype than the traditional primary myoblasts used in early transplant studies. Several reports have demonstrated that the MDSC will migrate from the vasculature to engraft in dystrophic muscle (Cao et al., 2003; Lee et al., 2000; Torrente et al., 2003, 2001). Embryonic and fetal stem cells have also shown the potential to rescue dystrophic muscle. Intra-arterial injection of wild-type mesoangioblasts (vessel associated fetal stem cells) in a murine model of limb girdle muscle dystrophy resulted in expression of the missing sarcoglycan in more than 50% of soleus muscle fibers. In addition, vascularly delivered mesoangioblasts have been shown to restore sarcolemmal integrity and results in functional recovery (Galvez et al., 2006; Sampaolesi et al., 2003). This restoration is especially important for the treatment of essential muscles, such as the diaphragm, where impairment results in severe respiratory problems in muscular dystrophy (Chen and Goldhamer, 2003). Overall, a number of studies have found that stem cells have therapeutic potential in skeletal muscle (Brussee et al., 1998; Qu-Petersen et al., 2002; Sampaolesi et al., 2003; Wernig et al., 2000) where transfer efficiencies depend on the cell type, with integrations ranging from 0.1 % to 50 % of a target tissue. In addition, the majority of studies have found that enhancement of chemotatic and mitogenic signals is necessary to increase the efficiency of cell transplantation (Skuk and Tremblay, 2003).

SPIO contrast agents can be utilized to monitor therapeutic muscle stem-cell transplants in a murine model of Duchenne/Becker's muscular dystrophy (*mdx* mice) (Walter et al., 2004). Early work utilized a dendrimer to shuttle SPIO into MDSC cells, prior to transplantation into *mdx* mice (Walter et al., 2004). Labeled MDSCs cells were transplanted into the gastrocnemius-plantaris-soleus muscle group of 6-week-old *mdx* mice. High-resolution MRIs were obtained 24 h, 2, 4, and 11 days post-injection. Distinct regions of signal hypo-intensity were identified in the posterior musculature of animals receiving labeled cell transplants at all time points. Control animals receiving unlabeled cell transplants displayed homogenous images, without the regions of hypo-intensity seen in the experimental animals. Following in vivo imaging, a comparative study was performed with conventional histochemical cell detection techniques. Engrafted cells were detected by analysis of β-galactosidase activity (LacZ), dystrophin expression, and iron content. LacZ expressing fibers were readily identified in regions corresponding to the hypo-intense regions in MR images. Additionally, Prussian blue staining of consecutive serial sections revealed the presence of iron accumulation in many of the LacZ positive fibers, confirming the correlation between the histological location of the cells and MR images. Immunostaining for mini-dystrophin indicated that the engrafted cells restored membrane dystrophin expression and were therefore potentially therapeutic. However, a small number of cells that appeared to be nonmuscle cells also displayed iron-containing granules. Due to the number of macrophages seen in regenerating muscle fibers, it was hypothesized these to be scavenger cells removing the remains of labeled cells that failed to engraft.

Muscle stem and progenitor cells can be efficiently and nontoxically labeled via simple in vitro incubation with SPIO complexes, consisting of commercially available ferumoxides, and the standard molecular biology transfection reagent poly-L-lysine (Cahill et al., 2004a). After 6 h of incubation, 100 % cellular labeling was achieved in MDSCs and C2C12 cells, primary myoblasts, and the multipotent human MSCs (Figure 6.7). Similar to magnetodendrimer-labeled cells, ferumoxide-labeled cells could be visualized by high-resolution MRI, following therapeutic transplantation into *mdx* mice (Cahill et al., 2004a). MRI can also visualize vascular delivery of SPIO MDSC cells (Cahill et al., 2004a). Following arterial delivery of either ferumoxides-labeled myoblasts or MDSCs into dystrophic mice, small, punctuate areas of decreased signal intensity were seen only in the limb musculature of the leg that received labeled cell infusion (Figure 6.8). Histological analyses of the leg musculature showed LacZ expressing muscle-derived stem cells within the vasculature, distributed in patterns corresponding to the MR images. X-gal staining confirmed the presence of stem cell integration in the soleus following vascular delivery (Figure 6.8).

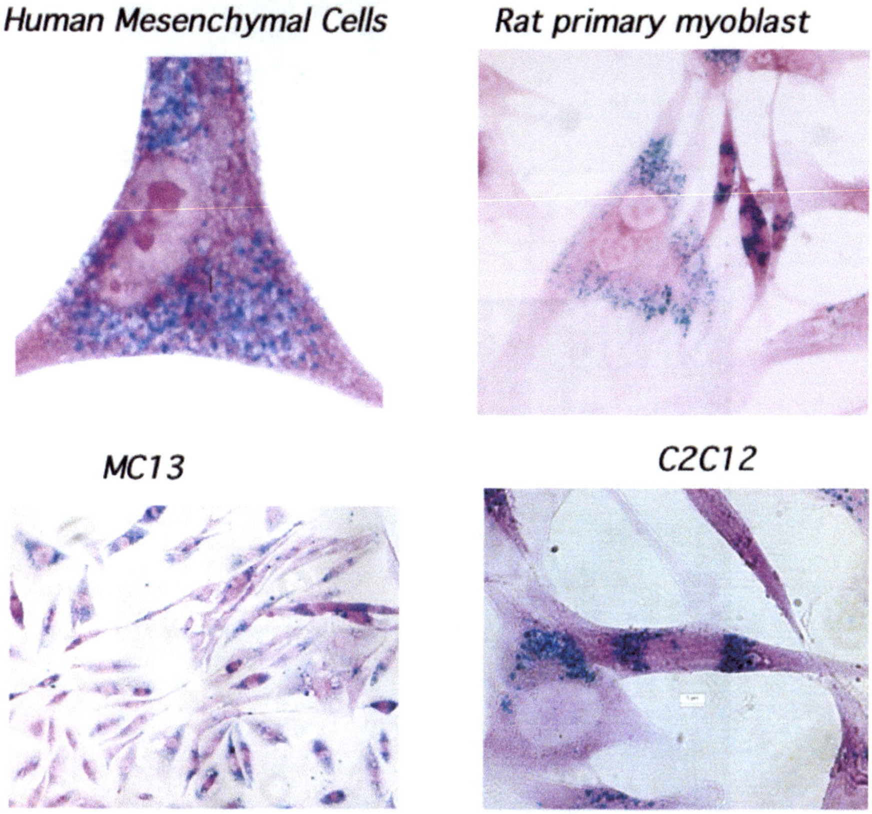

FIGURE 6.7. Prussian blue staining of fixed human mesenchymal, rat primary myoblast, muscle-derived stem cells, and C2C12 cells following a 6 h incubation with ferumoxide:poly-L-lysine.

SPIO labeling was found not to be toxic to MDSCs and does not alter the normal growth rate (Cahill et al., 2004a; Walter et al., 2004). The labeled cells differentiated to mature, multinucleated myotubes at rates comparable to unlabeled cells (Cahill et al., 2004b). The resulting myotubes displayed intracellular iron accumulation throughout the length of the myotubes and were otherwise morphologically indistinguishable from unlabeled myotubes. Immunofluorescent analysis of alpha-actinin and desmin expression also revealed that labeled myotubes contain normal sarcomeres (Walter et al., 2004). On transmission electron microscopy images, electron dense areas indicative of iron accumulation could be seen in the endosomal compartments (Cahill et al., 2004a). As previously suggested, trapping of iron oxide inside the endosome reduces the chance of Fenton-like reactions in the myoplasm and the containment of the iron until it can be metabolized (Bulte et al., 2001). In agreement with this hypothesis,

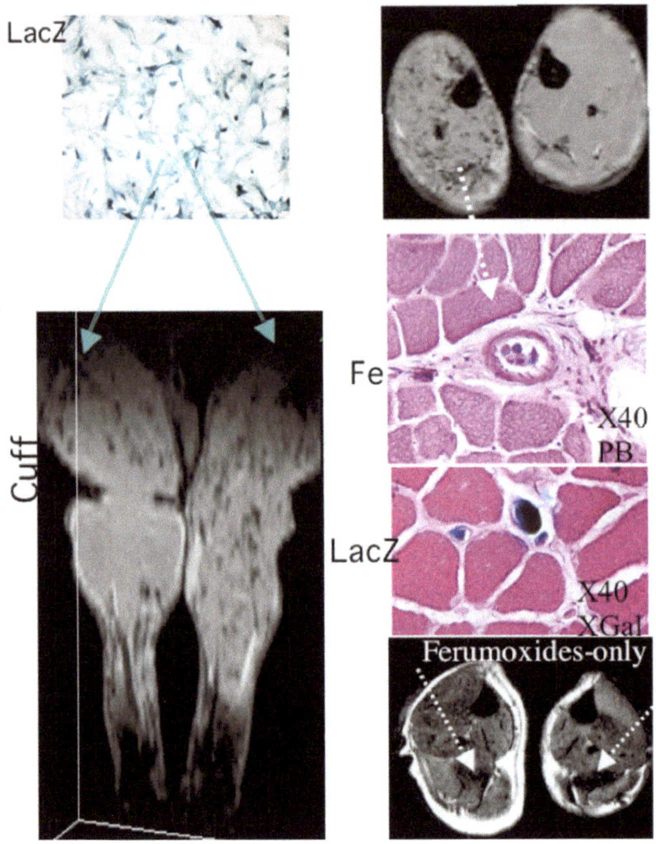

FIGURE 6.8. Imaging arterial delivery of SPIO-labeled muscle-derived stem cells to a normal mouse hindlimb. *Top*: 3D MRI of occluded leg (*left*) and contralateral limb (*right*) following arterial injection (1 h) of labeled muscle-derived stem cells. *Top-right*: Transaxial T_1 images of the same hindlimbs. *Right-middle*: Corresponding Prussian blue and XGal of gastrocnemius muscle showing the accumulation of labeled stem cells in major vessels and capillaries of the controlateral limb ($\times 40$) in contrast to Fermoxide alone (*right-bottom*). Modified from Cahill et al. (2004a).

ferumoxide accumulation did not affect cellular viability or alter the normal growth rate of labeled cells in vitro.

6.6.5. Renal

In a rat model of mesangiolysis, vascular delivery of SPIO-labeled MSCs could be detected in both the liver and the kidney (Bos et al., 2004; Hauger et al., 2006). In the case of intravenous injection, homing to the cortical area could be tracked using MR microscopy at 9.4 T (Figure 6.9). Unfortunately, despite detection of intravenously delivered cells in the liver, the cells could not be

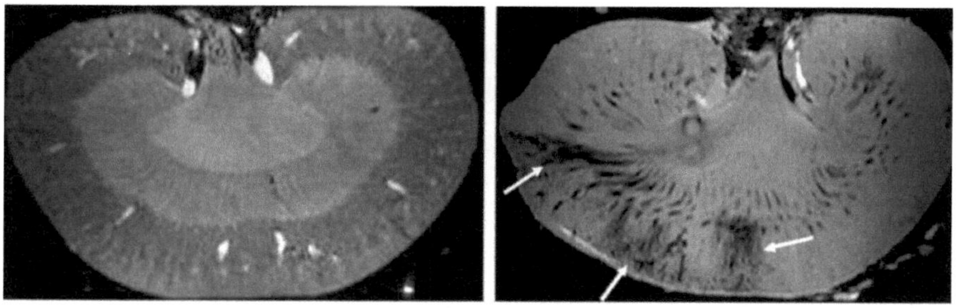

FIGURE 6.9. **a** Ex vivo sagittal three-dimensional T_2^*-weighted (110/15.7, flip angle of 30°) 9.4 T MR images of (a) control and (b) pathologic kidney 6 days after intravenous injection of 10^7-labeled MSCs (the upper pole is oriented left). In a, no signal intensity decrease is noted. In b, the corticomedullary differentiation is absent and distinct areas of cortical signal intensity decrease are present in the superior and superior midportion. **b** Ex vivo sagittal three-dimensional T_2^*-weighted (110/15.7, flip angle of 30°) 9.4 T MR images of (a) control and (b) pathologic kidney 6 days after intravenous injection of 10^7-labeled MSCs (the upper pole is oriented left). In a, no signal intensity decrease is noted. In b, the corticomedullary differentiation is absent and distinct areas of cortical signal intensity decrease are present in the superior and superior midportion (arrows) poles. Reprinted from Hauger et al. (2006). Reproduced with permission from the Radiological Society of North America.

detected in vivo in the kidney following arterial delivery (Hauger et al., 2006). This indicates that the higher signal to noise and resolution available from high-field MR scanners may be necessary to detect homing in some organ systems similar to the damaged kidney following intravenous delivery of SPIO-labeled cells (Hauger et al., 2006).

6.6.6. *Hematopoietic*

A major hurdle is the inability to track the migration of hematopoietic stem cells (HSCs) into the deep recesses of the bone marrow niche requiring extensive in vitro and histological analysis. The ability to noninvasively and longitudinally monitor the homing of HSCs to the different regions with the long bone in response to different chemokines would provide new insight into the mechanisms that result in stem cell targeting to hematopoietic niches. Daldrup-Link (Daldrup-Link et al., 2005b) found that the intravenous injection of SPIO-labeled hematopoietic bone marrow cells home to the bone marrow, whereas the intravenous injection of SPIO alone had very different distribution pattern. Targeting of negative contrast SPIO-labeled cells into the bone marrow and respective bone niches is extremely difficult to image due to the need to achieve high resolution in a tissue that displays inherently short T_2 and T_2^* (Mayer-Kuckuk et al., 2005). To avoid these relaxation problems, a multimodal contrast agent, Gadophrin-2 (a porphyrin-based contrast agent composed of an aromatic ring and covalently bound paramagnetic gadolinium chelates), (Daldrup-Link et al., 2004) has been used to image both optically and with MRI HSCs homing to the femoral and pelvic bone marrow. Further work needs to be done to determine if Gd-labeled cells can be imaged at enough high resolution to discriminate homing to different HSC niches within long bones.

6.7. Future Directions

Cell-based therapies represent a challenge for noninvasive monitoring due to the limited stem cell incorporation that occurs in the presence of massive cell death. In order to detect them in vivo, stem cells must be labeled with agents that will generate the greatest MR contrast possible in order to approach single cell sensitivity. This contrast can be accomplished currently by labeling cells with SPIO nanoparticles using standard molecular biology transfection techniques. Even at low magnetic fields, MRI provides a sensitive method in which picograms of contrast agent can be monitored in vivo. This methodology is not without limitations. Due to the bulk action and lack of specificity of the SPIO label, it is difficult to determine cell persistence and viability. This issue may be overcome by the use of smart probes that will be activated only in viable tissue, either by using new contrast mechanisms or gene markers. A second potential limitation is the dilution of label as the cell divides or is eliminated in vivo. It is possible that such cell division may dilute the label to the extent that it is no longer detectable. It has previously been estimated that with symmetric cell division and an initial cell labeling of 25 pg/cell, cells should be detected up to at least four cell divisions (Bulte and Kraitchman, 2004). The number of detectable cell divisions is expected to increase at ultra-high magnetic fields and with the development of more sensitive MR pulse sequences. Lastly, as is the case with all labels, there should be no effect on cell viability and its ability to differentiate, proliferate, and migrate in response to chemotatic signals.

Acknowledgment. Data presented in figures 6.4, 6.5, and 6.7 were acquired with support from NIH (R01 HL 78670) and NSF (NHFML; ECC0506560).

References

Aghi, M., Chiocca, E.A., 2005. Contribution of bone marrow-derived cells to blood vessels in ischemic tissues and tumors. Mol Ther 12, 994–1005.

Ahrens, E.T., Feili-Hariri, M., Xu, H., Genove, G., Morel, P.A., 2003. Receptor-mediated endocytosis of iron-oxide particles provides efficient labeling of dendritic cells for in vivo MR imaging. Magn Reson Med 49, 1006–1013.

Aime, S., Barge, A., Delli Castelli, D., Fedeli, F., Mortillaro, A., Nielsen, F.U., Terreno, E., 2002a. Paramagnetic lanthanide(III) complexes as pH-sensitive chemical exchange saturation transfer (CEST) contrast agents for MRI applications. Magn Reson Med 47, 639–648.

Aime, S., Calabi, L., Biondi, L., De Miranda, M., Ghelli, S., Paleari, L., Rebaudengo, C., Terreno, E., 2005a. Iopamidol: Exploring the potential use of a well-established x-ray contrast agent for MRI. Magn Reson Med 53, 830–834.

Aime, S., Delli Castelli, D., Fedeli, F., Terreno, E., 2002b. A paramagnetic MRI-CEST agent responsive to lactate concentration. J Am Chem Soc 124, 9364–9365.

Aime, S., Delli Castelli, D., Terreno, E., 2005b. Highly sensitive MRI chemical exchange saturation transfer agents using liposomes. Angew Chem Int Ed Engl 44, 5513–5515.

Alivisatos, A.P., 1996. Perspectives on the physical chemistry of semiconductor nanocrystals. J Phys Chem 100, 13226–13239.

Allen, M.J., MacRenaris, K.W., Venkatasubramanian, P.N., Meade, T.J., 2004. Cellular delivery of MRI contrast agents. Chem Biol 11, 301–307.

Allport, J.R., Weissleder, R., 2001. In vivo imaging of gene and cell therapies. Exp Hematol 29, 1237–1246.

Alvarez-Buylla, A., Herrera, D.G., Wichterle, H., 2000. The subventricular zone: source of neuronal precursors for brain repair. Prog Brain Res 127, 1–11.

Anderson, S.A., Glod, J., Arbab, A.S., Noel, M., Ashari, P., Fine, H.A., Frank, J.A., 2005. Noninvasive MR imaging of magnetically labeled stem cells to directly identify neovasculature in a glioma model. Blood 105, 420–425.

Anderson, S.A., Lee, K.K., Frank, J.A., 2006. Gadolinium-fullerenol as a paramagnetic contrast agent for cellular imaging. Invest Radiol 41, 332–338.

Anversa, P., Kajstura, J., Leri, A., Bolli, R., 2006. Life and death of cardiac stem cells: a paradigm shift in cardiac biology. Circulation 113, 1451–1463.

Aoki, I., Takahashi, Y., Chuang, K.H., Silva, A.C., Igarashi, T., Tanaka, C., Childs, R.W., Koretsky, A.P., 2006. Cell labeling for magnetic resonance imaging with the T1 agent manganese chloride. NMR Biomed 19, 50–59.

Arai, T., Kofidis, T., Bulte, J.W., de Bruin, J., Venook, R.D., Berry, G.J., McConnell, M.V., Quertermous, T., Robbins, R.C., Yang, P.C., 2006. Dual in vivo magnetic resonance evaluation of magnetically labeled mouse embryonic stem cells and cardiac function at 1.5 t. Magn Reson Med 55, 203–209.

Arbab, A.S., Bashaw, L.A., Miller, B.R., Jordan, E.K., Bulte, J.W., Frank, J.A., 2003. Intracytoplasmic tagging of cells with ferumoxides and transfection agent for cellular magnetic resonance imaging after cell transplantation: methods and techniques. Transplantation 76, 1123–1130.

Arbab, A.S., Liu, W., Frank, J.A., 2006a. Cellular magnetic resonance imaging: current status and future prospects. Expert Rev Med Devices 3, 427–439.

Arbab, A.S., Pandit, S.D., Anderson, S.A., Yocum, G.T., Bur, M., Frenkel, V., Khuu, H.M., Read, E.J., Frank, J.A., 2006b. Magnetic resonance imaging and confocal microscopy studies of magnetically labeled endothelial progenitor cells trafficking to sites of tumor angiogenesis. Stem Cells 24, 671–678.

Arbab, A.S., Yocum, G.T., Kalish, H., Jordan, E.K., Anderson, S.A., Khakoo, A.Y., Read, E.J., Frank, J.A., 2004a. Efficient magnetic cell labeling with protamine sulfate complexed to ferumoxides for cellular MRI. Blood 104, 1217–1223.

Arbab, A.S., Yocum, G.T., Rad, A.M., Khakoo, A.Y., Fellowes, V., Read, E.J., Frank, J.A., 2005. Labeling of cells with ferumoxides-protamine sulfate complexes does not inhibit function or differentiation capacity of hematopoietic or mesenchymal stem cells. NMR Biomed 18, 553–559.

Arbab, A.S., Yocum, G.T., Wilson, L.B., Parwana, A., Jordan, E.K., Kalish, H., Frank, J.A., 2004b. Comparison of transfection agents in forming complexes with ferumoxides, cell labeling efficiency, and cellular viability. Mol Imaging 3, 24–32.

Arriagada, F.J., Osseo-Asare, K., 1999. Synthesis of nanosize silica in a nonionic water-in-oil microemulsion: effects of the water/surfactant molar ratio and ammonia concentration. J Colloid Interface Sci 211, 210–220.

Arteaga, C., Revel, D., Zhao, S., Hadour, G., Forrat, R., Oksendal, A., Canet, E., 1999. Myocardial "low reflow" assessed by Dy-DTPA-BMA-enhanced first-pass MR imaging in a dog model. J Magn Reson Imaging 9, 679–684.

Artemov, D., Bhujwalla, Z.M., Bulte, J.W., 2004. Magnetic resonance imaging of cell surface receptors using targeted contrast agents. Curr Pharm Biotechnol 5, 485–494.

Beache, G.M., Kulke, S.F., Kantor, H.L., Niemi, P., Campbell, T.A., Chesler, D.A., Gewirtz, H., Rosen, B.R., Brady, T.J., Weisskoff, R.M., 1998. Imaging perfusion deficits in ischemic heart disease with susceptibility-enhanced T2-weighted MRI: preliminary human studies. Magn Reson Imaging 16, 19–27.

Beck, B., Plant, D.H., Grant, S.C., Thelwall, P.E., Silver, X., Mareci, T.H., Benveniste, H., Smith, M., Collins, C., Crozier, S., Blackband, S.J., 2002. Progress in high field MRI at the University of Florida. Magma 13, 152–157.

Benedetti, S., Pirola, B., Pollo, B., Magrassi, L., Bruzzone, M.G., Rigamonti, D., Galli, R., Selleri, S., Di Meco, F., De Fraja, C., Vescovi, A., Cattaneo, E., Finocchiaro, G., 2000. Gene therapy of experimental brain tumors using neural progenitor cells. Nat Med 6, 447–450.

Benveniste, H., Blackband, S.J., 2006. Translational neuroscience and magnetic-resonance microscopy. Lancet Neurol 5, 536–544.

Berens, S.A., Colvin, D.C., Yu, C.G., Yezierski, R.P., Mareci, T.H., 2005. Evaluation of the pathologic characteristics of excitotoxic spinal cord injury with MR imaging. AJNR Am J Neuroradiol 26, 1612–1622.

Bhorade, R., Weissleder, R., Nakakoshi, T., Moore, A., Tung, C.H., 2000. Macrocyclic chelators with paramagnetic cations are internalized into mammalian cells via a HIV-tat derived membrane translocation peptide. Bioconjug Chem 11, 301–305.

Billotey, C., Wilhelm, C., Devaud, M., Bacri, J.C., Bittoun, J., Gazeau, F., 2003. Cell internalization of anionic maghemite nanoparticles: quantitative effect on magnetic resonance imaging. Magn Reson Med 49, 646–654.

Blau, H.M., Brazelton, T.R., Weimann, J.M., 2001. The evolving concept of a stem cell: entity or function? Cell 105, 829–841.

Boockvar, J.A., Schouten, J., Royo, N., Millard, M., Spangler, Z., Castelbuono, D., Snyder, E., O'Rourke, D., McIntosh, T., 2005. Experimental traumatic brain injury modulates the survival, migration, and terminal phenotype of transplanted epidermal growth factor receptor-activated neural stem cells. Neurosurgery 56, 163–171; discussion 171.

Bos, C., Delmas, Y., Desmouliere, A., Solanilla, A., Hauger, O., Grosset, C., Dubus, I., Ivanovic, Z., Rosenbaum, J., Charbord, P., Combe, C., Bulte, J.W., Moonen, C.T., Ripoche, J., Grenier, N., 2004. In vivo MR imaging of intravascularly injected magnetically labeled mesenchymal stem cells in rat kidney and liver. Radiology 233, 781–789.

Boushel, R., Langberg, H., Olesen, J., Nowak, M., Simonsen, L., Bulow, J., Kjaer, M., 2000. Regional blood flow during exercise in humans measured by near-infrared spectroscopy and indocyanine green. J Appl Physiol 89, 1868–1878.

Brazelton, T.R., Blau, H.M., 2005. Optimizing techniques for tracking transplanted stem cells in vivo. Stem Cells 23, 1251–1265.

Brown, M.A., Semelka, R.C., NetLibrary Inc., 1999. MRI basic principles and applications, 2nd ed. Wiley-Liss, New York.

Bruchez, M., Jr., Moronne, M., Gin, P., Weiss, S., Alivisatos, A.P., 1998. Semiconductor nanocrystals as fluorescent biological labels. Science 281, 2013–2016.

Brussee, V., Merly, F., Tardif, F., Tremblay, J.P., 1998. Normal myoblast implantation in MDX mice prevents muscle damage by exercise. Biochem Biophys Res Commun 250, 321–327.

Bullok, K.E., Gammon, S.T., Violini, S., Prantner, A.M., Villalobos, V.M., Sharma, V., Piwnic-Worms, D., 2006. Permeation peptide conjugates for in vivo molecular imaging applications. Mol Imaging 5, 1–15.

Bulte J.W.M., Ben-Hur T., Miller B.R., Mizrachi-Kol R., Einstein O., Reinhartz E., Zywicke H.A., Douglas T., Frank J.A. MR microscopy of magnetically labeled neurospheres transplanted into the LewisEAE rat brain. Magn. Reson. Med. 2003; 50: 201–205.

Bulte, J.W., Douglas, T., Witwer, B., Zhang, S.C., Lewis, B.K., van Gelderen, P., Zywicke, H., Duncan, I.D., Frank, J.A., 2002a. Monitoring stem cell therapy in vivo using magnetodendrimers as a new class of cellular MR contrast agents. Acad Radiol 9 Suppl 2, S332–S335.

Bulte, J.W., Douglas, T., Witwer, B., Zhang, S.C., Strable, E., Lewis, B.K., Zywicke, H., Miller, B., van Gelderen, P., Moskowitz, B.M., Duncan, I.D., Frank, J.A., 2001. Magnetodendrimers allow endosomal magnetic labeling and in vivo tracking of stem cells. Nat Biotechnol 19, 1141–1147.

Bulte, J.W., Duncan, I.D., Frank, J.A., 2002b. In vivo magnetic resonance tracking of magnetically labeled cells after transplantation. J Cereb Blood Flow Metab 22, 899–907.

Bulte, J.W., Kostura, L., Mackay, A., Karmarkar, P.V., Izbudak, I., Atalar, E., Fritzges, D., Rodriguez, E.R., Young, R.G., Marcelino, M., Pittenger, M.F., Kraitchman, D.L., 2005. Feridex-labeled mesenchymal stem cells: cellular differentiation and MR assessment in a canine myocardial infarction model. Acad Radiol 12 Suppl 1, S2–S6.

Bulte, J.W., Kraitchman, D.L., 2004. Iron oxide MR contrast agents for molecular and cellular imaging. NMR Biomed 17, 484–499.

Bulte, J.W., Wu, C., Brechbiel, M.W., Brooks, R.A., Vymazal, J., Holla, M., Frank, J.A., 1998. Dysprosium-DOTA-PAMAM dendrimers as macromolecular T2 contrast agents. Preparation and relaxometry. Invest Radiol 33, 841–845.

Bulte, J.W.M., Zhang, S., van Gelderen, P., Herynek, V., Jordan, E.K., Duncan, I.D., Frank, J.A., 1999. Neurotransplantation of magnetically labeled oligodendrocyte progenitors: magnetic resonance tracking of cell migration and myelination. Proc Natl Acad Sci U S A 96, 15256–15261.

Cahill, K.S., Gaidosh, G., Huard, J., Silver, X., Byrne, B.J., Walter, G.A., 2004a. Noninvasive monitoring and tracking of muscle stem cell transplants. Transplantation 78, 1626–1633.

Cahill, K.S., Germain, S., Byrne, B.J., Walter, G.A., 2004b. Non-invasive analysis of myoblast transplants in rodent cardiac muscle. Int J Cardiovasc Imaging 20, 593–598.

Cao, B., Zheng, B., Jankowski, R.J., Kimura, S., Ikezawa, M., Deasy, B., Cummins, J., Epperly, M., Qu-Petersen, Z., Huard, J., 2003. Muscle stem cells differentiate into haematopoietic lineages but retain myogenic potential. Nat Cell Biol 5, 640–646.

Caravan, P., 2006. Strategies for increasing the sensitivity of gadolinium based MRI contrast agents. Chem Soc Rev 35, 512–523.

Caravan, P., Ellison, J.J., McMurry, T.J., Lauffer, R.B., 1999. Gadolinium(III) Chelates as MRI Contrast Agents: Structure, Dynamics, and Applications. Chem Rev 99, 2293–2352.

Caravan, P., Greenfield, M.T., Bulte, J.W., 2001. Molecular factors that determine Curie spin relaxation in dysprosium complexes. Magn Reson Med 46, 917–922.

Chakkalakal, J.V., Thompson, J., Parks, R.J., Jasmin, B.J., 2005. Molecular, cellular, and pharmacological therapies for Duchenne/Becker muscular dystrophies. Faseb J 19, 880–891.

Chan, W.C., Nie, S., 1998. Quantum dot bioconjugates for ultrasensitive nonisotopic detection. Science 281, 2016–2018.

Chang, G.Y., Xie, X., Wu, J.C., 2006. Overview of stem cells and imaging modalities for cardiovascular diseases. J Nucl Cardiol 13, 554–569.

Chen, J.C., Goldhamer, D.J., 2003. Skeletal muscle stem cells. Reprod Biol Endocrinol 1, 101–107.

Chien, K.R., 2006. Lost and found: cardiac stem cell therapy revisited. J Clin Invest 116, 1838–1840.

Ciobanu, L., Pennington, C.H., 2004. 3D micron-scale MRI of single biological cells. Solid State Nucl Magn Reson 25, 138–141.

Ciobanu, L., Seeber, D.A., Pennington, C.H., 2002. 3D MR microscopy with resolution 3.7 microm by 3.3 microm by 3.3 microm. J Magn Reson 158, 178–182.

Cogle, C.R., Yachnis, A.T., Laywell, E.D., Zander, D.S., Wingard, J.R., Steindler, D.A., Scott, E.W., 2004. Bone marrow transdifferentiation in brain after transplantation: a retrospective study. Lancet 363, 1432–1437.

Contag, C.H., Ross, B.D., 2002. It's not just about anatomy: in vivo bioluminescence imaging as an eyepiece into biology. J Magn Reson Imaging 16, 378–387.

Cossu, G., Mavilio, F., 2000. Myogenic stem cells for the therapy of primary myopathies: wishful thinking or therapeutic perspective? J Clin Invest 105, 1669–1674.

Crich, S.G., Biancone, L., Cantaluppi, V., Duo, D., Esposito, G., Russo, S., Camussi, G., Aime, S., 2004. Improved route for the visualization of stem cells labeled with a Gd-/Eu-chelate as dual (MRI and fluorescence) agent. Magn Reson Med 51, 938–944.

Cummings, B.J., Uchida, N., Tamaki, S.J., Salazar, D.L., Hooshmand, M., Summers, R., Gage, F.H., Anderson, A.J., 2005. Human neural stem cells differentiate and promote locomotor recovery in spinal cord-injured mice. Proc Natl Acad Sci U S A 102, 14069–14074.

Cunningham, C.H., Arai, T., Yang, P.C., McConnell, M.V., Pauly, J.M., Conolly, S.M., 2005. Positive contrast magnetic resonance imaging of cells labeled with magnetic nanoparticles. Magn Reson Med 53, 999–1005.

Daldrup-Link, H.E., Rudelius, M., Metz, S., Piontek, G., Pichler, B., Settles, M., Heinzmann, U., Schlegel, J., Oostendorp, R.A., Rummeny, E.J., 2004. Cell tracking with gadophrin-2: a bifunctional contrast agent for MR imaging, optical imaging, and fluorescence microscopy. Eur J Nucl Med Mol Imaging 31, 1312–1321.

Daldrup-Link, H.E., Rudelius, M., Oostendorp, R.A., Jacobs, V.R., Simon, G.H., Gooding, C., Rummeny, E.J., 2005a. Comparison of iron oxide labeling properties of hematopoietic progenitor cells from umbilical cord blood and from peripheral blood for subsequent in vivo tracking in a xenotransplant mouse model XXX. Acad Radiol 12, 502–510.

Daldrup-Link, H.E., Rudelius, M., Piontek, G., Metz, S., Brauer, R., Debus, G., Corot, C., Schlegel, J., Link, T.M., Peschel, C., Rummeny, E.J., Oostendorp, R.A., 2005b. Migration of iron oxide-labeled human hematopoietic progenitor cells in a mouse model: in vivo monitoring with 1.5-T MR imaging equipment. Radiology 234, 197–205.

Deans, A.E., Wadghiri, Y.Z., Bernas, L.M., Yu, X., Rutt, B.K., Turnbull, D.H., 2006. Cellular MRI contrast via coexpression of transferrin receptor and ferritin. Magn Reson Med 56, 51–59.

Deasy, B.M., Huard, J., 2002. Gene therapy and tissue engineering based on muscle-derived stem cells. Curr Opin Mol Ther 4, 382–389.

Dennig, J., Duncan, E., 2002. Gene transfer into eukaryotic cells using activated polyamidoamine dendrimers. J Biotechnol 90, 339–347.

Dick, A.J., Guttman, M.A., Raman, V.K., Peters, D.C., Pessanha, B.S., Hill, J.M., Smith, S., Scott, G., McVeigh, E.R., Lederman, R.J., 2003. Magnetic resonance fluoroscopy allows targeted delivery of mesenchymal stem cells to infarct borders in Swine. Circulation 108, 2899–2904.

Dimmeler, S., Zeiher, A.M., Schneider, M.D., 2005. Unchain my heart: the scientific foundations of cardiac repair. J Clin Invest 115, 572–583.

Dubois, J., Lethimonnier, F., Vaufrey, F., Robert, P., Le Bihan, D., 2005. Frequency-shift based detection of BMS contrast agents using SSFP: potential for MRA. Magn Reson Imaging 23, 453–462.

Dunning, M.D., Kettunen, M.I., Ffrench Constant, C., Franklin, R.J.M., Brindle, K.M., 2006. Magnetic resonance imaging of functional Schwann cell transplants labelled with magnetic microspheres. Neuroimage 31, 172–180.

Empedocles, S., Bawendi, M., 1999. Spectroscopy of single CdSe nanocrystallites. Acc Chem Res 32, 389–396.

Engler, A.J., Sen, S., Sweeney, H.L., Discher, D.E., 2006. Matrix elasticity directs stem cell lineage specification. Cell 126, 677–689.

Enzmann, G.U., Benton, R.L., Talbott, J.F., Cao, Q., Whittemore, S.R., 2006. Functional considerations of stem cell transplantation therapy for spinal cord repair. J Neurotrauma 23, 479–495.

Frangioni, J.V., 2003. In vivo near-infrared fluorescence imaging. Curr Opin Chem Biol 7, 626–634.

Frank, J.A., Miller, B.R., Arbab, A.S., Zywicke, H.A., Jordan, E.K., Lewis, B.K., Bryant, L.H., Jr., Bulte, J.W., 2003. Clinically applicable labeling of mammalian and stem cells by combining superparamagnetic iron oxides and transfection agents. Radiology 228, 480–487.

Frank, J.A., Zywicke, H., Jordan, E.K., Mitchell, J., Lewis, B.K., Miller, B., Bryant, L.H., Jr., Bulte, J.W., 2002. Magnetic intracellular labeling of mammalian cells by combining (FDA-approved) superparamagnetic iron oxide MR contrast agents and commonly used transfection agents. Acad Radiol 9 Suppl 2, S484–S487.

Franklin, R.J., Blaschuk, K.L., Bearchell, M.C., Prestoz, L.L., Setzu, A., Brindle, K.M., ffrench-Constant, C., 1999. Magnetic resonance imaging of transplanted oligodendrocyte precursors in the rat brain. Neuroreport 10, 3961–3965.

Fu, R., Brey, W.W., Shetty, K., Gor'kov, P., Saha, S., Long, J.R., Grant, S.C., Chekmenev, E.Y., Hu, J., Gan, Z., Sharma, M., Zhang, F., Logan, T.M., Bruschweller, R., Edison, A., Blue, A., Dixon, I.R., Markiewicz, W.D., Cross, T.A., 2005. Ultra-wide bore 900 MHz high-resolution NMR at the National High Magnetic Field Laboratory. J Magn Reson 177, 1–8.

Galvez, B.G., Sampaolesi, M., Brunelli, S., Covarello, D., Gavina, M., Rossi, B., Costantin, G., Torrente, Y., Cossu, G., 2006. Complete repair of dystrophic skeletal muscle by mesoangioblasts with enhanced migration ability. J Cell Biol 174, 231–243.

Gao, X., Chan, W.C., Nie, S., 2002. Quantum-dot nanocrystals for ultrasensitive biological labeling and multicolor optical encoding. J Biomed Opt 7, 532–537.

Gao, X., Cui, Y., Levenson, R.M., Chung, L.W., Nie, S., 2004. In vivo cancer targeting and imaging with semiconductor quantum dots. Nat Biotechnol 22, 969–976.

Gao, X., Nie, S., 2003. Molecular profiling of single cells and tissue specimens with quantum dots. Trends Biotechnol 21, 371–373.

Gao, X., Yang, L., Petros, J.A., Marshall, F.F., Simons, J.W., Nie, S., 2005. In vivo molecular and cellular imaging with quantum dots. Curr Opin Biotechnol 16, 63–72.

Gaponenko, S.V., 1998. Optical properties of semiconductor nanocrystals. Cambridge Unviersity Press, Cambridge, UK; New York, NY, USA.

Garden, O.A., Reynolds, P.R., Yates, J., Larkman, D.J., Marelli-Berg, F.M., Haskard, D.O., Edwards, A.D., George, A.J., 2006. A rapid method for labelling CD4(+) T cells with ultrasmall paramagnetic iron oxide nanoparticles for magnetic resonance imaging that preserves proliferative, regulatory and migratory behaviour in vitro. J Immunol Methods 314, 123–133.

Garot, J., Unterseeh, T., Teiger, E., Champagne, S., Chazaud, B., Gherardi, R., Hittinger, L., Gueret, P., Rahmouni, A., 2003. Magnetic resonance imaging of targeted catheter-based implantation of myogenic precursor cells into infarcted left ventricular myocardium. J Am Coll Cardiol 41, 1841–1846.

Ghostine, S., Carrion, C., Souza, L.C., Richard, P., Bruneval, P., Vilquin, J.T., Pouzet, B., Schwartz, K., Menasche, P., Hagege, A.A., 2002. Long-term efficacy of myoblast transplantation on regional structure and function after myocardial infarction. Circulation 106, I131–136.

Gillies, R.J., Raghunand, N., Garcia-Martin, M.L., Gatenby, R.A., 2004. pH imaging. A review of pH measurement methods and applications in cancers. IEEE Eng Med Biol Mag 23, 57–64.

Gilman, S., 2006. Pharmacologic management of ischemic stroke: Relevance to stem cell therapy. Exp Neurol 199, 28–36.

Gimi, B., Mori, N., Ackerstaff, E., Frost, E.E., Bulte, J.W., Bhujwalla, Z.M., 2006. Noninvasive MRI of endothelial cell response to human breast cancer cells. Neoplasia 8, 207–213.

Gupta, A.K., Gupta, M., 2005. Synthesis and surface engineering of iron oxide nanoparticles for biomedical applications. Biomaterials 26, 3995–4021.

Hadjantonakis, A.K., Dickinson, M.E., Fraser, S.E., Papaioannou, V.E., 2003. Technicolour transgenics: imaging tools for functional genomics in the mouse. Nat Rev Genet 4, 613–625.

Haraldseth, O., Jones, R.A., Muller, T.B., Fahlvik, A.K., Oksendal, A.N., 1996. Comparison of dysprosium DTPA BMA and superparamagnetic iron oxide particles as susceptibility contrast agents for perfusion imaging of regional cerebral ischemia in the rat. J Magn Reson Imaging 6, 714–717.

Hauger, O., Frost, E.E., van Heeswijk, R., Deminiere, C., Xue, R., Delmas, Y., Combe, C., Moonen, C.T., Grenier, N., Bulte, J.W., 2006. MR evaluation of the glomerular homing of magnetically labeled mesenchymal stem cells in a rat model of nephropathy. Radiology 238, 200–210.

He, G., Hao, Z., Wei, H., Yi, W., Zhang, X., Yue, T., Wei, Y., Hu, S., 2007. In vivo imaging of bone marrow mesenchymal stem cells transplanted into myocardium using magnetic resonance imaging: A novel method to trace the transplanted cells. Int J Cardiol 114, 4–10.

Hekmatyar, S.K., Hopewell, P., Pakin, S.K., Babsky, A., Bansal, N., 2005. Noninvasive MR thermometry using paramagnetic lanthanide complexes of 1,4,7,10-tetraazacyclodoecane-alpha,alpha′,alpha″,alpha‴-tetramethyl-1, 4,7,10-tetraacetic acid (DOTMA4-). Magn Reson Med 53, 294–303.

Heyn, C., Ronald, J.A., Mackenzie, L.T., MacDonald, I.C., Chambers, A.F., Rutt, B.K., Foster, P.J., 2006. In vivo magnetic resonance imaging of single cells in mouse brain with optical validation. Magn Reson Med 55, 23–29.

Himes, N., Min, J.Y., Lee, R., Brown, C., Shea, J., Huang, X., Xiao, Y.F., Morgan, J.P., Burstein, D., Oettgen, P., 2004. In vivo MRI of embryonic stem cells in a mouse model of myocardial infarction. Magn Reson Med 52, 1214–1219.

Himmelreich, U., Weber, R., Ramos-Cabrer, P., Wegener, S., Kandal, K., Shapiro, E.M., Koretsky, A.P., Hoehn, M., 2005. Improved stem cell MR detectability in animal models by modification of the inhalation gas. Mol Imaging 4, 104–109.

Hinds, K.A., Hill, J.M., Shapiro, E.M., Laukkanen, M.O., Silva, A.C., Combs, C.A., Varney, T.R., Balaban, R.S., Koretsky, A.P., Dunbar, C.E., 2003. Highly efficient endosomal labeling of progenitor and stem cells with large magnetic particles allows magnetic resonance imaging of single cells. Blood 102, 867–872.

Hoehn, M., Kustermann, E., Blunk, J., Wiedermann, D., Trapp, T., Wecker, S., Focking, M., Arnold, H., Hescheler, J., Fleischmann, B.K., Schwindt, W., Buhrle, C., 2002. Monitoring of implanted stem cell migration in vivo: a highly resolved in vivo magnetic resonance imaging investigation of experimental stroke in rat. Proc Natl Acad Sci U S A 99, 16267–16272.

Honma, T., Honmou, O., Iihoshi, S., Harada, K., Houkin, K., Hamada, H., Kocsis, J.D., 2006. Intravenous infusion of immortalized human mesenchymal stem cells protects against injury in a cerebral ischemia model in adult rat. Exp Neurol 199, 56–66.

Huard, J., Acsadi, G., Jani, A., Massie, B., Karpati, G., 1994. Gene transfer into skeletal muscles by isogenic myoblasts. Hum Gene Ther 5, 949–958.

Huard, J., Cao, B., Qu-Petersen, Z., 2003. Muscle-derived stem cells: potential for muscle regeneration. Birth Defects Res Part C Embryo Today 69, 230–237.

Huber, M.M., Staubli, A.B., Kustedjo, K., Gray, M.H., Shih, J., Fraser, S.E., Jacobs, R.E., Meade, T.J., 1998. Fluorescently detectable magnetic resonance imaging agents. Bioconjug Chem 9, 242–249.

Hudde, T., Rayner, S.A., Comer, R.M., Weber, M., Isaacs, J.D., Waldmann, H., Larkin, D.F., George, A.J., 1999. Activated polyamidoamine dendrimers, a non-viral vector for gene transfer to the corneal endothelium. Gene Ther 6, 939–943.

Ikezawa, M., Cao, B., Qu, Z., Peng, H., Xiao, X., Pruchnic, R., Kimura, S., Miike, T., Huard, J., 2003. Dystrophin delivery in dystrophin-deficient DMDmdx skeletal muscle by isogenic muscle-derived stem cell transplantation. Hum Gene Ther 14, 1535–1546.

Ittrich, H., Lange, C., Dahnke, H., Zander, A.R., Adam, G., Nolte-Ernsting, C., 2005. [Labeling of mesenchymal stem cells with different superparamagnetic particles of iron oxide and detectability with MRI at 3T]. Rofo 177, 1151–1163.

Jankowski, R.J., Deasy, B.M., Huard, J., 2002. Muscle-derived stem cells. Gene Ther 9, 642–647.

Jendelova, P., Herynek, V., DeCroos, J., Glogarova, K., Andersson, B., Hajek, M., Sykova, E., 2003. Imaging the fate of implanted bone marrow stromal cells labeled with superparamagnetic nanoparticles. Magn Reson Med 50, 767–776.

Jendelova, P., Herynek, V., Urdzikova, L., Glogarova, K., Kroupova, J., Andersson, B., Bryja, V., Burian, M., Hajek, M., Sykova, E., 2004. Magnetic resonance tracking of transplanted bone marrow and embryonic stem cells labeled by iron oxide nanoparticles in rat brain and spinal cord. J Neurosci Res 76, 232–243.

Jendelova, P., Herynek, V., Urdzikova, L., Glogarova, K., Rahmatova, S., Fales, I., Andersson, B., Prochazka, P., Zamecnik, J., Eckschlager, T., Kobylka, P., Hajek, M., Sykova, E., 2005. Magnetic resonance tracking of human CD34+ progenitor cells separated by means of immunomagnetic selection and transplanted into injured rat brain. Cell Transplant 14, 173–182.

Jiang, Q., Zhang, Z.G., Ding, G.L., Silver, B., Zhang, L., Meng, H., Lu, M., Pourabdillah-Nejed, D.S., Wang, L., Savant-Bhonsale, S., Li, L., Bagher-Ebadian, H., Hu, J., Arbab, A.S., Vanguri, P., Ewing, J.R., Ledbetter, K.A., Chopp, M., 2006. MRI detects white matter reorganization after neural progenitor cell treatment of stroke. Neuroimage 32, 1080–1089.

Jiang, Q., Zhang, Z.G., Ding, G.L., Zhang, L., Ewing, J.R., Wang, L., Zhang, R., Li, L., Lu, M., Meng, H., Arbab, A.S., Hu, J., Li, Q.J., Pourabdollah Nejad, D.S., Athiraman, H., Chopp, M., 2005. Investigation of neural progenitor cell induced angiogenesis after embolic stroke in rat using MRI. Neuroimage 28, 698–707.

Jiang, W., Papa, E., Fischer, H., Mardyani, S., Chan, W.C., 2004. Semiconductor quantum dots as contrast agents for whole animal imaging. Trends Biotechnol 22, 607–609.

Jobsis, F.F., 1977. Noninvasive, infrared monitoring of cerebral and myocardial oxygen sufficiency and circulatory parameters. Science 198, 1264–1267.

Johnson, K.M., Tao, J.Z., Kennan, R.P., Gore, J.C., 2000. Intravascular susceptibility agent effects on tissue transverse relaxation rates in vivo. Magn Reson Med 44, 909–914.

Ju, S., Teng, G., Zhang, Y., Ma, M., Chen, F., Ni, Y., 2006. In vitro labeling and MRI of mesenchymal stem cells from human umbilical cord blood. Magn Reson Imaging 24, 611–617.

Kellar, K.E., Fossheim, S.L., Koenig, S.H., 1998. Magnetic field dependence of solvent proton relaxation by solute dysprosium (III) complexes. Invest Radiol 33, 835–840.

Kessler, P.D., Byrne, B.J., 1999. Myoblast cell grafting into heart muscle: cellular biology and potential applications. Annu Rev Physiol 61, 219–242.

Kim, S., Lim, Y.T., Soltesz, E.G., De Grand, A.M., Lee, J., Nakayama, A., Parker, J.A., Mihaljevic, T., Laurence, R.G., Dor, D.M., Cohn, L.H., Bawendi, M.G., Frangioni, J.V., 2004. Near-infrared fluorescent type II quantum dots for sentinel lymph node mapping. Nat Biotechnol 22, 93–97.

Kircher, M.F., Allport, J.R., Graves, E.E., Love, V., Josephson, L., Lichtman, A.H., Weissleder, R., 2003. In vivo high resolution three-dimensional imaging of antigen-specific cytotoxic T-lymphocyte trafficking to tumors. Cancer Res 63, 6838–6846.

Kirik, D., Breysse, N., Bjorklund, T., Besret, L., Hantraye, P., 2005. Imaging in cell-based therapy for neurodegenerative diseases. Eur J Nucl Med Mol Imaging 32 Suppl 2, S417–S434.

Kraitchman, D.L., Heldman, A.W., Atalar, E., Amado, L.C., Martin, B.J., Pittenger, M.F., Hare, J.M., Bulte, J.W., 2003. In vivo magnetic resonance imaging of mesenchymal stem cells in myocardial infarction. Circulation 107, 2290–2293.

Kraitchman, D.L., Tatsumi, M., Gilson, W.D., Ishimori, T., Kedziorek, D., Walczak, P., Segars, W.P., Chen, H.H., Fritzges, D., Izbudak, I., Young, R.G., Marcelino, M., Pittenger, M.F., Solaiyappan, M., Boston, R.C., Tsui, B.M., Wahl, R.L., Bulte, J.W., 2005. Dynamic imaging of allogeneic mesenchymal stem cells trafficking to myocardial infarction. Circulation 112, 1451–1461.

Kukowska-Latallo, J.F., Bielinska, A.U., Johnson, J., Spindler, R., Tomalia, D.A., Baker, J.R., Jr., 1996. Efficient transfer of genetic material into mammalian cells using Starburst polyamidoamine dendrimers. Proc Natl Acad Sci U S A 93, 4897–4902.

Kustermann, E., Roell, W., Breitbach, M., Wecker, S., Wiedermann, D., Buehrle, C., Welz, A., Hescheler, J., Fleischmann, B.K., Hoehn, M., 2005. Stem cell implantation in ischemic mouse heart: a high-resolution magnetic resonance imaging investigation. NMR Biomed 18, 362–370.

Laflamme, M.A., Murry, C.E., 2005. Regenerating the heart. Nat Biotechnol 23, 845–856.

Lakshmipathy, U., Verfaillie, C., 2005. Stem cell plasticity. Blood Reviews 19, 29–38.

Lanza, G.M., Winter, P., Caruthers, S., Schmeider, A., Crowder, K., Morawski, A., Zhang, H., Scott, M.J., Wickline, S.A., 2004a. Novel paramagnetic contrast agents for molecular imaging and targeted drug delivery. Curr Pharm Biotechnol 5, 495–507.
Lanza, G.M., Winter, P.M., Caruthers, S.D., Morawski, A.M., Schmieder, A.H., Crowder, K.C., Wickline, S.A., 2004b. Magnetic resonance molecular imaging with nanoparticles. J Nucl Cardiol 11, 733–743.
Lapidot, T., Dar, A., Kollet, O., 2005. How do stem cells find their way home? Blood 106, 1901–1910.
Laywell, E.D., Kukekov, V.G., Steindler, D.A., 1999. Multipotent neurospheres can be derived from forebrain subependymal zone and spinal cord of adult mice after protracted postmortem intervals. Exp Neurol 156, 430–433.
Laywell, E.D., Rakic, P., Kukekov, V.G., Holland, E.C., Steindler, D.A., 2000. Identification of a multipotent astrocytic stem cell in the immature and adult mouse brain. Proc Natl Acad Sci U S A 97, 13883–13888.
Laywell, E.D., Steindler, D.A., 2002. Glial stem-like cells: implications for ontogeny, phylogeny, and CNS regeneration. Prog Brain Res 138, 435–450.
Lee, I.H., Bulte, J.W., Schweinhardt, P., Douglas, T., Trifunovski, A., Hofstetter, C., Olson, L., Spenger, C., 2004. In vivo magnetic resonance tracking of olfactory ensheathing glia grafted into the rat spinal cord. Exp Neurol 187, 509–516.
Lee, J.Y., Qu-Petersen, Z., Cao, B., Kimura, S., Jankowski, R., Cummins, J., Usas, A., Gates, C., Robbins, P., Wernig, A., Huard, J., 2000. Clonal isolation of muscle-derived cells capable of enhancing muscle regeneration and bone healing. J Cell Biol 150, 1085–1100.
Leor, J., Patterson, M., Quinones, M.J., Kedes, L.H., Kloner, R.A., 1996. Transplantation of fetal myocardial tissue into the infarcted myocardium of rat. A potential method for repair of infarcted myocardium? Circulation 94, II332–336.
Leor, J., Rozen, L., Zuloff-Shani, A., Feinberg, M.S., Amsalem, Y., Barbash, I.M., Kachel, E., Holbova, R., Mardor, Y., Daniels, D., Ocherashvilli, A., Orenstein, A., Danon, D., 2006. Ex vivo activated human macrophages improve healing, remodeling, and function of the infarcted heart. Circulation 114, I94–100.
Lepore, A.C., Walczak, P., Rao, M.S., Fischer, I., Bulte, J.W., 2006. MR imaging of lineage-restricted neural precursors following transplantation into the adult spinal cord. Exp Neurol 201, 49–59.
Leri, A., Kajstura, J., Anversa, P., 2005. Cardiac stem cells and mechanisms of myocardial regeneration. Physiol Rev 85, 1373–1416.
Lewin, M., Carlesso, N., Tung, C.H., Tang, X.W., Cory, D., Scadden, D.T., Weissleder, R., 2000. Tat peptide-derivatized magnetic nanoparticles allow in vivo tracking and recovery of progenitor cells. Nat Biotechnol 18, 410–414.
Lim, Y.T., Kim, S., Nakayama, A., Stott, N.E., Bawendi, M.G., Frangioni, J.V., 2003. Selection of quantum dot wavelengths for biomedical assays and imaging. Mol Imaging 2, 50–64.
Lin, Z., Cui, S., Zhang, H., Chen, Q., Yang, B., Su, X., Zhang, J., Jin, Q., 2003. Studies on quantum dots synthesized in aqueous solution for biological labeling via electrostatic interaction. Anal Biochem 319, 239–243.
Lindvall, O., Kokaia, Z., 2006. Stem cells for the treatment of neurological disorders. Nature 441, 1094–1096.
Liu, M., Guo, Y.M., Wu, Q.F., Yang, J.L., Wang, P., Wang, S.C., Guo, X.J., Qiang, Y.Q., Duan, X.Y., 2006. Paramagnetic particles carried by cell-penetrating peptide tracking of bone marrow mesenchymal stem cells, a research in vitro. Biochem Biophys Res Commun 347, 133–140.

Louie, A.Y., Huber, M.M., Ahrens, E.T., Rothbacher, U., Moats, R., Jacobs, R.E., Fraser, S.E., Meade, T.J., 2000. In vivo visualization of gene expression using magnetic resonance imaging. Nat Biotechnol 18, 321–325.

Lyons, A.B., Parish, C.R., 1994. Determination of lymphocyte division by flow cytometry. J Immunol Methods 171, 131–137.

Magnitsky, S., Watson, D.J., Walton, R.M., Pickup, S., Bulte, J.W., Wolfe, J.H., Poptani, H., 2005. In vivo and ex vivo MRI detection of localized and disseminated neural stem cell grafts in the mouse brain. Neuroimage 26, 744–754.

Mani, V., Briley-Saebo, K.C., Itskovich, V.V., Samber, D.D., Fayad, Z.A., 2006. Gradient echo acquisition for superparamagnetic particles with positive contrast (GRASP): sequence characterization in membrane and glass superparamagnetic iron oxide phantoms at 1.5T and 3T. Magn Reson Med 55, 126–135.

Mantyla, T., Hakumaki, J.M., Huhtala, T., Narvanen, A., Yla-Herttuala, S., 2006. Targeted magnetic resonance imaging of Scavidin-receptor in human umbilical vein endothelial cells in vitro. Magn Reson Med 55, 800–804.

Maronpot, R.R., Sills, R.C., Johnson, G.A., 2004. Applications of magnetic resonance microscopy. Toxicol Pathol 32 Suppl 2, 42–48.

Massoud, T.F., Gambhir, S.S., 2003. Molecular imaging in living subjects: seeing fundamental biological processes in a new light. Genes Dev 17, 545–580.

Masuda, C., Maki, Z., Morikawa, S., Morita, M., Inubushi, T., Matsusue, Y., Yamagata, S., Taguchi, H., Doi, Y., Shirai, N., Hirao, K., Tooyama, I., 2006. MR tracking of transplanted glial cells using poly-l-lysine-CF(3). Neurosci Res.

Mathur, A., Martin, J.F., 2004. Stem cells and repair of the heart. The Lancet 364, 183–192.

Matuszewski, L., Persigehl, T., Wall, A., Schwindt, W., Tombach, B., Fobker, M., Poremba, C., Ebert, W., Heindel, W., Bremer, C., 2005. Cell tagging with clinically approved iron oxides: feasibility and effect of lipofection, particle size, and surface coating on labeling efficiency. Radiology 235, 155–161.

Mayer-Kuckuk, P., Gade, T.P., Buchanan, I.M., Doubrovin, M., Ageyeva, L., Bertino, J.R., Boskey, A.L., Blasberg, R.G., Koutcher, J.A., Banerjee, D., 2005. High-resolution imaging of bone precursor cells within the intact bone marrow cavity of living mice. Mol Ther 12, 33–41.

Medintz, I.L., Uyeda, H.T., Goldman, E.R., Mattoussi, H., 2005. Quantum dot bioconjugates for imaging, labelling and sensing. Nat Mater 4, 435–446.

Menasche, P., Hagege, A.A., Scorsin, M., Pouzet, B., Desnos, M., Duboc, D., Schwartz, K., Vilquin, J.T., Marolleau, J.P., 2001. Myoblast transplantation for heart failure. Lancet 357, 279–280.

Metz, S., Bonaterra, G., Rudelius, M., Settles, M., Rummeny, E.J., Daldrup-Link, H.E., 2004. Capacity of human monocytes to phagocytose approved iron oxide MR contrast agents in vitro. Eur Radiol 14, 1851–1858.

Michalet, X., Pinaud, F.F., Bentolila, L.A., Tsay, J.M., Doose, S., Li, J.J., Sundaresan, G., Wu, A.M., Gambhir, S.S., Weiss, S., 2005. Quantum dots for live cells, in vivo imaging, and diagnostics. Science 307, 538–544.

Min, J.J., Gambhir, S.S., 2004. Gene therapy progress and prospects. Gene Ther 11, 115–125.

Mintz-Hittner, H.A., Semina, E.V., Frishman, L.J., Prager, T.C., Murray, J.C., 2004. VSX1 (RINX) mutation with craniofacial anomalies, empty sella, corneal endothelial changes, and abnormal retinal and auditory bipolar cells. Ophthalmology 111, 828–836.

Miyoshi, S., Flexman, J.A., Cross, D.J., Maravilla, K.R., Kim, Y., Anzai, Y., Oshima, J., Minoshima, S., 2005. Transfection of neuroprogenitor cells with iron nanoparticles for magnetic resonance imaging tracking: cell viability, differentiation, and intracellular localization. Mol Imaging Biol 7, 286–295.

Modo, M., Cash, D., Mellodew, K., Williams, S.C., Fraser, S.E., Meade, T.J., Price, J., Hodges, H., 2002. Tracking transplanted stem cell migration using bifunctional, contrast agent-enhanced, magnetic resonance imaging. Neuroimage 17, 803–811.

Modo, M., Hoehn, M., Bulte, J.W., 2005. Cellular MR imaging. Mol Imaging 4, 143–164.

Modo, M., Mellodew, K., Cash, D., Fraser, S.E., Meade, T.J., Price, J., Williams, S.C., 2004. Mapping transplanted stem cell migration after a stroke: a serial, in vivo magnetic resonance imaging study. Neuroimage 21, 311–317.

Montet-Abou, K., Montet, X., Weissleder, R., Josephson, L., 2005. Transfection agent induced nanoparticle cell loading. Mol Imaging 4, 165–171.

Moore, K.A., Lemischka, I.R., 2006. Stem cells and their niches. Science 311, 1880–1885.

Morawski, A.M., Lanza, G.A., Wickline, S.A., 2005. Targeted contrast agents for magnetic resonance imaging and ultrasound. Curr Opin Biotechnol 16, 89–92.

Morawski, A.M., Winter, P.M., Crowder, K.C., Caruthers, S.D., Fuhrhop, R.W., Scott, M.J., Robertson, J.D., Abendschein, D.R., Lanza, G.M., Wickline, S.A., 2004. Targeted nanoparticles for quantitative imaging of sparse molecular epitopes with MRI. Magn Reson Med 51, 480–486.

Mulder, W.J., Koole, R., Brandwijk, R.J., Storm, G., Chin, P.T., Strijkers, G.J., de Mello Donega, C., Nicolay, K., Griffioen, A.W., 2006a. Quantum dots with a paramagnetic coating as a bimodal molecular imaging probe. Nano Lett 6, 1–6.

Mulder, W.J., Strijkers, G.J., van Tilborg, G.A., Griffioen, A.W., Nicolay, K., 2006b. Lipid-based nanoparticles for contrast-enhanced MRI and molecular imaging. NMR Biomed 19, 142–164.

Muller, F.J., Snyder, E.Y., Loring, J.F., 2006. Gene therapy: can neural stem cells deliver? Nat Rev Neurosci 7, 75–84.

Ntziachristos, V., Ripoll, J., Wang, L.V., Weissleder, R., 2005. Looking and listening to light: the evolution of whole-body photonic imaging. Nat Biotechnol 23, 313–320.

Oshima, H., Payne, T.R., Urish, K.L., Sakai, T., Ling, Y., Gharaibeh, B., Tobita, K., Keller, B.B., Cummins, J.H., Huard, J., 2005. Differential myocardial infarct repair with muscle stem cells compared to myoblasts. Mol Ther 12, 1130–1141.

Othman, S.F., Xu, H., Royston, T.J., Magin, R.L., 2005. Microscopic magnetic resonance elastography (microMRE). Magn Reson Med 54, 605–615.

Palmer, G.D., Steinert, A., Pascher, A., Gouze, E., Gouze, J.N., Betz, O., Johnstone, B., Evans, C.H., Ghivizzani, S.C., 2005. Gene-induced chondrogenesis of primary mesenchymal stem cells in vitro. Mol Ther 12, 219–228.

Parungo, C.P., Colson, Y.L., Kim, S.W., Kim, S., Cohn, L.H., Bawendi, M.G., Frangioni, J.V., 2005. Sentinel lymph node mapping of the pleural space. Chest 127, 1799–1804.

Payne, T.R., Oshima, H., Sakai, T., Ling, Y., Gharaibeh, B., Cummins, J., Huard, J., 2005. Regeneration of dystrophin-expressing myocytes in the mdx heart by skeletal muscle stem cells. Gene Ther 12, 1264–1274.

Pearse, D.D., Bunge, M.B., 2006. Designing cell- and gene-based regeneration strategies to repair the injured spinal cord. J Neurotrauma 23, 438–452.

Pinaud, F., Michalet, X., Bentolila, L.A., Tsay, J.M., Doose, S., Li, J.J., Iyer, G., Weiss, S., 2006. Advances in fluorescence imaging with quantum dot bio-probes. Biomaterials 27, 1679–1687.

Pirko, I., Fricke, S.T., Johnson, A.J., Rodriguez, M., Macura, S.I., 2005. Magnetic resonance imaging, microscopy, and spectroscopy of the central nervous system in experimental animals. NeuroRx 2, 250–264.

Pittenger, M.F., Mosca, J.D., McIntosh, K.R., 2000. Human mesenchymal stem cells: progenitor cells for cartilage, bone, fat and stroma. Curr Top Microbiol Immunol 251, 3–11.

Qhobosheane, M., Santra, S., Zhang, P., Tan, W., 2001. Biochemically functionalized silica nanoparticles. Analyst 126, 1274–1278.

Qu-Petersen, Z., Deasy, B., Jankowski, R., Ikezawa, M., Cummins, J., Pruchnic, R., Mytinger, J., Cao, B., Gates, C., Wernig, A., Huard, J., 2002. Identification of a novel population of muscle stem cells in mice: potential for muscle regeneration. J Cell Biol 157, 851–864.

Quesenberry, P.J., Abedi, M., Aliotta, J., Colvin, G., Demers, D., Dooner, M., Greer, D., Hebert, H., Menon, M.K., Pimentel, J., Paggioli, D., 2004. Stem cell plasticity: an overview. Blood Cells, Molecules, and Diseases 32, 1–4.

Rando, T.A., 2006. Stem cells, ageing and the quest for immortality. Nature 441, 1080–1086.

Renshaw, P.F., Owen, C.S., Evans, A.E., Leigh, J.S., Jr., 1986. Immunospecific NMR contrast agents. Magn Reson Imaging 4, 351–357.

Riviere, C., Boudghene, F.P., Gazeau, F., Roger, J., Pons, J.-N., Laissy, J.-P., Allaire, E., Michel, J.-B., Letourneur, D., Deux, J.-F., 2005. Iron Oxide Nanoparticle-labeled Rat Smooth Muscle Cells: Cardiac MR Imaging for Cell Graft Monitoring and Quantitation. Radiology 235, 959–967.

Robinson, S.W., Cho, P.W., Levitsky, H.I., Olson, J.L., Hruban, R.H., Acker, M.A., Kessler, P.D., 1996. Arterial delivery of genetically labelled skeletal myoblasts to the murine heart: long-term survival and phenotypic modification of implanted myoblasts. Cell Transplant 5, 77–91.

Rudin, M., Weissleder, R., 2003. Molecular imaging in drug discovery and development. Nat Rev Drug Discov 2, 123–131.

Saeed, M., Saloner, D., Weber, O., Martin, A., Henk, C., Higgins, C., 2005. MRI in guiding and assessing intramyocardial therapy. Eur Radiol 15, 851–863.

Sakai, T., Ling, Y., Payne, T.R., Huard, J., 2002. The use of ex vivo gene transfer based on muscle-derived stem cells for cardiovascular medicine. Trends Cardiovasc Med 12, 115–120.

Sampaolesi, M., Torrente, Y., Innocenzi, A., Tonlorenzi, R., D'Antona, G., Pellegrino, M.A., Barresi, R., Bresolin, N., De Angelis, M.G., Campbell, K.P., Bottinelli, R., Cossu, G., 2003. Cell therapy of alpha-sarcoglycan null dystrophic mice through intra-arterial delivery of mesoangioblasts. Science 301, 487–492.

Santra, S., Bagwe, R.P., Dutta, D., Stanley, J.T., Walter, G., Tan, W., Moudgil, B., Mericle, R.A., 2005a. Synthesis and Characterization of Fluorescent, Radio-Opaque, and Paramagnetic Silica Nanoparticles for Multimodal Bioimaging Applications. Adv Mater 17, 2165–2169.

Santra, S., Dutta, D., Walter, G.A., Moudgil, B.M., 2005b. Fluorescent nanoparticle probes for cancer imaging. Technol Cancer Res Treat 4, 593–602.

Santra, S., Wang, K., Tapec, R., Tan, W., 2001a. Development of novel dye-doped silica nanoparticles for biomarker application. J Biomed Opt 6, 160–166.

Santra, S., Yang, H., Dutta, D., Stanley, J.T., Holloway, P.H., Tan, W., Moudgil, B.M., Mericle, R.A., 2004. TAT conjugated, FITC doped silica nanoparticles for bioimaging applications. Chem Commun (Camb), 2810–2811.

Santra, S., Yang, H., Stanley, J.T., Holloway, P.H., Moudgil, B.M., Walter, G., Mericle, R.A., 2005c. Rapid and effective labeling of brain tissue using TAT-conjugated CdS:Mn/ZnS quantum dots. Chem Commun (Camb), 3144–3146.

Santra, S., Zhang, P., Wang, K., Tapec, R., Tan, W., 2001b. Conjugation of biomolecules with luminophore-doped silica nanoparticles for photostable biomarkers. Anal Chem 73, 4988–4993.

Scheffler, B., Horn, M., Blumcke, I., Laywell, E.D., Coomes, D., Kukekov, V.G., Steindler, D.A., 1999. Marrow-mindedness: a perspective on neuropoiesis. Trends Neurosci 22, 348–357.

Scheffler, B., Walton, N.M., Lin, D.D., Goetz, A.K., Enikolopov, G., Roper, S.N., Steindler, D.A., 2005. Phenotypic and functional characterization of adult brain neuropoiesis. Proc Natl Acad Sci U S A 102, 9353–9358.

Schmieder, A.H., Winter, P.M., Caruthers, S.D., Harris, T.D., Williams, T.A., Allen, J.S., Lacy, E.K., Zhang, H., Scott, M.J., Hu, G., Robertson, J.D., Wickline, S.A., Lanza, G.M., 2005. Molecular MR imaging of melanoma angiogenesis with alphanubeta3-targeted paramagnetic nanoparticles. Magn Reson Med 53, 621–627.

Scorsin, M., Hagege, A.A., Marotte, F., Mirochnik, N., Copin, H., Barnoux, M., Sabri, A., Samuel, J.L., Rappaport, L., Menasche, P., 1997. Does transplantation of cardiomyocytes improve function of infarcted myocardium? Circulation 96, II-188–193.

Scorsin, M., Hagege, A., Vilquin, J.T., Fiszman, M., Marotte, F., Samuel, J.L., Rappaport, L., Schwartz, K., Menasche, P., 2000. Comparison of the effects of fetal cardiomyocyte and skeletal myoblast transplantation on postinfarction left ventricular function. J Thorac Cardiovasc Surg 119, 1169–1175.

Scorsin, M., Marotte, F., Sabri, A., Le Dref, O., Demirag, M., Samuel, J.L., Rappaport, L., Menasche, P., 1996. Can grafted cardiomyocytes colonize peri-infarct myocardial areas? Circulation 94, II337–340.

Sevick-Muraca, E.M., Houston, J.P., Gurfinkel, M., 2002. Fluorescence-enhanced, near infrared diagnostic imaging with contrast agents. Curr Opin Chem Biol 6, 642–650.

Shah, D.S., Sakthivel, T., Toth, I., Florence, A.T., Wilderspin, A.F., 2000. DNA transfection and transfected cell viability using amphipathic asymmetric dendrimers. Int J Pharm 208, 41–48.

Shapiro, E.M., Gonzalez-Perez, O., Manuel Garcia-Verdugo, J., Alvarez-Buylla, A., Koretsky, A.P., 2006a. Magnetic resonance imaging of the migration of neuronal precursors generated in the adult rodent brain. Neuroimage 32:1150–1157.

Shapiro, E.M., Sharer, K., Skrtic, S., Koretsky, A.P., 2006b. In vivo detection of single cells by MRI. Magn Reson Med 55, 242–249.

Shapiro, E.M., Skrtic, S., Koretsky, A.P., 2005. Sizing it up: cellular MRI using micron-sized iron oxide particles. Magn Reson Med 53, 329–338.

Sharma, P., Brown, S., Walter, G., Santra, S., Moudgil, B., 2006. Nanoparticles for bioimaging. Adv Colloid Interface Sci 16:123–126.

Sitharaman, B., Kissell, K.R., Hartman, K.B., Tran, L.A., Baikalov, A., Rusakova, I., Sun, Y., Khant, H.A., Ludtke, S.J., Chiu, W., Laus, S., Toth, E., Helm, L., Merbach, A.E., Wilson, L.J., 2005. Superparamagnetic gadonanotubes are high-performance MRI contrast agents. Chem Commun (Camb), 3915–3917.

Skuk, D., Tremblay, J.P., 2003. Cell therapies for inherited myopathies. Curr Opin Rheumatol 15, 723–729.

Smirnov, P., Gazeau, F., Beloeil, J.-C., Doan, B.T., Wilhelm, C., Gillet, B., 2006. Single-cell detection by gradient echo 9.4 T MRI: a parametric study. Contrast Media & Molecular Imaging 1, 165–174.

Smith, A.M., Gao, X., Nie, S., 2004. Quantum dot nanocrystals for in vivo molecular and cellular imaging. Photochem Photobiol 80, 377–385.

Smythe, G.M., Hodgetts, S.I., Grounds, M.D., 2000. Immunobiology and the future of myoblast transfer therapy. Mol Ther 1, 304–313.

Smythe, G.M., Hodgetts, S.I., Grounds, M.D., 2001. Problems and solutions in myoblast transfer therapy. J Cell Mol Med 5, 33–47.

Snoussi, K., Bulte, J.W., Gueron, M., van Zijl, P.C., 2003. Sensitive CEST agents based on nucleic acid imino proton exchange: detection of poly(rU) and of a dendrimer-poly(rU) model for nucleic acid delivery and pharmacology. Magn Reson Med 49, 998–1005.

Sosnovik, D.E., Schellenberger, E.A., Nahrendorf, M., Novikov, M.S., Matsui, T., Dai, G., Reynolds, F., Grazette, L., Rosenzweig, A., Weissleder, R., Josephson, L., 2005. Magnetic resonance imaging of cardiomyocyte apoptosis with a novel magneto-optical nanoparticle. Magn Reson Med 54, 718–724.

Steigerwald, M.L., Brus, L.E., 1990. Semiconductor Crystallites – a Class of Large Molecules. Accounts of Chemical Research 23, 183–188.

Steindler, D.A., 2002. Neural stem cells, scaffolds, and chaperones. Nat Biotechnol 20, 1091–1093.

Steindler, D.A., Kadrie, T., Fillmore, H., Thomas, L.B., 1996. The subependymal zone: "brain marrow". Prog Brain Res 108, 349–363.

Steindler, D.A., Pincus, D.W., 2002. Stem cells and neuropoiesis in the adult human brain. Lancet 359, 1047–1054.

Stroh, A., Faber, C., Neuberger, T., Lorenz, P., Sieland, K., Jakob, P.M., Webb, A., Pilgrimm, H., Schober, R., Pohl, E.E., Zimmer, C., 2005a. In vivo detection limits of magnetically labeled embryonic stem cells in the rat brain using high-field (17.6 T) magnetic resonance imaging. Neuroimage 24, 635–645.

Stroh, M., Zimmer, J.P., Duda, D.G., Levchenko, T.S., Cohen, K.S., Brown, E.B., Scadden, D.T., Torchilin, V.P., Bawendi, M.G., Fukumura, D., Jain, R.K., 2005b. Quantum dots spectrally distinguish multiple species within the tumor milieu in vivo. Nat Med 11, 678–682.

Stuckey, D.J., Carr, C.A., Martin-Rendon, E., Tyler, D.J., Willmott, C., Cassidy, P.J., Hale, S.J., Schneider, J.E., Tatton, L., Harding, S.E., Radda, G.K., Watt, S., Clarke, K., 2006. Iron particles for noninvasive monitoring of bone marrow stromal cell engraftment into, and isolation of viable engrafted donor cells from, the heart. Stem Cells 24, 1968–1975.

Sun, R., Dittrich, J., Le-Huu, M., Mueller, M.M., Bedke, J., Kartenbeck, J., Lehmann, W.D., Krueger, R., Bock, M., Huss, R., Seliger, C., Grone, H.J., Misselwitz, B., Semmler, W., Kiessling, F., 2005. Physical and biological characterization of superparamagnetic iron oxide- and ultrasmall superparamagnetic iron oxide-labeled cells: a comparison. Invest Radiol 40, 504–513.

Sykova, E., Jendelova, P., 2005. Magnetic resonance tracking of implanted adult and embryonic stem cells in injured brain and spinal cord. Ann N Y Acad Sci 1049, 146–160.

Sykova, E., Jendelova, P., 2006. Magnetic resonance tracking of transplanted stem cells in rat brain and spinal cord. Neurodegener Dis 3, 62–67.

Sykova, E., Jendelova, P., Urdzikova, L., Lesny, P., Hejcl, A., 2006. Bone marrow stem cells and polymer hydrogels-two strategies for spinal cord injury repair. Cell Mol Neurobiol 26:1113–1129.

Tallheden, T., Nannmark, U., Lorentzon, M., Rakotonirainy, O., Soussi, B., Waagstein, F., Jeppsson, A., Sjogren-Jansson, E., Lindahl, A., Omerovic, E., 2006. In vivo MR imaging of magnetically labeled human embryonic stem cells. Life Sci 79, 999–1006.

Tan, W.B., Zhang, Y., 2005. Surface modification of gold and quantum dot nanoparticles with chitosan for bioapplications. J Biomed Mater Res A 75, 56–62.

Terreno, E., Castelli, D.D., Cravotto, G., Milone, L., Aime, S., 2004. Ln(III)-DOTAMGly complexes: a versatile series to assess the determinants of the efficacy of paramagnetic chemical exchange saturation transfer agents for magnetic resonance imaging applications. Invest Radiol 39, 235–243.

Terreno, E., Geninatti Crich, S., Belfiore, S., Biancone, L., Cabella, C., Esposito, G., Manazza, A.D., Aime, S., 2006. Effect of the intracellular localization of a Gd-based imaging probe on the relaxation enhancement of water protons. Magn Reson Med 55, 491–497.

Thomson, J.A., Itskovitz-Eldor, J., Shapiro, S.S., Waknitz, M.A., Swiergiel, J.J., Marshall, V.S., Jones, J.M., 1998. Embryonic stem cell lines derived from human blastocysts. Science 282, 1145–1147.

Thorek, D.L., Chen, A.K., Czupryna, J., Tsourkas, A., 2006. Superparamagnetic iron oxide nanoparticle probes for molecular imaging. Ann Biomed Eng 34, 23–38.

Tomita, S., Li, R.K., Weisel, R.D., Mickle, D.A., Kim, E.J., Sakai, T., Jia, Z.Q., 1999. Autologous transplantation of bone marrow cells improves damaged heart function. Circulation 100, II247–256.

Torrente, Y., Camirand, G., Pisati, F., Belicchi, M., Rossi, B., Colombo, F., El Fahime, M., Caron, N.J., Issekutz, A.C., Constantin, G., Tremblay, J.P., Bresolin, N., 2003. Identification of a putative pathway for the muscle homing of stem cells in a muscular dystrophy model. J Cell Biol 162, 511–520.

Torrente, Y., Tremblay, J.P., Pisati, F., Belicchi, M., Rossi, B., Sironi, M., Fortunato, F., El Fahime, M., D'Angelo, M.G., Caron, N.J., Constantin, G., Paulin, D., Scarlato, G., Bresolin, N., 2001. Intraarterial injection of muscle-derived CD34(+)Sca-1(+) stem cells restores dystrophin in mdx mice. J Cell Biol 152, 335–348.

Toyoda, K., Tooyama, I., Kato, M., Sato, H., Morikawa, S., Hisa, Y., Inubushi, T., 2004. Effective magnetic labeling of transplanted cells with HVJ-E for magnetic resonance imaging. Neuroreport 15, 589–593.

Trokowski, R., Zhang, S., Sherry, A.D., 2004. Cyclen-based phenylboronate ligands and their Eu3+ complexes for sensing glucose by MRI. Bioconjug Chem 15, 1431–1440.

Tyszka, J.M., Fraser, S.E., Jacobs, R.E., 2005. Magnetic resonance microscopy: recent advances and applications. Curr Opin Biotechnol 16, 93–99.

Vander Elst, L., Roch, A., Gillis, P., Laurent, S., Botteman, F., Bulte, J.W., Muller, R.N., 2002a. Dy-DTPA derivatives as relaxation agents for very high field MRI: the beneficial effect of slow water exchange on the transverse relaxivities. Magn Reson Med 47, 1121–1130.

Vander Elst, L., Zhang, S., Sherry, A.D., Laurent, S., Botteman, F., Muller, R.N., 2002b. Dy-complexes as high field T2 contrast agents: influence of water exchange rates. Acad Radiol 9 Suppl 2, S297–S299.

van Tilborg, G.A., Mulder, W.J., Chin, P.T., Storm, G., Reutelingsperger, C.P., Nicolay, K., Strijkers, G.J., 2006. Annexin A5-conjugated quantum dots with a paramagnetic lipidic coating for the multimodal detection of apoptotic cells. Bioconjug Chem 17, 865–868.

Veiseh, O., Sun, C., Gunn, J., Kohler, N., Gabikian, P., Lee, D., Bhattarai, N., Ellenbogen, R., Sze, R., Hallahan, A., Olson, J., Zhang, M., 2005. Optical and MRI multifunctional nanoprobe for targeting gliomas. Nano Lett 5, 1003–1008.

Velardo, M.J., Burger, C., Williams, P.R., Baker, H.V., Lopez, M.C., Mareci, T.H., White, T.E., Muzyczka, N., Reier, P.J., 2004. Patterns of gene expression reveal a temporally orchestrated wound healing response in the injured spinal cord. J Neurosci 24, 8562–8576.

Villringer, A., Rosen, B.R., Belliveau, J.W., Ackerman, J.L., Lauffer, R.B., Buxton, R.B., Chao, Y.S., Wedeen, V.J., Brady, T.J., 1988. Dynamic imaging with lanthanide chelates in normal brain: contrast due to magnetic susceptibility effects. Magn Reson Med 6, 164–174.

Voura, E.B., Jaiswal, J.K., Mattoussi, H., Simon, S.M., 2004. Tracking metastatic tumor cell extravasation with quantum dot nanocrystals and fluorescence emission-scanning microscopy. Nat Med 10, 993–998.

Walczak, P., Kedziorek, D.A., Gilad, A.A., Lin, S., Bulte, J.W., 2005. Instant MR labeling of stem cells using magnetoelectroporation. Magn Reson Med 54, 769–774.

Walter, G.A., Cahill, K.S., Huard, J., Feng, H., Douglas, T., Sweeney, H.L., Bulte, J.W., 2004. Noninvasive monitoring of stem cell transfer for muscle disorders. Magn Reson Med 51, 273–277.

Wang, J.F., Yang, Y., Wang, G., Min, J., Sullivan, M.F., Ping, P., Xiao, Y.F., Morgan, J.P., 2002. Embryonic stem cells attenuate viral myocarditis in murine model. Cell Transplant 11, 753–758.

Wang, X., Ku, G., Wegiel, M.A., Bornhop, D.J., Stoica, G., Wang, L.V., 2004a. Noninvasive photoacoustic angiography of animal brains in vivo with near-infrared light and an optical contrast agent. Opt Lett 29, 730–732.

Wang, Y., Xing, D., Zeng, Y., Chen, Q., 2004b. Photoacoustic imaging with deconvolution algorithm. Phys Med Biol 49, 3117–3124.

Ward, K.M., Aletras, A.H., Balaban, R.S., 2000. A new class of contrast agents for MRI based on proton chemical exchange dependent saturation transfer (CEST). J Magn Reson 143, 79–87.

Ward, K.M., Balaban, R.S., 2000. Determination of pH using water protons and chemical exchange dependent saturation transfer (CEST). Magn Reson Med 44, 799–802.

Warlow, C., Sudlow, C., Dennis, M., Wardlaw, J., Sandercock, P., 2003. Stroke. The Lancet 362, 1211–1224.

Watanabe, E., Smith, D.M., Jr., Delcarpio, J.B., Sun, J., Smart, F.W., Van Meter, C.H., Jr., Claycomb, W.C., 1998. Cardiomyocyte transplantation in a porcine myocardial infarction model. Cell Transplant 7, 239–246.

Watson, D.J., Walton, R.M., Magnitsky, S.G., Bulte, J.W., Poptani, H., Wolfe, J.H., 2006. Structure-specific patterns of neural stem cell engraftment after transplantation in the adult mouse brain. Hum Gene Ther 17, 693–704.

Weber, A., Pedrosa, I., Kawamoto, A., Himes, N., Munasinghe, J., Asahara, T., Rofsky, N.M., Losordo, D.W., 2004a. Magnetic resonance mapping of transplanted endothelial progenitor cells for therapeutic neovascularization in ischemic heart disease. Eur J Cardiothorac Surg 26, 137–143.

Weber, A., Pedrosa, I., Kawamoto, A., Himes, N., Munasinghe, J., Asahara, T., Rofsky, N.M., Losordo, D.W., 2004b. Magnetic resonance mapping of transplanted endothelial progenitor cells for therapeutic neovascularization in ischemic heart disease. Eur J Cardiothorac Surg 26, 137–143.

Weber, R., Ramos-Cabrer, P., Hoehn, M., 2006. Present status of magnetic resonance imaging and spectroscopy in animal stroke models. J Cereb Blood Flow Metab 26, 591–604.

Weissleder, R., Ntziachristos, V., 2003. Shedding light onto live molecular targets. Nat Med 9, 123–128.
Weller, H., 1993a. Colloidal semiconductor Q-particles – chemistry in the transition region between solid-state and molecules. Angewandte Chemie-International Edition in English 32, 41–53.
Weller, H., 1993b. Quantized semiconductor particles – a novel state of matter for materials science. Adv Mater 5, 88–95.
Weller, H., 1998. Quantum size colloids: from size-dependent properties of discrete particles to self-organized superstructures. Current Opinion in Colloid & Interface Science 3, 194–199.
Wernig, A., Zweyer, M., Irintchev, A., 2000. Function of skeletal muscle tissue formed after myoblast transplantation into irradiated mouse muscles. J Physiol 522 Pt 2, 333–345.
Weston, S.A., Parish, C.R., 1990. New fluorescent dyes for lymphocyte migration studies. Analysis by flow cytometry and fluorescence microscopy. J Immunol Methods 133, 87–97.
Wirth, E.D., 3rd, Theele, D.P., Mareci, T.H., Anderson, D.K., Brown, S.A., Reier, P.J., 1992. In vivo magnetic resonance imaging of fetal cat neural tissue transplants in the adult cat spinal cord. J Neurosurg 76, 261–274.
Woessner, D.E., Zhang, S., Merritt, M.E., Sherry, A.D., 2005. Numerical solution of the Bloch equations provides insights into the optimum design of PARACEST agents for MRI. Magn Reson Med 53, 790–799.
Wu, J.C., Chen, I.Y., Sundaresan, G., Min, J.J., De, A., Qiao, J.H., Fishbein, M.C., Gambhir, S.S., 2003. Molecular imaging of cardiac cell transplantation in living animals using optical bioluminescence and positron emission tomography. Circulation 108, 1302–1305.
Xiao, D., Xi, L., Yang, W., Fu, H., Shuai, Z., Fang, Y., Yao, J., 2003. Size-tunable emission from 1,3-diphenyl-5-(2-anthryl)-2-pyrazoline nanoparticles. J Am Chem Soc 125, 6740–6745.
Xu, H., Othman, S.F., Hong, L., Peptan, I.A., Magin, R.L., 2006. Magnetic resonance microscopy for monitoring osteogenesis in tissue-engineered construct in vitro. Phys Med Biol 51, 719–732.
Yang, H., Holloway, P.H., Santra, S., 2004. Water-soluble silica-overcoated CdS:Mn/ZnS semiconductor quantum dots. J Chem Phys 121, 7421–7426.
Yang, H.I., Santra, S., Walter, G., Holloway, P.H., 2006. GdIII-functionalized fluorescent quantum dots as multimodal imaging probes. Advanced Materials 18:2890–2894.
Yin, T., Li, L., 2006. The stem cell niches in bone. J Clin Invest 116, 1195–1201.
Yoo, H., Juliano, R.L., 2000. Enhanced delivery of antisense oligonucleotides with fluorophore-conjugated PAMAM dendrimers. Nucleic Acids Res 28, 4225–4231.
Zaheer, A., Lenkinski, R.E., Mahmood, A., Jones, A.G., Cantley, L.C., Frangioni, J.V., 2001. In vivo near-infrared fluorescence imaging of osteoblastic activity. Nat Biotechnol 19, 1148–1154.
Zelivyanskaya, M.L., Nelson, J.A., Poluektova, L., Uberti, M., Mellon, M., Gendelman, H.E., Boska, M.D., 2003. Tracking superparamagnetic iron oxide labeled monocytes in brain by high-field magnetic resonance imaging. J Neurosci Res 73, 284–295.
Zhang, H.F., Maslov, K., Stoica, G., Wang, L.V., 2006. Functional photoacoustic microscopy for high-resolution and noninvasive in vivo imaging. Nat Biotechnol 24, 848–851.

Zhang, J.Z., 1997. Ultrafast studies of electron dynamics in semiconductor and metal colloidal nanoparticles: Effects of size and surface. Acc Chem Res 30, 423–429.

Zhang, R.L., Zhang, Z.G., Chopp, M., 2005. Neurogenesis in the adult ischemic brain: generation, migration, survival, and restorative therapy. Neuroscientist 11, 408–416.

Zhang, S., Merritt, M., Woessner, D.E., Lenkinski, R.E., Sherry, A.D., 2003a. PARACEST agents: modulating MRI contrast via water proton exchange. Acc Chem Res 36, 783–790.

Zhang, S., Trokowski, R., Sherry, A.D., 2003b. A paramagnetic CEST agent for imaging glucose by MRI. J Am Chem Soc 125, 15288–15289.

Zhang, Z., Jiang, Q., Jiang, F., Ding, G., Zhang, R., Wang, L., Zhang, L., Robin, A.M., Katakowski, M., Chopp, M., 2004. In vivo magnetic resonance imaging tracks adult neural progenitor cell targeting of brain tumor. Neuroimage 23, 281–287.

Zhang, Z.G., Jiang, Q., Zhang, R., Zhang, L., Wang, L., Zhang, L., Arniego, P., Ho, K.L., Chopp, M., 2003c. Magnetic resonance imaging and neurosphere therapy of stroke in rat. Ann Neurol 53, 259–263.

Zhang, Z.G., Zhang, L., Jiang, Q., Chopp, M., 2002. Bone marrow-derived endothelial progenitor cells participate in cerebral neovascularization after focal cerebral ischemia in the adult mouse. Circ Res 90, 284–288.

Zhao, M., Kircher, M.F., Josephson, L., Weissleder, R., 2002. Differential conjugation of tat peptide to superparamagnetic nanoparticles and its effect on cellular uptake. Bioconjug Chem 13, 840–844.

Zhao, M., Weissleder, R., 2004. Intracellular cargo delivery using tat peptide and derivatives. Med Res Rev 24, 1–12.

Zheng, Q., Dai, H., Merritt, M.E., Malloy, C., Pan, C.Y., Li, W.H., 2005. A new class of macrocyclic lanthanide complexes for cell labeling and magnetic resonance imaging applications. J Am Chem Soc 127, 16178–16188.

Zhou, J., Wilson, D.A., Sun, P.Z., Klaus, J.A., Van Zijl, P.C., 2004. Quantitative description of proton exchange processes between water and endogenous and exogenous agents for WEX, CEST, and APT experiments. Magn Reson Med 51, 945–952.

7
Micron-Sized Iron Oxide Particles (MPIOs) for Cellular Imaging: More Bang for the Buck

Erik M. Shapiro and Alan P. Koretsky

Abstract: Iron oxide particles have been used for molecular and cellular magnetic resonance imaging since the 1980s. Interestingly, early research in these fields used particles in the micron size range. After a long disappearance, the utility of micron-sized iron oxide particles, or MPIOs, has been rediscovered, largely due to their improved construction and stability. Due to the very high iron content of single MPIOs, their use as a cell-tracking agent has enabled the detection of single particles in single cells, the detection of single cells in vivo, and the ability to label cells directly in vivo. This chapter begins by discussing the physical and magnetic properties of MPIOs and compares them to the more commonly used iron oxide particles, USPIOs and SPIOs. Next, the unique properties of MPIOs as an MRI agent are illustrated, with examples of MRI experiments on the particles themselves, and on cells labeled with MPIOs. The chapter then reviews work on various applications of cell therapy, using cells labeled with MPIOs, with a section on the detection of single cells in vivo. Lastly, there is a presentation of work using MPIOs to directly label cells in vivo.

Keywords: MRI, stem cell, iron oxide, particle, contrast agent, transplant.

7.1. Introduction

Transplanted stem and post-mitotic cells show great promise as therapeutic agents for a number of diseases, including Alzheimer's disease, Parkinson's disease, and myocardial infarction, among many others. Recent work has even suggested the possible use of native immune or stem cells, to either report on or to treat diseases, such as cancer or stroke. The curative and/or reporter uses of these different cells would be aided greatly by using non-invasive techniques for visualizing the distribution and location of cells following transplantation, infusion, or native responses. Particularly in cases where widespread migration of delivered cells is necessary for success of treatment, longitudinal monitoring of cellular migration non-invasively could be crucial. Coupled with functional and metabolic assessments, non-invasive imaging of cell engraftment could have a profound effect on physicians' ability to serially monitor and adjust treatment.

Magnetic resonance imaging is a powerful tool for visualizing cellular trafficking, homing, and migration in intact animals. Particularly advantageous to cellular imaging in intact organisms is the ability of MRI examinations to be performed longitudinally and non-invasively, at image resolutions approaching 50 microns isotropic in live animals, all within appropriately rapid scan times. To distinguish cells in an MR image, cells are loaded with an MRI contrast agent prior to or during imaging. Historically, superparamagnetic iron oxide nanoparticles have been predominantly used to generate contrast based on magnetic susceptibility differences between the labeled cells and the tissue. Cellular detection is most often accomplished via T_2- or T_2^*-based mechanisms following incorporation of contrast agent either within the cell or attached to the cell.

Recently, micron-sized iron oxide particles, MPIOs, have received attention as an alternative cell labeling material. Due to the high quantity of iron oxide packaged into a single particle, MPIOs have some advantageous properties for molecular and cellular imaging. This chapter describes these special properties and discusses the use of these particles for magnetic cell labeling for cellular imaging by MRI.

7.2. Brief History

While molecular MRI and cellular MRI are recently coined terms, early experiments in the field were published in 1986 (Renshaw et al., 1986). These studies took their cues from the already well-developed field of targeting radiolabeled antibodies to various antigens, particularly cancer, in an effort to localize these antigens and detect the tumor. To accomplish the same by MRI, antibodies were coupled to homemade magnetite particles, varying in size from 5 nm to 1.5 micron, and injected into tumor bearing animals, which the antibodies would recognize (Renshaw et al., 1986). The antibody-coated particles bound to the antigens on the tumor and MRI experiments, sensitive to the presence of inhomogeneous magnetic fields caused by the presence of the magnetite, were able to detect the presence of the particles, and hence the tumor.

At the same time, several groups were experimenting with nanometer-sized coated iron oxide particles as an injectable contrast agent to delineate tumors in the liver or spleen. It had been discovered, quite expectedly, that resident macrophages of the RES system could endocytose iron oxide nanoparticles from the blood and that cells harboring these particles could be imaged, thus distinguishing tumor tissue devoid of the macrophages from normal tissue (Saini et al., 1987a,b). With the reliable commercial production of the nanometer-sized particles, in addition to the difficulty in working with, which were at the time ferromagnetic micron-sized iron oxide particles, the use of micron-sized particles was largely shelved in favor of the smaller particles.

The field of cell labeling for the purposes of observing these cells by MRI following transplantation evolved in the early 1990s. Experiments had already demonstrated that neurons could be labeled with fluorescent microspheres in culture and that following transplant into the brain, the intracellular label was maintained for months (Arbuthnott and Dunnett, 1990). Paul Lauterbur's group labeled fetal neural cells with virus encapsulated 11 nm iron oxide nanoparticles and transplanted these cells into host rats, where the iron oxide label was detected by MRI (Hawrylak et al., 1993). Concurrently, lymphocyte labeling was accomplished by two different ways in vitro. One method used biotinylated iron oxide nanoparticles, mediated by initial recognition of the lymphocytes by a biotinylated antibody and a streptavidin 'bridge' (Bulte et al., 1992), and the other method accomplished labeling by simple incubation with naked particles (Yeh et al., 1993).

Since then, advances have been accomplished in increasing the intracellular iron content with nanoparticles. These include the use of the TAT peptide (Josephson et al., 1999), lipofection agents (Frank et al., 2002), antibody- and transferrin-mediated incorporation (Ahrens et al., 2003; Hogemann et al., 2000) and the use of phagocytotic cell lines. These innovations have allowed the detection of single cells being labeled with nanoparticles, in vitro, both at high field (Dodd et al., 1999) and at clinical fields (Foster-Gareau et al., 2003). Furthermore, iron oxide nanoparticles are now in general use for labeling peripheral blood and tissue-resident macrophages for the tracking of their movements in response to inflammation and tissue damage (Corot et al., 2004). Still, using nanoparticles, difficulty remains in generating high-enough intracellular iron content to visualize single cells in vivo, as well as the need for high doses of injected nanoparticles to adequately label macrophages to monitor their movements. This is precisely the area where MPIOs possess an enormous potential advantage as cellular MRI contrast agents.

7.3. Micron-Sized Iron Oxide Particles

Nearly all MRI experiments aiming to monitor cellular migration use T_2- or T_2^*-weighted MRI to probe the location of the cells. This is because the contrast agent is most often iron oxide based. Thus, of central importance to the success of the experiment is loading the cells with sufficient iron oxide to render these relaxation processes extremely efficient when compared with the native tissue. The properties of some iron oxide particles for cell tracking by MRI are shown in Table 7.1. Common iron oxide nanoparticles consist of either a 5 nm (USPIO) or 7 nm (SPIO) iron oxide crystal covered by a dextran coating (Artemov et al., 2004), bringing the total hydrodynamic size of the particles to 30 or 150 nm, respectively (Benderbous et al., 1996; Raynal et al., 2004). Given their respective core diameters and the density of iron oxide of $\sim$5 g/cm^3, USPIO and SPIO have 3.3×10^{-19} and 9.0×10^{-19} grams of iron oxide per particle. This computes to 4.3×10^6 USPIOs and 1.5×10^6 SPIOs to achieve 1 pg of iron, given that iron

TABLE 7.1. Properties of iron oxide contrast agents for cell tracking. Overall particle diameter refers to hydrodynamic radius. The MPIO data is for a commercially available particle from Bangs Laboratories.

Particle	Core diameter	Overall particle diameter	No. of particles for 1 pg iron	Volume of 1 pg iron worth of particles
USPIO	5 nm	20–30 nm	4.3×10^6	(4.9 microns)^3
SPIO	7 nm	60–150 nm	1.5×10^6	(17.2 microns)^3
MPIO	47.5% magnetite	1630 nm	1	(1.3 microns)^3

oxide is 72% iron w/w. With the overall diameters of USPIO and SPIO particles at 30 and 150 nm, respectively, the volume of particle needed to deliver the 1 pg of iron would be 0.12 and 5.0 pL, respectively and encompass a cubic volume 4.9 and 17.2 microns on a side (based on simple packing of spheres), easily the same size, if not larger than many cells of interest for cellular MRI. The same volume of space necessary to encompass 1 pg of iron using an MPIO is a cubic volume1.3 microns on a side. Labeling efficiencies as high as 20 pg, reported with SPIO:poly-lysine complexes, would require a cubic volume 46.6 microns on a side (Frank et al., 2003)! For MPIOs, the volume required to deliver 20 pg of iron would be a cubic volume 4.34 microns on a side.

Thus, a major advantage of using MPIOs for cell labeling can be readily seen based on the volume taken up by the MPIOs with respect to smaller particles. This volume difference is largely due to the construction of the MPIO particles, which packs many smaller, superparamagnetic iron oxide cores into an inert polymer matrix. This approach limits the amount of coating of the particles that does not enhance the MRI signal. Importantly, because the volume of MPIOs is correspondingly smaller than the volume of USPIO or SPIO needed to acheive equivalent iron content, cells can be labeled with significantly more iron using MPIOs. Lastly, the inert coating of most MPIOs means that the labels remain intact for many months.

A useful comparison for evaluating contrast agents is the relaxivity of that agent. Relaxivity, expressed as $s^{-1}mM^{-1}$, reflects the ability of a standard concentration of material to alter relaxation rates and is variable with field strength. The R_2^* measurement is important because many cell tracking MRI experiments make use of gradient echo T_2^*-weighted images. The R_2^* relaxivity of SPIO and 0.76 micron MPIOs has been measured (Hinds et al., 2003). The relaxivities measured at 4.7 T were 240 ± 27 for SPIO and 356 ± 21 for MPIO, and at 11.7 T were 498 ± 19 and 851 ± 62 for SPIO and MPIO respectively. A major reason for the marked increase in relaxivity of MPIOs versus SPIOs is the phenomenon of clustering of magnetic centers to change relaxivity regimes from diffusion sensitive (small particles) to static dephasing regime (larger clusters) (Bowen et al., 2002; Tanimoto et al., 1994; Weisskoff et al., 1994). This is the major principle behind several studies using the clustering and unclustering of iron oxide particles to modulate MRI signal, reflecting underlying gene expression (Perez et al., 2002b; Perez et al., 2002a). Clustering of magnetic centers is effective for MPIOs as well, as 4.5 micron MPIOs from Dynal have three times higher R_2^* relaxivity than 1.63 micron MPIOs from Bangs (Shapiro et al., 2005).

7.4. MRI of MPIOs

In all MRI experiments, contrast is dependent on a multitude of MRI parameters such as pulse sequence, image resolution, and echo and repetition times. MPIOs and other iron oxide particles are particularly powerful contrast agents, taking advantage of differences in magnetic susceptibility by way of gradient echo–based T_2^*-weighted imaging. The magnetic core of the particles disturbs the otherwise uniform magnetic field water molecules experience in the MR magnet, causing dephasing of the spins of the water protons. The size of the effect from a particle is dependent on both the amount of iron in a particle and its distribution within the particle. Particles containing more iron possess a higher magnetic moment and perturb the field to a greater extent. This causes water protons further from the particle to experience magnetic field differences, and consequently the area of contrast is larger.

Because MPIOs pack so much iron into a single particle, MPIOs have unique MRI properties. Table 7.2 compares various MPIOs from several manufacturers, with the physical characteristics listed for each. The most advantageous MRI property of MPIOs is the ability to detect single particles, a feat first demonstrated in 1986 by Paul Lauterbur's group (Lauterbur et al., 1986). This was simply accomplished by suspending 3.5 micron diameter particles in agarose, acquiring images at 90 MHz and at very high resolution, $20 \times 20 \times 60$ micron voxels.

In a more recent study, several different commercially available MPIOs were imaged in agarose, in concentrations that placed individual particles in voxels. The goal was to quantitatively characterize the contrast generated for single

TABLE 7.2. Properties of various commercial micron-sized iron oxide particles. These are for 'off the shelf' preparations. For non-fluorescent particles, the functionalized coatings makes the introduction of chemically fluorescent tags facile.

Company	Diameter (Microns)	% ferrites, Pg Fe/ particle	Available fluorescent	Available functionalized
Bangs	0.76	46.5%, 0.13	Yes	Yes
	0.96	27.8%, 0.13		
	1.63	47.5%, 1.10		
	5.80	3.8%, 4.40		
Dynal	1.05	37%, 0.26	No	Yes
	2.79	15%, 1.50		
	4.50	29%, 10.3		
BioPAL	Ferrotrack 1.0	35%, 0.26	No	Yes
	BioMagic 1.0	97%, 1.60		
BioMag	1.5	90%, 2.70	No	Yes
Spherotech	1–2	70%, 1.30	Yes	Yes

MPIOs, using various imaging conditions (Shapiro et al., 2005). The results are summarized in Figure 7.1. It was shown that single 0.96 micron MPIOs from Bangs Labs, containing only 0.13 pg iron, were detectable using gradient echo T_2^*-weighted imaging, creating a hypointense contrast area 200 microns in size

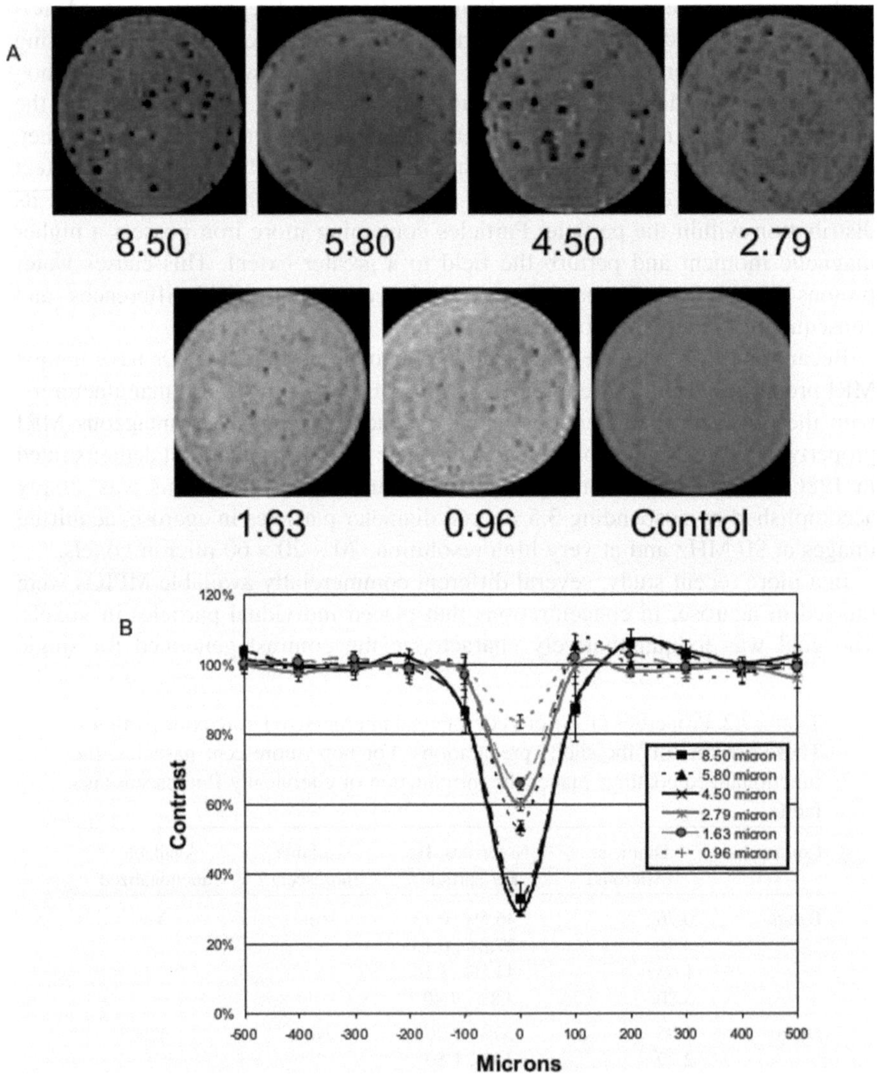

FIGURE 7.1. T_2^*-weighted GRE-MR images of single MPIOs of various sizes. **A** MR images of seven different samples, with different-sized MPIOs. Numbers below images indicates particle diameter in microns. MRI was performed at 7.0 T with a 10 ms TE, and at 100 micron isotropic resolution. **B** Average profile plots for hypointense contrast regions for each size MPIO. Reproduced from Shapiro et al. (2005).

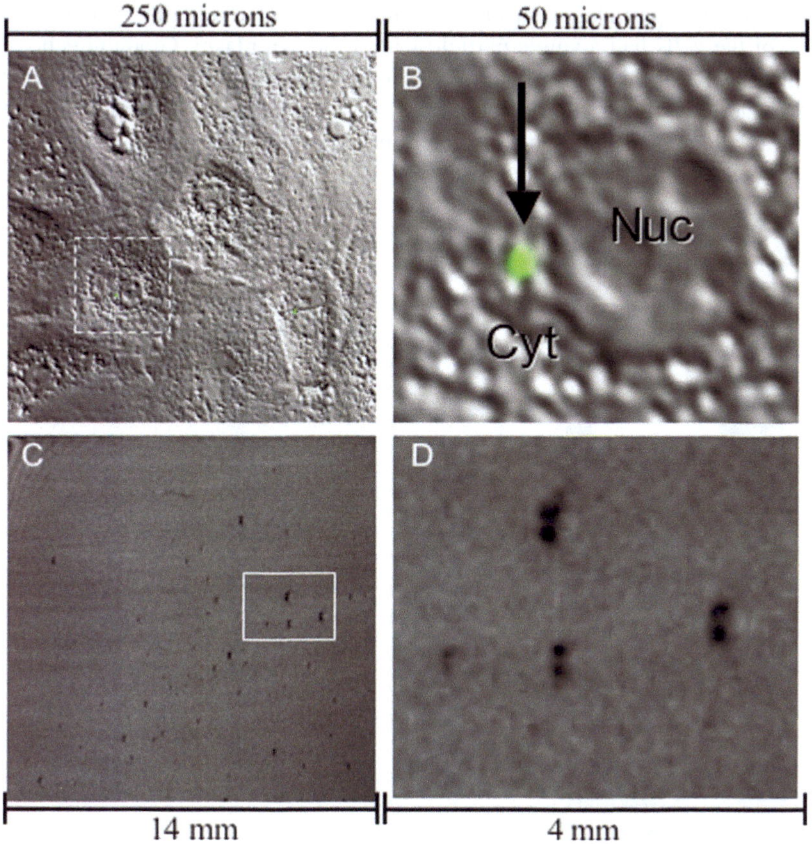

FIGURE 7.2. Optical and magnetic resonance images of single particles in single live mouse hepatocytes. **A** A wide-field confocal fluorescence image superimposed on a Nomarski optics image of a nearly confluent dish of live hepatocytes, showing few single-labeled cells. **B** An expansion of one of the singly labeled cells (dashed box in **A**). Note the single green fluorescent spot (arrow) in the cytoplasm (Cyt) bordering the nucleus (Nuc). **C** An MRI of the entire culture dish with singly labeled hepatocytes. **D** An expansion of the boxed region in **C**. Note the characteristic barbell-shaped contrast caused by single particles in four different cells. Reproduced from Shapiro et al. (2004).

and eliciting a 16% drop in signal intensity. This was accomplished at 7 T with 100 micron isotropic voxel resolution and a 10 ms echo time. A single 1.63 micron MPIO from Bangs containing 1.1 pg of iron elicited a signal drop of 34%, using identical imaging parameters. Most dramatically, a single 4.5 micron particle from Dynal containing 10.3 pg of iron caused a drop in signal intensity of 72% over 200 microns!

Single particles in cells have also been detected by MRI (Shapiro et al., 2004). Primary mouse hepatocytes were plated and maintained in culture. MPIOs of 0.96 micron from Bangs Laboratories were added to the cell culture medium at a ratio of 1 particle per 100 cells and incubated overnight. The next day ~1% of the cells had a single particle within the cytoplasm. Gradient echo, T_2^*-weighted MRI of this live cell culture at 50 microns isotropic resolution revealed a characteristic figure '8' dark contrast spot, indicative of a single, magnetic center (Callaghan, 1993). This is shown in Figure 7.2. The ability to distinguish

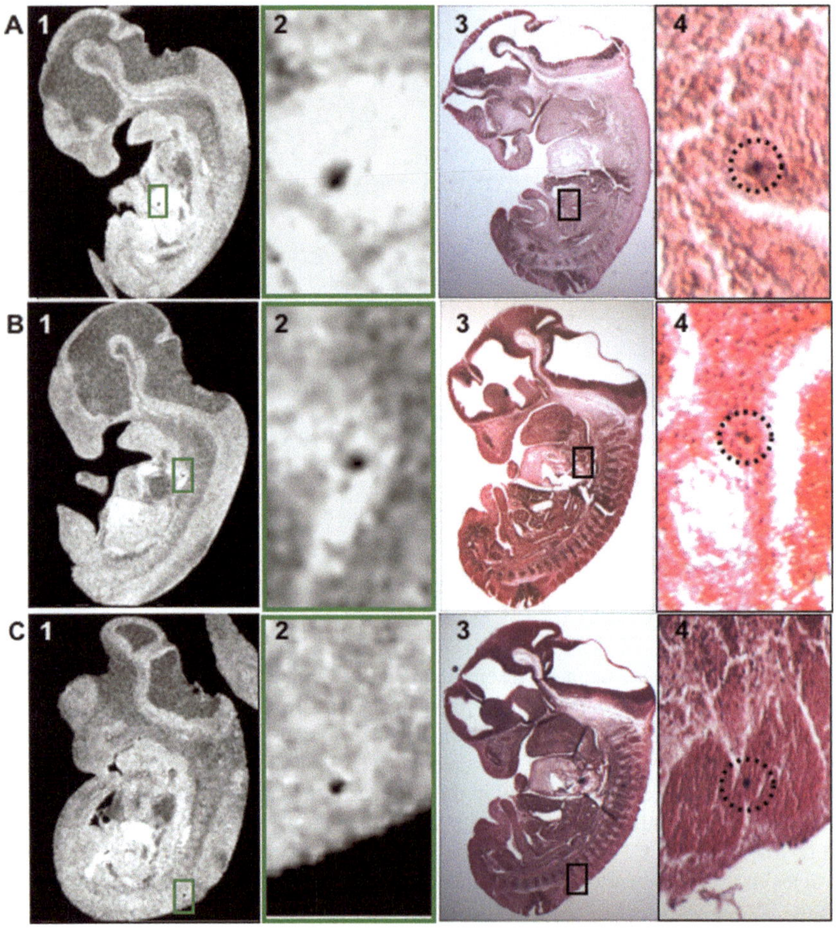

FIGURE 7.3. MRI and histology of Bangs particles in E11.5 mouse embryos. **A–C, A1–C1** Slices from 3D MRI data sets. **A2–C2** Expansions of the boxed regions in **A1–C1**, showing susceptibility-induced contrast regions in each image. **A3–C3** Matching histological sections that correspond to the same plane as the accompanying MRIs. **A4–C4**) A second set of images of the sections in **A3–C3**, acquired with a higher magnification, approximately representing the boxed regions. The stains from the individual particles are circled to assist visualization in the histology. Shown are single 1.63 micron particles registered to the areas of dark contrast in the MRIs. Each embryo image is 6.75 mm from head to tail, both in the MRI and in the histology. All *insets* are ×10 magnifications. Reproduced from Shapiro et al. (2004).

this figure '8' was due to the high resolution used and high signal-to-noise ratio attained. Images acquired at 100 microns isotropic resolution resulted in the collapse of the figure '8' into a more globular-shaped contrast region.

The goal of many magnetic cell-labeling experiments is the eventual transplant of the labeled cell into an animal model and to use MRI to follow the migration of the cell. Here then, it is key to maintain enough intracellular contrast agent to allow detection, even through cellular dilution due to migration and cell division. For cells with little mitotic activity, lymphocytes for example, the challenge for detection is maximizing labeling capacity. For cancer cells, which undergo many cell divisions to form tumors, a larger premium is placed upon retention of sufficient contrast agent for detection. Because individual MPIOs can be detected by MRI, using resolutions suitable for studying entire animals, cell division cannot dilute the label beyond detectability, the way it can when using nanoparticles. It can be estimated that at current maximal reported labeling efficiencies for SPIO (30–50 pg), cellular iron concentrations drop below 1 pg of iron after five cell divisions and below 0.1 pg of iron after nine cell divisions. To demonstrate continued detection of MPIOs even through tremendous cell division, fifty 1.63 micron MPIOs were injected into a single-cell fertilized mouse egg, and re-implanted in a pseudo-pregnant female (Shapiro et al., 2004). At embryonic day 14.5, the embryos were removed and imaged. As shown in Figure 7.3, dark contrast regions in the embryo could be correlated to Prussian blue iron stains in matching histological sections, indicating that MPIOs were distributed during development and could be imaged, even with a tremendous amount of cell division. While only ~30 dark contrast regions were detected within the whole embryos, representing only 30 cells, this experiment performed with nanoparticles would have resulted in zero detection.

7.5. MRI of MPIO-Labeled Cells

7.5.1. Cell Labeling

Cell labeling with MPIOs has been achieved with many different sized particles, as shown in Figure 7.4. For example, it has been demonstrated that mouse hepatocytes can be labeled with as many as thirty-five 4.50 micron MPIOs from Dynal (Shapiro et al., 2005), embryonic stem cell derived neural precursors can be labeled with up to five 5.80 micron MPIOs from Bangs (Shapiro et al., 2005), and rat Schwann cells can be labeled with numerous 1.63 micron MPIOs, also from Bangs (Dunning et al., 2006). In most studies, cell labeling has been achieved by the simple addition of particles to the cell culture medium, in a desired ratio for labeling, i.e. 100 particles per cell. Particles can then be washed away from the medium the next day, leaving labeled cells. Additional steps to remove all free particles may be necessary for some experiments, and in these situations density centrifugation through Ficoll or Percoll has been shown to work. The drawback, however, is the loss of some cells. Suspension-borne cells, such as T lymphocytes, have been more difficult to label. Success has been

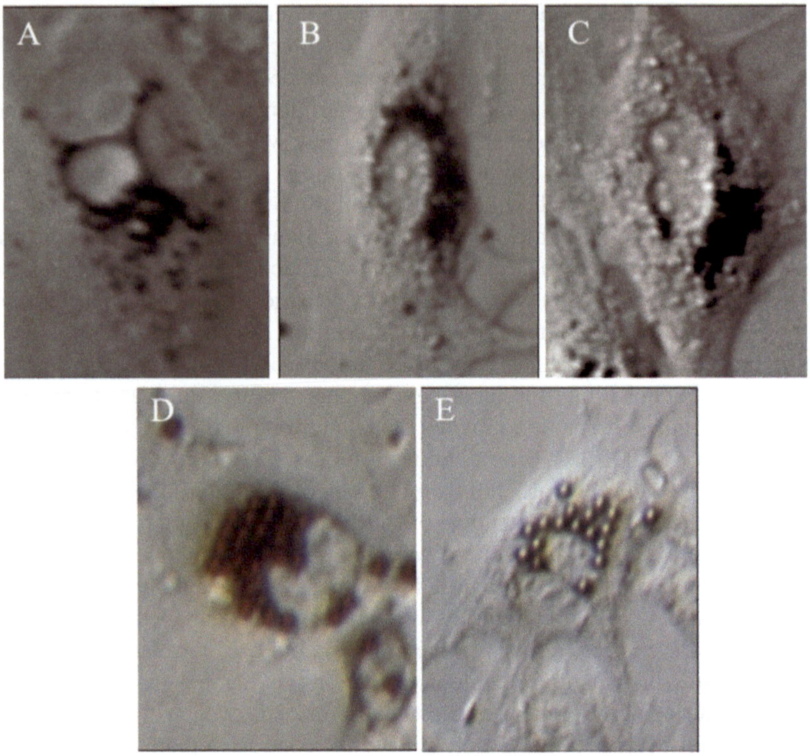

FIGURE 7.4. Light microscopy images of hepatocytes labeled with (**A**) 0.96, (**B**) 1.63, (**C**) 2.79, (**D**) 4.50, and (**E**) 5.80 micron particles. Dark perinuclear spheres are individual MPIOs. Particle numbers per cell ranged from several hundred for the smaller particles to tens for the largest particles. **D** and **E** Thirty-five 4.5 micron and 17 5.8 micron particles, yielding iron contents of 387 and 75 pg iron, respectively. Reproduced from Shapiro et al. (2005).

achieved by pre-targeting the cells with biotinylated antibodies, and then labeling the cells with streptavidin-coated particles (Shapiro et al., 2006b). This is a modification of earlier work by Bulte et al. (1992). Preliminary results showed that as many as 50 beads can be attached to the surface of the T cells safely, with evidence of some internalization of particles. Whether internalization is necessary for proper cell trafficking still needs to be determined.

The choice for particle type is partly dependent on the experiment. Again, one of the most important factors for cell tracking by MRI is to accumulate enough iron within the cells for visualization. If a cell is not expected to undergo multiple cell divisions, the size of the particle is not particularly important, just the capacity for labeling. If a cell is expected to undergo many cell divisions, it is advisable to use 1 pg particles, rather than larger particles. The reason for this is that higher quantities of smaller, rather than larger, particles can be loaded into a cell. Cell division will dilute the particles to one or zero particles faster in cells carrying a fewer number of larger particles. Therefore, since single 1 pg particles can be robustly detected, it may be better to use them in this case.

7.5.2. Cardiac Studies

The use of cell transplants to repair damaged heart tissue is attractive due to the heart's lack of regenerative ability. To this end, it would be useful to be able to monitor cellular therapy in the heart using a non-invasive analytical tool. In an effort to magnetically label CD34+ hematopoietic stem cells and mesenchymal stem cells (MSCs) for MRI cell tracking, Hinds et al. (2003) demonstrated near 100% labeling of both cell types in culture by simple overnight co-incubation of 0.96 micron diameter MPIOs from Bangs Laboratories. It was determined that particles were present in intracellular endosomes by co-labeling with an endosomal labeling dye and co-localizing fluorescence from the particles and the dye (Figure 7.5). Additionally, differentiation capacity and enzymatic assays for labeled mesenchymal stem cells were equivalent to unlabeled cells (Hill et al., 2003; Hinds et al., 2003). Furthermore, it was quantified using flow cytometry that the particles were evenly distributed to daughter cells following cell division. The iron labeling was sufficient to enable single cell detection by MRI of labeled cells growing in culture. This was manifested as isolated, hypointense signal voids approximately 200 microns in diameter in a gradient echo, T_2^*-weighted image.

Exploring the hypothesis that MSC delivery could repair damaged heart tissue, labeled MSCs were then injected into animal models of myocardial infarction (Dick et al., 2003; Hill et al., 2003). Following x-ray–guided, catheter-mounted injections into beating hearts, both with and without infarctions, cells were detectable by MRI as dark contrast as long as 12 weeks post-injection (Hill et al., 2003). Upon histological examination, labeled cells were observed at the injection site with cells having adopted an elongated appearance, suggesting the beginnings of in vivo cellular differentiation, despite the high number of incorporated MPIOs (Figure 7.6). The use of dual magnetic and fluorescent particles facilitated the use of fluorescent immunohistochemistry.

7.5.3. Neural Stem Cell Studies

Similar to the heart, the brain has a limited capacity to repair damaged tissue, making cellular therapies an attractive treatment option for CNS diseases. In one of the first studies using MPIOs, Zhang et al., (2003) successfully demonstrated that neural precursor cells derived from the subventricular zone (SVZ) of adult rats could be labeled in culture with 0.76 micron diameter MPIOs. Instead of simple incubation, they used a biolistic gene gun method to confer intracellular labeling. This produced >95% cell labeling. Following cell labeling, these cells were injected into the adult striatum where a mass of dark contrast was observed immediately after injection, by gradient echo T_2^*-weighted MRI. After 28 days of injection, the mass of dark contrast expanded compared to day 0 for live

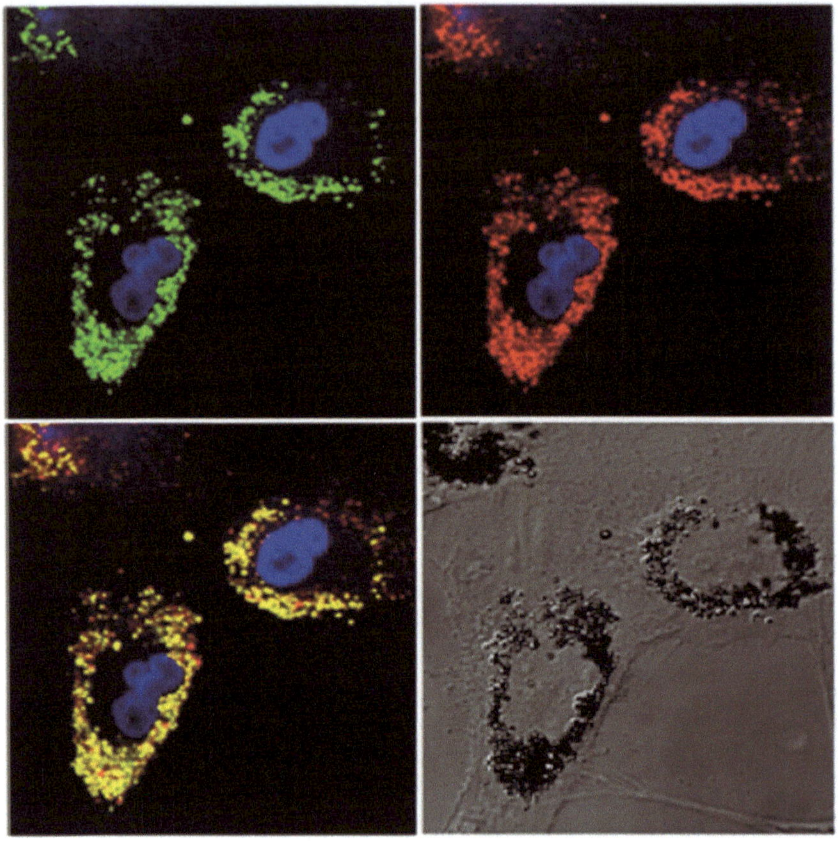

FIGURE 7.5. Confocal fluorescent microscopy of porcine mesenchymal stem cells. A subconfluent monolayer of primary porcine marrow MSCs were exposed to 10 μL/mL of the iron oxide fluorescent microparticles overnight. Excess particles were washed off with PBS. **A** Green fluorescence of beads. **B** Red fluorescence of endosomal marker CM-DiI. **C** Colocalization of the two colors confirming endosomal particle uptake. **D** Nomarski optics view revealing the outlines of the fibroblastic cells and the iron particles clearly clustered in perinuclear organelles. ×100 original magnification for all panels. Reproduced from Hinds et al. (2003).

injected cells, whereas the contrast for dead injected cells was reduced. This was interpreted as live-labeled cells migrating throughout the striatum, carrying the contrast generating particles with them. Dead cells were presumed to have been cleared.

This labeling method was subsequently used to monitor the migration of both neural precursor cells and MSCs to tumors in rat brains (Zhang et al., 2004). Here, it was demonstrated that MPIO-labeled neural precursor cells as well as labeled MSCs, transplanted into the brain, homed to 9L-gliosarcoma tumors (Figure 7.7). In a further study, it was demonstrated that labeled SVZ-derived

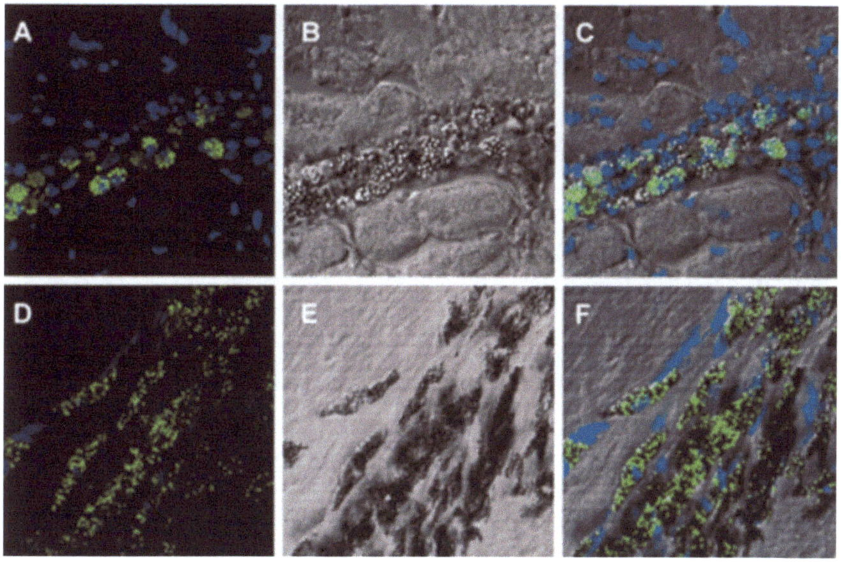

FIGURE 7.6. Histological confirmation of MPIO-labeled cells after endomyocardial injection. **A–C** Immediately after injection. **A** Confocal fluorescence micrograph of round fluorescent green MSCs with DAPI nuclear counterstaining of surrounding myocardium. The cells appear intact immediately after injection with few exogenous particles in the surrounding myocardium. **B** Corresponding section using differential interference microscopy. **C** Overlay of images **A** and **B**. **D–F** Animal shown 21 days after injection. **D** Endomyocardial engraftment of fluorescent green MSCs with DAPI nuclear counterstaining. MSCs appear elongated and aligned with the host myocardium. **E** Corresponding section using differential interference microscopy. **F** Overlay of images **D** and **E**. Cells appear intact with maintenance of DAPI nuclear staining with surrounding perinuclear MPIO. Reproduced from Hill et al. (2003).

adult neural precursor cells, transplanted into the brain, migrated to areas of ischemia (Jiang et al., 2005).

7.5.4. Cancer Studies

Various cancer cell lines have been labeled with MPIOs. Rodriguez et al. (2006) have demonstrated the labeling of human and mouse prostate and breast cancer cells. All cell types were labeled by simple overnight co-incubation of cells in culture with MPIOs. Figure 7.8 shows data on MPIO-labeled mouse prostate cancer cell xenografts. Importantly, upon injection of MPIO-labeled cancer cells in vivo, tumors formed that were identical in shape to unlabeled tumors, and tumors from labeled cells had identical immuohistochemical staining profiles to tumors that formed from unlabeled cells. This demonstrated that MPIOs do not affect the overall function of the tumor cells. Lastly, as expected, labeled tumors elicited dark contrast on MR images, compared to tumors from unlabeled cells.

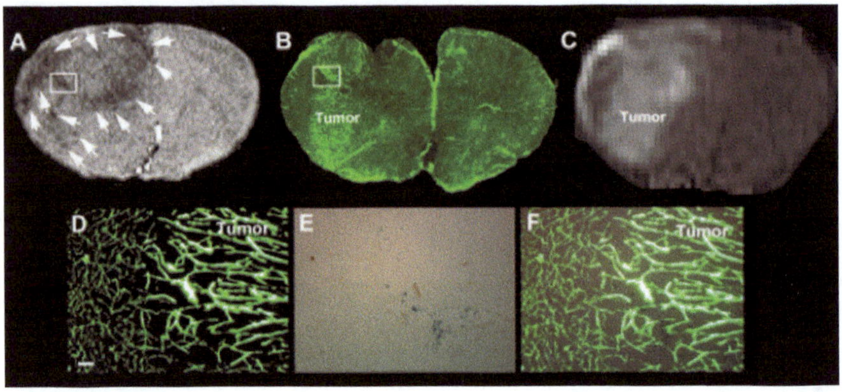

FIGURE 7.7. Transplanted superparamagnetic particle-labeled MSCs target brain tumor. A coronal section of three-dimensional gradient echo image shows loss of MRI signals around and within the tumor (**A**, arrows). Panel **B** shows enlarged cerebral vessels in the tumor, which matches the outline of tumor on 3D gradient echo image (**A**) and the extent of the tumor on Gd–DTPA contrast image (**C**, tumor). Panels **D–F** are high magnification of the box area in panel **B**, which matches the box area of panel **A**. Panels **D** and **E** show enlarged abnormal tumor vessels along with adjacent relatively normal cerebral vessels and Prussian blue-positive cells, respectively. Panel **F** is a merged image of panels **D** and **E**, showing grafted superparamagnetic particle-labeled MSCs around and within the tumor, which matches signal losses observed on 3D gradient echo image (**A**, box). Bar = 100 μm. Reproduced from Zhang et al. (2004), copyright 2004, with permission from Elsevier.

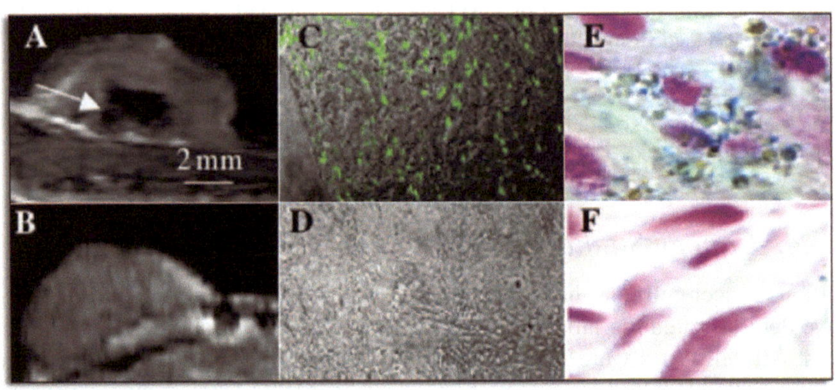

FIGURE 7.8. Imaging of mouse prostate cancer xenografts. MPIO-labeled (**A**) and unlabeled (**B**) TRAMP-C1 mouse prostate xenografts were imaged in vivo by T_2^*-weighted MRI. **C** and **D** Frozen tumor sections imaged by confocal fluorescent microscopy. Perl's Prussian blue staining of MPIO positive cells (**E**) or MPIO-negative cells (**F**). The arrow in (**A**) identifies the negative contrast MR image created by the superparamagnetic MPIOs. Reproduced from Rodriguez et al. (2006).

7.6. In Vivo Detection of Single MPIO-Labeled Cells

Truly, setting MRI apart from other imaging modalities has been the recent demonstration of the detection of single cells in living animals. In most cases, this has been enabled by the use of MPIOs to safely deliver large amounts of iron to cells. For example, a single hepatocyte labeled with iron contents of ~100 pg iron (Shapiro et al., 2005) presents a cellular iron concentration of 1–2 M! When placed into a 100 micron isotropic voxel, the iron concentration is 1.8 mM in that voxel.

One example of in vivo single cell detection using MPIOs as the magnetic label was demonstrated using a hepatocyte transplant model in mice (Shapiro et al., 2006c). One possible treatment for various liver disorders is the delivery of hepatocytes. In the case of a liver that cannot function as a host to cell transplantation, hepatocytes can be injected into the spleen. These cells then migrate to the liver and engraft as single cells (Ponder et al., 1991). This system was used to investigate whether MRI could detect single cells in vivo. Hepatocytes were

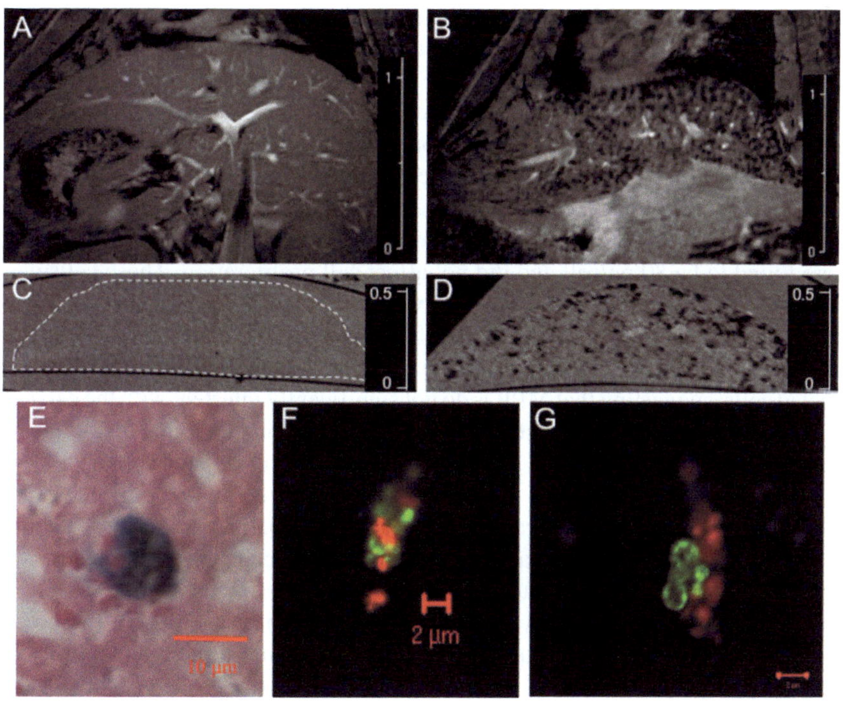

FIGURE 7.9. In vivo MRI slice of (**A**) a control liver and (**B**) a liver from an animal whose spleen was injected with labeled hepatocytes 1 month prior to imaging. **C** and **D** In vitro MRI slices of the same samples shown in **A** and **B**. The dotted outline in **C** delineates the liver boundary. Scale bars are in centimeters. **E** Prussian blue stained cell showing particles surrounding the nucleus of a grafted cell in the liver. **F** and **G** Two dual channel confocal fluorescence images of grafted hepatocytes in the liver showing green MPIOs and red cell tracker dye. Reproduced from Shapiro et al. (2006c).

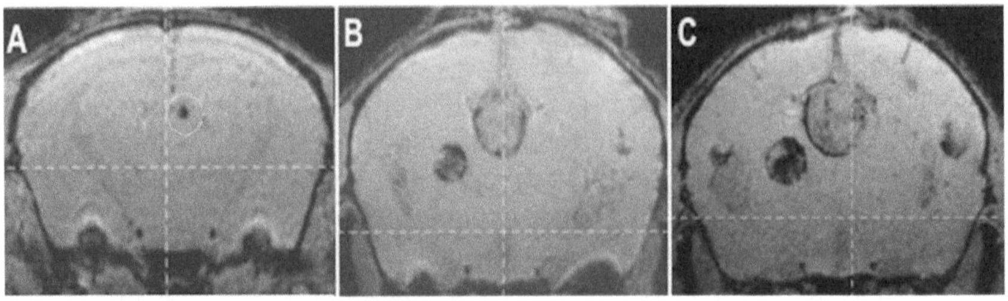

FIGURE 7.10. MR images following the progression of brain metastasis in an untreated mouse 2 weeks (**A**), 4 weeks (**B**), and 5 weeks (**C**) after tumor cell inoculation.

labeled with 1.63 micron MPIOs and injected into the spleens of mice. One month post-injection, mice were scanned. The results are summarized in Figure 7.9. MR images showed isolated, hypointense contrast regions scattered throughout the livers of these mice. Histological and immunohistochemical evaluation of tissue samples showed that most cells had grafted to the liver as individual cells. Comparing the MRI characteristics of the livers with agarose phantoms containing single isolated cells revealed that the sizes of the contrast regions were identical, indicating the presence of single cells. On average, the dark contrast generated from a single engrafted cell occupied six to seven 100 micron isotropic voxels in three-dimensional space.

Two other reports have applied the detection of single cells by MRI to cancer studies. In one study, cells derived from a 'brain-seeking' clone of a breast cancer cell line were labeled with MPIOs in culture (Heyn et al., 2006). Iron contents were found to be 80 pg/cell and did not affect cell proliferation. These cells were then injected into the left ventricle of live nu/nu mice. MRI was used to follow several punctate dark contrast regions forming larger, dark tumors. Metastatic burden was not statistically different for animals injected with unlabeled versus labeled cells, suggesting that MPIOs do not significantly interfere with the capacity of the cells to form tumors. Lastly, while MPIOs could not be found histologically in large tumors, they were seen in micrometastases and in isolated cells.

Another study investigated the growth of single metastatic prostate cancer cells to form tumors in the brain (Tracy et al., 2006). These results are shown in Figure 7.10. A highly metastatic prostate cancer cell line was labeled with 1.63 micron MPIOs and injected i.v. Some cells were retained in the brain and were visible as single cells by MRI. Similar to Heyn et al. (2006), these cells could then be monitored to quantify how many became tumorogenic and what the longitudinal characteristics of tumor formation were.

7.7. In Vivo Labeling of Cells with MPIOs

While most cell labeling for MRI is performed in cell culture, only peripheral macrophages have been consistently labeled in vivo with iron oxide particles of any size. This has made the study of endogenous cell migration and homing

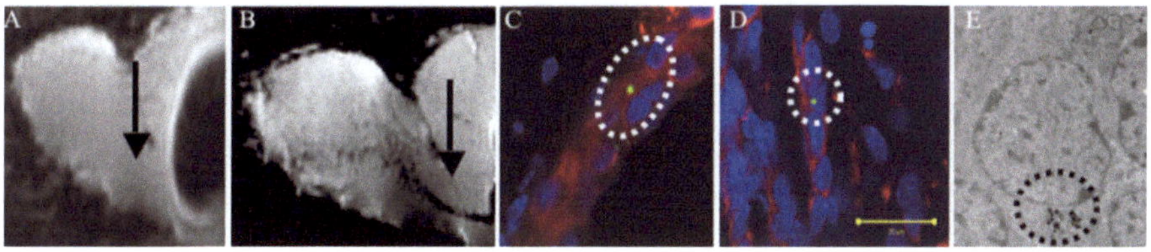

FIGURE 7.11. **A** MRI of RMS (arrow) and OB immediately post-injection. **B** MRI of RMS and OB 5 weeks post-injection. Dark contrast from neuroblasts carrying MPIOs is seen along the RMS, from the ventricle into the olfactory bulb. **C** Green particles in GFAP+(red) NSCs at the ventricle. **D** Green particles in Doublecortin+(red) neural precursors in the RMS. **E** Electron microscopy identification of intracellular particles in granule cell neuron in olfactory bulb. Reproduced from Shapiro et al. (2006a).

by MRI nearly impossible. Recently, Shapiro et al. (2006a) have demonstrated that neural stem cells in the subventricular zone of adult rats can be labeled by directly injecting MPIOs in the lateral cerebral ventricles. These neural stem cells divide to produce neural progenitor cells which migrate exclusively to the olfactory bulb along a very well defined route (Doetsch and Alvarez-Buylla, 1996). Following endocytosis by the stem cell, the particle can be transferred to the daughter neural progenitor which can carry it while it migrates. The results are shown in Figure 7.11. Hypointense contrast was visualized along the rostral migratory stream leading from the ventricle into the olfactory bulbs. Immunohistochemistry showed that particles had labeled neural stem cells in the SVZ, migrating neuroblasts in the rostral migratory stream, and mature neurons in the olfactory bulb. Furthermore, this study established the ability of MRI to longitudinally monitor this migration in the same animal over several weeks. A similar study has since been successfully performed in adult mice (Vreys et al., 2006), and the migration of endogenous neural progenitor cells has also been detected in embryonic mice (Deans et al., 2006).

In a different study, the phagocytotic ability of peripheral macrophages was taken advantage of (Wu et al., 2006). MPIOs were injected i.v. into animals that

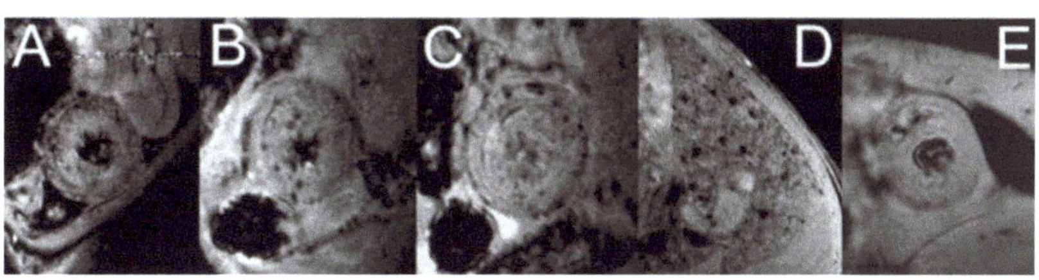

FIGURE 7.12. In vivo MRI of allograft hearts and lungs, 1 day after i.v. injection of MPIO particles. **A** Allograft heart on POD 5. **B** and **C** Allograft heart on POD 6. **D** Allograft lung on POD 6. **E** Isograft heart on POD 6. Shown with 156 μm in-plane resolution at 4.7 T by using a Bruker Biospec AVANCE-DBX MRI instrument. Reproduced from Wu et al. (2006), copyright 2006, with permission from National Academy of Sciences, USA.

had received organ transplantations. While MPIOs have an extremely fast half-life in the blood (less than 5 minutes, Shapiro et al., 2007), macrophages need only ingest one or a few particles to contain enough iron for single cell detection. After 24 hours of injection, dark contrast spots were observed in the transplanted organs, due to the immune response of the host to the transplant (Figure 7.12). Interestingly, the number of dark spots gradually increased, suggesting that perhaps macrophages resident in other tissues had endocytosed particles and were homing to the rejected organ.

7.8. Conclusions

MPIOs offer distinct advantages over nanometer-sized iron oxide particles for cellular MRI. Their high iron contents allow individual particles to be detected. Cell labeling is often accomplished by simple overnight incubation of cell cultures with stoichiometric numbers of particles. Cells can be safely labeled with large amounts of iron, facilitating the detection of single cells in living animals. Most unique, however, is the principle that because MPIOs pack so much iron into a single mass, adequate in vivo cell labeling for MRI-based cell tracking becomes possible, because only a few particles are required for detection.

While most cell tracking studies using MPIOs take advantage of gradient echo, susceptibility-weighted imaging, MPIOs may yet still offer new advantages over nanosized particles for pulse sequences which attempt to create bright contrast (Cunningham et al., 2005; Zurkiya and Hu, 2006). Further improvements may also come from the design of MPIOs with coatings that prolong their half-life in blood, such as PEG. For eventual clinical applications, it may be desirable to construct MPIOs with dextran coatings.

References

Ahrens, E.T., Feili-Hariri, M., Xu, H., Genove, G., Morel, P.A., 2003. Receptor-mediated endocytosis of iron-oxide particles provides efficient labeling of dendritic cells for in vivo MR imaging. Magn Reson Med 49, 1006–1013.

Arbuthnott, G.W., Dunnett, S.B., 1990. Indentification of grafted neurons with fluorescent-labelled microspheres. In: Dunnett, S.B., Richards, S.J. (Eds.), Progress in Brain Research. Elsevier Science Publisher B.V., Amsterdam, pp. 385–390.

Artemov, D., Bhujwalla, Z.M., Bulte, J.W., 2004. Magnetic resonance imaging of cell surface receptors using targeted contrast agents. Curr Pharm Biotechnol 5, 485–494.

Benderbous, S., Corot, C., Jacobs, P., Bonnemain, B., 1996. Superparamagnetic agents: physicochemical characteristics and preclinical imaging evaluation. Acad Radiol 3 Suppl 2, S292–S294.

Bowen, C.V., Zhang, X., Saab, G., Gareau, P.J., Rutt, B.K., 2002. Application of the static dephasing regime theory to superparamagnetic iron-oxide loaded cells. Magn Reson Med 48, 52–61.

Bulte, J.W., Hoekstra, Y., Kamman, R.L., Magin, R.L., Webb, A.G., Briggs, R.W., Go, K.G., Hulstaert, C.E., Miltenyi, S., The, T.H., 1992. Specific MR imaging of human lymphocytes by monoclonal antibody-guided dextran-magnetite particles. Magn Reson Med 25, 148–157.

Callaghan, P.T., 1993. Principles of nuclear magnetic resonance microscopy. Clarendon Press, Oxford, England.

Corot, C., Petry, K.G., Trivedi, R., Saleh, A., Jonkmanns, C., Le Bas, J.F., Blezer, E., Rausch, M., Brochet, B., Foster-Gareau, P., Baleriaux, D., Gaillard, S., Dousset, V., 2004. Macrophage imaging in central nervous system and in carotid atherosclerotic plaque using ultrasmall superparamagnetic iron oxide in magnetic resonance imaging. Invest Radiol 39, 619–625.

Cunningham, C.H., Arai, T., Yang, P.C., McConnell, M.V., Pauly, J.M., Conolly, S.M., 2005. Positive contrast magnetic resonance imaging of cells labeled with magnetic nanoparticles. Magn Reson Med 53, 999–1005.

Deans, A.E., Wadghiri, Y.Z., Turnbull, D.H., 2006. MRI of neural progenitor migration in the developing mouse brain. Proc 14th ISMRM, abstract #355.

Dick, A.J., Guttman, M.A., Raman, V.K., Peters, D.C., Pessanha, B.S., Hill, J.M., Smith, S., Scott, G., McVeigh, E.R., Lederman, R.J., 2003. Magnetic resonance fluoroscopy allows targeted delivery of mesenchymal stem cells to infarct borders in Swine. Circulation 108, 2899–2904.

Dodd, S.J., Williams, M., Suhan, J.P., Williams, D.S., Koretsky, A.P., Ho, C., 1999. Detection of single mammalian cells by high-resolution magnetic resonance imaging. Biophys J 76, 103–109.

Doetsch, F., Alvarez-Buylla, A., 1996. Network of tangential pathways for neuronal migration in adult mammalian brain. Proc Natl Acad SciUSA 93, 14895–14900.

Dunning, M.D., Kettunen, M.I., Ffrench, C.C., Franklin, R.J., Brindle, K.M., 2006. Magnetic resonance imaging of functional Schwann cell transplants labelled with magnetic microspheres. Neuroimage 31, 172–180.

Foster-Gareau, P., Heyn, C., Alejski, A., Rutt, B.K., 2003. Imaging single mammalian cells with a 1.5 T clinical MRI scanner. Magn Reson Med 49, 968–971.

Frank, J.A., Miller, B.R., Arbab, A.S., Zywicke, H.A., Jordan, E.K., Lewis, B.K., Bryant, L.H., Jr., Bulte, J.W., 2003. Clinically applicable labeling of mammalian and stem cells by combining superparamagnetic iron oxides and transfection agents. Radiology 228, 480–487.

Frank, J.A., Zywicke, H., Jordan, E.K., Mitchell, J., Lewis, B.K., Miller, B., Bryant, L.H., Jr., Bulte, J.W., 2002. Magnetic intracellular labeling of mammalian cells by combining (FDA-approved) superparamagnetic iron oxide MR contrast agents and commonly used transfection agents. Acad Radiol 9 Suppl 2, S484–S487.

Hawrylak, N., Ghosh, P., Broadus, J., Schlueter, C., Greenough, W.T., Lauterbur, P.C., 1993. Nuclear magnetic resonance (NMR) imaging of iron oxide-labeled neural transplants. Exp Neurol 121, 181–192.

Heyn, C., Ronald, J.A., Mackenzie, L.T., MacDonald, I.C., Chambers, A.F., Rutt, B.K., Foster, P.J., 2006. In vivo magnetic resonance imaging of single cells in mouse brain with optical validation. Magn Reson Med 55, 23–29.

Hill, J.M., Dick, A.J., Raman, V.K., Thompson, R.B., Yu, Z.X., Hinds, K.A., Pessanha, B.S., Guttman, M.A., Varney, T.R., Martin, B.J., Dunbar, C.E., McVeigh, E.R., Lederman, R.J., 2003. Serial cardiac magnetic resonance imaging of injected mesenchymal stem cells. Circulation 108, 1009–1014.

Hinds, K.A., Hill, J.M., Shapiro, E.M., Laukkanen, M.O., Silva, A.C., Combs, C.A., Varney, T.R., Balaban, R.S., Koretsky, A.P., Dunbar, C.E., 2003. Highly efficient endosomal labeling of progenitor and stem cells with large magnetic particles allows magnetic resonance imaging of single cells. Blood 102, 867–872.

Hogemann, D., Josephson, L., Weissleder, R., Basilion, J.P., 2000. Improvement of MRI probes to allow efficient detection of gene expression. Bioconjug Chem 11, 941–946.

Jiang, Q., Zhang, Z.G., Ding, G.L., Zhang, L., Ewing, J.R., Wang, L., Zhang, R., Li, L., Lu, M., Meng, H., Arbab, A.S., Hu, J., Li, Q.J., Pourabdollah Nejad, D.S., Athiraman, H., Chopp, M., 2005. Investigation of neural progenitor cell induced angiogenesis after embolic stroke in rat using MRI. Neuroimage 28, 698–707.

Josephson, L., Tung, C.H., Moore, A., Weissleder, R., 1999. High-efficiency intracellular magnetic labeling with novel superparamagnetic-Tat peptide conjugates. Bioconjug Chem 10, 186–191.

Lauterbur, P.C., Bernardo, M.L., Jr., Mendonca Dias, M.H., Heldman, A.W., 1986. Microscopic NMR imaging of the magnetic fields around magnetite particles. Proc 5th SMRM, 229.

Perez, J.M., Josephson, L., O'Loughlin, T., Hogemann, D., Weissleder, R., 2002a. Magnetic relaxation switches capable of sensing molecular interactions. Nat Biotechnol 20, 816–820.

Perez, J.M., O'Loughin, T., Simeone, F.J., Weissleder, R., Josephson, L., 2002b. DNA-based magnetic nanoparticle assembly acts as a magnetic relaxation nanoswitch allowing screening of DNA-cleaving agents. J Am Chem Soc 124, 2856–2857.

Ponder, K.P., Gupta, S., Leland, F., Darlington, G., Finegold, M., DeMayo, J., Ledley, F.D., Chowdhury, J.R., Woo, S.L., 1991. Mouse hepatocytes migrate to liver parenchyma and function indefinitely after intrasplenic transplantation. Proc Natl Acad Sci USA 88, 1217–1221.

Raynal, I., Prigent, P., Peyramaure, S., Najid, A., Rebuzzi, C., Corot, C., 2004. Macrophage endocytosis of superparamagnetic iron oxide nanoparticles: mechanisms and comparison of ferumoxides and ferumoxtran-10. Invest Radiol 39, 56–63.

Renshaw, P.F., Owen, C.S., Evans, A.E., Leigh, J.S., Jr., 1986. Immunospecific NMR contrast agents. Magn Reson Imaging 4, 351–357.

Rodriguez, O., Fricke, S., Chien, C., Dettin, L., Vanmeter, J., Shapiro, E., Dai, H.N., Casimiro, M., Ileva, L., Dagata, J., Johnson, M.D., Lisanti, M.P., Koretsky, A., Albanese, C., 2006. Contrast-enhanced in vivo imaging of breast and prostate cancer cells by MRI. Cell Cycle 5, 113–119.

Saini, S., Stark, D.D., Hahn, P.F., Bousquet, J.C., Introcasso, J., Wittenberg, J., Brady, T.J., Ferrucci, J.T., Jr., 1987a. Ferrite particles: a superparamagnetic MR contrast agent for enhanced detection of liver carcinoma. Radiology 162, 217–222.

Saini, S., Stark, D.D., Hahn, P.F., Wittenberg, J., Brady, T.J., Ferrucci, J.T., Jr., 1987b. Ferrite particles: a superparamagnetic MR contrast agent for the reticuloendothelial system. Radiology 162, 211–216.

Shapiro, E.M., Gonzalez-Perez, O., Garcia-Verdugo, J.M., Alvarez-Buylla, A., Koretsky, A.P., 2006a. Magnetic resonance imaging of the migration of neuronal precursors generated in the adult rodent brain. Neuroimage. 32, 1150–1157.

Shapiro, E.M., Medford-Davis, L.N., Dunbar, C.E., Koretsky, A.P., 2006b. Antibody mediated cell labeling of peripheral T cells with micron sized iron oxide particles (MPIOs) allows single cell detection by MRI. Proc 14th ISMRM, abstract # 1868.

Shapiro, E.M., Medford-Davis, L.N., Fahmy, T., Dunbar, C.E., Koretsky, A.P., 2007. Antibody-red cell labeling of peripheral T cells with micron-Sized iron oxide particles

(MPIOs) allows single cell detection by MRI, Contest Media and Molecular Imaging, 2, 53.

Shapiro, E.M., Sharer, K., Skrtic, S., Koretsky, A.P., 2006c. In vivo detection of single cells by MRI. Magn Reson Med 55, 242–249.

Shapiro, E.M., Skrtic, S., Koretsky, A.P., 2005. Sizing it up: cellular MRI using micron-sized iron oxide particles. Magn Reson Med 53, 329–338.

Shapiro, E.M., Skrtic, S., Sharer, K., Hill, J.M., Dunbar, C.E., Koretsky, A.P., 2004. MRI detection of single particles for cellular imaging. Proc Natl Acad Sci USA 101, 10901–10906.

Tanimoto, A., Pouliquen, D., Kreft, B.P., Stark, D.D., 1994. Effects of spatial distribution on proton relaxation enhancement by particulate iron oxide. J Magn Reson Imaging 4, 653–657.

Tracy, K., Yin, J.J., Munasinghe, J., Shapiro, E.M., Koretsky, A.P., Kelly, K., 2006. The role of VEGF in growth of brain metastases from single cells as visualized by contrast enhanced MRI. Proc 14th ISMRM, abstract # 466.

Vreys, R., Peleman, C., Geraerts, M., De Cuyper, M., Debyser, Z., Baekelandt, V., Ven der Linden, A., 2006. Validation of magnetoliposomes as MR contrast agents for in situ labeling of endogenous neural progenitor cells in the mouse brain. Proc 14th ISMRM, abstract #356.

Weisskoff, R.M., Zuo, C.S., Boxerman, J.L., Rosen, B.R., 1994. Microscopic susceptibility variation and transverse relaxation: theory and experiment. Magn Reson Med 31, 601–610.

Wu, Y.L., Ye, Q., Foley, L.M., Hitchens, T.K., Sato, K., Williams, J.B., Ho, C., 2006. In situ labeling of immune cells with iron oxide particles: an approach to detect organ rejection by cellular MRI. Proc Natl Acad Sci USA 103, 1852–1857.

Yeh, T.C., Zhang, W., Ildstad, S.T., Ho, C., 1993. Intracellular labeling of T-cells with superparamagnetic contrast agents. Magn Reson Med 30, 617–625.

Zhang, R.L., Zhang, L., Zhang, Z.G., Morris, D., Jiang, Q., Wang, L., Zhang, L.J., Chopp, M., 2003. Migration and differentiation of adult rat subventricular zone progenitor cells transplanted into the adult rat striatum. Neuroscience 116, 373–382.

Zhang, Z., Jiang, Q., Jiang, F., Ding, G., Zhang, R., Wang, L., Zhang, L., Robin, A.M., Katakowski, M., Chopp, M., 2004. In vivo magnetic resonance imaging tracks adult neural progenitor cell targeting of brain tumor. Neuroimage. 23, 281–287.

Zurkiya, O., Hu, X., 2006. Off-resonance saturation as a means of generating contrast with superparamagnetic nanoparticles. Magn Reson Med 56, 726–732.

8
Molecular MR Imaging with Paramagnetic Perfluorocarbon Nanoparticles

Patrick M. Winter, Shelton D. Caruthers, Anne H. Schmieder, Anne M. Neubauer, Gregory M. Lanza, and Samuel A. Wickline

Abstract: Targeted contrast agents, such as perfluorocarbon (PFC) nanoparticles, have been developed to allow conventional imaging modalities, including MRI, to detect and characterize specific pathological biomarkers of early disease rather than simply observe the anatomical manifestations occurring at very late stages. PFC nanoparticles are typically 200–300 nm in diameter and are encapsulated in a phospholipid shell, which provides an ideal surface for the incorporation of targeting ligands and/or imaging agents. Through chemical modification of the paramagnetic chelates incorporated on the particle surface, nanoparticle relaxivity as well as stability can be increased to improve the efficacy of MR molecular imaging. PFC nanoparticles can be targeted to a number of different biological epitopes, including fibrin, an abundant marker of ruptured atherosclerotic plaques; $\alpha_v\beta_3$-integrin, an endothelial biomarker of angiogenesis associated with atherosclerosis, tumor growth, and vascular injury; collagen III, a component of the extracellular matrix that is exposed after balloon angioplasty; and tissue factor, a vascular smooth muscle cell (VSMC) marker that is overexpressed following vascular injury. In addition to paramagnetic nanoparticles for ^{1}H MRI, the PFC core has a high fluorine content that can be detected with ^{19}F MRI, providing unambiguous and quantitative mapping of the contrast agent distribution. Another distinctive advantage of PFC nanoparticles for molecular imaging applications is their compatibility with several imaging modalities, including MRI, ultrasound, nuclear imaging, and CT.

Keywords: molecular imaging, MRI, nanoparticle, angiogenesis, atherosclerosis, cancer.

8.1. Introduction

Perfluorocarbon (PFC) nanoparticle emulsions are similar to but distinct from liposomes, a popular structure used in contrast agent design and drug delivery (Banerjee, 2001). PFC nanoparticles consist of a heterogeneous mixture containing at least one immiscible liquid intimately dispersed in another, forming droplets with diameters around 200 nm. PFCs have a similar structure as hydrocarbons, except that all the hydrogen atoms are replaced with fluorine. They are clear, colorless, and odorless compounds that do not conduct electricity and are not flammable. They are volatile, biologically inert, chemically stable, nontoxic, and do not become degraded in the body through metabolism (Flaim, 1994). The PFCs typically used for molecular imaging with MRI, perfluorooctylbromide and perfluoro-15-crown-5-ether, are liquid at room temperature and atmospheric pressure, but other PFC species are gaseous under these conditions.

PFC nanoparticles are stabilized for use in vivo with surfactants, the most common being phospholipids, which restrict the ability of the perfluorcarbon cores to coalesce with one another. The phospholipid surface provides an ideal location for the incorporation of specialized components, such as targeting ligands or imaging agents. The phospholipid monolayer constrains these compounds to only one orientation on the nanoparticles: the hydrophilic end (paramagnetic chelate or targeting molecule) points out into the surrounding biological environment, while the hydrophobic end intercolates into the surface of the particle. Because they are chemically inert, highly biocompatible, and stable in vivo, perfluorocarbon emulsions have been investigated for a number of medical uses, including artificial blood (Riess, 2005), liquid ventilation (Hamilton et al., 2005), and x-ray contrast agents (Mattrey, 1994; Mattrey and Long, 1988).

To reach and interact with molecular markers, targeted contrast agents must have appropriate pharmacokinetic and pharmacodynamic profiles. This requires proper design of the surface chemistry and particle size to achieve acceptable in vivo biodistribution and clearance. Nanoparticles clear from the blood stream according to a bi-exponential function with an estimated circulatory half-life greater than 1 hour, which may represent a nearly ideal retention time. Approximately 2–3 hours are required to saturate a target in contact with the blood pool, corresponding to the time needed for all blood to make a single pass through a remote vascular bed. Although higher doses might increase the observed signal, they could also produce a significant blood pool signal, obscuring whether the agent is actually binding to the desired epitope or merely circulating through the region of interest. Minimizing the blood pool signal is particularly important when considering agents that have a very high signal intensity and/or a prolonged half-life.

8.2. Paramagnetic Payload

The amount of MRI enhancement achieved with paramagnetic PFC nanoparticles depends upon the number of nanoparticles bound and the paramagnetic payload of each particle. One way to improve the MRI enhancement is to increase the relaxivity per nanoparticle by attaching more gadolinium agents to each individual particle. In this way, the relaxivity at 1.5 T can be increased from 180,000 (s mM)$^{-1}$ (10 mol%) to 540,000 (s mM)$^{-1}$ (40 mol%), in terms of particle concentration, without reaching a saturation plateau (Flacke et al., 2001). Increasing the payload, however, can lead to impaired particle stability and might also interfere with the binding affinity of the targeting ligands. Therefore, the paramagnetic chelate is usually limited to 30–40 mol% of the total surfactant, yielding 50,000–100,000 gadolinium agents per particle.

Another avenue of increasing nanoparticle relaxivity is to maximize the T_1 shortening effect of the paramagnetic chelate. The relaxivity of a paramagnetic contrast agent depends upon the effective correlation time (τ_c), a composite function of the rotational correlation time (τ_r), the electron spin relaxation time (τ_s), and the water exchange time (τ_M), with the overall effect being dominated by the fastest process. The rotational correlation time is largely independent of particle size in the range of 50–400 nm (Burnell et al., 1980); therefore, efforts to increase relaxivity by altering particle size are ineffective (Tilcock et al., 1992). However, manipulation of the water exchange time could lead to significant increases in the relaxivity of paramagnetic nanoparticles.

Variable field relaxometry allows evaluation of paramagnetic chelates with differing chemistries, such as different chelating structures or lipid conjugation strategies. These measurements show that the r_1 dispersion of paramagnetic nanoparticles is dominated by the long rotational correlation time (τ_r) of the slowly tumbling complex (Figure 8.1) (Winter et al., 2003b), as a consequence of the relatively large particle size. This produces an r_1 dispersion curve with a gradual increase at low frequency to a "peak" relaxivity around 30 MHz followed by a gradual decrease at higher frequencies. Nanoparticles formulated with Gd-DTPA-phosphatidyl ethanolamine (Gd-DTPA-PE) have a much higher magnitude of relaxivity compared to Gd-DTPA-bis-oleate (Gd-DTPA-BOA) nanoparticles, illustrating the increased T_1 shortening ability of this formulation. The r_2 values initially follow a similar pattern as r_1, but do not decrease at higher fields, in accordance with expectations based on the Solomon-Bloembergen equations (Wood et al., 1993), due to the nondispersive term involving τ_c.

For Gd-DTPA-BOA nanoparticles, the value of r_1 is largely independent of temperature, whereas r_2 increases slightly at the lower temperature. In contrast, r_1 decreases at the lower temperature in the Gd-DTPA-PE emulsion, while r_2 remains independent of temperature. These temperature-dependence curves show that the water exchange rate is greater for nanoparticles formulated with Gd-DTPA-PE compared to Gd-DTPA-BOA. At the higher temperature, the r_1 of Gd-DTPA-PE nanoparticles increased due to the faster water exchange and increased kinetic activity. The Gd-DTPA-BOA nanoparticles, however, have somewhat restricted water access and do not benefit from the increased kinetic activity of water at the higher temperature. Structurally, Gd-DTPA-PE elevates the chelate away from the phosphate headgroups on the particle surface (Grant et al., 1989) and provides better interaction with the surrounding water molecules in comparison to Gd-DTPA-BOA.

Increasing the relaxivity of paramagnetic nanoparticles, however, would provide no practical benefit if the improved chelate interferes with the targeting efficacy. Therefore, T_1 shortening must be evaluated with the nanoparticles bound

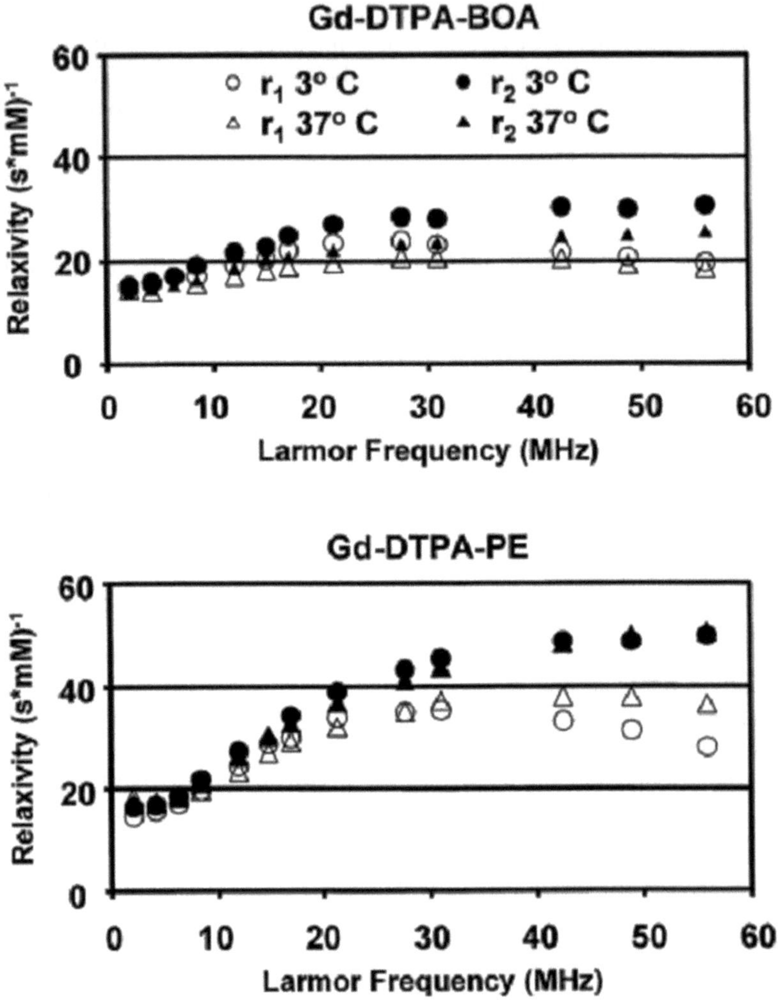

FIGURE 8.1. Relaxivity of PFC nanoparticles measured over a range of magnetic field strengths and at two different temperatures. Nanoparticles formulated with Gd-DTPA-PE as the paramagnetic chelate had a higher r_1 (open symbols) and r_2 (closed symbols) at all fields compared to Gd-DTPA-BOA nanoparticles. The increase in r_1 at the higher temperature shows that Gd-DTPA-PE nanoparticles have ample interaction between the paramagnetic chelate and the surrounding water molecules. On the other hand, Gd-DTPA-BOA nanoparticles do not show an increase in r_1 at the higher temperature, indicating reduced water interaction probably due to the close proximity between the paramagnetic chelate and the phospholipid particle surface. Reproduced with permission from Winter et al. (2003b).

to a biological target, such as fibrin incorporated into thrombus. Relaxation maps collected at 1.5 T show higher r_1 values at the clot surface, demonstrating the paramagnetic influence of the fibrin-targeted nanoparticles (Figure 8.2) (Winter et al., 2003b). Compared with the clot interior, the surface layer of

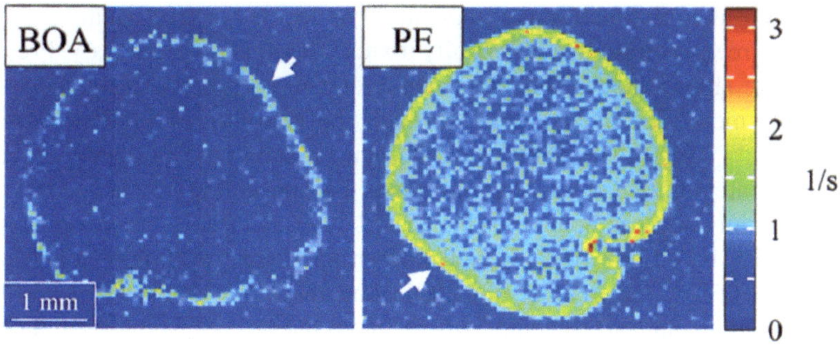

FIGURE 8.2. Color-coded maps of r_1 relaxation rates from human plasma clots imaged on a clinical 1.5 T scanner. Fibrin-targeted Gd-DTPA-BOA nanoparticles (*left*) induced significant T_1 shortening at the clot surface (white arrow) relative to the clot interior. Gd-DTPA-PE nanoparticles, however, produced a 72% larger change in r_1 compared to Gd-DTPA-BOA. Reproduced with permission from Winter et al. (2003b).

Gd-DTPA-BOA nanoparticles increased r_1 at the clot margin by 48% ($\Delta r_1 = 0.45 \pm 0.02$ 1/s), while the Gd-DTPA-PE agent increased r_1 by 72% ($\Delta r_1 = 0.77 \pm 0.02$ 1/s). No significant differences were observed between Gd-DTPA-BOA and Gd-DTPA-PE nanoparticles in terms of the r_1 values obtained inside the clot or in the saline surrounding the clot.

The density of Gd-DTPA-BOA and Gd-DTPA-PE nanoparticles, and their respective gadolinium concentrations over the clot surfaces were compared and did not differ. On average, 0.22 ± 0.01 μmol Gd^{3+} bound to the surface of Gd-DTPA-BOA-treated clots, which had a surface area of 175 ± 17 mm^2. The Gd-DTPA-PE-treated clots had 0.22 ± 0.02 μmol Gd^{3+} bound to a surface area of $190 + 17$ mm^2. Therefore, the increased r_1 observed for Gd-DTPA-PE-treated clots is a direct result of the improved relaxivity of this chelate and not caused by any differences in nanoparticle binding characteristics.

Similar to the improved relaxivity with Gd-DTPA-PE (Winter et al., 2003b), Gd-MeO-DOTA-PE and Gd-MeO-DOTA-triglycine-PE nanoparticles were shown to have much higher particulate relaxivities (1,470,000 and 1,780,000 $(s\ mM)^{-1}$, respectively) than Gd-DTPA-BOA (1,260,000 $(s\ mM)^{-1}$) (Winter et al., 2005a). The MeO-DOTA chelate is a cyclic chelate, which offers better stability and relaxivity compared to linear chelates such as DTPA. Adding a tri-peptide spacer between the chelate and the phospholipid results in the Gd-MeO-DOTA-triglycine-PE structure. The improved relaxivity likely reflects improved water exchange due to displacing the Gd-MeO-DOTA complex beyond the lipid surface. The enhanced relaxivity of Gd-MeO-DOTA-triglycine-PE compared to Gd-MeO-DOTA-PE (~20%) is presumably due to slight further improvements in water exchange because of the triglycine spacer.

Not only does the DOTA chelate offer better relaxivity, but it also provides more stable binding to the paramagnetic metal. Highly stable agents are particularly important for molecular imaging applications. Targeted nanoparticles remain in circulation much longer than small molecule contrast agents and must not release free gadolinium, a potentially toxic compound, into the body. Previous studies have shown that DTPA transmetallation increases proportionately with the loss of metal coordinate bonds through the coupling of lipophilic linkers (Sherry et al., 1988). The DTPA chelate can be stabilized by using functionalization strategies that do not compromise the coordinate bonds securing the gadolinium ion, such as benzyl-DTPA (Sharkey et al., 1990). However, macrocyclic DOTA-based chelates are well known to retain gadolinium much more avidly than DTPA (Laurent et al., 2001) and therefore may be the preferred avenue for molecular imaging applications given the extended biological half-life of targeted contrast agents. The use of MeO-DOTA preserves the coordination of gadolinium to the chelate while providing a reactive isothiocyanate side group for reaction with the terminal amines of the PE anchor.

Using zinc as a competing species (Laurent et al., 2001), the transmetallation of Gd-DTPA-BOA, Gd-MeO-DOTA-PE, and Gd-MeO-DOTA-triglycine-PE nanoparticles was measured (Winter et al., 2005a). Gd-DTPA-BOA nanoparticles displayed rapid transmetallation as expected with the linear chemistry of the chelate and the compromise of two coordination bonds imparted by the lipid modification (Sherry et al., 1988). Nanoparticles formulated with the DOTA-based chelates showed markedly less transmetallation, which was consistent with improved stability of the Gd-MeO-DOTA complex. The retained gadolinium at equilibrium was much higher for the two Gd-MeO-DOTA ligands (91%) compared to the Gd-DTPA chelate (75%).

8.3. Detection of Sparse Epitopes

Some biological epitopes are present in relatively high concentrations, allowing targeted contrast agents to achieve very high contrast-to-noise levels, even in vivo. For instance, fibrin is an abundant component of thrombi and an early marker of ruptured atherosclerotic plaques. Plaque rupture is the proximate cause of myocardial ischemia and stroke, which represent the most prevalent causes of death in America today (Thom et al., 2006). Fibrin-targeted PFC nanoparticles can bind to fibrin fibrils and form a dense layer on the surface of thrombi (Figure 8.3) (Flacke et al., 2001). Contrast-to-noise levels greater than 100 can be achieved with in vivo MRI of stenotic clots formed in experimental animals (Figure 8.4) (Flacke et al., 2001). Many other targets of interest in cardiovascular disease and cancer, such as cell surface receptors, are present in much lower concentrations. Consequently, the signal enhancement observed for these targets tends to be much lower.

The $\alpha_v\beta_3$-integrin is a general marker of angiogenesis that can serve as a biomarker for numerous diseases, including atherosclerosis, inflammation, and cancer (Brooks et al., 1995; Kerr et al., 2001). The $\alpha_v\beta_3$-integrin is a heterodimeric adhesion molecule widely expressed by endothelial cells, monocytes, fibroblasts, and vascular smooth muscle cells (VSMC). It plays

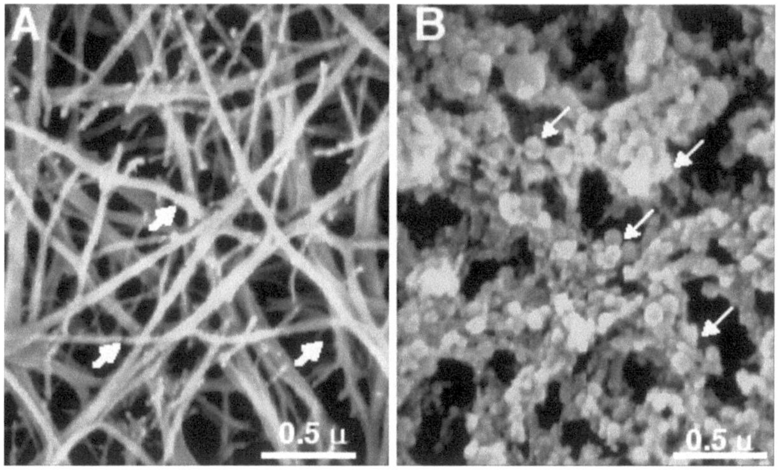

FIGURE 8.3. Scanning electron microscope images of (**A**) untreated fibrin clot and (**B**) clot after application of fibrin-targeted PFC nanoparticles. **A** Fibrin fibrils (arrows) form a tight weave in these acellular clots. **B** Targeted nanoparticles densely bind to the exposed fibrin epitopes. Reproduced with permission from Flacke et al. (2001).

a critical part in smooth muscle cell migration and cellular adhesion (Bishop et al., 2001; Corjay et al., 1999), which are required for the formation of new blood vessels. The $\alpha_v\beta_3$-integrin is expressed on the luminal surface of activated endothelial cells, but not on mature quiescent cells (Brooks et al., 1995). Nanoparticles can be specifically targeted to angiogenesis by covalently coupling peptidomimetic $\alpha_v\beta_3$-integrin antagonists onto the particle surface. $\alpha_v\beta_3$-Integrin-targeted nanoparticles have been used to detect and characterize angiogenesis associated with growth factor expression (Anderson et al., 2000), atherosclerosis (Winter et al., 2003c), and tumor growth (Schmieder et al., 2005; Sipkins et al., 1998; Winter et al., 2003a).

8.3.1. Atherosclerosis

Angiogenesis plays a key role in the development of atherosclerotic plaques. The neovasculature can originate in the adventitia and extend into the thickening intimal layer of the atheroma as well as deriving from the primary arterial lumen (Gossl et al., 2003a,b,c). Increased proliferation of angiogenic vessels is associated with atherosclerotic plaques that cause clinical events, such as unstable angina, myocardial infarction, and stroke (Khurana et al., 2004; Moreno et al., 2004; Tenaglia et al., 1998). These immature vessels may initiate hemorrhaging and instability within the plaque, leading to rupture and thrombosis (Kolodgie et al., 2004).

Cholesterol feeding induces hyperlipidemia in rabbits, serving as a model of early atherosclerosis. Dynamic T_1-weighted MR imaging of the abdominal aorta before and up to 2 hours after intravenous injection of $\alpha_v\beta_3$-targeted nanoparticles

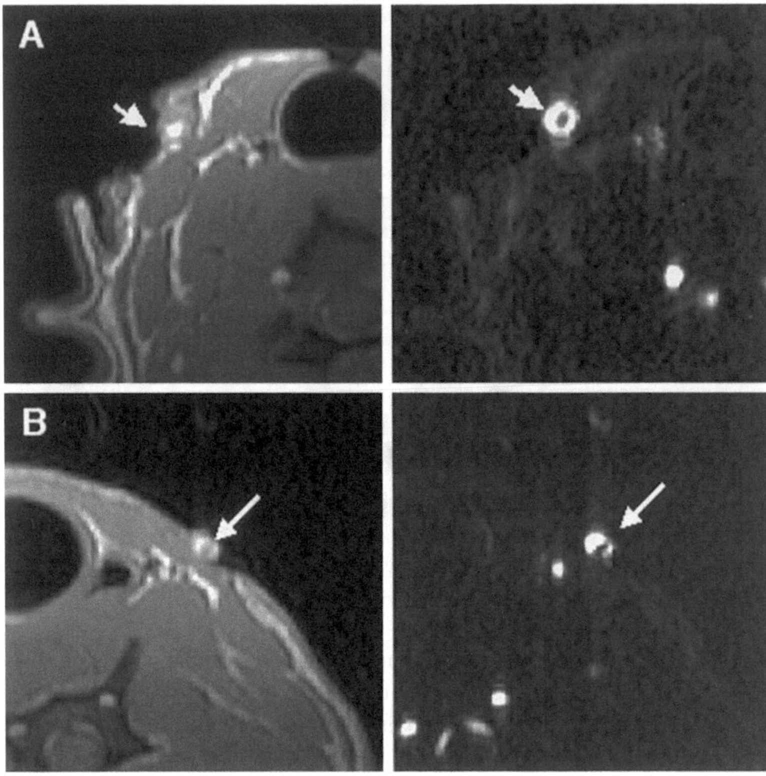

FIGURE 8.4. In vivo molecular imaging of thrombus conducted at 1.5 T with fibrin-targeted PFC nanoparticles. **A** Thrombus in the external jugular vein treated with fibrin-targeted nanoparticles shows clear T_1-weighted contrast enhancement (*left*). Phase-contrast angiogram (*right*) shows flow defect corresponding to the location of the experimentally induced thrombus. **B** Untreated thrombus in the contralateral jugular vein (*left*) had a T_1-weighted MRI signal intensity similar to adjacent muscle. Angiography confirmed the presence of a flow limiting thrombus (*right*). Reproduced with permission from Flacke et al. (2001).

allows mapping the extent and magnitude of angiogenesis (Winter et al., 2003c). Image contrast within the aortic wall showed dramatic heterogeneity along the circumference and length of the aorta (Figure 8.5) indicating diffuse disease progression with focal sites of active plaque formation.

When averaged across all images, aortic signal enhancement among cholesterol-fed rabbits receiving $\alpha_v\beta_3$-targeted paramagnetic nanoparticles showed an increase of $26 \pm 4\%$ and $47 \pm 5\%$ over baseline at 15 and 120 minutes, respectively. In cholesterol-fed rabbits receiving nontargeted nanoparticles, the aortic wall enhanced by $19 \pm 1\%$ within 15 minutes, and remained at $26 \pm 1\%$ from 60 to 120 minutes, probably due to delayed particle washout from the dysmorphic neovasculature. Competitive blockade of angiogenic $\alpha_v\beta_3$-integrins with targeted nonparamagnetic nanoparticles reduced the signal enhancement

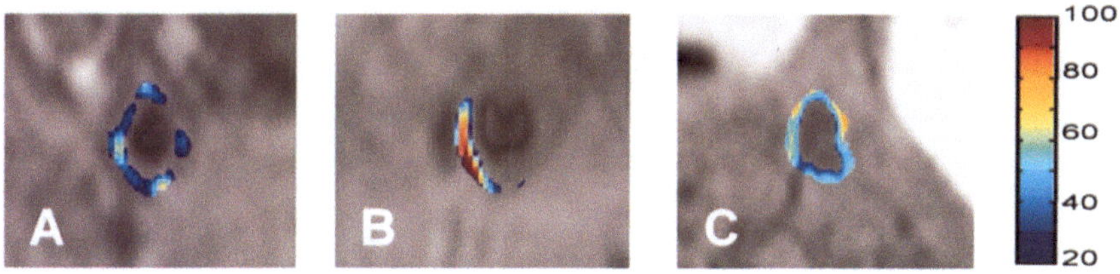

FIGURE 8.5. Molecular imaging of angiogenesis in the aortic wall of a hypercholesteremic rabbit using $\alpha_v\beta_3$-integrin–targeted PFC nanoparticles. Images obtained at three anatomical levels: (**A**) near the renal arteries, (**B**) mid-abdominal aorta, and (**C**) diaphragm, with color-coded overlays depicting the percent enhancement in MRI signal 2 hours after nanoparticle injection. These images illustrate the heterogeneous development of early-stage atherosclerotic plaques, both across slices and within a single slice. Reproduced with permission from Winter et al. (2003c).

of $\alpha_v\beta_3$-targeted paramagnetic nanoparticles by at least 50%. In control-diet rabbits, aortic wall enhancement with $\alpha_v\beta_3$-targeted nanoparticles was much lower than for cholesterol-fed animals. The signal enhancement observed in the adjacent skeletal muscle due to nanoparticles in all treatment groups was negligible relative to that observed in the aortic wall.

Histology of cholesterol-fed rabbit aortas (Figure 8.6) showed mild intimal thickening that was not present in control-diet animals. Immunohistochemistry with a general vascular stain, PECAM, revealed increased capillary density among atherosclerotic rabbits in comparison with the controls, most notably along the adventitia-media interface of the aortic wall. Angiogenic vessels were identified by co-localized staining of $\alpha_v\beta_3$-integrin and PECAM. In hyperlipidemic rabbits, abundant neovasculature was typically observed in close proximity of intimal lesions, while only a very sparse distribution of angiogenic vessels was apparent in control animals.

8.3.2. Vessel Injury

Lumenal expression of $\alpha_v\beta_3$-integrin is limited to angiogenic vasculature, but vessel injury exposes integrins associated with smooth muscle cells. Other epitopes, such as collagen III in the extracellular matrix, also become available for binding to PFC nanoparticles after rupture of the vessel wall. Targeted paramagnetic PFC nanoparticles have been used to visualize vascular damage in the carotid arteries of pigs following balloon injury (Cyrus et al., 2006). Both collagen III and $\alpha_v\beta_3$-integrin were exposed by stretch fracture of the vessel wall and were available immediately after injury for MR molecular imaging. Three-dimensional reconstruction of the MRI enhancement pattern was similar for both molecular targets, indicating sufficient distribution throughout the vessel wall. Treatment with either nontargeted nanoparticles or saline produced no MR contrast enhancement and therefore provided no information about the extent of vessel injury.

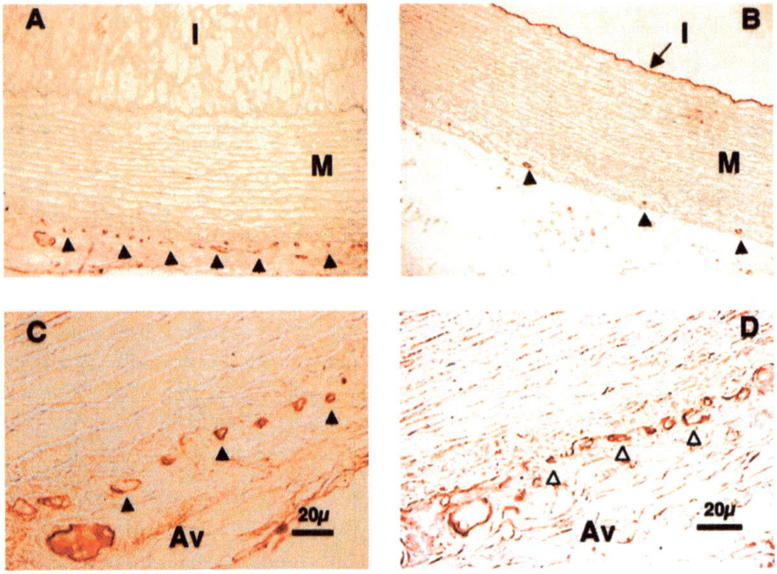

FIGURE 8.6. Immunohistochemistry of $\alpha_v\beta_3$-integrin expression in the abdominal aorta of hyper-cholesteremic (**A, C, D**) and control (**B**) rabbits. A thickened intima (I) is evident at ×200 magnification in the cholesterol-fed rabbit (**A**), but not in the animal fed a control diet (**B**). Higher power magnification (×600) sections stained for $\alpha_v\beta_3$-integrin (**C**) and PECAM (**D**) reveal increased angiogenesis at the interface between the media (M) and the adventitia (Av). Co-localization of $\alpha_v\beta_3$-integrin (solid arrows) and the general vascular marker PECAM (open arrows) demonstrates the presence of angiogenic vessels in the aortic wall. Angiogenic vessels (solid arrows) are much less prevalent in the section from the control animal (**B**). Reproduced with permission from Winter et al. (2003c).

The contrast-to-noise ratio achieved with $\alpha_v\beta_3$-integrin-targeted nanoparticles was much higher (13.8 ± 5.2) compared to collagen III–targeted nanoparticles (3.3 ± 0.3), probably reflecting the relative density of each biomarker and/or the relative avidity of each targeting probe. The low molecular weight (~1000 d) of the $\alpha_v\beta_3$-integrin peptidomimetic allows 250–300 homing ligands to be incorporated onto the surface of each nanoparticle, whereas only 25–50 of the larger (50,000 d) collagen III fragments can be attached to each particle. Morphological analysis of the vessel injury was similar for both $\alpha_v\beta_3$-integrin-targeted and collagen III-targeted nanoparticles. The injury length was ~31 mm, which exceeded the actual balloon length (20 mm) by 50%, and the total injury volume was ~950 mm^3.

Histology of the carotid arteries showed irregular loss of endothelium and disruptions in the media and internal elastic lamina as well as fractures reaching into the adventitia. Morphometric analyses from vessels treated with $\alpha_v\beta_3$-targeted, collagen III-targeted, nontargeted nanoparticles or saline were not significantly different. The presence and distribution of $\alpha_v\beta_3$-integrin and collagen III in the media were confirmed histologically.

8.3.3. Cancer

Molecular imaging of angiogenesis also provides a unique opportunity to study the growth of solid tumors. High-resolution imaging of $\alpha_v\beta_3$-integrin expression on the neovasculature of minute human melanoma tumors implanted in nude mice was demonstrated using targeted paramagnetic nanoparticles and a clinical 1.5 T MRI scanner (Schmieder et al., 2005). The average xenograft volume 12 days post tumor implantation was similar for animals receiving $\alpha_v\beta_3$-targeted (33.1 ± 7.1 mm^3), nontargeted (26.8 ± 5.4 mm^3), and competitively inhibited (29.5 ± 8.3 mm^3) PFC nanoparticles. Prior to nanoparticle treatment, these minute tumors appeared isointense with respect to skeletal muscle, making it difficult to distinguish them from surrounding tissues (Figure 8.7a). Injection of paramagnetic $\alpha_v\beta_3$-targeted nanoparticles produced abundant MRI signal enhancement, allowing clear identification of the tumor location (Figure 8.7b).

MRI enhancement was observed in a heterogeneous pattern along the tumor periphery among mice treated with $\alpha_v\beta_3$-targeted paramagnetic nanoparticles. The signal contrast in animals receiving $\alpha_v\beta_3$-targeted nanoparticles was 122.8% at 30 minutes post-injection and continued to increase up to 120 minutes (173.1%). Contrast enhancement in animals receiving nontargeted paramagnetic nanoparticles was approximately 50% lower than in the targeted group, presumably due to slow nanoparticle washout from the hyperpermeable neovasculature. Moreover, the enhancing region of angiogenesis was fourfold greater in animals treated with $\alpha_v\beta_3$-targeted nanoparticles than in the nontargeted control. Pretreatment with $\alpha_v\beta_3$-targeted, nonparamagnetic nanoparticles occupied a significant portion of the $\alpha_v\beta_3$-integrin receptors and greatly diminished the MRI contrast compared with $\alpha_v\beta_3$-targeted paramagnetic particles. The reduced signal enhancement observed with nontargeted particles and competitive inhibition illustrates the in vivo targeting specificity of this molecular imaging agent. The signal intensity from nearby muscle showed no significant change in any experimental group.

Routine histologic examination of melanoma xenografts 12 days post-implantation typically revealed a small, consolidated tumor. The tumor size assessed microscopically corroborated the minute dimensions measured by MRI. The tumors were confluent sheets of melanoma cells surrounded peripherally by loose connective tissues and a modest intermingled host microvasculature. The majority of co-localized immunohistochemical staining of $\alpha_v\beta_3$-integrin and PECAM was observed heterogeneously along the outer connective tissue border of the tumors near dermal vasculature or muscle fascia. The neovascular spatial distribution seen microscopically, which was consistent across all experimental groups, was analogous to the contrast enhancement patterns of angiogenesis detected with molecular imaging via $\alpha_v\beta_3$-targeted paramagnetic particles. The neovasculature, as depicted histologically, occupied a smaller fraction of the tumor volume than suggested by the MR images, which presumably reflects the high molecular relaxivity of the nanoparticles and partial volume effects.

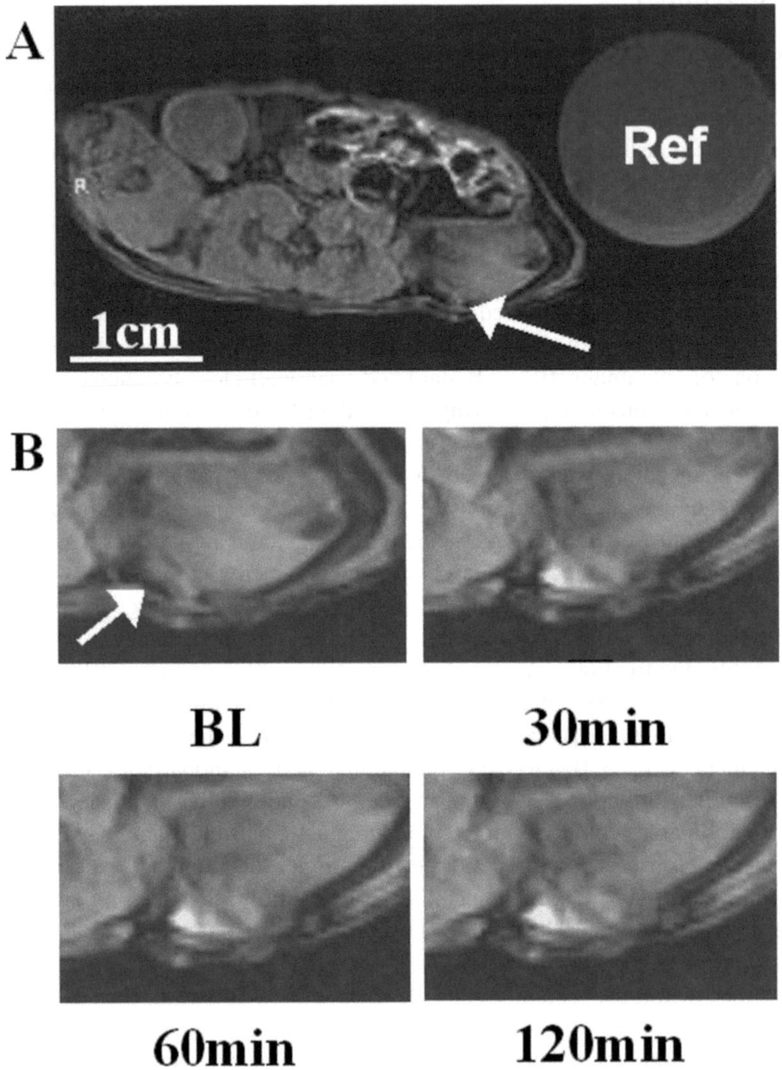

FIGURE 8.7. T_1-weighted MRI of an athymic nude mouse bearing a subcutaneous human melanoma tumor collected on a clinical 1.5 T scanner. **A** The 2 mm tumor (arrow) is impossible to recognize on the image obtained prior to contrast agent injection. A 10 cc syringe filled with Gd-doped water (Ref) is included in the field of view for spatial comparison. **B** Enlarged view of melanoma tumor before (BL = baseline) and 30, 60, and 120 minutes after injection of $\alpha_v\beta_3$-integrin–targeted PFC nanoparticles. Progressive MRI signal enhancement indicates presence of angiogenic vasculature forming in order to support tumor growth. Reproduced with permission from Schmieder et al. (2005).

8.3.4. Tissue Factor

Another sparse molecular epitope of interest in cardiovascular disease is tissue factor. Tissue factor is a transmembrane glycoprotein that is expressed on vascular smooth muscle cells. It is involved in a variety of normal and pathological processes, including thrombosis, hemostasis, angiogenesis, cell signaling,

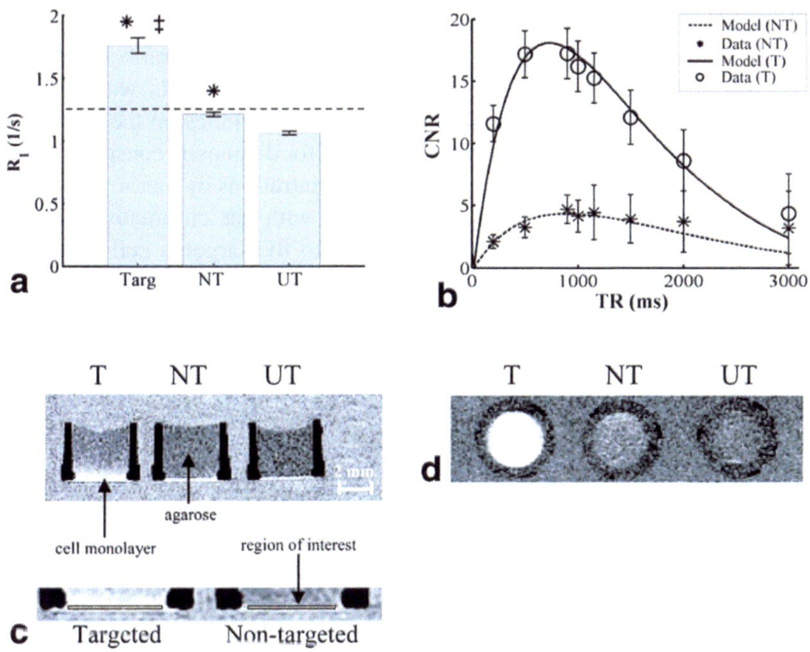

FIGURE 8.8. Molecular imaging of tissue factor expressed on cultured vascular smooth muscle cells (VSMC). **A** Relaxation rates measured at 1.5 T of VSMC monolayers treated with tissue factor–targeted PFC nanoparticles (Targ), nontargeted nanoparticles (NT), and untreated (UT). The dotted line represents the relaxation rate required to discriminate between treated and untreated cells, i.e. produce a contrast-to-noise ratio >5. Nontargeted nanoparticles do not reach this threshold, while tissue factor-targeted particles induce sufficient T_1 shortening to distinguish from native cell cultures. **B** Contrast-to-noise ratio for VSMC treated with tissue factor–targeted and nontargeted nanoparticles imaged with spin echo pulse sequences utilizing a range of repetition times (TR). Experimental data (individual points) agree very closely with a mathematical model of MRI signal enhancement (solid lines). Again, nontargeted nanoparticles do not produce a contrast-to-noise ratio exceeding 5, while tissue factor–targeted nanoparticles can induce a contrast-to-noise ratio over 15. **C** Representative spin echo image collected with a repetition time of 900 ms, displaying cell culture wells. Magnified view indicates regions of interest (black boxes) used to measure signal intensities. **D** A maximum intensity projection through a three-dimensional set of T_1-weighted images obtained parallel to the cell monolayers demonstrating the sensitive detection of VSMC surface markers with tissue factor–targeted PFC nanoparticles. Reproduced with permission from Morawski et al. (2004a).

and mitogenesis. In particular, VSMC are exposed following angioplasty and tissue factor expression is upregulated in order to initiate vascular repair. Molecular imaging of tissue factor expression on cultured VSMC was performed at 1.5 T and the experimental results were compared to a mathematical model of MRI signal enhancement (Morawski et al., 2004a).

Cells treated with tissue factor-targeted nanoparticles displayed a higher longitudinal relaxation rate (R_1) than VSMC incubated with nontargeted or no nanoparticles (Figure 8.8). The concentration of nanoparticles calculated from the measured changes in T_1 for targeted cells (468 $\pm$ 30 pM) substantially exceeded the predicted threshold required to produce a contrast-to-noise ratio greater than 5 (113 pM). The contrast-to-noise ratio of the targeted cells was 17.7 with respect to untreated cells. The concentration of nanoparticles on the nontargeted cell cultures (88 $\pm$ 3 pM) fell below the threshold for diagnostic conspicuity.

We confirmed the predicted nanoparticle concentrations by measuring the total perfluorocarbon content of the VSMC samples with gas chromatography. The absolute concentration of nanoparticles bound to the targeted cell monolayers substantially exceeded that for the nontargeted layers (530 $\pm$ 42 vs 111 $\pm$ 25 pM, respectively). These values did not differ significantly from those predicted from T_1 measurements. The relationship between cellular nanoparticle concentrations determined by gas chromatography and MRI from T_1 measurements was linear with a slope of 1.09.

The MRI contrast between the cell layers varied substantially at different TR values, with all other scanning parameters held constant. This quantitative dependence of contrast on TR agreed closely with the mathematical model for both the targeted and the nontargeted cell layers (Figure 8.8b). The greatest contrast between the targeted and the untreated cell monolayers was observed for a TR of 900 ms. Using a standard three-dimensional imaging technique, high contrast was observed between the targeted cell monolayers and the other treatment groups, demonstrating the ability of tissue factor-targeted nanoparticles to image a single layer of cells at 1.5 T (Figure 8.8d).

8.4. ^{19}F Imaging

The liquid perfluorocarbon core represents 98% of the total nanoparticle volume, leading to a ^{19}F concentration of 100 M within perfluorooctylbromide particles. Like hydrogen, fluorine is a spin 1/2 nucleus and can be detected with MR imaging and spectroscopy. Fluorine has a very high gyromagnetic ratio, 83% relative to ^{1}H, and is 100% naturally abundant, generating a high signal intensity. In addition, there is no background ^{19}F signal in the body, allowing ^{19}F MRI to provide definitive identification and localization of PFC nanoparticles.

Because fluorine has seven outer-shell electrons, in contrast to one for hydrogen, the chemical shift range is very wide. The large separation between ^{19}F resonances allows selective excitation and/or detection of individual signals, either from a single compound or different chemical species of PFCs. The ^{19}F signal from site-targeted nanoparticles has been measured at 1.5 T with spectroscopy as well as with imaging utilizing a steady-state pulse sequence (Caruthers et al., 2006). Both imaging and spectroscopy could detect and distinguish nanoparticles containing either perfluorooctylbromide (PFOB) or perfluoro-15-crown-5-ether (CE) as the core material. The signal-to-noise ratio for PFOB was lower than CE (10 vs 25, respectively), presumably due to the single CE peak (20 equivalent fluorine atoms) compared to the multiple PFOB peaks (17 fluorine atoms distributed over 5 peaks). A clear linear relationship between the ^{19}F signal intensity and perfluorocarbon concentration was demonstrated for both PFOB and CE using both imaging and spectroscopy.

Given their high fluorine content, fibrin-targeted nanoparticles may provide quantification of exposed fibrin, indicating the size and/or number of ruptures in the fibrous cap. Imaging human carotid endarterectomy samples at 4.7 T, multi-slice ^{1}H MRI showed high levels of signal enhancement along the luminal surface due to binding of targeted paramagnetic nanoparticles to fibrin deposits (Morawski et al., 2004b). A ^{19}F projection image of the artery, acquired in less than 5 minutes, shows the asymmetric distribution of fibrin-targeted nanoparticles around the vessel wall corroborating the ^{1}H signal enhancement (Figure 8.9). Spectroscopic quantification of nanoparticle binding allowed calibration of the ^{19}F MRI signal intensity. Co-registration of the quantitative nanoparticle map with the ^{1}H image permits visualization of anatomical and pathological information in a single image. Combining information from ^{1}H and ^{19}F MRI could allow prediction of subsequent occlusion or distal embolization, and aid clinical decision-making for acute invasive intervention versus pharmaceutical therapies.

In addition to molecular imaging of fibrin, ^{19}F MRI has been used to track stem cells labeled with PFC nanoparticles (Ahrens et al., 2005). Stem cell tracking by MRI has typically relied upon iron oxide (Ahrens et al., 2003; Bulte et al., 2004;

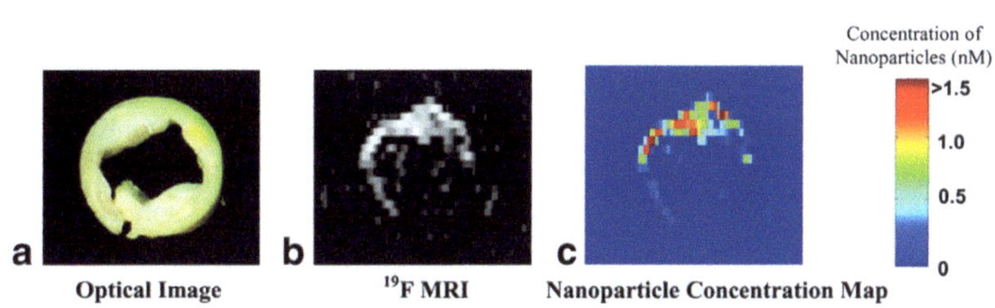

FIGURE 8.9. **A** Cross-sectional photograph of 5 mm section of a human carotid endarterectomy sample showing moderate lumenal narrowing and several atherosclerotic lesions. **B** ^{19}F MRI performed at 4.7 T showing high signal intensity along the lumenal border due to binding of fibrin-targeted PFC nanoparticles to ruptured plaques. **C** Color-coded concentration map of bound nanoparticles based upon the ^{19}F image intensity. Reproduced with permission from Morawski et al. (2004b).

Kircher et al., 2003; Yeh et al., 1993) or gadolinium agents (Jacobs et al., 1994; Modo et al., 2004), but definitive identification of labeled cells can be difficult due to innate susceptibility artifacts and variations in ^{1}H signal intensity within the body. ^{19}F MRI provides unambiguous identification of labeled cells because there is no confounding background signal. After labeling with CE nanoparticles, the ^{19}F spectrum from pelleted cells showed a single sharp resonance, 150 Hz linewidth, which is ideal for imaging with standard pulse sequences (Ahrens et al., 2005). Spectroscopic quantitation revealed that each labeled cell contained ~0.25 ng of PFC. Intracellular localization of the nanoparticles was confirmed by electron microscopy, revealing accumulation in vacuoles. Labeling with nanoparticles had minimal effect on cellular viability or phenotype. No cellular toxicity or altered proliferation was observed and the expression of cell surface markers, including CD80 and MHC, was not effected. Labeled cells delivered via intramuscular or intravenous injection could be detected and tracked with ^{19}F MRI. Due to the efficient cellular labeling and the high fluorine content of the nanoparticles, in vivo ^{19}F images were obtained with a SNR ~10 and a minimum detection limit of 2×10^5 labeled cells per voxel.

8.5. Multi-Modality Imaging

Another distinctive advantage of PFC nanoparticles for molecular imaging applications is their compatibility with several different imaging modalities. Nanoparticles can be used with ultrasound imaging because the speed of sound is much slower in the perfluorocarbon core compared to tissue, imparting inherent acoustic contrast when bound to biological epitopes (Hall et al., 2001; Lanza et al., 2000a,b,1996; Marsh et al., 2002). As described previously, nanoparticles can be formulated with paramagnetic chelates on the outer surface for MRI detection (Flacke et al., 2001) and the high ^{19}F content of the core provides a unique signal with no competing background in the body (Caruthers et al., 2006; Morawski et al., 2004b). Similar to paramagnetic chelates used for MRI, the nanoparticles can be formulated to carry radioisotopes, such as indium and technetium, for SPECT imaging (Hu et al., 2007). The PFC core is also inherently radio-opaque, producing modest contrast on x-ray or CT images (Mattrey, 1994; Mattrey and Long, 1988). The x-ray contrast can be increased by replacing the PFC in the core with a more highly attenuating material, such as iodinated oils, without altering the particle size or targeting (Winter et al., 2005b). The ability to image nanoparticles with MRI, ultrasound, SPECT, and CT provides the opportunity to choose the modality best suited for the particular needs of a given application, such as high resolution, high sensitivity, quantitation, etc. The nanoparticles may also be uniquely suitable for use with hybrid imaging systems, such as SPECT/CT or SPECT/MRI. In addition, the contrast agent can be specifically designed to be compatible with traditional clinical indices, such as calcium scoring by CT.

8.6. Conclusions

Traditional clinical imaging techniques provide gross anatomical characterization of diseases at very late stages of progression, where treatment options are limited and clinical prognosis is poor. Targeted contrast agents, such as PFC nanoparticles, are helping to shift the focus from imaging the anatomical manifestations of disease to the detection and characterization of specific pathological biomarkers. The nanoparticle platform allows the assembly of vast numbers of paramagnetic agents for reliable detection of very sparse epitopes with T_1-weighted MRI. This technology has been demonstrated in numerous animal models for the detection of early stage tumor growth, atherosclerosis, thrombosis, and vascular injury. In addition, the core material provides a unique ^{19}F signal that can be exploited for confirmation, quantitation, or differentiation of the signal enhancement observed by ^{1}H MRI. These approaches may provide a means to diagnose disease at the earliest and most easily treatable stages, which is expected to improve clinical prognosis.

References

Ahrens, E.T., Feili-Hariri, M., Xu, H., Genove, G., Morel, P.A., 2003. Receptor-mediated endocytosis of iron-oxide particles provides efficient labeling of dendritic cells for in vivo MR imaging. Magn Reson Med 49, 1006–1013.

Ahrens, E.T., Flores, R., Xu, H., Morel, P.A., 2005. In vivo imaging platform for tracking immunotherapeutic cells. Nat Biotechnol 23, 983–987.

Anderson, S.A., Rader, R.K., Westlin, W.F., Null, C., Jackson, D., Lanza, G.M., Wickline, S.A., Kotyk, J.J., 2000. Magnetic resonance contrast enhancement of neovasculature with alpha(v)beta(3)-targeted nanoparticles. Magn Reson Med 44, 433–439.

Banerjee, R., 2001. Liposomes: applications in medicine. J Biomater Appl 16, 3–21.

Bishop, G.G., McPherson, J.A., Sanders, J.M., Hesselbacher, S.E., Feldman, M.J., McNamara, C.A., Gimple, L.W., Powers, E.R., Mousa, S.A., Sarembock, I.J., 2001. Selective alpha(v)beta(3)-receptor blockade reduces macrophage infiltration and restenosis after balloon angioplasty in the atherosclerotic rabbit. Circulation 103, 1906–1911.

Brooks, P.C., Stromblad, S., Klemke, R., Visscher, D., Sarkar, F.H., Cheresh, D.A., 1995. Antiintegrin alpha v beta 3 blocks human breast cancer growth and angiogenesis in human skin. J Clin Invest 96, 1815–1822.

Bulte, J.W., Arbab, A.S., Douglas, T., Frank, J.A., 2004. Preparation of magnetically labeled cells for cell tracking by magnetic resonance imaging. Methods Enzymol 386, 275–299.

Burnell, E.E., Cullis, P.R., de Kruijff, B., 1980. Effects of tumbling and lateral diffusion on phosphatidylcholine model membrane 31P-NMR lineshapes. Biochim Biophys Acta 603, 63–69.

Caruthers, S.D., Neubauer, A.M., Hockett, F.D., Lamerichs, R., Winter, P.M., Scott, M.J., Gaffney, P.J., Wickline, S.A., Lanza, G.M., 2006. In vitro demonstration using 19F magnetic resonance to augment molecular imaging with paramagnetic perfluorocarbon nanoparticles at 1.5 Tesla. Invest Radiol 41, 305–312.

Corjay, M.H., Diamond, S.M., Schlingmann, K.L., Gibbs, S.K., Stoltenborg, J.K., Racanelli, A.L., 1999. alphavbeta3, alphavbeta5, and osteopontin are coordinately upregulated at early time points in a rabbit model of neointima formation. J Cell Biochem 75, 492–504.

Cyrus, T., Abendschein, D.R., Caruthers, S.D., Harris, T.D., Glattauer, V., Werkmeister, J.A., Ramshaw, J.A., Wickline, S.A., Lanza, G.M., 2006. MR three-dimensional molecular imaging of intramural biomarkers with targeted nanoparticles. J Cardiovasc Magn Reson 8, 535–541.

Flacke, S., Fischer, S., Scott, M.J., Fuhrhop, R.J., Allen, J.S., McLean, M., Winter, P., Sicard, G.A., Gaffney, P.J., Wickline, S.A., Lanza, G.M., 2001. Novel MRI contrast agent for molecular imaging of fibrin: implications for detecting vulnerable plaques. Circulation 104, 1280–1285.

Flaim, S.F., 1994. Pharmacokinetics and side effects of perfluorocarbon-based blood substitutes. Artif Cells Blood Substit Immobil Biotechnol 22, 1043–1054.

Gossl, M., Malyar, N.M., Rosol, M., Beighley, P.E., Ritman, E.L., 2003a. Impact of coronary vasa vasorum functional structure on coronary vessel wall perfusion distribution. Am J Physiol Heart Circ Physiol 285, H2019–2026.

Gossl, M., Rosol, M., Malyar, N.M., Fitzpatrick, L.A., Beighley, P.E., Zamir, M., Ritman, E.L., 2003b. Functional anatomy and hemodynamic characteristics of vasa vasorum in the walls of porcine coronary arteries. Anat Rec A Discov Mol Cell Evol Biol 272, 526–537.

Gossl, M., von Birgelen, C., Mintz, G.S., Bose, D., Eggebrecht, H., Baumgart, D., Haude, M., Erbel, R., 2003c. Volumetric assessment of ulcerated ruptured coronary plaques with three-dimensional intravascular ultrasound in vivo. Am J Cardiol 91, 992–996, A997.

Grant, C.W., Karlik, S., Florio, E., 1989. A liposomal MRI contrast agent: phosphatidylethanolamine-DTPA. Magn Reson Med 11, 236–243.

Hall, C.S., Marsh, J.N., Scott, M.J., Gaffney, P.J., Wickline, S.A., Lanza, G.M., 2001. Temperature dependence of ultrasonic enhancement with a site-targeted contrast agent. J Acoust Soc Am 110, 1677–1684.

Hamilton, M.C., Peek, G.J., Dux, A.E., 2005. Partial liquid ventilation. Pediatr Radiol 35, 1152–1156.

Hu, G., Lijowski, M., Zhang, H., Partlow, K.C. Caruthers, S.D., Kiefer, G., Gulyas, G., Athey, P., Slott, M.J., Wickline, S.A., Lanza, G.M., 2007. Imaging of Vx-2 rabbit tumors with alpha(nu) beta3-integrin-targeted ^{111}In nanoparticles. Int. J. Cancer 120:1951–1957.

Jacobs, R.E., Fraser, S.E., 1994. Magnetic resonance microscopy of embryonic cell lineages and movements. Science 263, 681–684.

Kerr, J.S., Mousa, S.A., Slee, A.M., 2001. Alpha(v)beta(3) integrin in angiogenesis and restenosis. Drug News Perspect 14, 143–150.

Khurana, R., Zhuang, Z., Bhardwaj, S., Murakami, M., De Muinck, E., Yla-Herttuala, S., Ferrara, N., Martin, J.F., Zachary, I., Simons, M., 2004. Angiogenesis-dependent and independent phases of intimal hyperplasia. Circulation 110, 2436–2443.

Kircher, M.F., Allport, J.R., Graves, E.E., Love, V., Josephson, L., Lichtman, A.H., Weissleder, R., 2003. In vivo high resolution three-dimensional imaging of antigen-specific cytotoxic T-lymphocyte trafficking to tumors. Cancer Res 63, 6838–6846.

Kolodgie, F.D., Virmani, R., Burke, A.P., Farb, A., Weber, D.K., Kutys, R., Finn, A.V., Gold, H.K., 2004. Pathologic assessment of the vulnerable human coronary plaque. Heart 90, 1385–1391.

Lanza, G.M., Abendschein, D.R., Hall, C.S., Marsh, J.N., Scott, M.J., Scherrer, D.E., Wickline, S.A., 2000a. Molecular imaging of stretch-induced tissue factor expression in carotid arteries with intravascular ultrasound. Invest Radiol 35, 227–234.
Lanza, G.M., Abendschein, D.R., Hall, C.S., Scott, M.J., Scherrer, D.E., Houseman, A., Miller, J.G., Wickline, S.A., 2000b. In vivo molecular imaging of stretch-induced tissue factor in carotid arteries with ligand-targeted nanoparticles. J Am Soc Echocardiogr 13, 608–614.
Lanza, G.M., Wallace, K.D., Scott, M.J., Cacheris, W.P., Abendschein, D.R., Christy, D.H., Sharkey, A.M., Miller, J.G., Gaffney, P.J., Wickline, S.A., 1996. A novel site-targeted ultrasonic contrast agent with broad biomedical application. Circulation 94, 3334–3340.
Laurent, S., Elst, L.V., Copoix, F., Muller, R.N., 2001. Stability of MRI paramagnetic contrast media: a proton relaxometric protocol for transmetallation assessment. Invest Radiol 36, 115–122.
Marsh, J.N., Hall, C.S., Wickline, S.A., Lanza, G.M., 2002. Temperature dependence of acoustic impedance for specific fluorocarbon liquids. J Acoust Soc Am 112, 2858–2862.
Mattrey, R.F., 1994. The potential role of perfluorochemicals (PFCs) in diagnostic imaging. Artif Cells Blood Substit Immobil Biotechnol 22, 295–313.
Mattrey, R.F., Long, D.C., 1988. Potential role of PFOB in diagnostic imaging. Invest Radiol 23 Suppl 1, S298–S301.
Modo, M., Mellodew, K., Cash, D., Fraser, S.E., Meade, T.J., Price, J., Williams, S.C., 2004. Mapping transplanted stem cell migration after a stroke: a serial, in vivo magnetic resonance imaging study. Neuroimage 21, 311–317.
Morawski, A.M., Winter, P.M., Crowder, K.C., Caruthers, S.D., Fuhrhop, R.W., Scott, M.J., Robertson, J.D., Abendschein, D.R., Lanza, G.M., Wickline, S.A., 2004a. Targeted nanoparticles for quantitative imaging of sparse molecular epitopes with MRI. Magn Reson Med 51, 480–486.
Morawski, A.M., Winter, P.M., Yu, X., Fuhrhop, R.W., Scott, M.J., Hockett, F., Robertson, J.D., Gaffney, P.J., Lanza, G.M., Wickline, S.A., 2004b. Quantitative "magnetic resonance immunohistochemistry" with ligand-targeted (19)F nanoparticles. Magn Reson Med 52, 1255–1262.
Moreno, P.R., Purushothaman, K.R., Fuster, V., Echeverri, D., Truszczynska, H., Sharma, S.K., Badimon, J.J., O'Connor, W.N., 2004. Plaque neovascularization is increased in ruptured atherosclerotic lesions of human aorta: implications for plaque vulnerability. Circulation 110, 2032–2038.
Riess, J.G., 2005. Understanding the fundamentals of perfluorocarbons and perfluorocarbon emulsions relevant to in vivo oxygen delivery. Artif Cells Blood Substit Immobil Biotechnol 33, 47–63.
Schmieder, A.H., Winter, P.M., Caruthers, S.D., Harris, T.D., Williams, T.A., Allen, J.S., Lacy, E.K., Zhang, H., Scott, M.J., Hu, G., Robertson, J.D., Wickline, S.A., Lanza, G.M., 2005. Molecular MR imaging of melanoma angiogenesis with alphanubeta3-targeted paramagnetic nanoparticles. Magn Reson Med 53, 621–627.
Sharkey, R.M., Motta-Hennessy, C., Gansow, O.A., Brechbiel, M.W., Fand, I., Griffiths, G.L., Jones, A.L., Goldenberg, D.M., 1990. Selection of a DTPA chelate conjugate for monoclonal antibody targeting to a human colonic tumor in nude mice. Int J Cancer 46, 79–85.
Sherry, A.D., Cacheris, W.P., Kuan, K.T., 1988. Stability constants for Gd3+ binding to model DTPA-conjugates and DTPA-proteins: implications for their use as magnetic resonance contrast agents. Magn Reson Med 8, 180–190.

Sipkins, D.A., Cheresh, D.A., Kazemi, M.R., Nevin, L.M., Bednarski, M.D., Li, K.C., 1998. Detection of tumor angiogenesis in vivo by alphaVbeta3-targeted magnetic resonance imaging. Nat Med 4, 623–626.

Tenaglia, A.N., Peters, K.G., Sketch, M.H., Jr., Annex, B.H., 1998. Neovascularization in atherectomy specimens from patients with unstable angina: implications for pathogenesis of unstable angina. Am Heart J 135, 10–14.

Thom, T., Haase, N., Rosamond, W., Howard, V.J., Rumsfeld, J., Manolio, T., Zheng, Z.J., Flegal, K., O'Donnell, C., Kittner, S., Lloyd-Jones, D., Goff, D.C., Jr., Hong, Y., Adams, R., Friday, G., Furie, K., Gorelick, P., Kissela, B., Marler, J., Meigs, J., Roger, V., Sidney, S., Sorlie, P., Steinberger, J., Wasserthiel-Smoller, S., Wilson, M., Wolf, P., 2006. Heart disease and stroke statistics–2006 update: a report from the American Heart Association Statistics Committee and Stroke Statistics Subcommittee. Circulation 113, e85–e151.

Tilcock, C., Ahkong, Q.F., Koenig, S.H., Brown, R.D., 3rd, Davis, M., Kabalka, G., 1992. The design of liposomal paramagnetic MR agents: effect of vesicle size upon the relaxivity of surface-incorporated lipophilic chelates. Magn Reson Med 27, 44–51.

Winter, P., Athey, P., Kiefer, G., Gulyas, G., Fuhrhop, R., Robertson, D., Wickline, S., Lanza, G., 2005a. Improved paramagnetic chelate for molecular imaging with MRI. J Magn Magn Mater 293, 540–545.

Winter, P.M., Caruthers, S.D., Kassner, A., Harris, T.D., Chinen, L.K., Allen, J.S., Lacy, E.K., Zhang, H., Robertson, J.D., Wickline, S.A., Lanza, G.M., 2003a. Molecular imaging of angiogenesis in nascent Vx-2 rabbit tumors using a novel alpha(nu)beta3-targeted nanoparticle and 1.5 tesla magnetic resonance imaging. Cancer Res 63, 5838–5843.

Winter, P.M., Caruthers, S.D., Yu, X., Song, S.K., Chen, J., Miller, B., Bulte, J.W., Robertson, J.D., Gaffney, P.J., Wickline, S.A., Lanza, G.M., 2003b. Improved molecular imaging contrast agent for detection of human thrombus. Magn Reson Med 50, 411–416.

Winter, P.M., Morawski, A.M., Caruthers, S.D., Fuhrhop, R.W., Zhang, H., Williams, T.A., Allen, J.S., Lacy, E.K., Robertson, J.D., Lanza, G.M., Wickline, S.A., 2003c. Molecular imaging of angiogenesis in early-stage atherosclerosis with alpha(v)beta3-integrin-targeted nanoparticles. Circulation 108, 2270–2274.

Winter, P.M., Shukla, H.P., Caruthers, S.D., Scott, M.J., Fuhrhop, R.W., Robertson, J.D., Gaffney, P.J., Wickline, S.A., Lanza, G.M., 2005b. Molecular imaging of human thrombus with computed tomography. Acad Radiol 12 Suppl 1, S9–S13.

Wood, M.L., Hardy, P.A., 1993. Proton relaxation enhancement. J Magn Reson Imaging 3, 149–156.

Yeh, T.C., Zhang, W., Ildstad, S.T., Ho, C., 1993. Intracellular labeling of T-cells with superparamagnetic contrast agents. Magn Reson Med 30, 617–625.

9 Magnetic Nanosensors for Probing Molecular Interactions

J. Manuel Perez and Charalambos Kaittanis

Abstract: Magnetic nanosensors exhibiting high specificity and biocompatibility have been synthesized for the in vitro and in vivo detection of molecular interactions. Upon target-induced nanoassembly formation, a sensitive and dose-dependent decrease in the spin-spin relaxation time (T_2) of adjacent water molecules was observed. Various molecular targets ranging from DNA-oligonucleotides, proteins, small molecules, and even whole viruses have been detected using this technology, with sensitivity in the low femtomole range (0.5–30 fmol). Furthermore, the engymatic activity of restriction endonucleases and proteases, among others, has also been detected using an alternate method, based on the enzyme-induced dispersion of nanoparticles from a pre-formed nanoassembly. Furthermore, this technique was used to monitor disease-associated enzymatic activity, such as up-regulated activity of telomerase, myeloperoxidase and caspase. The observed changes in T_2 can be easily detected in water suspension of the magnetic nanosensors by existing magnetic resonance (NMR/MRI) techniques with minimal sample preparation and permitting high-throughput screening. Homogenous and high-throughput assays to selectively detect a variety of molecular targets in solution can be designed using this magnetic nanosensor technology. Finally, the presence of a target in whole blood or turbid media can be performed as the magnetic relaxation signal is independent of the optical properties of the solution.

9.1. Introduction

The use of superparamagnetic nanoparticles as molecular probes for in vitro and in vivo applications has gained considerable attention in recent years (Thorek et al., 2006). Typically, these magnetic nanoprobes are comprised of a superparamagnetic core surrounded by a polymeric coating of dextran or silica, among others. The coating not only helps in stabilizing the nanoparticle colloid, but also facilitates the conjugation of various ligands and macromolecules for targeting applications. These nanoparticles are composed of a superparamagnetic core of magnetite (Fe_3O_4), maghemite (Fe_2O_3), or a mixture of both magnetite/maghemite. However, other metals in various compositions have been used such as cobalt oxide, spinel cobalt or manganese ferrite ($CoFe_2O_4$, $MnFe_2O_4$) (Song and Zhang, 2004, 2006), and gadolinium oxide (McDonald and Watkin, 2006). Magnetic nanoparticles are ideal candidates for highly sensitive biosensing of molecular targets. In this case, the classical label is replaced by a superparamagnetic nanoparticle, which can be detected magnetically. Magnetic labeling complements other labeling techniques that make use of optical reporters such as fluorescent probes, gold nanoparticles, and quantum dots. However, most of the magnetic sensors so far have taken advantage of the intrinsic magnetic properties of the magnetic core. Examples include superconductive quantum interference device (SQUID)-based detection (Chemla et al., 2000; Grossman et al., 2004), handheld magnetic readers (Kriz et al., 1998), and magneto-resistive detectors (Baselt et al., 1998; Edelstein et al., 2000), among others.

This review will focus on an alternative detection technology that relies on the effect that iron oxide–based nanoparticles have on the T_2 relaxation time of neighboring water molecules. In the presence of an external magnetic field, these superparamagnetic nanoparticles become magnetized and generate significant susceptibility changes that result in a strong decrease in T_2 relaxation time (Bulte et al., 1998; Bulte and Kraitchman, 2004a,b). This remarkable effect has been utilized successfully in various applications in magnetic resonance imaging (MRI), from cell tracking (Bulte and Kraitchman, 2004a,b; Frank et al., 2003; Modo et al., 2005), liver and lymph node imaging (Harisinghani et al., 2003; Harisinghani and Weissleder, 2004), to tissue specific contrast agent (Kelly et al., 2004, 2005; Weissleder et al., 2005). The observed T_2 contrast is dependent on the amount of nanoparticle present in the particular tissue. In solution, as the concentration of superparamagentic nanoparticles increases, the T_2 relaxation time becomes shorter. More recently, it was found that when a set amount of monodispersed magnetic nanoparticles was allowed to cluster or self-assemble, a similar T_2 contrast enhancement was observed, as determined by a decrease in the T_2 relaxation times (Perez et al., 2002a). This observation was utilized in the development of magnetic nanosensors to facilitate the detection of molecular interactions through magnetic resonance (NMR/MRI) techniques. Upon target binding, the nanoparticles form stable nanoassemblies, which cause a decrease in the spin-spin relaxation time (T_2) of water molecules in the nanoassembly's vicinity. Sensitive and homogenous assays have been developed to detect and monitor a plethora of molecular interactions in vitro, requiring limited or no sample preparation. For example, reversible and irreversible biomolecular interactions between DNA–DNA, DNA–protein, enzyme–substrate, protein–protein, and protein–small molecule have been detected with this technique. These magnetic nanosensors can be engineered to be selective towards a biological target in a complex medium, like GFP mRNA in cell lysate (Perez et al., 2002a) or virions (HSV-1) in serum samples (Perez et al., 2003). Additionally, this approach has been used for the detection of enzymatic processes, catalyzed by restriction endonucleases and proteases. This method is highly sensitive considering its detection thresholds for DNA and proteins being at the low femtomole level (0.5 fmols) in non-purified samples. Additionally, they can serve as a powerful tool in the quest to answer many of the molecular biology's riddles in the fields of proteomics, signal transduction, and system biology, and to facilitate an enhanced in vivo imaging vital for medical decision-making.

9.2. Magnetic Nanosensor Composition

These magnetic nanosensors are composed of a 3–5 nm core of superparamagnetic iron oxide core with an inverse spinel structure (close cubic packed) of $(Fe_2O_3)_n(Fe_3O_4)_m$ surrounded by a 10 nm thick 10 kDa dextran coating (Shen et al., 1993). In order to enhance their stability and functionality, the dextran coating is cross-linked with epichlorohydrin and followed by ammonia treatment to incorporate accessible functional amino groups (Josephson et al., 1999). The average particle size of the resulting aminated nanoparticles was 30–40 nm, each nanoparticle bearing about 40 amino groups on its surface. Notably, these nanoparticles exhibited great stability even under extremely harsh conditions, such as a 30-minute incubation at 120 °C, with neither size nor coating alterations. Attachment of biomolecules containing sulfhydryl moieties to the amino groups on the amino magnetic nanosensor was facilitated by the use of *N*-Succinimidyl-3-(2-PyridylDithio)-Propionate (SPDP), a bifunctional cross-linking agent (Figure 9.1). This treatment allows the direct

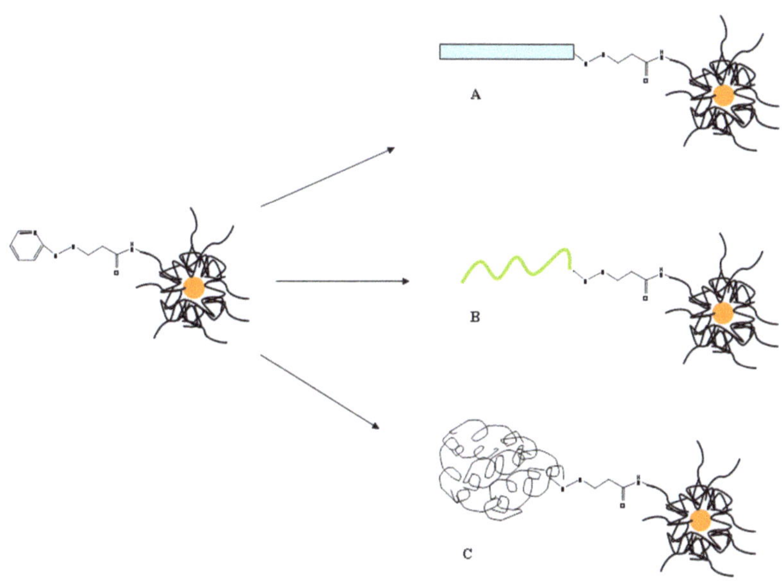

FIGURE 9.1. Magnetic nanosensor bioconjugates. An SPDP-nanoparticle conjugate is utilized as starting material to synthesize a variety of magnetic nanoparticles bioconjugates containing (**A**) peptides, (**B**) DNA oligonucleotides, or (**C**) protein, among others.

conjugation of biomolecules such as peptides, proteins, and oligonucleotides to the nanoparticles, thereby incorporating target specificity and increasing their biological utility in contrast to simple polymer-coated iron oxide nanoparticles. Additionally, fluorescent dyes have been conjugated to the magnetic nanosensors allowing for multimodal, magnetic and optical sensing.

9.3. Mechanism of Magnetic Relaxation Switching

The magnetic nanosensors carry a crystalline superparamagnetic iron oxide core, which becomes magnetized due to induction by the application of an external magnetic field (Brooks, 2002; Gillis et al., 2002). The spins of the iron atoms that compose the crystal lattice collectively produce a single large magnetic dipole, which facilitates the formation of a localized magnetic field gradient. Consequently, the presence of the induced magnetic gradient creates heterogeneity that alters the external magnetic field. Within these magnetic heterogeneities, protons from the surrounding water molecules diffuse at an off resonance frequency. Protons in the water molecule dephase their spins, increasing the relaxation rate ($1/T_2$) of water as explained by the outer sphere theory, which postulates that the relaxation rate ($1/T_2$) is directly proportional to the nanoparticle's cross-sectional area (Brooks, 2002; Gillis et al., 2002). Considering this, it is plausible to speculate that clustering of the individual superparamagnetic nanoparticles increases the corresponding effective cross-sectional area. Therefore, the nanoassembly more efficiently can alter the spins of adjacent water protons enhancing the relaxation rates ($1/T_2$), and justifying the characterization of magnetic nanosenors as magnetic relaxation switches (Brooks, 2002; Gillis et al., 2002).. Evidently, the relaxation rate can facilitate the assessment of concentration in both nanoassembled and dispersed states within the same solution.

9.4. Model System: Sensing DNA

To test our hypothesis, we first investigated the ability of oligonucleotide-modified magnetic nanoparticles to cluster in the presence of a target oligonucleotide (Josephson et al., 2001). In these studies, alkanethiololigonucleotides were conjugated to the aminated magnetic nanoparticles via SPDP conjugation, forming a disulfide bond link as shown in Figure 9.1A. Two nanoparticle preparations (P1 and P2) were synthesized per target sequence. These nanoparticles contain a 12-base oligonucleotide, each recognizing two adjacent sequences on a target DNA oligonucleotide. Upon addition of the target, induction of nanoassembly formation took place (Figure 9.2A), facilitating a rapid and significant decrease in the spin-spin relaxation times (T_2) of proximal water molecules, which was recorded with a 0.47 T NMR bench-top relaxometer at 40 °C (Figure 9.2B). In contrast, T_2 did not change when a non-complementary oligonucleotide aliquot was added to the preparations, supporting the superior

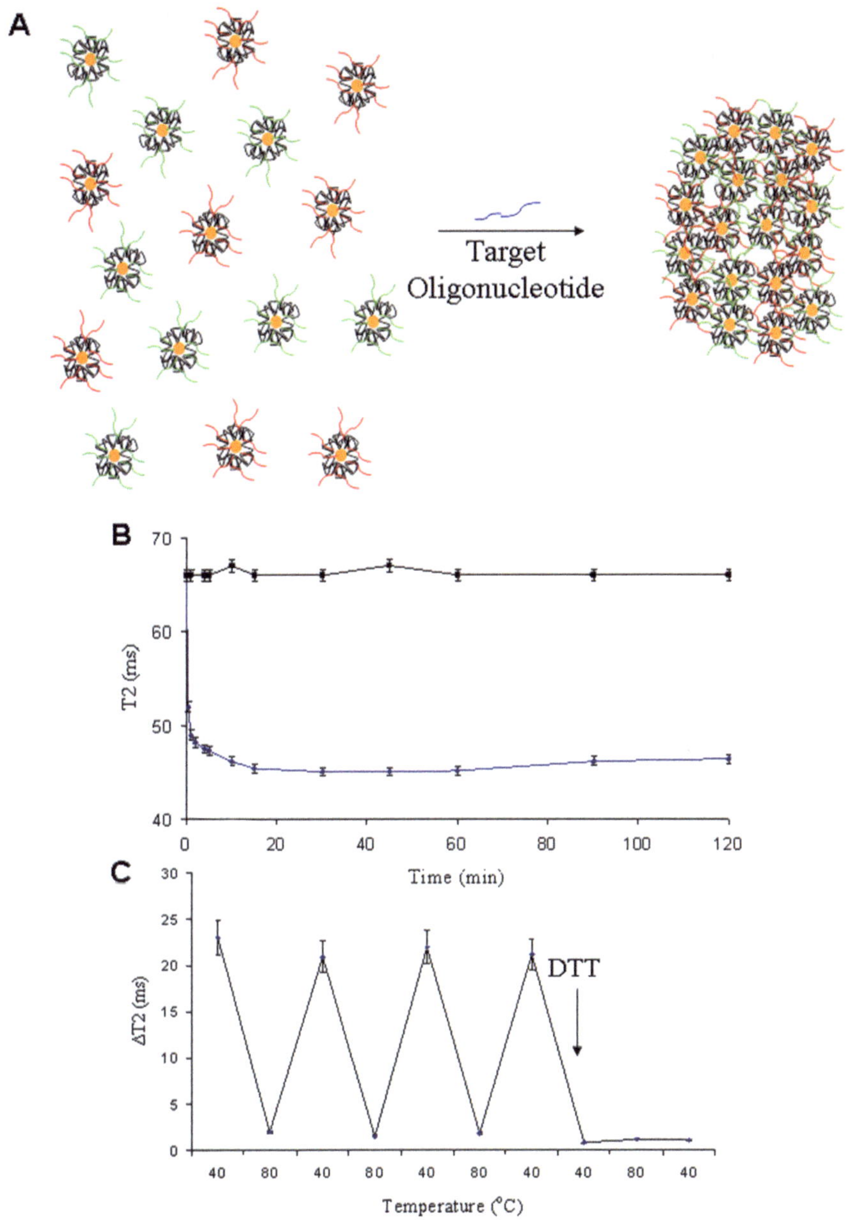

FIGURE 9.2. Oligonucleotide-induced magnetic nanoparticle assembly. Two distinct nanoparticle population, **P1** and **P2,** self-assemble upon addition of a complementary oligonucleotide (**A**), reporting fast and dose-dependent sensing of the oligonucleotide sequence (**B**). Changes in T_2 values are dependent on temperature as the oligonucleotides linking the nanoparticle together hydridize at 40 °C and de-hybridize at 80 °C (**C**). Addition of DTT cleaves the disulfide bond linking the nanoparticle together and making the nanoparticles insensitive to hybridization-induced changes in T_2. Reproduced with permission from Josephson et al. (2001).

target specificity of the nanosensors. Notably, there was a linear relationship between the average T_2 decrease and the solution's DNA concentration. Studies in turbid solutions revealed an analogous decrease in T_2 relaxation, which was not affected by the solutions' cloudiness, or optical properties. Furthermore, the magnetic nanosensors were sensitive to temperature cycles, as at 80 °C the nanoparticles modestly hybridized with complementary oligonucleotides, resulting in minor variations in T_2. Figure 9.2C illustrates the effect of multiple heating (80 °C) and cooling (40 °C) cycles on δT_2. Treatment of the nanoparticles with dithiothreitol (DTT), cleaved the disulfide bond between the nanoparticle and the oligonucleotide, making the nanoparticle solution insensitive to hybridization-induced changes in T_2. Determination of the nanosensors detection threshold and selectivity toward a given DNA target was achieved in a higher magnetic field strength (1.5 T). In these studies, a 24-base sequence complementary to a GFP mRNA target sequence was used. At 1.5 T, samples in the low femtomole range (0.5–2.7 fmol) depicted significant variations in T_2 that were visible by MRI[27]. Furthermore, the magnetic nanosensors could identify single and multiple nucleotide polymorphisms, or single and multiple additions.

9.4.1. Sensing Restriction Endonuclease Activity

Monitoring the catalytic activity of an enzyme in biological media and biological systems is critical in molecular biology. In particular, the development of magnetic nanosensors capable of measuring enzymatic activities, such as of restriction endonucleases, methylases, and proteases, is important. Considering the aforementioned nanoassembly between P1 and P2 upon DNA binding, it was speculated that cleavage of the P1–P2 oligonucleotide linker by a restriction endonuclease, such as *Bam*HI, would disrupt the nanoassembly, dispersing the nanoparticles and increasing the sample's T_2 (Perez et al., 2002b). To investigate this hypothesis, two magnetic nanoparticle preparations, P1 and P2, were synthesized so that the two nanoparticles could inter-hybridize when mixed together, forming a *Bam*HI sensitive nanostructure (Figure 9.3A). As expected, when P1 and P2 were mixed, a decrease in T_2 in the solution was observed. Upon incubation with *Bam*HI, a time- and dose-dependent increase in the sample's T_2 was observed (Figure 9.3B). The nanoassembly's *Bam*HI specificity was confirmed after treatment with several other endonucleases (*Eco*RI, *Hind*III, and *Dpn*I) that did not manage to affect the nanoassembly's T_2 values. Further experimentation with *dam* methylase demonstrated the potential to methylate the DNA linker that holds the monomeric units of the nanoassembly. Subsequently, adding a methylation-sensitive restriction endonuclease facilitated the detection of this methylation event (Perez et al., 2002b).

9.4.2. Sensing Protease Activity

To further investigate the application in other classes of enzymes, magnetic nanosensors were designed to detect protease activity. In these experiments,

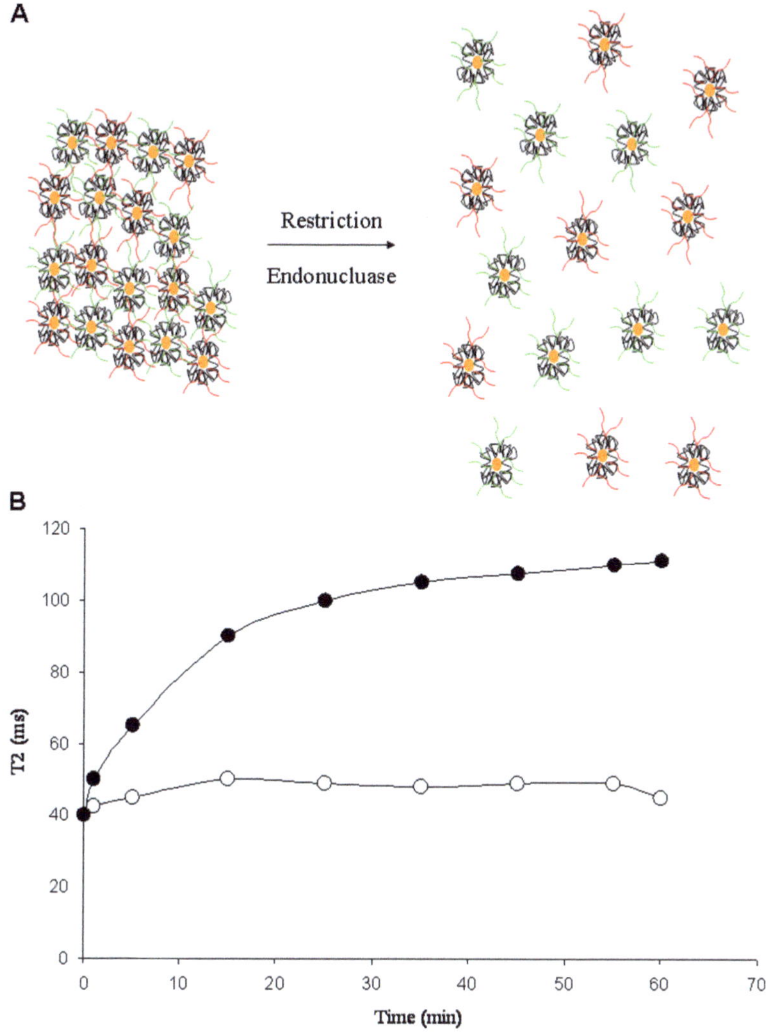

FIGURE 9.3. Restriction endonuclease. Sequence-specific cleavage of a magnetic nanoparticle assembly results in separation of the nanoparticles (**A**) and a corresponding time-dependent increase in the T_2 relaxation times of a *Bam HI* specific magnetic nanoassembly upon incubation with *BamHI* (**B**). Reproduced by permission from Perez et al. (2002b).

two nanoparticle preparations (P1 and P2) were synthesized: one was conjugated with avidin, whereas the other was conjugated to a biotinylated DEVD peptides. Therefore, incubating P1 and P2 led to the formation of a caspase 3–sensitive magnetic nanoassembly (Perez et al., 2002a). Caspase 3, which is an effector caspase in the cell apoptotic pathway, recognizes the DEVD amino acid sequence. Hence, this proteolytic cleavage by caspase 3 can be sensed via a change in the P1–P2 nanoassembly's T_2. Upon incubation with caspase 3, the T_2 relaxation time increased in a dose-dependent manner, exhibiting similar kinetic characteristics to those reported with fluorogenic substrates. In another report, an assay was developed that relies on the use of peptides with a protease recognition motif, containing a pair of biotin moieties on the *N*- and *C*-termini respectively (Zhao et al., 2003). This peptide then facilitates the assembly of avidin-conjugated nanoparticles, causing a decrease in T_2. Pretreatment of such a peptide with an active protease cleaves the oligopeptide to two biotin peptide fragments that are unable to facilitate formation of a nanoparticle assembly and therefore no change in T_2 is observed. This strategy was used to monitor the activity of proteases, such as trypsin, renin, and matrix metalloprotease-2, via engineered magnetic nanosensors.

9.4.3. Antibody-Mediated Sensing

9.4.3.1. Sensing Proteins

Apart from nucleic acid detection, magnetic nanosensors can probe the presence of a target protein in solution using antigen–antibody interactions. In order to decorate the nanosensor with an antibody, avidin-P1 conjugates were prepared. For probing of GFP, a biotinylated anti-GFP polyclonal antibody was attached to the avidin P1. The interaction between the GFP-specific nanoparticle and its target caused significant changes in T_2 that were time and dose dependent. As a negative control, the GFP nanosensors were incubated with bovine serum albumin (BSA) causing no changes in T_2 relaxation times. Furthermore, in order to investigate the aforementioned findings in other model systems, magnetic nanosensors were designed to probe for CA-125, a protein present in high levels in the blood of women affected with ovarian cancer. In these studies, magnetic nanoparticles carrying an anti-CA-125 antibody were used. Results showed that significant and reproducible T_2 values where observed in these experiments allowing the detection of the target protein in the low femtomol (1–10 fmol) range. Finally, GFP transcripts could be sensed from a group of GFP transfected cell lines, in a mixture of isolated total RNA and in cell lysate. In addition, there was a linear trend between T_2 and GFP fluorescence in the GFP expressing cells.

9.4.3.2. Sensing Viral Particles

The use of magnetic nanosensors for the identification of more complex targets, like viral particles in serum, was also reported (Perez et al., 2003). Considering that the identification of protein targets with this novel method had been possible and that viral particles contain multiple proteins on their surface, the application of this technology for viral detection in solution sounded feasible. Additionally, the presence of multiple copies of a target protein on the viral coat facilitates multivalent interactions between the multi-epitope virions and the magnetic nanosensors (Figure 9.4). This feature enhances nanoassembly formation, promoting high sensitivity for viral detection in biological media. Having this in mind, magnetic nanoparticles were designed capable to sense

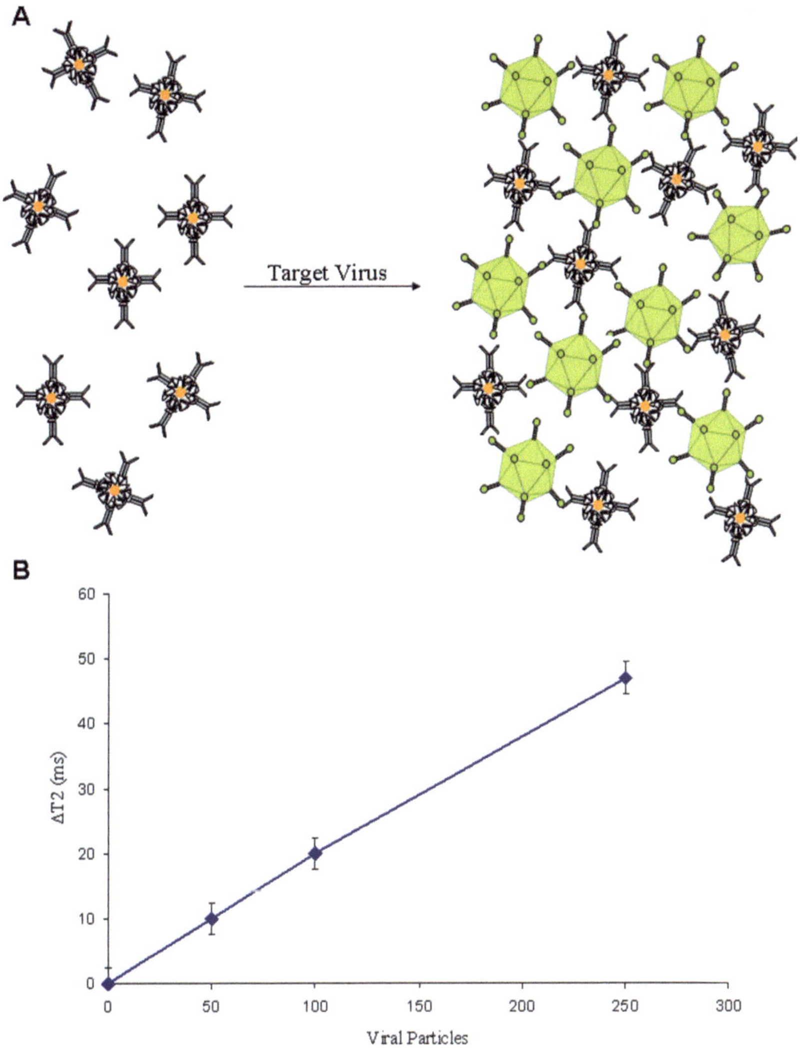

FIGURE 9.4. Viral-induced assembly of magnetic nanoparticles. Upon addition of the target virus to a suspension of magnetic nanoparticles conjugated with anti-virus antibodies, self-assembly of the nanoparticles occurs (**A**), allowing sensitive detection of the virus in solution (**B**). Reproduced with permission from Perez et al. (2003).

low levels (5 virions per 10 μL) of Herpes simplex virus-1 (HSV-1) and adenovirus-5 (ADV-5) in serum. The identification of viral particles with magnetic nanosensors outperforms the detection of viruses with contemporary PCR techniques, providing quick and easy-to-read results with minimal artifacts, and without the need for protein removal or sample amplification. In the future, acknowledging the potential and utility of this assay, viral-specific nanoparticles can be designed for the detection and determination of the localization of viruses in vivo, serving as viral-specific MRI agents.

9.4.4. Sensing Disease-Related Targets

9.4.4.1. Telomerase Activity

Our previous results demonstrating the detection of restriction endonuclease activity and caspase activity motivated us to pursue other enzymatic activities associated with disease. Telomerase is a reverse transcriptase enzyme found in high levels in many tumors (Avilion et al., 1996; Shay and Bacchetti, 1997). High telomerase activity is believed to play a critical role in tumorigenesis; hence, detecting telomerase activity in tumor extracts would be highly beneficial in the prognosis of the disease. Telomerase assays typically use PCR-based amplification that, although highly sensitive, is susceptible to amplification-related errors. Thus, there is a need for an easier telomerase activity assay that does not rely on PCR amplification and can be performed in crude cell or tissue extract. For these studies, we designed oligonucleotide-based magnetic nanoparticle conjugates that will specifically hybridize with telomeric repeats, (TTAGGG)n (Grimm et al., 2004). These telomeric repeats are generated after incubation of a specific DNA primer with tumor extracts containing active telomerase. Upon incubation of the elongated primer with telomerase-specific nanosensors, self-assembly along the telomeric repeats occurs, with a corresponding change in the T2 relaxation time (Figure 9.5a). This effect was not observed with control nonsense sequences or heat-inactivated tumor samples. However, telomerase activity in a panel of tumor cell lysate was detected and correlated well with their reported telomerase activity (Figure 9.5b). The magnetic nanosensor assay was sensitive enough to detect 300 attomols of telomerase and specific to tissue samples containing active telomerase. Interestingly, normal human melanocytes reported minimal telomerase activity. Furthermore, the nanosensor assay was utilized in the study of the phosphorylation-dependent activation of telomerase in various tissue samples in addition to the study of various telomerase inhibitors in high throughput using MRI.

9.4.4.2. Myeloperoxidase

Myeloperoxidase is a critical enzyme found in phagocytic white blood cells and is associated with inflammatory pathologies such as atherosclerosis, cystic fibrosis, and rheumatoid arthritis. This enzyme is particularly generated when arteries become inflamed and have rupture-prone fatty deposits, which could potentially lead to a blood clot and eventually to a heart attack or stroke. Therefore, by measuring myeloperoxidase levels in blood or even imaging myeloperoxidase activity in vivo, it could be possible to predict whether a person is at risk of heart attack (Brennan et al., 2003; Zhang et al., 2001). In the pathogenesis of vascular disease, myeloperoxidase facilitates the formation of *o,o*-dityrosine cross-links in LDL-associated proteins via the enzymatic generation of tyrosyl radicals.

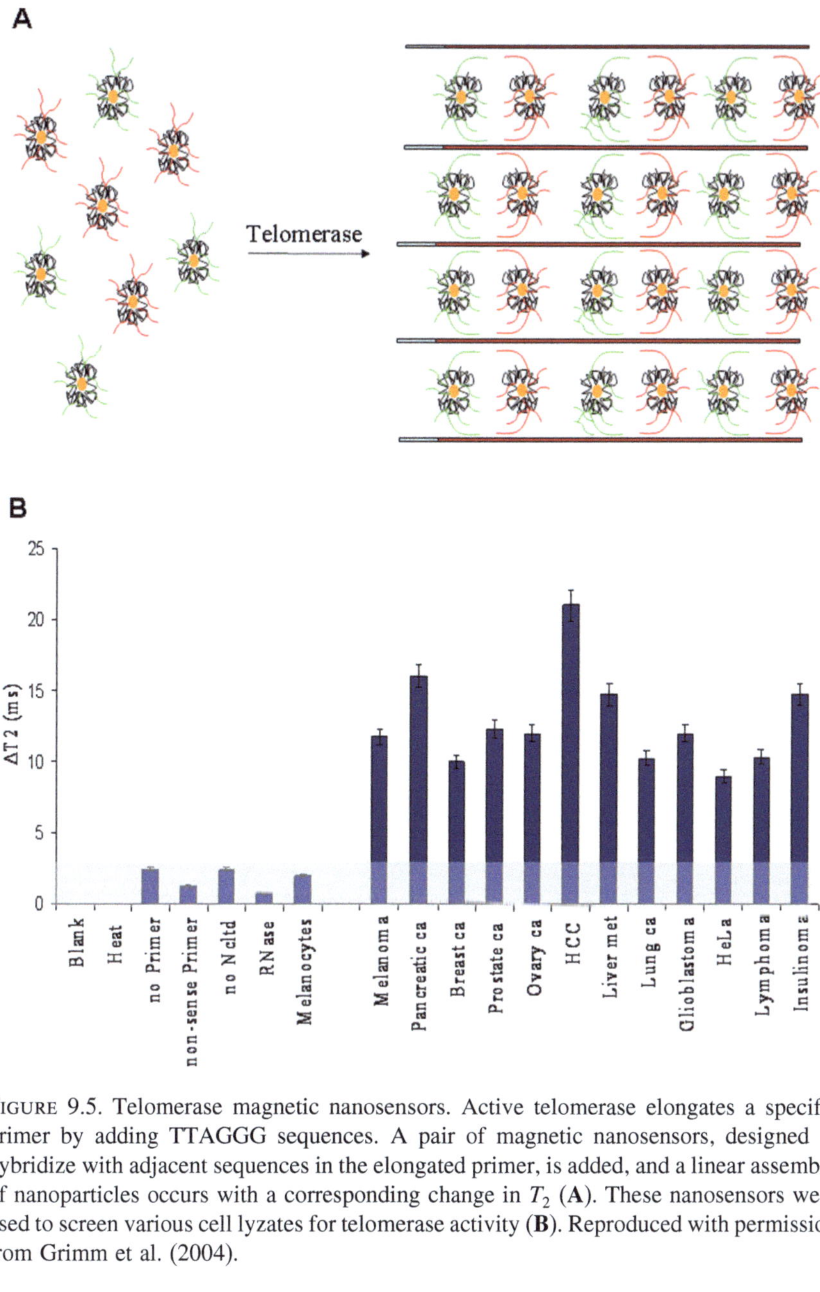

FIGURE 9.5. Telomerase magnetic nanosensors. Active telomerase elongates a specific primer by adding TTAGGG sequences. A pair of magnetic nanosensors, designed to hybridize with adjacent sequences in the elongated primer, is added, and a linear assembly of nanoparticles occurs with a corresponding change in T_2 (**A**). These nanosensors were used to screen various cell lyzates for telomerase activity (**B**). Reproduced with permission from Grimm et al. (2004).

We therefore hypothesized that phenolic compounds, such as tyrosine, could be attached to the magnetic nanoparticles and create a myeloperoxidase-active nanosensor. For attachment to the magnetic nanoparticles, serotonin was chosen since it has been found to be a better substrate for myeloperoxidase compared to tyrosine. Results show that upon myeloperoxidase incubation, self-assembly of these nanosensors occurred in solution along with a detectable change in T2 (Perez et al., 2004). These changes in T2 increased as a function of enzyme concentration and were abrogated by the addition of myeloperoxidase inhibitors or in the absence of hydrogen peroxide. In addition, it was demonstrated that the magnetic sensor assay was sensitive enough to measure low amounts of myeloperoxidase (0.0031 units/μL)

9.4.5. Small Molecule Detection

Highly sensitive and selective magnetic nanosensors have been developed to detect small molecules. This idea was first investigated using steptavidin-coated nanoparticles for the detection of biotin in solution (Perez et al., 2002a). As expected, a dose-dependent T_2 response was observed that was abrogated by preincubation of the nanoparticles with free streptavidin. Most recently, a magnetic nanosensor was designed to detect phenylalanine in solution (Tsourkas et al., 2004). This nanosensor was sensitive enough to detect an enantiomeric form of phenylalanine (D-Phe) in a racemic mixture of D-Phe and L-Phe. For this particular application, a magnetic nanoparticle was first conjugated to D-Phe. Addition of an antibody raised against D-Phe caused the nanoparticles to assemble with a concomitant decrease in T_2 relaxation times (Figure 9.6). This magnetic nanoparticle configuration allows for the highly sensitive detection of

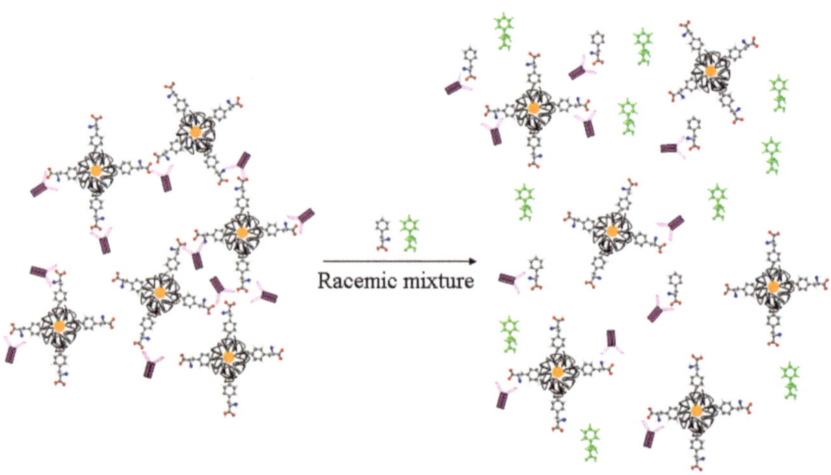

FIGURE 9.6. Magnetic nanosensor for the detection of enantiomeric impurities. A magnetic nanoparticle assembly is created by the addition of anti-D-phenylalanine antibodies to a suspension of phenylalanine-containing nanoparticles. Addition of a racemic mixture (containing both D- and L-phenylalanine) result in competition between free D-Phe and nanoparticle bound D -Phe for binding to the antibody, resulting in dispersion of the nanoassembly and increase in the T_2 of the solution. Reproduced with permission from Tsourkas et al. (2004).

D-phenylalanine in a racemic mixture of D- and L-phenylalanine by a one-step competitive magnetic immunoassay. The presence of D-Phe impurities in solution competed for antibody binding with the D-Phe on the magnetic nanoparticles, leading to dispersion of nanoparticles and causing an increase in T_2 as expected. This assay was reported to detect enantiomeric impurities at levels far below the requirements of the FDA.

9.5. Outlook and New Directions

The developed magnetic nanosensors can facilitate rapid detection of a molecular target without extensive sample preparation or target amplification. The fact that the assay relies on changes in the spin-spin relaxation time (T_2) of water molecules allows for the use of turbid samples, as light interference and scattering do not affect the sensitivity of the assay. Complex and opaque media such as blood, cell suspensions, culture media, lipid emulsions, and even whole tissue can be used. Additionally, there is no need for sample immobilization onto a flat surface, such as a microarray, facilitating faster hybridization and monitoring of binding kinetics. Magnetic nanosensors can detect subcellular or extracellular molecular interactions, such as mRNA and protein, as well as the presence of enzymatic activity in a sample. Using high-throughput MRI and NMR, hundreds of samples can be screened at once, drastically speeding up screening of a library of nanoparticles and or various molecular targets. Finally, most recent advances in portable NMR will undoubtedly expand the potential of magnetic resonance–based nanosensors.

Acknowledgment. We gratefully acknowledge the support from the National Cancer Institute by a Career Award to JMP (CA101781).

References

Avilion, A.A., Piatyszek, M.A., Gupta, J., Shay, J.W., Bacchetti, S., Greider, C.W., 1996. "Human telomerase RNA and telomerase activity in immortal cell lines and tumor tissues". Cancer Res 56, 645–650.

Baselt, D.R., Lee, G.U., Natesan, M., Metzger, S.W., Sheehan, P.E., Colton, R.J., 1998. "A biosensor based on magnetoresistance technology". Biosens Bioelectron 13, 731–739.

Brennan, M.L., Penn, M.S., Van Lente, F., Nambi, V., Shishehbor, M.H., Aviles, R.J., Goormastic, M., Pepoy, M.L., McErlean, E.S., Topol, E.J., Nissen, S.E., Hazen, S.L., 2003. "Prognostic value of myeloperoxidase in patients with chest pain". N Engl J Med 349, 1595–1604.

Brooks, R. A., 2002. "T(2)-shortening by strongly magnetized spheres: a chemical exchange model". Magn Reson Med 47, 388–391.

Bulte, J.W., Brooks, R.A., Moskowitz, B.M., Bryant, L.H. Jr., Frank, J.A., 1998. "T1 and T2 relaxometry of monocrystalline iron oxide nanoparticles (MION-46L): theory and experiment". Acad Radiol 5 Suppl 1, S137–140.

Bulte, J.W., Kraitchman, D.L., 2004a. "Iron oxide MR contrast agents for molecular and cellular imaging". NMR Biomed 17, 484–499.

Bulte, J.W., Kraitchman, D.L., 2004b. "Monitoring cell therapy using iron oxide MR contrast agents". Curr Pharm Biotechnol 5, 567–584.

Chemla, Y.R., Grossman, H.L., Poon, Y., McDermott, R., Stevens, R., Alper, M.D., Clarke, J., 2000. "Ultrasensitive magnetic biosensor for homogeneous immunoassay". Proc Natl Acad Sci U S A 97, 14268–14272.

Edelstein, R.L., Tamanaha, C.R., Sheehan, P.E., Miller, M.M., Baselt, D.R., Whitman, L.J., Colton, R.J., 2000. "The BARC biosensor applied to the detection of biological warfare agents". Biosens Bioelectron 14, 805–813.

Frank, J.A., Miller, B.R., Arbab, A.S., Zywicke, H.A., Jordan, E.K., Lewis, B.K., Bryant, L.H. Jr., Bulte, J.W., 2003. "Clinically applicable labeling of mammalian and stem cells by combining superparamagnetic iron oxides and transfection agents". Radiology 228, 480–487.

Gillis, P., Moiny, F., Brooks, R.A., 2002. "On T(2)-shortening by strongly magnetized spheres: a partial refocusing model". Magn Reson Med 47, 257–263.

Grimm, J., Perez, J.M., Josephson, L., Weissleder, R., 2004. " Novel nanosensors for rapid analysis of telomerase activity". Cancer Res 64, 639–643.

Grossman, H.L., Myers, W.R., Vreeland, V.J., Bruehl, R., Alper, M.D., Bertozzi, C.R., Clarke, J., 2004. "Detection of bacteria in suspension by using a superconducting quantum interference device". Proc Natl Acad Sci U S A 101, 129–134.

Harisinghani, M.G., Barentsz, J., Hahn, P.F., Deserno, W.M., Tabatabaei, S., van de Kaa, C.H., de la Rosette, J., Weissleder, R., 2003. "Noninvasive detection of clinically occult lymph-node metastases in prostate cancer". N Engl J Med 348, 2491–2499.

Harisinghani, M.G., Weissleder, R., 2004. "Sensitive, noninvasive detection of lymph node metastases". PLoS Med 1, e66, 202–209.

Josephson, L., Perez, J.M., Weissleder, R., 2001. "Magnetic nanosensors for the detection of oligonucleotide sequences". Angewandte Chemie-International Edition 40, 3204–3206.

Josephson, L., Tung, C.H., Moore, A., Weissleder, R., 1999. "High-efficiency intracellular magnetic labeling with novel superparamagnetic-Tat peptide conjugates". Bioconjug Chem 10, 186–191.

Kelly, K., Alencar, H., Funovics, M., Mahmood, U., Weissleder, R., 2004. "Detection of invasive colon cancer using a novel, targeted, library-derived fluorescent peptide". Cancer Res 64, 6247–6251.

Kelly, K.A., Allport, J.R., Tsourkas, A., Shinde-Patil, V.R., Josephson, L., Weissleder, R., 2005. "Detection of vascular adhesion molecule-1 expression using a novel multimodal nanoparticle". Circ Res 96, 327–336.

Kriz, K., Gehrke, J., Kriz, D., 1998. "Advancements toward magneto immunoassays". Biosens Bioelectron 13, 817–823.

McDonald, M.A., Watkin, K.L., 2006. "Investigations into the physicochemical properties of dextran small particulate gadolinium oxide nanoparticles". Acad Radiol 13, 421–427.

Modo, M., Hoehn, M., Bulte, J.W., 2005. "Cellular MR imaging". Mol Imaging 4, 143–164.

Perez, J.M., Josephson, L., O'Loughlin, T., Hogemann, D., Weissleder, R., 2002a. "Magnetic relaxation switches capable of sensing molecular interactions". Nat Biotechnol 20, 816–820.

Perez, J.M., O'Loughin, T., Simeone, F.J., Weissleder, R., Josephson, L., 2002b. "DNA-based magnetic nanoparticle assembly acts as a magnetic relaxation nanoswitch allowing screening of DNA-cleaving agents". J Am Chem Soc 124, 2856–2857.

Perez, J.M., Simeone, F.J., Saeki, Y., Josephson, L., Weissleder, R., 2003. "Viral-induced self-assembly of magnetic nanoparticles allows the detection of viral particles in biological media". J Am Chem Soc 125, 10192–10193.

Perez, J.M., Simeone, F.J., Tsourkas, A., Josephson, L., Weissleder, R., 2004. "Peroxidase substrate nanosensors for MR imaging". Nano Letters 4, 119–122.

Shay, J.W., Bacchetti, S., 1997. "A survey of telomerase activity in human cancer". Eur J Cancer 33, 787–791.

Shen, T., Weissleder, R., Papisov, M., Bogdanov, A. Jr., Brady, T.J., 1993. "Monocrystalline iron oxide nanocompounds (MION): physicochemical properties". Magn Reson Med 29, 599–604.

Song, Q., Zhang, Z.J., 2004. "Shape control and associated magnetic properties of spinel cobalt ferrite nanocrystals". J Am Chem Soc 126, 6164–6168.

Song, Q., Zhang, Z.J., 2006. "Correlation between spin-orbital coupling and the superparamagnetic properties in magnetite and cobalt ferrite spinel nanocrystals". J Phys Chem B Condens Matter Mater Surf Interfaces Biophys 110, 11205–11209.

Thorek, D.L., Chen, A.K., Czupryna, J., Tsourkas, A., 2006. "Superparamagnetic iron oxide nanoparticle probes for molecular imaging". Ann Biomed Eng 34, 23–38.

Tsourkas, A., Hofstetter, O., Hofstetter, H., Weissleder, R., Josephson, L., 2004. "Magnetic relaxation switch immunosensors detect enantiomeric impurities". Angewandte Chemie-International Edition 43, 2395–2399.

Weissleder, R., Kelly, K., Sun, E.Y., Shtatland, T., Josephson, L., 2005. "Cell-specific targeting of nanoparticles by multivalent attachment of small molecules". Nat Biotechnol 23, 1418–1423.

Zhang, R., Brennan, M.L., Fu, X., Aviles, R.J., Pearce, G.L., Penn, M.S., Topol, E.J., Sprecher, D.L., Hazen, S.L., 2001. "Association between myeloperoxidase levels and risk of coronary artery disease". Jama 286, 2136–2142.

Zhao, M., Josephson, L., Tang, Y., Weissleder, R., 2003. "Magnetic sensors for protease assays". Angewandte Chemie-International Edition 42, 1375–1378.

10 Magnetoptical Probes

Eyk Schellenberger

Abstract: The necessity of probe validation of widely used iron oxide particles in molecular magnetic resonance imaging (mMRI) has called for a growing demand for simple detection and quantification methods in vitro and ex vivo. A simple option to adapt these particles to standard methods of biological research such as microscopy and flow cytometry is the conjugation of fluorescent dyes. With the rapid developing optical imaging techniques such magnetoptical probes could not only be used for mMRI, but also for in vivo fluorescence imaging. Furthermore, the option of screening magnetoptical particles for their binding affinity to target cells has been employed to optimize particle designs or to screen libraries of particles with unknown binding properties for their binding affinity to target cells. Finally, the conjugation of fluorescence dyes via specifically cleavable linkers to MR detectable nanoparticles provides the option to image probe distribution by MRI and single or even multiple enzyme functions by fluorescence imaging.

Thus, fluorescent dyes do not only add functionality for improving probes for mMRI, but also increase potential information content by enabling multimodal imaging methods.

Keywords: Iron oxide particles, fluorescence, molecular imaging, magnetic resonance imaging, optical imaging.

10.1. Why Magnetoptical Probes?

With the introduction of magnetic resonance imaging (MRI) in diagnostic radiology, some researchers believed that the high soft tissue contrast makes the development of contrast media for MRI unnecessary. But the emerging paramagnetic and later superparamagnetic contrast agents could demonstrate impressively the gain of information in many diagnostic fields, e.g. in the detection of cancer, inflammation, and angiography. The sensible detectability especially of super-paramagnetic iron oxide nanoparticles, based on the high relaxivity, stimulated the development of ligand-coupled target-specific iron oxide particles (Bulte et al., 1992, Weissleder et al., 1992, Bulte and Kraitchman, 2004) for the in vivo characterization and measurement of biologic processes at the cellular and molecular level, known as 'molecular imaging' (Weissleder and Mahmood, 2001). With the introduction of multivalent metal chelates based, for instance, on dendrimers (Wiener et al., 1994), gadolinium-based paramagnetic particles proved to be useful for the design of molecular MRI probes as well (Kobayashi and Brechbiel, 2003). Although the MRI-detection limits of iron oxide particles are currently in the micro molar range, and thus still several orders of magnitude less sensible than nuclear imaging techniques, impressive results have been already achieved in molecular MR imaging. The future will show whether new technologies like magnetic particle imaging (MPI) can push the detection limits to new levels of sensitivity—competitive with nuclear imaging (Gleich and Weizenecker, 2005).

The increasing use of magnetic particles in molecular imaging and the necessity to validate particle uptake in the target tissues demanded the development of simple and sensitive detection methods for histology. While traditional Prussian blue staining is time-consuming and has limited sensitivity, fluorescent dyes coupled to antibodies or other affinity ligands have proven to be very useful as labels for in vitro and in vivo detection. Since one precondition of magnetic nanoparticles usable for molecular imaging is the possibility to couple compounds like target-specific ligands, the labeling with fluorescence dyes was an obvious choice. Fluorescently labeled magnetoptical probes are directly detectable by fluorescence microscopy, flow cytometry, and in vivo fluorescence imaging, which open the possibility to evaluate target binding with minimal effort. Figure 10.1 shows some prominent uses and detection methods for magnetoptical probes.

10.2. MR Imaging with Target-Specific Magnetoptical Probes

Central to the use of magnetic imaging probes for molecular imaging with MRI is the coupling of target-specific affinity ligands to superparamagnetic iron oxide particles. Examples for such affinity ligands are antibodies, proteins, peptides (Moore et al., 2004), and nucleotides (Bulte and Kraitchman, 2004). Frequently, the coupling procedures partially or completely hamper the binding affinity of the ligands, making it necessary to go through several optimization steps. Fluorescently labeled magnetic particles not only proved to be useful for histology after in vivo applications, but also for these optimization steps, since the particles can be readily detected with flow cytometry. Cytometry is an excellent tool for obtaining quantitative results of the binding behavior of magnetic particles to cells, which express the desired target-receptor. This approach was used, for instance, for the optimization of the MR-apoptosis probe Annexin V-CLIO-Cy5.5 (Schellenberger et al., 2004b) (model shown in Figure 10.2).

Since the applied linker chemistry for these particles was directed against the lysines of the protein, which are predominantly distributed at the membrane-binding site of Annexin V, the binding affinity was lost if more than one linker

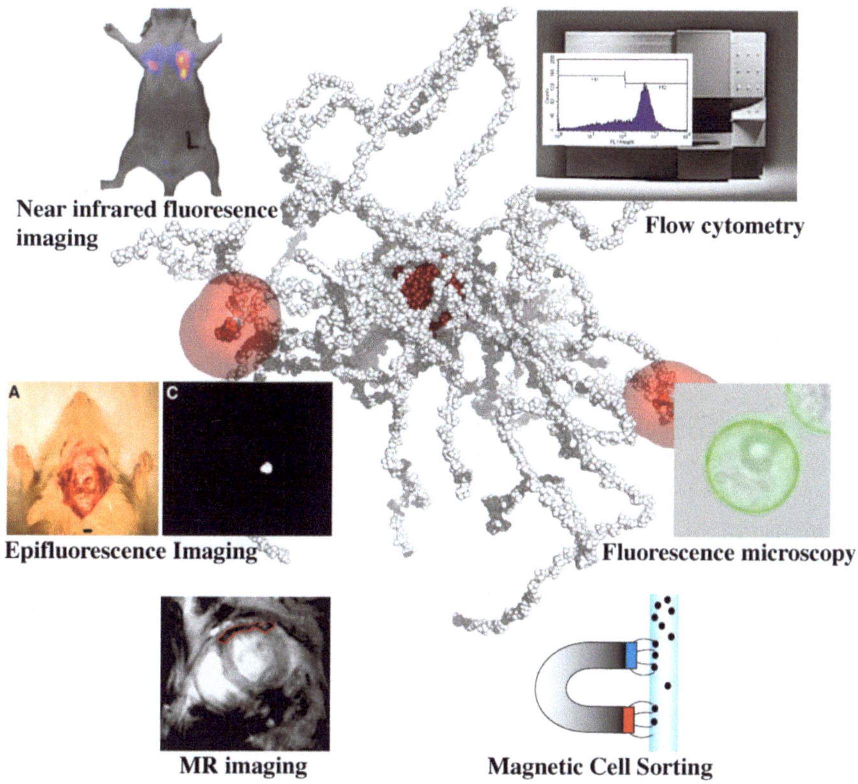

FIGURE 10.1. Applications for magnetoptical probes

or fluorescent dye per protein was coupled, but could be preserved with only one compound coupled to the protein (Schellenberger et al., 2004c). With the help of flow cytometry, we could develop a highly optimized magnetoptical apoptosis probe, which was later used for in vivo imaging apoptosis in mice. The objective of this study was to test Anx-CLIO-Cy5.5 probe for MR-imaging of apoptosis as a result of ischemia/reperfusion injuries in mouse hearts (Sosnovik et al., 2005). For that purpose, the fluorescence feature of the particles was used to verify the specificity for apoptotic heart muscle cells by fluorescence microscopy and was compared with unspecific CLIO-Cy5.5 particles. For the in vivo imaging, ischemia/reperfusion injuries were caused by a transient (30 min) ligation of the left anterior descending coronary artery (LAD). This procedure led to an induction of apoptosis of up to 20 % of the cardiac muscle cells in the anterior wall of the left ventricles. During the reperfusion time, Anx-CLIO-Cy5.5 or CLIO-Cy5.5 as control was injected intravenously and the animals were imaged by MR 24 h later. MRI revealed a strong signal loss in the muscle of the anterior wall for Anx-CLIO-Cy5.5 but not for CLIO-Cy5.5 (Figure 10.3). To further confirm that the signal loss was due to the accumulation of the specific iron

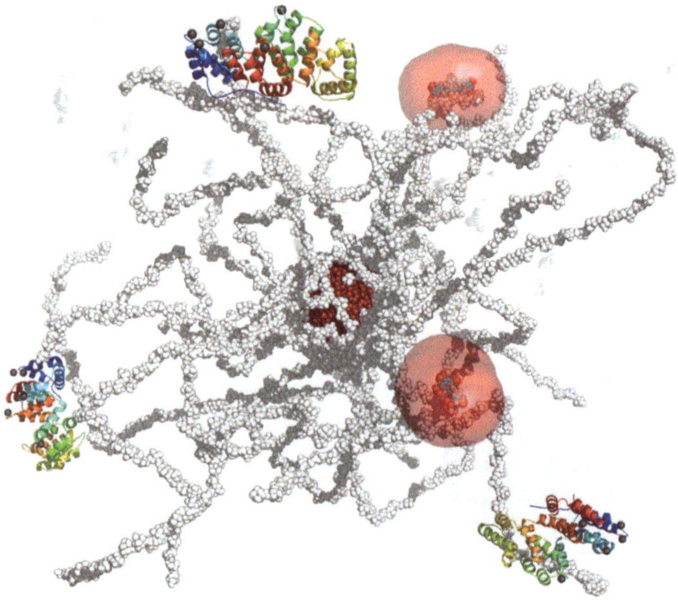

FIGURE 10.2. Molecular model of AnnexinV-CL10-Cy5.5

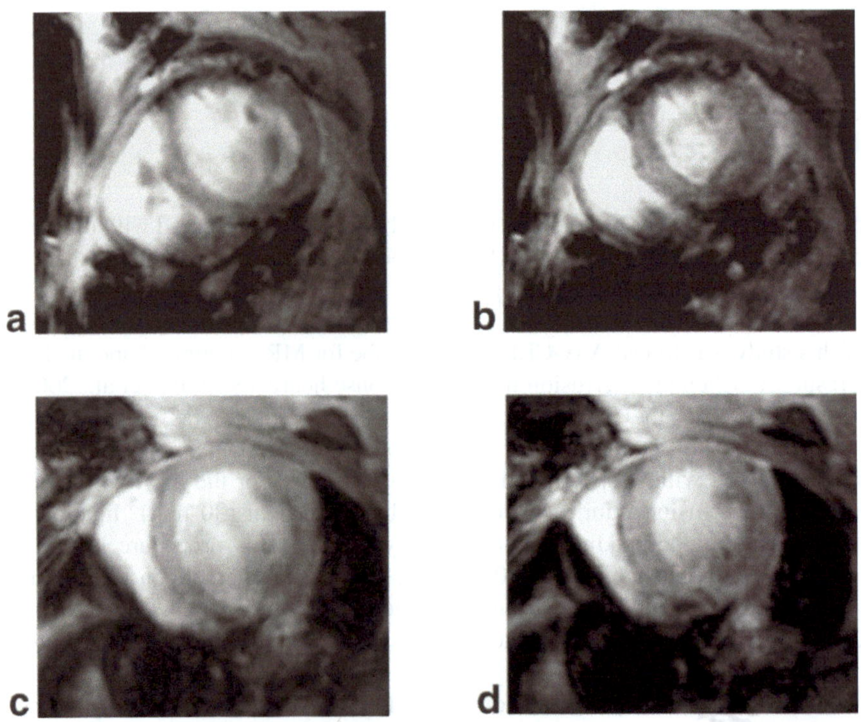

FIGURE 10.3. MR Imaging of ischemia/reperfusion injuries in mice

oxide particles, heart sections were committed to ex vivo near infrared imaging, which directly confirmed the accumulation of the particles in the anterior wall with a strong fluorescence signal.

10.3. MR and Epifluorescence Imaging

Another use for magnetoptical particles is the combination of MR imaging for pre-operative planning and intra-operative epifluorescence imaging for improvement in delineation and quality control, as suggested by Kircher et al. (2003). The authors used CLIO particles labeled with the near infrared fluorescence dye Cy5.5 similar to the particle in Figure 10.1, as an unspecific marker for MR-detection of glioblastomas in rats. Partially caused by the disruption of the blood brain barrier, the particles accumulate substantially in glioblastomas allowing a good delineation of the tumors by pre-operative MRI and subsequently an exact planning of the operational procedure. After 24 h of i.v. injection of the particles, a craniotomy was performed. Because of the relatively long-lasting intracellular sequestration of the particles by microglia cells, the particles could still be imaged during the surgery by noninvasive epifluorescence imaging one day later (Figure 10.4c). The accumulation of the fluorescence dyes in the tumors due to their particle character could be of great help to guide surgeons in the difficult task to resect all tumor tissue, at the same time avoiding unnecessary resections of normal brain structures. Even highly accurate, stereotactic planning of the surgery proves frequently useless, e.g. if the brain tissue dislocates after opening the skull due to increased intracranial pressure. Therefore, this method for identification of the tumor tissue by fluorescence imaging could be of great help.

10.4. Magnetoptical Probes for Combined Imaging of Probe Distribution and Enzyme Function

The attachment of fluorescent dyes to iron oxide nanoparticles provides not only the basis to localize these particles in vivo by MRI and optical imaging,

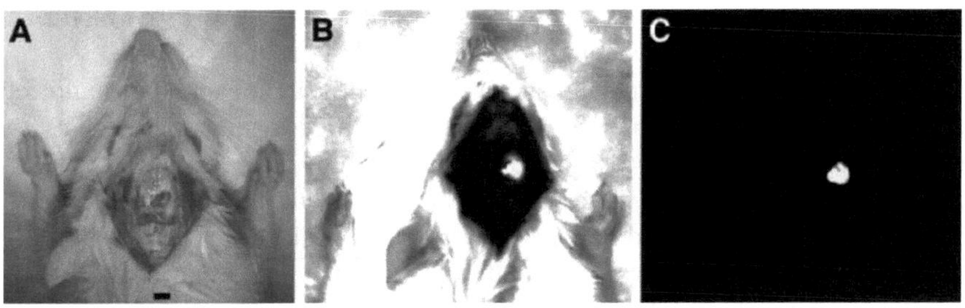

FIGURE 10.4. Combination of MR imaging for pre-operative planning and intraoperative epifluorescence imaging

but gives also the opportunity to design 'smart' nanoparticles. Lee Josephson et al. discovered that similar to activatable smart fluorescence probes (Weissleder et al., 1999, Bremer et al., 2001) the attachment of fluorescent dyes to iron oxide particles leads to substantial quenching of the fluorescence (e.g., 12-fold), which is reversed, when the dyes are released by cleavage (Josephson et al., 2002). This effect is at least partially due to interactions of the dyes with the iron oxide core itself and not only between the fluorescent dyes like in the conventional activatable fluorescence probes (Tung et al., 1999). For this, the near infrared dye Cy5.5 was coupled via peptide bridges to the particles, which contained a sequence of four arginines recognized as cleavage side by the protease trypsin. Thus, T_2^*-weighted imaging with MRI could reveal the in vivo distribution of these particles, while optical imaging displayed information about the enzyme activity in the environment of the particles.

An extension of this principle was presented by Kircher et al. (2004), who coupled two distinct near infrared fluorescent dyes (Cy5.5 and Cy7) to the CLIO particles. Again, the Cy5.5 dye was coupled to the particles via a protease-sensitive peptide containing a cleavage motive for trypsin. The second dye (Cy7) was directly attached—resistant to proteases. Therefore, this marker could be used for imaging the in vivo distribution of the particles. Applying the feature of simultaneous multi-wavelength imaging, ratio imaging of both dyes was used to correct the signal derived by protease activation for probe distribution and implantation depth, which otherwise causes strong attenuation of fluorescence signal due to tissue absorption. The described method is a way to obtain more accurate, robust results in optical imaging in addition to the possibility to localize the probe distribution by MR.

10.5. Biomimetic Magnetoptical Nanoparticles

Traditionally, target-specific magnetic nanoparticles are developed according the following scheme. First, affinity ligands like antibodies, peptides, or proteins for the desired target have to be searched for, selected, and tested. Then, a coupling procedure for attachment of the ligand to the magnetic nanoparticles has to be chosen. After coupling the binding affinity for the targets needs to be verified. Frequently, the affinity of the ligands is attenuated or even lost due to the coupling. In these cases, the coupling procedure needs to be modified, optimized, or sometimes even another affinity ligand has to be selected.

The principle of biomimetic nanoparticles is to circumvent this multistep procedure by creating libraries of surface-modified nanoparticles that are modified by a combination of simple chemical compounds, peptides, or sugars that can be matched to the target cells without knowing the binding receptors. The size of commonly used nanoparticles is between 4 nm for electrostatically stabilized particles (e.g., VSOP (Taupitz et al., 2000)) and 50 nm for sterically stabilized nanoparticles. This is in the size range of proteins and should, with a suitable surface, mimic the binding properties of proteins. Additionally, the possibility to detect magnetoptical nanoparticles by a broad selection of fluores-

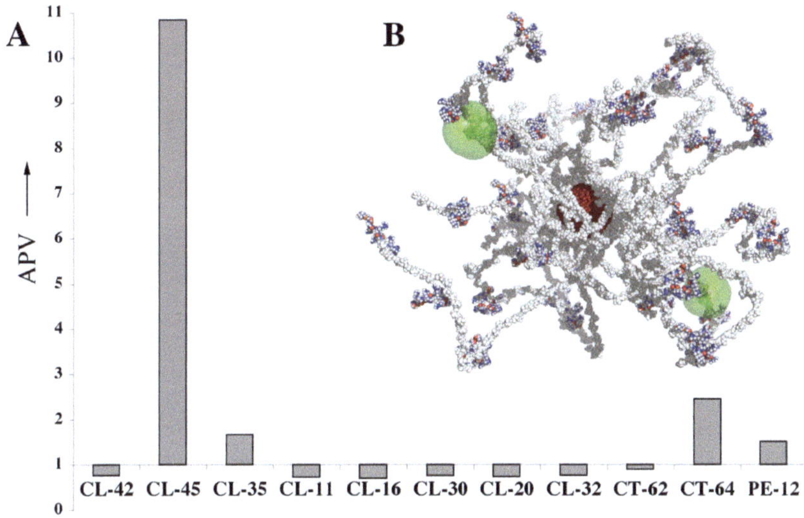

FIGURE 10.5. Screening of a magnetoptical nanoparticle library for selectivity for apoptotic cells

cence methods (e.g., by flow cytometry) provides a very efficient and accurate handle to screen a library for affinity to the desired target.

To test this strategy we developed a small library of 11 CLIO particles modified with simple chemical compounds or short peptides and screened them by flow cytometry for their binding affinity to apoptotic Jurkat T cells (Schellenberger et al., 2004a). On average, we coupled about 43 positive, negative, or neutral surface modifiers per fluorescein-labeled particle. As a result, we found one particle that was modified with positively charged arginine peptides that bound about 11-fold stronger to apoptotic cells compared with healthy cells. The result of the library screening and a model of the particle with the highest affinity to apoptotic cells are shown in Figure 10.5.

Weissleder et al. (2005) extended this principle by using a robotic system to synthesize a library of 146 surface-modified CLIO nanoparticles and screening them against five different cell targets (Weissleder et al., 2005). Out of this library, a series of nanoparticles with high specificity for endothelial cells, activated human macrophages, or pancreatic cancer cells was discovered. The particles with high affinity for pancreatic cancer cells also showed target specific uptake in pancreatic cancer in vivo.

10.6. Conclusions

Initially, fluorescent dyes of magnetoptical probes have been a tool for histology and probe optimization with the additional option to perform in vivo optical imaging. More recently, the optical imaging became an integral tool for in vitro screening procedures of, for instance, biomimetic nanoparticles. With the design of magnetoptical probes, where the optical part is activated by enzyme activity, even more versatile imaging probes for multimodality imaging are on the horizon, which can be located by MRI with high spatial resolution and at the same time specifically display one or more enzyme activities.

References

Bremer, C., Tung, C.H., Weissleder, R., 2001. In vivo molecular target assessment of matrix metalloproteinase inhibition. Nat Med 7, 743–748.

Bulte, J.W., Hoekstra, Y., Kamman, R.L., Magin, R.L., Webb, A.G., Briggs, R.W., Go, K.G., Hulstaert, C.E., Miltenyi, S., The, T.H., et al., 1992. Specific MR imaging of human lymphocytes by monoclonal antibody-guided dextran-magnetite particles. Magn Reson Med 25, 148–157.

Bulte, J.W., Kraitchman, D.L., 2004. Iron oxide MR contrast agents for molecular and cellular imaging. NMR Biomed 17, 484–499.

Gleich, B., Weizenecker, J., 2005. Tomographic imaging using the nonlinear response of magnetic particles. Nature 435, 1214–1217.

Josephson, L., Kircher, M.F., Mahmood, U., Tang, Y., Weissleder, R., 2002. Near-Infrared Fluorescent Nanoparticles as Combined MR/Optical Imaging Probes. Bioconjug Chem 13, 554–560.

Kircher, M.F., Mahmood, U., King, R.S., Weissleder, R., Josephson, L., 2003. A multimodal nanoparticle for preoperative magnetic resonance imaging and intraoperative optical brain tumor delineation. Cancer Res 63, 8122–8125.

Kircher, M.F., Weissleder, R., Josephson, L., 2004. A dual fluorochrome probe for imaging proteases. Bioconjug Chem 15, 242–248.

Kobayashi, H., Brechbiel, M.W., 2003. Dendrimer-based macromolecular MRI contrast agents: characteristics and application. Mol Imaging 2, 1–10.

Moore, A., Medarova, Z., Potthast, A., Dai, G., 2004. In vivo targeting of underglycosylated MUC-1 tumor antigen using a multimodal imaging probe. Cancer Res 64, 1821–1827.

Schellenberger, E.A., Reynolds, F., Weissleder, R., Josephson, L., 2004a. Surface-functionalized nanoparticle library yields probes for apoptotic cells. Chem Biochem 5, 275–279.

Schellenberger, E.A., Sosnovik, D., Weissleder, R., Josephson, L., 2004b. Magneto/optical annexin V, a multimodal protein. Bioconjug Chem 15, 1062–1067.

Schellenberger, E.A., Weissleder, R., Josephson, L., 2004c. Optimal modification of annexin V with fluorescent dyes. Chem Biochem 5, 271–274.

Sosnovik, D.E., Schellenberger, E.A., Nahrendorf, M., Novikov, M.S., Matsui, T., Dai, G., Reynolds, F., Grazette, L., Rosenzweig, A., Weissleder, R., Josephson, L., 2005. Magnetic resonance imaging of cardiomyocyte apoptosis with a novel magneto-optical nanoparticle. Magn Reson Med 54, 718–724.

Taupitz, M., Schnorr, J., Abramjuk, C., Wagner, S., Pilgrimm, H., Hunigen, H., Hamm, B., 2000. New generation of monomer-stabilized very small superparamagnetic iron oxide particles (VSOP) as contrast medium for MR angiography: preclinical results in rats and rabbits. J Magn Reson Imaging 12, 905–911.

Tung, C.H., Bredow, S., Mahmood, U., Weissleder, R., 1999. Preparation of a cathepsin D sensitive near-infrared fluorescence probe for imaging. Bioconjug Chem 10, 892–896.
Weissleder, R., Kelly, K., Sun, E.Y., Shtatland, T., Josephson, L., 2005. Cell-specific targeting of nanoparticles by multivalent attachment of small molecules. Nat Biotechnol 23, 1418–1423.
Weissleder, R., Lee, A.S., Khaw, B.A., Shen, T., Brady, T.J., 1992. Antimyosin-labeled monocrystalline iron oxide allows detection of myocardial infarct: MR antibody imaging. Radiology. 182, 381–385.
Weissleder, R., Mahmood, U., 2001. Molecular imaging. Radiology 219, 316–333.
Weissleder, R., Tung, C.H., Mahmood, U., Bogdanov, A., Jr., 1999. In vivo imaging of tumors with protease-activated near-infrared fluorescent probes. Nat Biotechnol 17, 375–378.
Wiener, E.C., Brechbiel, M.W., Brothers, H., Magin, R.L., Gansow, O.A., Tomalia, D.A., Lauterbur, P.C., 1994. Dendrimer-based metal chelates: a new class of magnetic resonance imaging contrast agents. Magn Reson Med 31, 1–8.

Part 2
Radiolabeled Nanoparticles: Applications in Nuclear Medicine

11
Use of Radiolabeled Liposomes for Tumor Imaging

Tamer Elbayoumi and Vladimir Torchilin

Abstract: Liposomes or microscopic phospholipid vesicles have a long history of medical applications as drug carriers. Among other pharmaceutical agents, liposomes have been used as carrier systems for various imaging agents including those used in gamma-scintigraphy. This chapter provides a brief description of the preparation and application of radiolabeled liposomes. Various methods of liposome labeling with gamma-emitting isotopes are considered as well as specific problems associated with obtaining the optimal preparations and results of both animal experiments with radiolabeled liposomes and early clinical data. The examples of using the radiolabeled liposomes for the development of the liposomal anti-cancer drugs are also presented.

11.1. Introduction

Successful medical imaging requires that a sufficient intensity of a signal from the area of interest is achieved to differentiate this area from surrounding tissue, which usually involves the use of specific contrast agents for early detection and localization of numerous pathologies in different tissues. According to the type of the reporter agent, and sensitivity and resolution of the registration method, various amounts of the contrast agent have to be delivered to the area of interest. The common objective, however, is to accumulate a sufficient quantity of the contrast moiety in the area of interest and keep its presence in normal tissues on a minimal level. One possible approach is the use of contrast-loaded pharmaceutical nanocarriers as contrast materials providing an enhanced accumulation of reporter moieties in pathological areas. For almost three decades now, liposomes have been identified as promising vehicles to deliver a wide range of encapsulated and/or membrane-incorporated therapeutic or diagnostic agents into the required areas (Gregoriadis et al., 1974b; Papahadjopoulos, 1988; Lasic and Papahadjopoulos, 1998). In particular, these nanocarriers have been utilized for carrying the diagnostic moieties used with all imaging modalities: gamma-scintigraphy, magnetic resonance imaging (MRI), computed tomography (CT) imaging, and sonography (Torchilin, 1995, 1996).

Liposomes can be loaded with gamma-imaging agents that permit visual tracking through scintigraphic imaging (Phillips, 1999). Scintigraphic nuclear imaging offers several advantages over higher resolution cross-sectional imaging modalities like CT and MRI (Wolf, 1995), including its rapid ability to image and screen the whole body for sites of abnormal uptake along with the ability to provide a temporal visual tracking of the radionuclide movement in the body over a course of time. Another advantage is the ability of scintigraphic imaging to non-invasively quantitate the amount of injected activity that reaches a particular location in the body (Goins and Phillips, 2001). In addition, scintigraphy requires only a small amount of contrast agent to be used (usually in the nanogram range), which does not interfere with the biodistribution of the labeled liposomes and physiological processes underlying this distribution. In contrast, other imaging modalities, such as MRI and CT, require the administration of a significantly higher amount of the contrast agent to obtain clinically sound images (milligrams for MRI and grams for CT) (Torchilin, 2000b). The small quantity of the contrast agent required for gamma-scintigraphy decreases also the risk for an adverse reaction induced by the contrast agent itself (Goins and Phillips, 2001).

One can say that while imaging techniques such as CT and MRI provide anatomical information, scintigraphic images of the in vivo distribution of a radiopharmaceutical agent obtained with a gamma camera depict changes in physiological processes (Boerman et al., 2000; Goins and Phillips, 2001). It has been shown that scintigraphic images can be positive for the presence of tumor lesions or infectious foci before any anatomical abnormalities are apparent (Boerman et al., 2000; Laverman et al., 2000).

To date, clinicians have generally utilized MRI and CT contrast agents to demonstrate changes in vascular permeability and blood flow. On the other hand, scintigraphic imaging has been used to depict a wide variety of physiological processes, ranging from changes in glucose, protein, and fatty acid metabolism to the demonstration of gene expression and detection of changes in the concentration of cell signaling receptors (Strauss, 1997; Herschman et al., 2000).

11.2. Radiolabeling of Liposomes

Many individual lipids and their mixtures, when suspended in an aqueous phase, spontaneously form bilayered structures in which the hydrophobic parts of their molecules face inwards, and the hydrophilic parts are exposed to the aqueous phase surrounding them. These vesicles are called liposomes. Several different types of liposomes exist; each type has specific characteristics and can be prepared by specific methods (Lasic, 1993; Torchilin, 1996; Torchilin and Weissig, 2003). Usual classification of liposomes is based on their size and number of concentric bilayers forming a single vesicle. When liposomes are prepared by the hydration of a dried lipid mixture, they spontaneously form relatively large particles (around 1 μm in size) containing multiple lipid bilayers alternating with aqueous interphases. Such liposomes are called multilamellar vesicles (MLVs). Upon being subjected to certain other procedures, MLVs form small unilamellar vesicles (SUVs) with a diameter of around or even less than 100 nm, or large unilamellar vesicles (LUVs), which are normally sized between 200 and 800 nm. The preparation methods can be easily scaled up for industrial production of large batches of liposomes with a predictable size and a narrow size distribution (Lasic, 1993).

Since their discovery by Alec Bangham about 40 years ago, liposomes became a pharmaceutical carrier of choice for numerous practical applications. Different liposomes have been under extensive investigation as pharmaceutical carriers for the improved delivery of a broad spectrum of agents, including chemotherapeutic drugs, antibiotics, fungicides, antigens, chelating compounds, radionuclides, and other contrast agents (Lasic, 1993; Torchilin, 2005).

This broad applicability of liposomes is based on the following reasons: (1) Liposomes are completely biocompatible; (2) liposomes can entrap practically any drug or diagnostic agent into either their internal water compartment or into the membrane itself depending on the physicochemical properties of the drug; (3) liposome-incorporated pharmaceuticals are protected from the inactivating effect of external conditions, yet at the same time do not cause undesirable side-reactions; (4) liposomes modified with certain specific ligands also provide a unique opportunity to deliver pharmaceuticals into cells or even inside individual cellular compartments (Philippot and Puyal, 1995). Pursuing different in vivo delivery purposes, the size, charge, and surface properties of liposomes can be easily controlled by varying ingredients and/or preparation methods (Lasch et al., 2003). In particular, it was also recognized that liposomes when labeled with gamma radiation–emitting radionuclides might serve as efficient radiopharmaceuticals to visualize a broad variety of pathological processes (Phillips and Goins, 1995).

There are two classes of radioisotopes available for scintigraphic imaging: single photon radioisotopes and positron-emitting radioisotopes (Table 11.1). Single photon radionuclides are the most widely used radioisotopes in the current clinical nuclear medicine practice, these radionuclides are readily available and relatively inexpensive. The single photon radionuclides have varying activity half-lives and photon energies, with ^{99m}Tc being the most clinically used one because its photon energy characteristics and half-life allow for generating high-quality images. In addition, ^{99m}Tc is readily available and inexpensive. ^{99m}Tc-loaded/labeled liposomes also appear to be the most attractive for clinical diagnostics and research studies. On the other hand, positron-emitting radionuclides tend to have much shorter half-lives (Goins and Phillips, 2003).

A large variety of techniques have been developed to load/label liposomes with radioisotopes. The following were found to be the main characteristics of a successful liposome labeling method: (1) sufficient shelf stability of the liposomal formulation to be labeled, (2) high labeling efficiency, (3) convenient and efficient labeling at room temperature, (4) use of readily available isotopes that have good physical imaging characteristics, dosimetry, and half-life, (5) reasonable in vitro and in vivo stability of label association with liposomes,

TABLE 11.1. Different radionuclides used in scintigraphy and their applications in liposome labeling

Radionuclide	Half life	γ-Ray energy (Kev)	Methods to label liposomes
		Single photon radionuclides	
^{99m}Tc	6h	140	Encapsulation, Stannous chloride/tin(II) dioxinate, DTPA-PE–/HYNIC-PE–containing liposomes, PAPs, ^{99m}Tc–HMPAO/^{99m}Tc–DISIDA/^{99m}Tc–BMEDA with GSH–liposomes
^{111}In	2.8 d	171 & 245	DTPA-PE–containing liposomes, PAPs, ^{111}In-oxine with NTA/defuroxamine-liposomes, ionophore A23187–containing liposomes
^{201}Tl	3.0 d	68–62, 167	
^{67}Ga	3.2 d	93, 185, 300, 394	DTPA-PE–containing liposomes, ^{67}Ga-oxine with NTA/defuroxamine-liposomes
^{133}Xe	5.3 d	81	
^{123}I	13 h	160	Encapsulation of Na ^{123}I, using iodinated phospholipid
^{131}I	8 d	364	
		Positron-emmiting radionuclides	
^{15}O	2 min		Binding to haemoglobin in performed liposomes
^{13}N	10 min		
^{11}C	20 min	511	
^{68}Ga	68 min		
^{18}F	110 min		Encapsulation of 18F-FDG

(6) universal applicability to all types of liposomes, regardless of size, surface charge, or lipid composition (Phillips, 1999; Goins and Phillips, 2003).

Several general approaches have been developed for the labeling of liposomes. They are overviewed here in order of complexity and progression from the poorer to higher in vivo stability (Goins and Phillips, 2003).

11.2.1. Label Encapsulation During Liposome Preparation

Liposomes can be labeled by the incorporation of the gamma-imaging agent into the lipid phase (Hnatowich et al., 1981) or encapsulation in the liposome aqueous interior during formulation (Caride et al., 1976; Torchilin et al., 1995a). This method has many disadvantages including low encapsulation efficiency, high radiation exposures to personnel, difficulty in maintaining liposome sterility, and the necessity of making liposomes within a very short period of the intended clinical use, due to the overall instability of the formulation (Phillips, 1999). This is particularly true with radioisotopes with half-lives under 24 h, which includes ^{99m}Tc and ^{18}F. The only method reported to date for labeling liposomes with a PET radionuclide is the encapsulation of 18-fluorodeoxyglucose

(18F-FDG) (110 min half-life) within liposomes during liposome manufacture (Oku, 1999). The development of methods to label liposomes with PET radionuclides after their manufacture will be very valuable. A better approach is the labeling of liposomes after their manufacture and just prior to use (Goins and Phillips, 2001).

11.2.2. Surface Labeling After Liposome Preparation

Liposomes can be labeled by attaching the radioisotope to the surface of preformed liposomes. This method in fact offers an improvement over the encapsulation method as the preformed liposomes can be made in a different laboratory than where they were labeled, where the labeling process can occur just prior to the clinical use. This surface labeling employs the basic electrostatic interactions between the charged isotope and the liposome surface. Labeling using this method has the problems of inconsistent labeling efficiencies and insufficient in vivo stability due to exposure of the external surface of liposomes to plasma proteins in the circulation (Phillips, 1999).

Initial methods described for labeling post-manufactured liposomes with ^{99m}Tc relied on the labeling of the liposome membrane with reduced ^{99m}Tc. This method proved to produce ^{99m}Tc-labeled liposomes that were not stable when injected in vivo, with frequent dissociation of the ^{99m}Tc from the liposome after administration (Patel et al., 1984).

11.2.3. Label Incorporation into the Lipid Bilayer of Preformed Liposomes

Preformed liposomes can be labeled by incubation with a lipophilic chelator into which the radioisotope has been chelated prior to the incubation with the liposome. Although this method provides high labeling efficiency, its in vivo stability is questionable, because most of these lipophilic labels can readily exchange with lipoprotein components in blood (Damen et al., 1981).

Morgan and co-workers (1981) were among the first to label liposomes (MLVs) with ^{99m}Tc by reducing pertechnetate with stannous chloride in the presence of the liposomes. In animal studies, the liposomal preparation labeled according to this "stannous chloride method" showed relatively high uptake of the radiolabel in kidneys and bladder, suggesting extensive in vivo release of the radiolabel from the liposomes. This in fact was confirmed by other groups (Love et al., 1989; Ahkong and Tilcock, 1992). Several methods to label liposomes with ^{99m}Tc have been developed. Alafandy and co-workers (1996) developed the oxido-reduction method, as an improved version of the stannous chloride method. Liposomes were prepared with tin(II)dioxinate complexes inserted in the bilayers. However, this method required a series of extensive washing in order to remove the free pertechnetate from the liposomes (Boerman et al., 2000).

11.2.4. Surface Labeling of Preformed Liposome by Incorporating Lipid Chelator Conjugate During Preparation

Another approach is to incorporate a lipid chelator into the liposome during the formulation stage, more specifically during the preparation of the lipid film (Phillips, 1999). These amphiphilic chelators are composed of diethylene triamine pentaacetic acid (DTPA) conjugated to the head group region of a long-chain fatty acid or phospholipid molecule. The chelator is incorporated into the lipid components of the bilayer, prior to the actual preparation of the liposomes. After hydration of the liposomal lipids the chelator is exposed on the outer surface of the liposomes, thus being easily accessible for the radionuclide. The liposomes are then incubated with the radioisotope, which becomes bound to the chelator present on the external surface of the liposome. Still, lipid-chelator systems described to date suffer from decreased in vivo stability, mainly due to the exposure of the radiolabel on the surface of liposome (Boerman et al., 2000).

Historically, Hnatowich et al. (1981) described the synthesis of a long hydrocarbon chain covalently coupled to DTPA. These liposomes could be labeled with ^{67}Ga or ^{99m}Tc. However, labeling efficiencies did not exceed 60 %. During the incubation in 50 % human serum, approximately 30 % of the ^{99m}Tc activity was released from the liposomes, indicating instability of these radiolabeled formulations. Later, Goto et al. and co-workers (1989) developed liposomes containing stearylamine-DTPA conjugates to label liposomes with ^{99m}Tc. These preparations showed good in vitro stability in 50 % human serum.

A second method has been recently developed by Laverman et al. (1999). With this method, the highly stable ^{99m}Tc-chelator, hydrazino nicotinamide (HYNIC), is attached to distearoyl phosphatidylethanolamine (DSPE), which is added to the liposome formulation during manufacture. Reduced ^{99m}Tc is then incubated with the HYNIC liposomes where it becomes chelated by the HYNIC on the liposome surface.

A new approach aimed at the amplification of the diagnostic signal using poly-chelating polymers, for higher loading capacity of reporter molecules was suggested later by Torchilin et al. (Khaw et al., 1991; Slinkin et al., 1991; Torchilin et al., 1991). Since it is rather difficult to enhance the signal intensity from a given reporter metal in liposomes, one may attempt to increase the quantity of carrier-associated reporter metals (such as Gd for MRI or ^{111}In for scintigraphy) (Torchilin, 2000b). In this design, a series of single terminus-activated polychelating polymers were synthesized using poly-L-lysine (PLL) as a main chain and chelating moieties (such as DTPA) as side groups. These polychelating polymers were additionally modified with a hydrophobic residue to assure their firm incorporation into the liposome membrane. As a result, polychelating amphiphilic polymers (PAP) have been obtained. These constitute a family of co-polymers containing a hydrophilic fragment with multiple chelating groups and a relatively short, but highly hydrophobic, phospholipid fragment on one end of a polymeric chain suitable for incorporation into hydrophobic liposome membrane or a micelle core (Torchilin, 1997). The chemistry was based on the use of carbobenzoxy (CBZ)-protected PLL with free terminal amino-group, which was used to modify the PLL chain with the phospholipid anchor, with subsequent CBZ deprotection and incorporation of DTPA residues at lysine aminogroups in PLL. This polychelator was found to easily incorporate into the liposomal membrane in the process of liposome preparation, and sharply increases the number of chelated heavy metal atoms attached to a single lipid anchor (Figure 11.1A).

Liposomes loaded with ^{111}In via the liposome-incorporated PAP demonstrated significantly higher labeling efficiency than liposomes loaded with ^{111}In via the monomeric DTPA-PE (Figure 11.1B). This technique allowed to increase the number of bound reporter metal atoms per vesicle and to decrease the dosage of an administered total liposomal lipid without compromising the image signal intensity (Erdogan et al., 2006a).

11.2.5. Aqueous Phase Loading of Preformed Liposomes

Several methods have been developed for loading the radioisotope into the aqueous phase of preformed liposomes. This approach generally uses a lipophilic chelator capable of carrying the radioisotope through the lipid membrane into the aqueous phase, where the isotope is re-chelated into a more firm complex with an entrapped compound and remains inside the liposome. The success of this method depends mainly on the high affinity of the radioisotope for the agent entrapped in the aqueous phase of the liposome. Otherwise, the radioisotope will shuttle back and forth across the lipid membrane. This method has shown high labeling efficiency (more than 90 %). Moreover, it is likely to have the greatest in vivo stability, because of the protected location of the radioisotope inside the liposome (Goins and Phillips, 2001).

Initially, the approach was demonstrated with ^{67}Ga transported through the liposomal bilayer as the lipophilic ^{67}Ga–oxine complex, and then irreversibly trapped inside liposomes via the presence of an aqueous interior-loaded hydrophilic chelator in preformed liposomes (Hwang et al., 1982). The same procedure was suggested for labeling liposomes with ^{111}In as well (Gabizon et al., 1988). In the last decade, two stable methods of labeling post-manufactured liposomes with ^{99m}Tc have been reported. In the first method, ^{99m}Tc is chelated into the lipophilic carrier, hexamethylpropyleneamine oxime (HMPAO) (Phillips et al., 1992). The ^{99m}Tc–HMPAO complex is then added to liposomes with encapsulated glutathione (GSH) and incubated with the GSH-containing liposomes for 30 min during which the ^{99m}Tc–HMPAO migrates across the lipid membrane inside liposomes, where it is converted to a hydrophilic complex by reductive partial decomposition via the interaction with GSH. The hydrophilic ^{99m}Tc–HMPAO complex remains trapped within the liposome. This method proved reliable, yielding in vivo stable radiolabeled liposomes with high labeling efficiency (approximately 85 %).

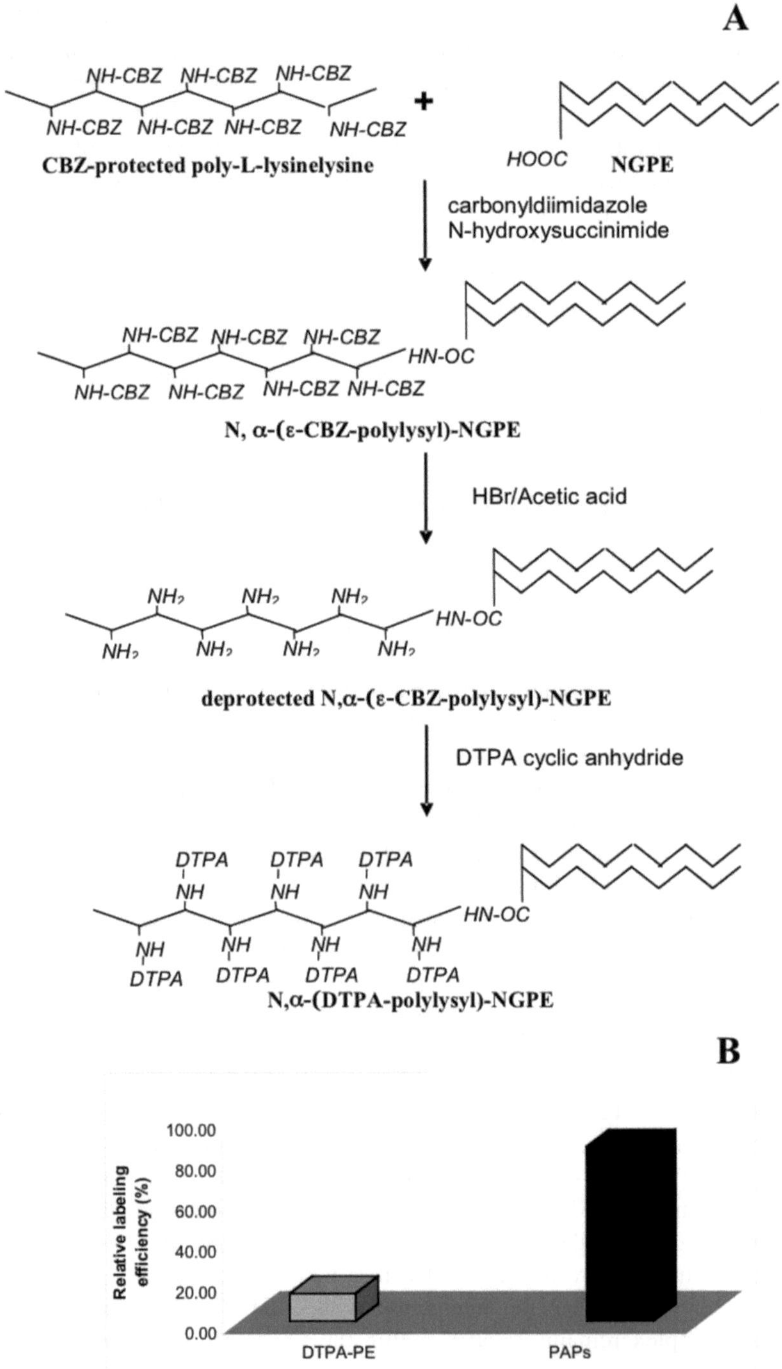
A
NH-CBZ NH-CBZ NH-CBZ
NH-CBZ NH-CBZ NH-CBZ NH-CBZ
+
HOOC
CBZ-protected poly-L-lysinelysine
NGPE
carbonyldiimidazole
N-hydroxysuccinimide
HN-OC
N, α-(ε-CBZ-polylysyl)-NGPE
HBr/Acetic acid
NH2 NH2 NH2
NH2 NH2 NH2 NH2
deprotected N,α-(ε-CBZ-polylysyl)-NGPE
DTPA cyclic anhydride
DTPA DTPA DTPA
NH NH NH
DTPA DTPA DTPA DTPA
N,α-(DTPA-polylysyl)-NGPE
B
Relative labeling efficiency (%)
100.00
80.00
60.00
40.00
20.00
0.00
DTPA-PE
PAPs

In a similar way, Suresh and Cao (1998) proposed the use of the hepatobiliary imaging agent ^{99m}Tc–diisopropyl iminodiacetic acid (^{99m}Tc–DISIDA) to label preformed liposomes with encapsulated GSH. As in the previously mentioned study, the ^{99m}Tc–DISIDA complex undergoes the reductive decomposition to more hydrophilic species by reaction with encapsulated GSH, yielding the incorporation of the ^{99m}Tc–DISIDA into the liposomes with 90 % efficiency. The labeled liposomes were stable in saline and 90 % FBS for up to 24 h. Biodistribution studies in mice showed that ^{99m}Tc-labeled liposomes accumulated in liver and spleen at 24 h post-injection. The use of ^{99m}Tc–DISIDA was preferred compared to HMPAO, since the ^{99m}Tc–DISIDA is much cheaper and has a longer shelf life when used for liposome labeling.

Phillips et al. (1992) introduced a novel method for radiolabeling preformed liposomes with ^{99m}Tc–SNS/S [*N*-ethyl-substituted bis(2-mercaptoethyl)amine and phenylthiolate coligand, S] complexes, which can be applied to liposome radiolabeling with therapeutic radionuclides, ^{186}Re and ^{188}Re as well (Bao et al., 2003). In their technique, liposomes containing GSH were incubated with ^{99m}TC–SNS/S complexes having a neutral core coordinate structure. These ^{99m}TC–SNS/S complexes can change into hydrophilic complexes following their entry into the inner space of liposomes and reaction with pre-encapsulated GSH, cysteine, or other hydrophilic chemicals containing a thiol group. It was also found that the higher the lipophilicity of the ^{99m}TC–SNS/S complex, the easier it can pass through the double layer of the liposome membrane. Yet, after the complex enters into the inner space and changes into another complex with GSH, its lipophilic side chain compromises the retention of the ^{99m}Tc within the liposomes, leading to decreased in vitro and in vivo stability compared to less-lipophilic ^{99m}TC–SNS/S complexes (^{99m}Tc–BMEDA, ^{99m}Tc–*N*,*N*-bis(2-mercaptoethyl)-*N*′,*N*′-diethyl-ethylenediamine). This method resulted in a good labeling efficiency (about 50 %–70 %) and high in vitro stability of ^{99m}Tc–BMEDA liposomes. Moreover, these ^{99m}Tc–GSH liposomes also showed the slow blood pool clearance and low spleen accumulation typical for PEG-coated liposomes.

In interpreting and analyzing scintigraphic images with radiolabeled liposomes, one has to recognize that these radionuclide labels track the biodistribution of liposomes and do not always track the disposition of encapsulated therapeutic agents (Barenholz et al., 1993). The particular chelate with which the radionuclide is bound may also influence the behavior of the radionuclide after the liposome leaves the blood and becomes metabolized. Thus, Ogihara-Umeda et al. (1992) have shown that when liposome-encapsulated ^{67}Ga is in the form of ^{67}Ga–nitrilotriacetic acid (NTA) complex, the ^{67}Ga accumulation in a tumor is twice as high as when ^{67}Ga is used as ^{67}Ga-deferoxamine, even at similar blood

FIGURE 11.1. Synthesis of amphiphilic polychelate DTPA-PLL-NGPE consisting of hydrophilic DTPA-polylysyl (DTPA-PLL) moiety and hydrophobic *N*-glutaryl phosphatidyl ethanolamine (NGPE) moiety (**A**). Labeling efficiency of liposomes containing the same molar fraction of PAP or DTPA-PE (**B**).

level of ^{67}Ga-liposome activities at 24 h. It was speculated that this difference was due to the high affinity of the unbound ^{67}Ga for the tumor tissue, which resulted in ^{67}Ga tumor accumulation after metabolism of the liposome in the region of the tumor and release of ^{67}Ga from the low-affinity nitrilotriacetic acid chelate. Hence, the characteristics of the radionuclide and the radionuclide ligand complex used in liposome labeling must always be carefully considered when analyzing the scintigraphic images, particularly after the beginning of liposome metabolism (Goins and Phillips, 2001).

11.3. Some General Considerations for Target Imaging with Radiolabeled Liposomes

11.3.1. Effect of Liposome Properties on the In Vivo Behavior of the Radiolabeled Formulation

It was demonstrated earlier that mean vesicle size is a major factor affecting the circulation time of "plain" liposomes (Gregoriadis et al., 1974). A non-linear pharmacokinetics of MLV was observed, and the variability in the MLV sizes was assumed to be the underlying reason. The effect of particle size was later studied in mice (Senior et al., 1985) to reveal the circulation times of liposomes varying in mean size (200 vs 400 nm), but of identical lipid composition. The larger liposomes cleared much faster ($t_{1/2} = 0.2$ h) than the smaller vesicles ($t_{1/2} = 1.5$ h). The liposomes' in vitro and in vivo properties depend also on their composition. The lipid composition and the preparation technique determine the characteristics of the liposomes. Overall, the use of phospholipids with longer saturated acyl chains and the inclusion of cholesterol (Chol) yield more stable vesicles. Therefore, for application as a radiopharmaceutical (when the radionuclide is incorporated in the aqueous phase), the liposomes should preferably contain such components to prevent leakage of the radionuclide from the liposomes (Senior, 1987). The incorporation of Chol into the liposomal bilayers, and the use of sphingomyelin (SM) and/or phospholipids with long and saturated acyl chains prolong also the vascular residence of liposomes (Kirby et al., 1980a,b; Senior and Gregoriadis, 1982). Moreover, liposomes are cleared from the circulation by the reticulo-endothelial system (RES) in a dose-dependent manner (Abra and Hunt, 1982; Bosworth et al., 1982). It was shown in mice that SUVs (80 nm; SM:Chol = 2:1) administered at the dose of 6 mg lipid/kg body weight show high liver uptake (about 60 % of injected dose 23 h post-injection.), while at doses exceeding 120 mg/kg, liver uptake is much lower and reaches a plateau at 20% of injected dose (Beaumier et al., 1983).

The surface charge and the charge density of liposomes can be controlled by introducing charged components in the lipid bilayers. In general, the presence of a high electrostatic surface charge promotes the interaction of liposomes with biomolecules (e.g., serum proteins) and cells. It has been shown that neutral and positively charged liposomes acquire a negative charge in the circulation, probably due to adsorption of plasma components (e.g., α2-macroglobulin) (Goins et al., 1994). However, regardless of charge acquired on contact with plasma, tissue uptake of negatively charged liposomes after i.v. injection is more rapid than of neutral or positively charged vesicles as indicated by their faster clearance from the circulation (Gregoriadis et al., 1974a; Senior, 1987).

11.3.2. Background Clearance

To clearly visualize the tumor, it has to accumulate a sufficient amount of radioactivity above the background level of activity demonstrated by normal tissues. Generally, a target-to-background ratio of at least 1.5 to 2 is necessary for detecting a pathological site and providing diagnostically significant images (Phillips and Goins, 1995). As the radioactivity of agents remaining in the blood form the main source for the background, proper identification and selection of the radiopharmaceutical plasma half-life is essential. Long circulation is important for higher tumor accumulation via the enhanced permeability and retention (EPR) effect (Maeda et al., 2000; Maeda, 2001). At the same time, the remaining blood activity decreases the target-to-background ratio. On the other hand, a too short plasma half-life of a contrast agent results in great reduction in tumor accumulation, although providing low background activity. Thus, a half-life of about 6–12 h appears to be ideal to allow for sufficient accumulation of liposomes in inflamed or cancerous targets, yet provide enough background clearance to permit early identification of the target (Gabizon et al., 1990; Phillips, 1999; Harrington et al., 2000b).

11.3.3. Active Background Removal

The target-to-background ratio can also be increased by developing special protocols for decreasing the background signal. Thus, the method was described for the rapid removal of liposomes from the circulation prior to the imaging (Ogihara-Umeda et al., 1993). In this method, biotin-coated liposomes are injected intravenously and allowed to accumulate in the tumor. Then, after 4 h, avidin is injected to induce the aggregation of the biotin-coated liposomes due to strong avidin–biotin interaction. This resulted in rapid removal of aggregated liposomes from the circulation by liver and spleen. Using this technique, the tumor-to-blood ratio increased to 13.8 compared to about 0.9 in the control. This method has great potential for rapid imaging using liposomes.

11.4. Experimental and Clinical Gamma Imaging of Tumors with Radiolabeled Surface-Modified Liposomes

11.4.1. Long-Circulating PEG Liposomes

"Plain" liposomal preparations are rapidly sequestered after i.v. administration by phagocytic cells of the RES, primarily those located in liver and spleen.

This efficient targeting to RES cells was exploited for liver and spleen imaging with radiolabeled liposomes (Kabalka et al., 1991). Thus, intra-hepatic and intra-splenic metastases could be imaged as with radiolabeled liposomes (Jaggi et al., 1991). For this application, almost any conventional liposomal formulation can be used.

The manipulation of liposome membrane surface properties provided a new approach to produce RES-evading liposomes, and the development of long-circulating liposomes represented a major step forward toward the successful application of liposomes in vivo. Current formulations of long-circulating liposomes contain lipid derivatives of the hydrophilic polymer polyethylene glycol (PEG), such as PEG distearyl phosphatidyl-ethanolamine (PEG-DSPE) (Klibanov et al., 1990; Papahadjopoulos et al., 1991; Huang et al., 1992). The presence of bulky hydrophilic surface groups on the liposome surface appears to sterically hinder the interaction of plasma proteins (opsonins) with the liposomal membranes. Consequently, the liposomes are protected from recognition by cells of the RES (Torchilin, 1994; Torchilin et al., 1995b). The prolonged circulation of such liposomes in the blood enhances also their tumor accumulation due to the EPR effect (Gabizon et al., 1990; Noguchi et al., 1998) through the tumor vasculature known for its increased permeability (Yuan et al., 1994). In a murine tumor model, Ishida et al. (1999) observed that if PEG liposomes were found intracellularly then they had been engulfed by tumor macrophages, but not by tumor cells. The ability of various PEG-liposomal formulations labeled with ^{67}Ga, ^{111}In, or ^{99m}Tc to localize in tumors has been demonstrated in a series of preclinical studies (Goins and Phillips, 2003). In fact, numerous successful tumor detection studies were done with contrast-loaded long-circulating liposomes using human tumor xenografts in nude mice (Harrington et al., 2000a, b). Moreover, the clinical data on ^{111}In-labeled long-circulating liposomes is already available for the visualization of lung cancer (Koukourakis et al., 1999; Harrington et al., 2001b), head and neck cancers (Koukourakis et al., 1999; Harrington et al., 2001b), AIDS-related Kaposi sarcoma (Harrington et al., 2001b), skin cancer (Harrington et al., 2001a), glioblastomas and metastatic brain tumors (Koukourakis et al., 2000a), soft tissue sarcomas (Koukourakis et al., 2000b), and other malignancies (Harrington et al., 2001b).

^{111}In-DTPA-labeled PEGylated liposomes (prepared using the ^{111}In-oxine method) have also been used to study the biodistribution and the pharmacokinetics of long-circualting liposomes and to assess their applicability for the radioimaging of tumor localization and evaluation of the different therapeutic treatment strategies against various types of cancers (Harrington et al., 2000c; Bao et al., 2004). Clinically, successful tumor imaging was achieved in 15 of 17 patients with cancer (4 of 5 breasts, 5 of 5 heads and necks, 3 of 4 bronchi, 2 of 2 gliomas, and 1 of 1 cervical cancers) (Harrington et al., 2001b). The greatest level of liposome uptake by the tumor was seen in the patients with head and neck cancers [33.0 $\pm$ 15.8 % ID/kg (percentage of injected dose/kg)]. The uptake in lung tumors was intermediate (18.3 $\pm$ 5.7 % ID/kg), and the breast cancers showed relatively low levels of uptake (5.3 $\pm$ 2.6 % ID/kg). These liposome uptake values mirrored the estimated tumor volumes of the various tumor types, and the use of such liposomes was considered very promising for the detection and imaging of tumor growth (Figure 11.2).

Koukorakis and colleagues(2000a) investigated the accumulation of ^{99m}Tc-DTPA-radiolabelled long-circulating liposomal doxorubicin (Doxil™) in ten patients with metastatic brain tumors and five patients with brain glioblastoma undergoing radiotherapy. Patients with metastatic brain lesions were treated with ten consecutive fractions of radiotherapy (whole brain, 3 Gy/fraction, day 1–12) followed by a booster dose of 9 Gy (3 Gy/fraction, day 21–23). Doxil™ at a dose of 25 mg/m^2 was given on day 1 and on day 21. Radiolabeled Doxil™ accumulation was 13–19 times higher in the glioblastomas and 7–13 times higher in the metastatic lesions, as compared to the normal brain. The drug accumulation in the tumoral areas was 40–60 % of the accumulation in the bone marrow of the skull bones. The normal brain radioactivity was <4 % of the bone marrow, confirming an important shielding effect of the blood–brain barrier in the normal but not in the tumoral tissue. As a result, 4 of 10 patients with metastatic lesions showed a complete recovery in CT-scan performed 2 months following the therapy. There was no severe toxicity related to radiotherapy or to chemotherapy noted. It was

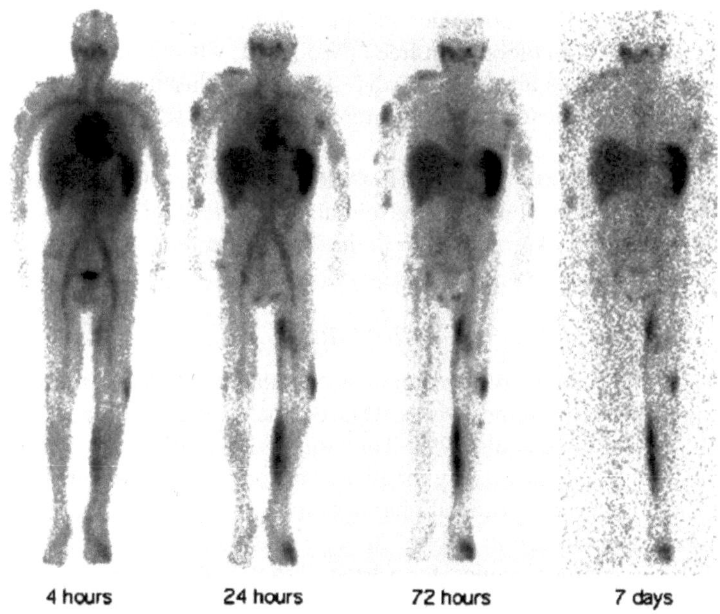

FIGURE 11.2. Serial whole-body gamma camera images over 7 days of a patient with stage T1I1S1 AIDS-KS. Multiple areas of uptake of radiolabeled liposomes are seen in the left foot and leg, right arm, and face. Each of these areas corresponded with a typical Kaposi sarcoma lesion. Prolonged retention of the radiolabel is seen despite significant clearance of circulating liposomes, as demonstrated by disappearance of the cardiac blood pool image Reproduced from Harrington et al. (2001b).

concluded that long-circulating liposomal drugs selectively overcome the blood–brain barrier in the tumoral areas. The clinical importance of this observation is now under investigation.

A new type of radiolabeled liposomes was suggested by Zheng et al. (2004) for the antisense imaging of human colon cancer in mice. In this work, 15-mer single-stranded amino-linked phosphorothioate antisense oligonucleotides of c-myc mRNA were labeled with ^{99m}Tc-pertechnetate and entrapped in liposomes. Interestingly, when looking at the in vivo data, the biodistributions showed that the tumor accumulation ($13.46 \pm 0.20\,\%$) of injected was the highest at 2 h post-injection with liposome-entrapped ^{99m}Tc-labeled antisense oligos, yet decreased sharply to $4.58 \pm 0.45\,\%$ at 4 h. In general, the tumor was shown clearly in the whole-body scan at 2 h post-injection and the ratio of the tumor-specific:non-specific (contralateral equivalent region of abdomen) signal count was about 1.8 at 2 h post-injection. The authors concluded that this strategy with liposome-entrapped ^{99m}Tc-labeled antisense oligos shows a promise toward being developed for colon cancer imaging.

Belhaj-Tayeb et al. (2003) have recently suggested an original method to encapsulate ^{99m}Tc-MIBI in preformed PEG liposomes. They used an exchange with an efflux of K+ through the valinomycin ionophores, i.e. an active encapsulation, which resulted in 50 % encapsulation efficiency of ^{99m}Tc-MIBI in PEG liposomes. By one hour post-injection in rats, PEG liposomes showed a 10 times higher activity in blood than free ^{99m}Tc-MIBI, whereas the activity of free ^{99m}Tc-MIBI in kidneys and bladder was markedly higher than that of encapsulated ^{99m}Tc-MIBI, indicating faster clearance of the free radiotracer. In the breast cancer (MCF7-ras)-bearing nude mice, PEG-liposome uptake in tumors was 2 times of that for the free ^{99m}Tc-MIBI. Overall, the ^{99m}Tc-MIBI-PEG liposomes demonstrated not only longer blood circulation time, but clearly showed an improved tumor-to-background ratio in the in vivo imaging.

11.4.2. Ligand-Targeted Liposomes

To increase the quantity of liposomes accumulating in the "required" areas, targeted liposomes (liposomes with a surface-attached specific ligand) have been frequently used (Blume et al., 1993; Torchilin, 1994; Phillips and Goins, 1995). Liposomes with a specific affinity for an affected organ or tissue were believed to increase the efficacy of liposomal pharmaceutical agents, including the imaging ones. Immunoglobulins, primarily of the IgG class, are the most promising and widely used targeting moieties for various drugs and drug carriers including liposomes (in this case we are talking about immunoliposomes). Numerous methods for antibody coupling to liposomes are reviewed in Torchilin (2000a). Usually, as much as 50–200 antibody molecules can be bound per single 100–250 nm liposome (Torchilin, 2000a). Despite evident success in the development of antibody-to-liposome coupling technique and improvements in the targeting efficacy, the majority of immunoliposomes still ended in the liver, which was usually a consequence of insufficient time for the interaction between the target and the targeted liposome. This is especially true in cases when a target of choice has diminished blood supply (ischemic or necrotic areas), and only a small number of liposomes pass through the target with the blood during the time period when liposomes are present in the circulation. The same insufficient targeting occurs if the concentration of the target antigen is very low, and the number of productive collisions between target antigens and immunoliposomes is too small. So, it was a natural development to combine the properties of targeted and long-circulating liposomes in one preparation (Torchilin, 1994, 2000a). Clearly, such combinations can significantly enhance the specificity and intensity of radiologic signal localization in the target tumor. As long-circulating liposomes are now widely used in biomedical in vitro and in vivo studies and even found their way into clinical practice (Gabizon et al., 2003; Swenson et al., 2003; Gabizon et al., 2004), the use of ligand-targeted PEGylated liposomes as carriers for contrast agents represents now an important area of research.

11.4.2.1. Antibody-Targeted Radilabeled Immunoliposomes

The EPR-dependent passive tumor-targeting of long-circulating liposomes for the delivery of imaging agents was further improved by adding a specific targeting moiety to the liposomal composition (Elbayoumi and Torchilin, 2006). This was achieved via the attachment to the liposomes of the anti-cancer monoclonal antibody 2C5 (mAb 2C5) with nucleosome (NS)-restricted activity that can recognize the surface of various tumor but not normal cells (Iakoubov and Torchilin, 1997; Chakilam, 2004) and specifically targets pharmaceutical carriers to tumor cells in vitro and in vivo (Torchilin et al., 2003; Lukyanov et al., 2004; Gupta et al., 2005). These mAb 2C5-modified PEG liposomes demonstrated 3-to 8-fold increase in the in vitro–specific cell binding with various cancer cell lines of diverse origins compared to non-modified PEG liposomes. In addition, using ^{111}In-labeled liposomes, the biodistribution data showed almost doubled tumor-specific accumulation of the mAb 2C5-modified liposomes compared to non-specific controls. At the same time, the gamma-imaging, starting 6 h and up to 12 h post-injection, of tumor-bearing mice clearly demonstrate the far superior tumor-accumulation of the mAb 2C5-modified liposomes compared to both non-targeted controls, in both 4T1 and LLC tumor models. Moreover, the fast and distinct tumor visualization acquired with the 2C5-liposomes at 6 h compared to 12–24 h with PEG liposomes suggests that the EPR effect responsible for tumor accumulation of nanocarriers through the leaky vasculature is much less significant for the accumulation of 2C5-liposomes, which proceeds via a more specific mechanism. These results clearly demonstrate an evident potential of specific antibody (including mAb 2C5)-modified long-circulating liposomes as targeted vehicles capable of specifically delivering diagnostic moieties to a variety of tumors decreasing the image acquisition time. It may also allow for the non-invasive quantification and characterization of tumor delivery and therapeutic efficacy of various nano-particulate formulations of anti-cancer agents.

Recently, the "targeting" approach have also been applied to PAP-bearing liposomes (Erdogan et al., 2006b). The authors combined in a single preparation of long-circulating liposomes an increased load with diagnostic radiometal (^{111}In) by using liposome-incorporated PAPs and tumor specificity by additionally decorating liposomes with monoclonal antibody 2C5 (mAb 2C5) possessing the nucleosome-restricted specificity that is capable of specific recognition of a broad variety of live cancer cells via the cancer cell–surface-bound NSs. The tumor-specific accumulation of mAb 2C5-modified PAP-liposomes prepared in this study was significantly (3- to 5-fold) higher than in the neighboring muscle tissue at all times after administration (6, 24, and 48 h) in mice bearing murine Lewis lung carcinoma (LLC) and human HT-29 tumors (Table 11.2), which should allow for distinctive tumor detection by the planar gamma-scintigraphy. Furthermore, the whole-body direct gamma-imaging of LLC tumor-bearing mice at different times has demonstrated the superior in vivo tumor accumulation of the targeted mAb 2C5-modified PAP-containing PEGylated liposomes compared to non-targeted liposomal control formulations, which resulted in better and faster tumor imaging as shown with LLC-bearing mice, already after 6 h (Figure 11.3).

11.4.2.2. VIP-Targeted ^{99m}Tc Liposomes for Imaging Breast Cancer

Based on the fact that vasoactive intestinal peptide receptors (VIP-R) are approximately 5 times more expressed in human breast cancer compared to normal breast tissue, Onyuksel et al. (Dagar et al., 2003) used VIP, a 28 amino acid mammalian neuropeptide, as a breast cancer targeting moiety for targeted imaging of breast cancer. VIP was covalently attached to the surface of long-circulating liposomes that encapsulated the ^{99m}Tc–HMPAO complex. It was found that the presence of VIP did not affect the size and ^{99m}Tc–HMPAO encapsulation ability of the liposomes as well as did not alter the pharmacokinetic profile of PEGylated liposomal formulation. Liposomes with and without VIP on their surface accumulated at significantly higher quantities in breast cancer when compared to normal breast in rats, indicating the EPR-dependent passive targeting of these formulations to cancer tissues. Still, in breast cancer, ^{99m}Tc–HMPAO liposomes modified with VIP showed significantly more accumulation than analogs without VIP.

TABLE 11.2. Tumor/Muscle ratios for LLC and HT-29 tumor–bearing mice at different time points

Time (h)	2C5-PAP liposomes		IgG-PAP liposomes		PAP liposomes	
	LLC	HT-29	LLC	HT-29	LLC	HT-29
3	10.7	–	3.0	–	2.3	–
6	11.3	3.8	2.5	–	2.8	1.8
24	13.9	7.2	4.3	–	3.0	1.9
48	9.5	–	2.7	–	2.2	–

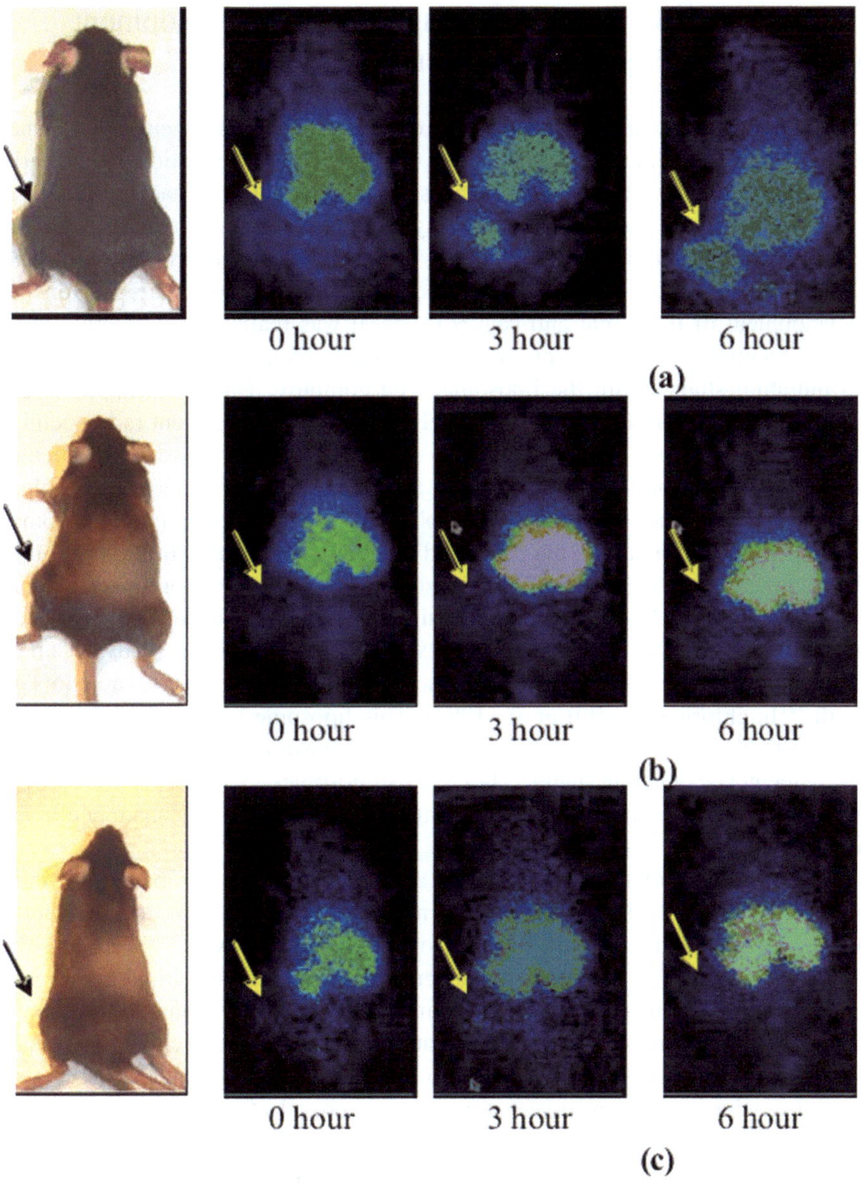

FIGURE 11.3. Whole-body imaging of LLC tumor-bearing mice at different time points after the injection of ^{111}In-labeled PAP-containing PEGylated liposomes: (**a**) 2C5-modified; (**b**) control IgG-modified, and (**c**) unmodified. Arrows indicate tumor locations.

The tumor-to-non-tumor ratio was also significantly higher for ^{99m}Tc–HMPAO liposomes modified with VIP than for VIP-free ^{99m}Tc–HMPAO liposomes, suggesting active targeting of VIP-liposomes to breast cancer.

11.5. Use of Scintigraphic Imaging in the Development of Liposome-Based Anti-cancer Drugs

Scintigraphic imaging has been used more often in the development of liposomal drugs for the treatment of cancer than for any other application. Very early after the discovery of liposomes, it was suggested that they had the potential to effectively treat tumors by delivering a significant amount of drug to the tumors while avoiding the side effects observed when the drug was injected alone (Janoff, 1992). Since most of these early efforts were directed on the development of liposomal anti-cancer drugs, it was realized that determination of the kinetics and elimination of the liposome vehicle alone was needed prior to initiating studies with the liposome–drug complex. This, in turn, led to the development of various methods to label liposomes with different radionuclides (Goins and Phillips, 2003). Richardson et al. (1979) were the first to attempt to show the feasibility of scintigraphic imaging in cancer patients by labeling phosphatidylcholine, cholesterol, and phosphatidic acid containing liposomes with ^{99m}Tc. Unfortunately, the ^{99m}Tc did not stay associated with the liposomes long enough, as little of the ^{99m}Tc liposomes reached the tumor site. Later on, other and improved labeling methods using ^{111}In and ^{67}Ga have been developed (Proffitt et al., 1983; Gabizon et al., 1988). The first studies used an "after-loading" technique to radiolabel liposomes containing nitrilotriacetic acid (NTA) with ^{111}In-chloride (Proffitt et al., 1983). This liposome formulation was found to effectively target tumors so that neutral ^{111}In liposomes began to be developed as a diagnostic imaging agent. These ^{111}In liposomes (VesCan^{TM}) were under commercial development as a tumor-imaging agent until being abandoned in the early 1990s.

Liposome formulation studies were also performed to compare the tumor uptake of neutral liposomes versus negatively charged liposomes (Goins et al., 1994). Both liposome formulations were labeled with ^{99m}Tc using the glutathione–HMPAO method. The percentage of injected dose in the tumor was 3 times higher for neutral liposomes compared to negatively charged liposomes, indicating that neutral liposome formulations may be a better choice for tumor drug carriers. Scintigraphic imaging can also be helpful during the process of deciding what indication to seek for during approval of a liposomal drug. For example, it was shown that Kaposi's sarcoma and malignant lymphoma could be the most successfully imaged using ^{111}In-NTA-liposomes (Presant et al., 1990). Kaposi's sarcoma was the first approved indication for both liposomal daunorubicin and doxorubicin products (Tulpule et al., 1998; Cheung et al., 1999). VesCan^{TM} (^{111}In-liposomes) has also been used to demonstrate the feasibility of liposomal chemotherapeutic agents as an alternative therapy for patients with recurrent malignant glioma (Khalifa et al., 1997). SPECT brain imaging demonstrated the accumulation of ^{111}In liposomes in the brain tumors in seven of eight patients, showing that liposome delivery of encapsulated anti-cancer agents has much promise for brain cancer therapy.

Gabizon et al. (1990) used radionuclide-labeled liposomes in animal studies to monitor the pharmacokinetics and distribution of liposomes incorporating either ganglioside GM1 or phosphatidylinositol. Subsequently, it was found that these longer-circulating liposome formulations resulted in higher accumulation of the liposomes at the tumor site and improved therapeutic efficacy (Gabizon, 1995), although later it was demonstrated that PEG-phospholipid was much better than ganglioside or phosphatidylinositol in extending the circulation half-life of liposomes (Allen et al., 1991, Mori et al., 1991). Papahadjopoulos et al. (1991) used the ^{67}Ga-deferoxamine labeling method to demonstrate that the incorporation of PEG-phospholipid significantly prolonged the circulation time of liposomes. These studies resulted in the development of the clinically approved PEG-liposome formulation containing doxorubicin, Doxil™ (Lasic and Martin, 1995). In certain cases, the methods currently used to label liposomes with diagnostic radionuclides could be easily adapted for the labeling of liposomes with therapeutic radionuclides (Goins and Phillips, 2003).

Lately, Phillips et al. (1992) employed the neutral and lipophilic ^{99m}Tc–BMEDA complex to directly label the commercially available Doxil™ (Bao et al., 2004). ^{99m}Tc–BMEDA contains amine groups, so it can be trapped within a liposome using a pH or ammonium gradient mechanism. These mechanisms enable the direct radiolabeling of the commercial Doxil™ with ^{99m}Tc–BMEDA. The labeling efficiency of ^{99m}Tc–Doxil™ was 70 % without changes in the original size of Doxil™ (approximately 100 nm) and showed good in vitro–labeling stability (70 %) up to 24 h. The in vivo biodisribution studies

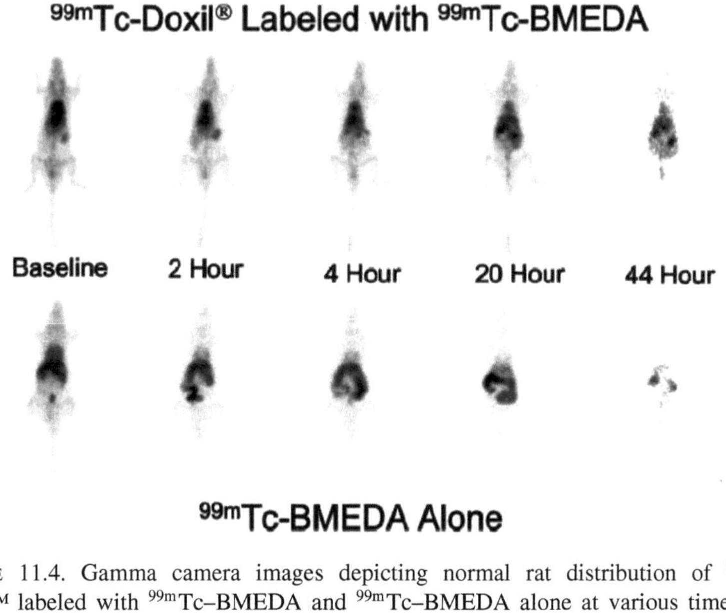

FIGURE 11.4. Gamma camera images depicting normal rat distribution of ^{99m}Tc–Doxil™ labeled with ^{99m}Tc–BMEDA and ^{99m}Tc–BMEDA alone at various times after injection. Reproduced from Bao et al. (2004).

revealed a two-phase blood clearance with half-clearance times of 2.2 and 26.2 h after bolus i.v. injection in normal rats, similar to the pharmacokinetics of the original Doxil™. In addition, the biodistribution of ^{99m}Tc–Doxil™ at 44 h after injection had 19.8 % of injected dose in blood, 14.1 % in liver, 3 % in spleen and, most importantly, up to 6 % in skin, and 15.3 % in bowel, which in fact confirmed the hypothesized liposomal side-accumulation in these tissues as a cause for the mucocutaneous adverse effects of Doxil™ therapy. Based on these results, the authors suggested using this ^{99m}Tc-labeling method for non-invasive scintigraphic imaging in the elucidation and determination of biodistribution and pharmacokinetics of drug-loaded liposomes prepared using the ammonium gradient protocol (Figure 11.4). Additionally, this labeling method may also be extended to ^{186}Re and ^{188}Re labeling to combine chemotherapy and radionuclide therapy for tumor treatment.

11.6. Summary and Conclusions

Soon after their discovery, it was recognized that liposomes labeled with gamma emitters could become valuable radiopharmaceuticals to visualize pathological sites such as tumors and infections. Efficient and convenient methods have been devised to label both plain and long-circulating PEGylated liposomes with various gamma emitters, opening the way for the development of liposome-based radiopharmaceuticals. Preformed liposomes can be easily labeled with ^{67}Ga, ^{111}In, or ^{99m}Tc, using the so-called “after-loading” procedures. With these methods, radiolabeled-liposomal preparations can be obtained with high in vivo stability. Non-targeted long-circulating liposomes preferentially localize in tumors due to the EPR effect. Clinical studies have confirmed the potential utility of ^{111}In-labeled long-circulating liposomes in tumor imaging. A wide range of tumors could also be imaged with good sensitivity and specificity by using radiolabeled long-circulating liposomes additionally modified with tumor-specific ligands, such as anti-tumor monoclonal antibodies. The application of radiolabeled liposomes seems to show a good promise for both clinical tumor imaging and control over anti-tumor drug development and delivery.

References

Abra, R.M., Hunt, C.A., 1982. Liposome disposition in vivo IV: the interaction of sequential doses of liposomes having different diameters. Res Commun Chem Pathol Pharmacol 36, 17–31.

Ahkong, Q.F., Tilcock, C., 1992. Attachment of 99mTc to lipid vesicles containing the lipophilic chelate dipalmitoylphosphatidylethanolamine-DTTA. Int J Rad Appl Instrum B 19, 831–840.

Alafandy, M., Goffinet, G., Umbrain, V., D'Haese, J., Camu, F., Legros, F.J., 1996. 99mTechnetium-stannous oxinate as marker of liposome formulations. Nucl Med Biol 23, 881–887.

Allen, T.M., Hansen, C., Martin, F., Redemann, C., Yau-Young, A., 1991. Liposomes containing synthetic lipid derivatives of poly(ethylene glycol) show prolonged circulation half-lives in vivo. Biochim Biophys Acta 1066, 29–36.

Bao, A., Goins, B., Klipper, R., Negrete, G., Mahindaratne, M., Phillips, W.T., 2003. A novel liposome radiolabeling method using 99mTc-"SNS/S" complexes: in vitro and in vivo evaluation. J Pharm Sci 92, 1893–1904.

Bao, A., Goins, B., Klipper, R., Negrete, G., Phillips, W.T., 2004. Direct 99mTc labeling of pegylated liposomal doxorubicin (Doxil) for pharmacokinetic and non-invasive imaging studies. J Pharmacol Exp Ther 308, 419–425.

Barenholz, Y., Amselem, S., Goren, D., Cohen, R., Gelvan, D., Samuni, A., Golden, E.B., Gabizon, A., 1993. Stability of liposomal doxorubicin formulations: problems and prospects. Med Res Rev 13, 449–491.

Beaumier, P.L., Hwang, K.J., Slattery, J.T., 1983. Effect of liposome dose on the elimination of small unilamellar sphingomyelin/cholesterol vesicles from the circulation. Res Commun Chem Pathol Pharmacol 39, 277–289.

Belhaj-Tayeb, H., Briane, D., Vergote, J., Kothan, S., Leger, G., Bendada, S.E., Tofighi, M., Tamgac, F., Cao, A., Moretti, J.L., 2003. In vitro and in vivo study of 99mTc-MIBI encapsulated in PEG-liposomes: a promising radiotracer for tumour imaging. Eur J Nucl Med Mol Imaging 30, 502–509.

Blume, G., Cevc, G., Crommelin, M.D., Bakker-Woudenberg, I.A., Kluft, C., Storm, G., 1993. Specific targeting with poly(ethylene glycol)-modified liposomes: coupling of homing devices to the ends of the polymeric chains combines effective target binding with long circulation times. Biochim Biophys Acta 1149, 180–184.

Boerman, O.C., Laverman, P., Oyen, W.J., Corstens, F.H., Storm, G., 2000. Radiolabeled liposomes for scintigraphic imaging. Prog Lipid Res 39, 461–475.

Bosworth, M.E., Hunt, C.A., 1982. Liposome disposition in vivo II: Dose dependency. J Pharm Sci 71, 100–104.

Caride, V.J., Taylor, W., Cramer, J.A., Gottschalk, A., 1976. Evaluation of liposome-entrapped radioactive tracers as scanning agents. Part 1: Organ distribution of liposome (99mTc-DTPA) in mice. J Nucl Med 17, 1067–1072.

Chakilam, A.R., Pabba, S., Mongayt, D., Iakoubov, L.Z., Torchilin, V.P., 2004. A single monoclonal antinuclear autoantibody with nucleosome-restricted specificity inhibits the growth of diverse human tumors in nude mice. Cancer Therapy 2, 353–364.

Cheung, T.W., Remick, S.C., Azarnia, N., Proper, J.A., Barrueco, J.R., Dezube, B.J., 1999. AIDS-related Kaposi's sarcoma: a phase II study of liposomal doxorubicin. The TLC D-99 Study Group. Clin Cancer Res 5, 3432–3437.

Dagar, S., Krishnadas, A., Rubinstein, I., Blend, M.J., Onyuksel, H., 2003. VIP grafted sterically stabilized liposomes for targeted imaging of breast cancer: in vivo studies. J Control Release 91, 123–133.

Damen, J., Regts, J., Scherphof, G., 1981. Transfer and exchange of phospholipid between small unilamellar liposomes and rat plasma high density lipoproteins. Dependence on cholesterol content and phospholipid composition. Biochim Biophys Acta 665, 538–545.

Elbayoumi, T.A., Torchilin, V.P., 2006. Enhanced accumulation of long-circulating liposomes modified with the nucleosome-specific monoclonal antibody 2C5 in various tumours in mice: gamma-imaging studies. Eur J Nucl Med Mol Imaging 33, 1196–1205.

Erdogan, S., Roby, A., Sawant, R., Hurley, J., Torchilin, V.P., 2006a. Gadolinium-loaded polychelating polymer-containing cancer cell-specific immunoliposomes. J Liposome Res 16, 45–55.

Erdogan, S., Roby, A., Torchilin, V.P., 2006b. Enhanced tumor visualization by gamma scintigraphy with ^{111}In-labeled polychelating polymer-containing immunoliposomes. Molecular Pharmaceutics. (In press).

Gabizon, A.A., 1995. Liposome circulation time and tumor targeting: implications for cancer therapy. Adv Drug Deliv Rev 16, 285–294.

Gabizon, A.A., Lyass, O., Berry, G.J., Wildgust, M., 2004. Cardiac safety of pegylated liposomal doxorubicin (Doxil/Caelyx) demonstrated by endomyocardial biopsy in patients with advanced malignancies. Cancer Invest 22, 663–669.

Gabizon, A., Huberty, J., Straubinger, R.M., Price, D.C., Papahadjopoulos, D., 1988. An improved method for in vivo tracking and imaging of liposomes using a gallium-67-deferoxamine complex. J Liposome Res 1, 123–135.

Gabizon, A., Price, D.C., Huberty, J., Bresalier, R.S., Papahadjopoulos, D., 1990. Effect of liposome composition and other factors on the targeting of liposomes to experimental tumors: biodistribution and imaging studies. Cancer Res. 50: 6371–6378.

Gabizon, A., Shmeeda, H., Barenholz, Y., 2003. Pharmacokinetics of pegylated liposomal Doxorubicin: review of animal and human studies. Clin Pharmacokinet 42, 419–436.

Goins, B., Klipper, R., Rudolph, A.S., Phillips, W.T., 1994. Use of technetium-99m-liposomes in tumor imaging. J Nucl Med 35, 1491–1498.

Goins, B.A., Phillips, W.T., 2001. The use of scintigraphic imaging as a tool in the development of liposome formulations. Prog Lipid Res 40, 95–123.

Goins, B., Phillips, T., 2003. Radiolabeled liposomes for imaging and biodistribution studies. In: *Liposomes: A Practical Approach*, Vol. 1, eds V. Torchilin, V. Weissig, 2nd edn, pp. 319–336. Oxford University Press, London.

Goto, R., Kubo, H., Okada, S., 1989. Liposomes prepared from synthetic amphiphiles. I. Their technetium labeling and stability. Chem Pharm Bull (Tokyo) 37, 1351–1354.

Gregoriadis, G., Neerunjun, D.E., 1974. Control of the rate of hepatic uptake and catabolism of liposome-entrapped proteins injected into rats. Possible therapeutic applications. Eur J Biochem 47, 179–185.

Gregoriadis, G., Putman, D., Louis, L., Neerunjun, D., 1974a. Comparative effect and fate of non-entrapped and liposome-entrapped neuraminidase injected into rats. Biochem J 140, 323–330.

Gregoriadis, G., Wills, E.J., Swain, C.P., Tavill, A.S., 1974b. Drug-carrier potential of liposomes in cancer chemotherapy. Lancet 1, 1313–1316.

Gupta, B., Levchenko, T.S., Mongayt, D.A., Torchilin, V.P., 2005. Monoclonal antibody 2C5-mediated binding of liposomes to brain tumor cells in vitro and in subcutaneous tumor model in vivo. J Drug Target 13, 337–343.

Harrington, K.J., Lewanski, C., Northcote, A.D., Whittaker, J., Peters, A.M., Vile, R.G., Stewart, J.S., 2001a. Phase II study of pegylated liposomal doxorubicin (Caelyx) as induction chemotherapy for patients with squamous cell cancer of the head and neck. Eur J Cancer 37, 2015–2022.

Harrington, K.J., Mohammadtaghi, S., Uster, P.S., Glass, D., Peters, A.M., Vile, R.G., Stewart, J.S., 2001b. Effective targeting of solid tumors in patients with locally advanced cancers by radiolabeled pegylated liposomes. Clin Cancer Res 7, 243–254.

Harrington, K.J., Rowlinson-Busza, G., Syrigos, K.N., Abra, R.M., Uster, P.S., Peters, A.M., Stewart, J.S., 2000a. Influence of tumour size on uptake of(111)ln-DTPA-labelled pegylated liposomes in a human tumour xenograft model. Br J Cancer 83, 684–688.

Harrington, K.J., Rowlinson-Busza, G., Syrigos, K.N., Uster, P.S., Abra, R.M., Stewart, J.S., 2000b. Biodistribution and pharmacokinetics of 111In-DTPA-labelled

pegylated liposomes in a human tumour xenograft model: implications for novel targeting strategies. Br J Cancer 83, 232–238.
Harrington, K.J., Rowlinson-Busza, G., Syrigos, K.N., Uster, P.S., Vile, R.G., Stewart, J.S., 2000c. Pegylated liposomes have potential as vehicles for intratumoral and subcutaneous drug delivery. Clin Cancer Res 6, 2528–2537.
Herschman, H.R., MacLaren, D.C., Iyer, M., Namavari, M., Bobinski, K., Green, L.A., Wu, L., Berk, A.J., Toyokuni, T., Barrio, J.R., Cherry, S.R., Phelps, M.E., Sandgren, E.P., Gambhir, S.S., 2000. Seeing is believing: non-invasive, quantitative and repetitive imaging of reporter gene expression in living animals, using positron emission tomography. J Neurosci Res 59, 699–705.
Hnatowich, D.J., Friedman, B., Clancy, B., Novak, M., 1981. Labeling of preformed liposomes with Ga-67 and Tc-99m by chelation. J Nucl Med 22, 810–814.
Huang, S.K., Mayhew, E., Gilani, S., Lasic, D.D., Martin, F.J., Papahadjopoulos, D., 1992. Pharmacokinetics and therapeutics of sterically stabilized liposomes in mice bearing C-26 colon carcinoma. Cancer Res 52, 6774–6781.
Hwang, K.J., Merriam, J.E., Beaumier, P.L., Luk, K.F., 1982. Encapsulation, with high efficiency, of radioactive metal ions in liposomes. Biochim Biophys Acta 716, 101–109.
Iakoubov, L.Z., Torchilin, V.P., 1997. A novel class of antitumor antibodies: nucleosome-restricted antinuclear autoantibodies (ANA) from healthy aged nonautoimmune mice. Oncol Res 9, 439–446.
Ishida, O., Maruyama, K., Sasaki, K., Iwatsuru, M., 1999. Size-dependent extravasation and interstitial localization of polyethyleneglycol liposomes in solid tumor-bearing mice. Int J Pharm 190, 49–56.
Jaggi, M., Khar, R., Chauhan, U., Gangal, S., 1991. Liposomes as carriers of technetium-99m glucoheptonate for liver imaging. Int J Pharm 69, 77–79.
Janoff, A.S., 1992. Lipids, liposomes, and rational drug design. Lab Invest 66, 655–658.
Kabalka, G.W., Davis, M.A., Moss, T.H., Buonocore, E., Hubner, K., Holmberg, E., Maruyama, K., Huang, L., 1991. Gadolinium-labeled liposomes containing various amphiphilic Gd-DTPA derivatives: targeted MRI contrast enhancement agents for the liver. Magn Reson Med 19, 406–415.
Khalifa, A., Dodds, D., Rampling, R., Paterson, J., Murray, T., 1997. Liposomal distribution in malignant glioma: possibilities for therapy. Nucl Med Commun 18, 17–23.
Khaw, B.A., Klibanov, A., O'Donnell, S.M., Saito, T., Nossiff, N., Slinkin, M.A., Newell, J.B., Strauss, H.W., Torchilin, V.P., 1991. Gamma imaging with negatively charge-modified monoclonal antibody: modification with synthetic polymers. J Nucl Med 32, 1742–1751.
Kirby, C., Clarke, J., Gregoriadis, G., 1980a. Cholesterol content of small unilamellar liposomes controls phospholipid loss to high density lipoproteins in the presence of serum. FEBS Lett 111, 324–328.
Kirby, C., Clarke, J., Gregoriadis, G., 1980b. Effect of the cholesterol content of small unilamellar liposomes on their stability in vivo and in vitro. Biochem J 186, 591–598.
Klibanov, A.L., Maruyama, K., Torchilin, V.P., Huang, L., 1990. Amphipathic polyethyleneglycols effectively prolong the circulation time of liposomes. FEBS Lett 268, 235–237.
Koukourakis, M.I., Koukouraki, S., Fezoulidis, I., Kelekis, N., Kyrias, G., Archimandritis, S., Karkavitsas, N., 2000a. High intratumoural accumulation of stealth liposomal doxorubicin (Caelyx) in glioblastomas and in metastatic brain tumours. Br J Cancer 83, 1281–1286.

Koukourakis, M.I., Koukouraki, S., Giatromanolaki, A., Archimandritis, S.C., Skarlatos, J., Beroukas, K., Bizakis, J.G., Retalis, G., Karkavitsas, N., Helidonis, E.S., 1999. Liposomal doxorubicin and conventionally fractionated radiotherapy in the treatment of locally advanced non-small-cell lung cancer and head and neck cancer. J Clin Oncol 17, 3512–3521.

Koukourakis, M.I., Koukouraki, S., Giatromanolaki, A., Kakolyris, S., Georgoulias, V., Velidaki, A., Archimandritis, S., Karkavitsas, N.N., 2000b. High intratumoral accumulation of stealth liposomal doxorubicin in sarcomas–rationale for combination with radiotherapy. Acta Oncol 39, 207–211.

Lasch, J., Weissig, V., Brandl, M., 2003. Preparation of liposomes. In: *Liposomes: A Practical Approach*, eds V. P. Torchilin, V. Weissig, 2nd edn, pp. 3–29. Oxford University Press, Oxford, New York.

Lasic, D.D., 1993. *Liposomes from physics to applications*, 1st edn. Elsevier Science publishers, Amsterdam.

Lasic, D.D., Martin, F., eds, 1995. *Stealth liposomes*. CRC Press, Boca Raton.

Lasic, D.D., Papahadjopoulos, D., eds, 1998. *Medical applications of liposomes*. Elsevier, New York.

Laverman, P., Dams, E.T., Oyen, W.J., Storm, G., Koenders, E.B., Prevost, R., van der Meer, J.W., Corstens, F.H., Boerman, O.C., 1999. A novel method to label liposomes with 99mTc by the hydrazino nicotinyl derivative. J Nucl Med 40, 192–197.

Laverman, P., Zalipsky, S., Oyen, W.J., Dams, E.T., Storm, G., Mullah, N., Corstens, F.H., Boerman, O.C., 2000. Improved imaging of infections by avidin-induced clearance of 99mTc-biotin-PEG liposomes. J Nucl Med 41, 912–918.

Love, W.G., Amos, N., Williams, B.D., Kellaway, I.W., 1989. Effect of liposome surface charge on the stability of technetium (99mTc) radiolabelled liposomes. J Microencapsul 6, 105–113.

Lukyanov, A.N., Elbayoumi, T.A., Chakilam, A.R., Torchilin, V.P., 2004. Tumor-targeted liposomes: doxorubicin-loaded long-circulating liposomes modified with anti-cancer antibody. J Control Release 100, 135–144.

Maeda, H., 2001. The enhanced permeability and retention (EPR) effect in tumor vasculature: the key role of tumor-selective macromolecular drug targeting. Adv Enzyme Regul 41, 189–207.

Maeda, H., Wu, J., Sawa, T., Matsumura, Y., Hori, K., 2000. Tumor vascular permeability and the EPR effect in macromolecular therapeutics: a review. J Control Release 65, 271–284.

Morgan, J.R., Williams, K.E., Davies, R.L., Leach, K., Thomson, M., Williams, L.A., 1981. Localisation of experimental staphylococcal abscesses by 99MTC-technetium-labelled liposomes. J Med Microbiol 14, 213–217.

Mori, A., Klibanov, A.L., Torchilin, V.P., Huang, L., 1991. Influence of the steric barrier activity of amphipathic poly(ethyleneglycol) and ganglioside GM1 on the circulation time of liposomes and on the target binding of immunoliposomes in vivo. FEBS Lett 284, 263–266.

Noguchi, Y., Wu, J., Duncan, R., Strohalm, J., Ulbrich, K., Akaike, T., Maeda, H., 1998. Early phase tumor accumulation of macromolecules: a great difference in clearance rate between tumor and normal tissues. Jpn J Cancer Res 89, 307–314.

Ogihara-Umeda, I., Sasaki, T., Nishigori, H., 1992. Development of a liposome-encapsulated radionuclide with preferential tumor accumulation–the choice of radionuclide and chelating ligand. Int J Rad Appl Instrum B 19, 753–757.

Ogihara-Umeda, I., Sasaki, T., Nishigori, H., 1993. Active removal of radioactivity in the blood circulation using biotin-bearing liposomes and avidin for rapid tumour imaging. Eur J Nucl Med 20, 170–172.

Oku, N., 1999. Delivery of contrast agents for positron emission tomography imaging by liposomes. Adv Drug Deliv Rev 37, 53–61.

Papahadjopoulos, D., 1988. Liposome formation and properties: an evolutionary profile. Biochem Soc Trans 16, 910–912.

Papahadjopoulos, D., Allen, T.M., Gabizon, A., Mayhew, E., Matthay, K., Huang, S.K., Lee, K.D., Woodle, M.C., Lasic, D.D., Redemann, C., Martin, F.J., 1991. Sterically stabilized liposomes: improvements in pharmacokinetics and antitumor therapeutic efficacy. Proc Natl Acad Sci U S A 88, 11460–11464.

Patel, H.M., Boodle, K.M., Vaughan-Jones, R., 1984. Assessment of the potential uses of liposomes for lymphoscintigraphy and lymphatic drug delivery. Failure of 99m-technetium marker to represent intact liposomes in lymph nodes. Biochim Biophys Acta 801, 76–86.

Philippot, J.R., Puyal, C.O., 1995. In: *Liposomes, New Systems and New Trends in Their Application*, Vol. 1, eds F. Puisieux, P. Couvreur, J. Delattre, J. P. Devissaguet, pp. 193–215. Editions de Sante, Paris.

Phillips, T.W., Goins, B., 1995. Targeted delivery of imaging agents by liposomes. In: *Handbook of targeted delivery of imaging agents*, ed V. P. Torchilin. CRS Press, Boca Raton.

Phillips, W.T., 1999. Delivery of gamma-imaging agents by liposomes. Adv Drug Deliv Rev 37, 13–32.

Phillips, W.T., Rudolph, A.S., Goins, B., Timmons, J.H., Klipper, R., Blumhardt, R., 1992. A simple method for producing a technetium-99m-labeled liposome which is stable in vivo. Int J Rad Appl Instrum B 19 539–547.

Presant, C.A., Blayney, D., Proffitt, R.T., Turner, A.F., Williams, L.E., Nadel, H.I., Kennedy, P., Wiseman, C., Gala, K., Crossley, R.J., et al., 1990. Preliminary report: imaging of Kaposi sarcoma and lymphoma in AIDS with indium-111-labelled liposomes. Lancet 335, 1307–1309.

Proffitt, R.T., Williams, L.E., Presant, C.A., Tin, G.W., Uliana, J.A., Gamble, R.C., Baldeschwieler, J.D., 1983. Tumor-imaging potential of liposomes loaded with In-111-NTA: biodistribution in mice. J Nucl Med 24, 45–51.

Richardson, V.J., Ryman, B.E., Jewkes, R.F., Jeyasingh, K., Tattersall, M.N., Newlands, E.S., Kaye, S.B., 1979. Tissue distribution and tumour localization of 99m-technetium-labelled liposomes in cancer patients. Br J Cancer 40, 35–43.

Senior, J., Crawley, J.C., Gregoriadis, G., 1985. Tissue distribution of liposomes exhibiting long half-lives in the circulation after intravenous injection. Biochim Biophys Acta 839, 1–8.

Senior, J., Gregoriadis, G., 1982. Stability of small unilamellar liposomes in serum and clearance from the circulation: the effect of the phospholipid and cholesterol components. Life Sci 30, 2123–2136.

Senior, J.H., 1987. Fate and behavior of liposomes in vivo: a review of controlling factors. Crit Rev Ther Drug Carrier Syst 3, 123–193.

Slinkin, M.A., Klibanov, A.L., Torchilin, V.P., 1991. Terminal-modified polylysine-based chelating polymers: highly efficient coupling to antibody with minimal loss in immunoreactivity. Bioconjug Chem 2, 342–348.

Strauss, L.G., 1997. Positron emission tomography: Current role for diagnosis and therapy monitoring in oncology. Oncologist 2, 381–388.

Suresh, M., Cao, Y., 1998. A simple and efficient method for radiolabeling of preformed liposomes. J Pharm Pharm Sci 1, 31–37.

Swenson, C.E., Bolcsak, L.E., Batist, G., Guthrie, T.H., Jr., Tkaczuk, K.H., Boxenbaum, H., Welles, L., Chow, S.C., Bhamra, R., Chaikin, P., 2003. Pharmacokinetics of doxorubicin administered i.v. as Myocet (TLC D-99; liposome-encapsulated doxorubicin citrate) compared with conventional doxorubicin when given in combination with cyclophosphamide in patients with metastatic breast cancer. Anticancer Drugs 14, 239–246.

Torchilin, V.P., 1994. Immunoliposomes and PEGylated immunoliposomes: possible use for targeted delivery of imaging agents. Immunomethods 4, 244–258.

Torchilin, V.P., ed 1995. *Handbook of targeted delivery of imaging agents*. CRS Press, Boca Raton.

Torchilin, V.P., 1996. Liposomes as delivery agents for medical imaging. Mol Med Today 2, 242–249.

Torchilin, V.P., 1997. Pharmacokinetic considerations in the development of labeled liposomes and micelles for diagnostic imaging. Q J Nucl Med 41, 141–153.

Torchilin, V.P., 2000a. Drug targeting. Eur J Pharm Sci 11 Suppl 2, S81–S91.

Torchilin, V.P., 2000b. Polymeric contrast agents for medical imaging. Curr Pharm Biotechnol 1, 183–215.

Torchilin, V.P., 2005. Recent advances with liposomes as pharmaceutical carriers. Nat Rev Drug Discov 4, 145–160.

Torchilin, V.P., Klibanov, A.L., 1991. The antibody-linked chelating polymers for nuclear therapy and diagnostics. Crit Rev Ther Drug Carrier Syst 7, 275–308.

Torchilin, V.P., Lukyanov, A.N., Gao, Z., Papahadjopoulos-Sternberg, B., 2003. Immunomicelles: Targeted pharmaceutical carriers for poorly soluble drugs. Proc Natl Acad Sci U S A 100, 6039–6044.

Torchilin, V.P., Omelyanenko, V.G., Papisov, M.I., Bogdanov, A.A., Jr., Trubetskoy, V.S., Herron, J.N., Gentry, C.A., 1994. Poly(ethylene glycol) on the liposome surface: on the mechanism of polymer-coated liposome longevity. Biochim Biophys Acta 1195, 11–20.

Torchilin, V.P., Trubetskoy, V.S., 1995a. In vivo visualizing of organs and tissues with liposomes. J Liposome Research 5, 795–812.

Torchilin, V.P., Trubetskoy, V.S., 1995b. Which polymers can make nanoparticulate drug carriers long-circulating? Adv Drug Deliv Rev 16, 141–155.

Torchilin, V.P., Weissig, V., 2003. *Liposomes: a practical approach*. 2nd edn. Oxford University Press, Oxford, New York.

Tulpule, A., Yung, R.C., Wernz, J., Espina, B.M., Myers, A., Scadden, D.T., Cabriales, S., Ilaw, M., Boswell, W., Gill, P.S., 1998. Phase II trial of liposomal daunorubicin in the treatment of AIDS-related pulmonary Kaposi's sarcoma. J Clin Oncol 16, 3369–3374.

Wolf, G.L., 1995. Targeted delivery of imaging agents: An over view. In: *Handbook of targeted delivery of imaging agents*, ed V.P. Torchilin. CRS Press, Boca Raton.

Yuan, F., Leunig, M., Huang, S.K., Berk, D.A., Papahadjopoulos, D., Jain, R.K., 1994. Microvascular permeability and interstitial penetration of sterically stabilized (stealth) liposomes in a human tumor xenograft. Cancer Res 54, 3352–3356.

Zheng, J.G., Tan, T.Z., 2004. Antisense imaging of colon cancer-bearing nude mice with liposome-entrapped 99m-technetium-labeled antisense oligonucleotides of c-myc mRNA. World J Gastroenterol 10, 2563–2566.

12
Use of Radiolabeled Liposomes for Imaging of Infection and Inflammation

Peter Laverman, Gert Storm, and Otto C. Boerman

Abstract: Liposomes are small vesicles consisting of one or more concentric lipid bilayers enclosing discrete aqueous spaces. Liposomes potentially serve as carriers for radiolabels to visualize pathological processes scintigraphically. Research performed during the last two decades has revealed insight on the characteristics of liposomes with regard to their tissue distribution after i.v. injection. Various approaches have been developed to radiolabel liposomes with gamma-emitting radionuclides, such as ^{67}Ga, ^{111}In, and ^{99m}Tc. Preferably, preformed liposomes are labeled using a so-called 'after-loading' method. The radionuclide can either be encapsulated in the aqueous interior or coupled to the lipid bilayer. The potential of radiolabeled liposomes to image infectious and inflammatory foci has been demonstrated in a variety of preclinical studies. In addition, two clinical studies in patients suspected of having infectious or inflammatory disease have been performed, demonstrating the potential of ^{99m}Tc-PEG-liposomes to image infection and inflammation.

Keywords: Liposomes, radiolabeling, scintigraphic imaging, experimental models of infection and inflammation.

12.1. Introduction

Liposomes are small vesicles consisting of one or more concentric lipid bilayers enclosing discrete aqueous spaces. Liposomes are formed spontaneously when (phospho)lipids are suspended in aqueous media. Lipophilic parts of the molecules face inward. Hydrophilic parts are exposed to the aqueous phase surrounding them. Liposomes can vary in size considerably. Usually the diameter varies between 0.1 and 5.0 μm (Bangham and Horne, 1964). After hydration of an organic lipid mixture, a liposome dispersion is formed which is usually very heterogeneous in size. In such a dispersion, large multilamellar vesicles (MLVs) are the major component. Ultrasonification and high pressure filtration are commonly used techniques to obtain vesicles of uniformal size, consisting of one phospholipid bilayer. Liposomes with a large aqueous phase and with a diameter between 100 and 1000 nm are called large unilamellar vesicles (LUVs), while liposomes smaller than 100 nm are usually called small unilamellar vesicles (SUVs).

Because of their structural versatility in terms of size, surface charge, composition, bilayer fluidity, and their ability to encapsulate almost any drug, regardless of their solubility, it was recognized already long ago that liposomes could serve as systems to carry drugs. Soon thereafter, it was also recognized that liposomes potentially could serve as carriers for radiolabels in order to visualize pathological processes scintigraphically. Unfortunately, the results of initial attempts to target liposomes to pathologic sites, such as tumors, inflammation, and infections, were disappointing (Morgan et al., 1981).

Upon i.v. injection, liposomes are recognized as 'foreign particles' by the mononuclear phagocytic system (MPS). They are rapidly cleared from the circulation to organs rich in phagocytic cells like liver, spleen, and bone marrow (Senior, 1987). This rapid clearance has been a major obstacle in the use of liposomes as (radio)pharmaceuticals. However, research during the last two decades has revealed insight into the impact of various characteristics of liposomes on their clearance rate and tissue distribution after i.v. injection (Woodle and Lasic, 1992). These characteristics include vesicle size, lipid composition, surface charge, and dose. Optimized liposomes have been widely studied for their potential as radiopharmaceutical to image infection and inflammation. In this chapter various preclinical studies as well as two clinical studies will be discussed.

12.2. Radiolabeling of Liposomes

In order to use liposomes for scintigraphic applications, they need to be labeled with gamma radiation–emitting radionuclides. The commonly used isotopes for scintigraphic imaging—gallium-67 (^{67}Ga), technetium-99m (^{99m}Tc), indium-111 (^{111}In), and iodine-123 (^{123}I)—are all widely available. However, generally ^{99m}Tc is preferred over the other isotopes, due to its optimal imaging characteristics with ideal photon energy. Moreover, in a nuclear medicine department setting, ^{99m}Tc is readily available since it can be eluted daily from a commercially available ^{99}Mo/^{99m}Tc generator. Since ^{67}Ga, ^{111}In, and ^{123}I are cyclotron products, they are more expensive and not always available in every nuclear medicine department.

One option to prepare radiolabeled liposomes is the addition of the radionuclide of choice to the aqueous medium used to hydrate the lipids (Figure 12.1A). This approach has been used to label liposomes with ^{99m}Tc (Andreopoulos and Kasi, 1997; Oku et al., 1993). However, this approach had two major disadvantages. Firstly, only part of the radiolabel was incorporated in the inner aqueous phase, and the labeling efficiency in general did not exceed 30 % (Caride, 1990). Secondly, this approach was very laborious, requiring the preparation of

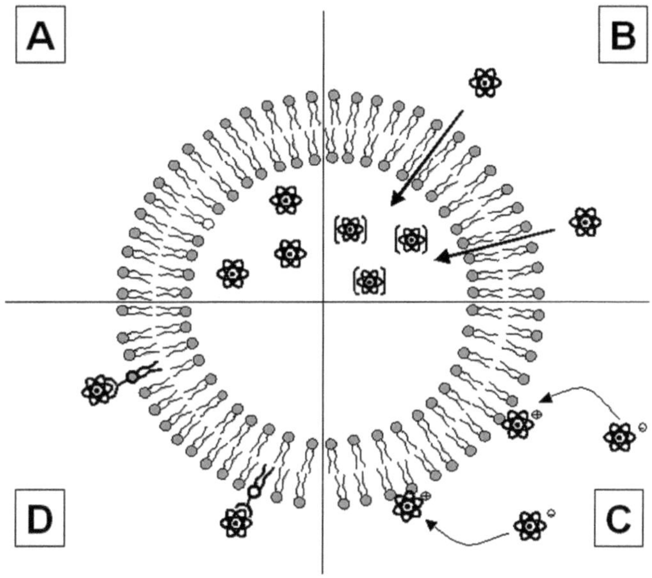

FIGURE 12.1. Overview of the main liposome labeling techniques. **A** encapsulation during manufacturing. When preparing the liposomes, radioactivity is added to the buffer in which the lipids are hydrated. **B** After-loading method. A chelator is entrapped in the liposomes during manufacturing. Radioactivity is added and will pass the lipid bilayer by various mechanisms and will be trapped. **C.** Tin technetium reduction method. ^{99m}Tc is reduced with $SnCl_2$ in the presence of liposomes, resulting in association of the ^{99m}Tc with the lipid bilayer. **D.** Chelation method. A chelator is conjugated to a (phospho)lipid and embedded in the lipid bilayer. Adding of the radiolabel will lead to binding to the surface exposed chelator.

new liposomes for each and every experiment. Therefore, strategies have been developed to radiolabel preformed liposomes with high efficiency.

Methods to label preformed liposomes—so-called 'after-loading' methods—can be divided into two categories. Firstly, the radionuclide can be transported through the lipid bilayer and then be trapped in the aqueous interior (Figure 12.1B). Secondly, liposomes can be labeled by coupling the radiolabel to the lipid bilayer, either directly or via the use of a chelator (Figure 12.1C and 12.1D, respectively).

In 1981, Morgan et al. and co-workers (1981) were among the first who labeled large multilamellar liposomes with ^{99m}Tc by reducing pertechnetate with stannous chloride in the presence of the liposomes (the so-called 'stannous chloride' method). In animal studies, liposomes labeled according to this method showed high uptake of the radiolabel in kidneys and bladder, suggesting extensive in vivo release of radiolabel from the liposomes. Love et al. (1989b) reported the instability of such a labeling method, especially when using neutral or positively charged liposomes. Similarly, Ahkong and Tilcock described instability, although less severe, in the case of negatively charged liposomes

(Ahkong and Tilcock, 1992). To overcome these disadvantages, several alternative methods to label liposomes with ^{99m}Tc have been investigated. One of these alternatives is to encapsulate the radionuclide in the aqueous internal space of preformed liposomes. To transport the radionuclide to the aqueous interior of the liposomes, the radionuclide has to be in a lipophilic form to pass through the lipid bilayer. In order to label liposomes with ^{67}Ga, liposomes were hydrated in phosphate buffered saline (PBS) containing nitrilotriacetic acid (NTA) (Hwang et al., 1982) or deferoxamine (Gabizon et al., 1988). Subsequently, ^{67}Ga is transported through the bilayer in the form of ^{67}Ga-oxine and then irreversibly trapped in the internal aqueous phase by the encapsulated NTA or deferoxamine. Boerman and colleagues used this method to investigate the feasabilty of ^{111}In-labeled pegylated liposomes to image focal infections in rats (Boerman et al., 1995). Similar to these methods, diethylene triamine pentaacetic acid (DTPA) can be used as a chelator in the aqueous phase of the liposomes for labeling with ^{111}In-oxine (Corvo et al., 1999).

A very elegant method for labeling liposomes with ^{99m}Tc according to the same principle has been developed by Phillips et al. and co-workers in 1992 (Phillips et al., 1992). Their method is based on the ^{99m}Tc-labeling method of erythrocytes, using hexamethylpropylene amine oxime (HMPAO). The ^{99m}Tc is transported as a lipophilic ^{99m}Tc–HMPAO complex through the bilayer of preformed liposomes containing glutathione in the internal phase. As a result, the enclosed ^{99m}Tc–HMPAO complex is reduced first to its hydrophilic form by the glutathione and then irreversibly trapped in the internal aqueous phase. This method proved reliable, yielding relatively stable radiolabeled liposomes in vivo (Goins et al., 1993; Phillips et al., 1992).

A second approach to label liposomes with radionuclides is to attach chelating agents like DTPA or hydrazino nicotinamide (HYNIC) to the lipid bilayer. In this method, the chelator is conjugated to a lipid component of the bilayer, prior to the actual preparation of the liposomes. After preparation of the liposomes, the chelator is exposed on the outer surface of the liposome, thus being easily accessible to the radionuclide. In 1981, Hnatowich et al. (1981) described the synthesis of a long hydrocarbon chain covalently coupled to DTPA. The lipophilic hydrocarbon anchored the molecule in the lipid bilayer, exposing the DTPA groups on the surface for chelation of the radiolabel. These liposomes could be labeled with ^{67}Ga or ^{99m}Tc. However, efficiencies did not exceed 60 % and stability was poor (Hnatowich et al., 1981). In contrast, Goto et al. and co-workers (1989) developed liposomes prepared with synthetic amphiphiles containing stearylamine–DTPA conjugates to label liposomes with ^{99m}Tc. These preparations showed good in vitro stability after incubation in 50 % human serum. Recently, our group developed a ^{99m}Tc-labeling method based on the bifunctional chelator HYNIC (Laverman et al., 1999). When conjugated to peptides or proteins, this chelator yields radioconjugates that could be labeled rapidly and with high efficiency (Abrams et al., 1990; Rennen et al., 2000). We conjugated HYNIC to distearoylphosphatidyl-ethanolamine (DSPE), a component of the lipid bilayer. When this conjugate was incorporated in the lipid bilayer, liposomes could be labeled with ^{99m}Tc rapidly and with a labeling efficiency generally higher than 95 %. ^{99m}Tc-HYNIC-PEG liposomes showed high in vitro and in vivo stability (Laverman et al., 1999).

12.3. Radiolabeled Liposomes in Experimental Infection Models

The development of radiopharmaceuticals that will rapidly and reliably visualize foci of infection and inflammation with high sensitivity and specificity form a major research topic in nuclear medicine (Rennen et al., 2001). One of these new radiopharmaceuticals for infection imaging could be radiolabeled liposomes. In the past decades, liposomes have been extensively studied in many different animal models of infection or inflammation in order to investigate their potential application in clinical practice. Liposomes of various compositions labeled with either ^{99m}Tc or ^{111}In have been studied in models of bacterial infections, sterile inflammation, arthritis, infected lungs, colitis, and osteomyelitis. In addition, clinical studies have been performed.

12.3.1. Bacterial Infections

A widely used infection model is the *Staphylococcus aureus* hind limb infection model in rats. The first studies on radiolabeled liposomes for imaging infection date back to the early 1980s. Morgan et al. and co-workers (1981) studied ^{99m}Tc-labeled liposomes in albino rats with an intramuscular *S. aureus* abscess. Anionic, cationic, and neutral MLVs were labeled with ^{99m}Tc according to the stannous chloride technique. The abscesses were clearly visualized with the anionic liposomes. The liposomes were rapidly cleared from the blood ($t^1/_2ß$ < 2 h). Due to instability, high uptake in liver and spleen was observed. Maximum abscess uptake was acquired at 30 min post-injection (p.i.). With the neutral and cationic liposomes enhanced uptake in the infectious foci was observed (Morgan et al., 1981).

Goins and colleagues reported improved performance of radiolabeled long circulating liposomes for imaging focal infection and proposed that a longer circulation time would enhance the abscess uptake (Goins et al., 1993).

In their study in rats with a focal *S. aureus* abscess, they used small (185 nm), unilamellar, cholesterol-rich liposomes. The liposomes were labeled according to the ^{99m}Tc-HMPAO method (Phillips et al., 1992). The performance of the ^{99m}Tc-liposomes was compared to that of two conventional radiopharmaceuticals: ^{67}Ga-citrate and ^{99m}Tc-HSA (human serum albumin). Abscess-to-muscle ratios with the liposomes were much higher than with the conventional agents at 24 h p.i. (35.3 vs 8.0 vs 4.1; ^{99m}Tc-liposomes vs ^{99m}Tc-HSA vs ^{67}Ga-citrate, respectively). Splenic uptake of the liposomes remained rather high (39 % ID/g, 24 h p.i.) (Goins et al., 1993). It has been proposed that the use of polyethyleneglycol (PEG)-liposomes could possibly further optimize the in vivo characteristics for scintigraphic imaging. Because of their 'stealth' characteristics, they were not detected by the MPS, thus avoiding uptake by these cells.

We have evaluated the potential of radiolabeled PEG-liposomes for infection imaging. The biodistribution of both ^{111}In-Desferal and ^{99m}Tc-HMPAO labeled PEG-liposomes ($\sim$100 nm) was studied in rats with focal *S. aureus* infection (Boerman et al., 1995; Oyen et al., 1996). In both studies, radiolabeled non-specific polyclonal IgG served as a control agent. The abscess uptake of ^{111}In-liposomes was higher than that obtained with ^{111}In-IgG (1.93 %ID/g vs 0.97 %ID/g, respectively). Abscess-to-muscle ratios were lower when compared to the ratios obtained with non-PEGylated liposomes as reported by Goins et al. (1993). However, splenic uptake of the PEG-liposomes was significantly lower (11.7 %ID/g at 24 h p.i.) than that of the non-PEG liposomes. Uptake in other non-target organs was relatively low and decreased with time. In a subsequent study, PEG-liposomes labeled with ^{99m}Tc-HMPAO were compared with ^{99m}Tc-IgG in the same animal model. The biodistribution of these ^{99m}Tc-liposomes was similar to the biodistribution of the ^{111}In-liposomes. Awasthi et al. (1998b) reported a study with dual-labeled liposomes (^{99m}Tc and ^{111}In) in rats with *S. aureus* infection. Target-to-background ratios were in the same range for both radiolabels. Uptake in spleen and bone marrow of the ^{111}In-liposomes was approximately 2.5 times higher than that of ^{99m}Tc-labeled liposomes.

Biodistribution was also studied in a model of infection in Wistar rats with *Escherichia coli* infection (Oyen et al., 1996). Uptake and retention of the ^{99m}Tc-labeled liposomes appeared to correlate well with the intensity of the inflammatory response. *E. coli*—inducing a relatively mild inflammatory response—showed significantly lower abscess uptake (0.6 %ID/g vs 1.0 %ID/g, 24 h p.i.) and abscess-to-muscle ratio (24 vs 42, 24 hr p.i.) when compared to the *S. aureus* infection.

In a subsequent study, we investigated the biodistribution of PEG-liposomes labeled with ^{99m}Tc via HYNIC, in rats with *S. aureus* infection (Laverman et al., 1999). Abscess uptake and abscess-to-muscle ratios were similar to the ^{99m}Tc-HMPAO-PEG liposomes (Figure 12.2). Fortunately, uptake of the radiolabel in the kidneys was threefold lower, indicating a much more stable radiolabeled liposome formulation, with better potential for use in routine clinical practice.

12.3.2. Uptake and Retention Mechanism

To further investigate the uptake and retention mechanism of PEG-liposomes in infected tissue, we studied the localization of gold-labeled PEG-liposomes within a *S. aureus* abscess in rats at the cellular and the subcellular level using both light microscopy and transmission electron microscopy (Laverman et al., 2001). It was hypothesized that the mechanism of uptake of PEG-liposomes in a *S. aureus* abscess is similar to that in tumor tissue, i.e. extravasation of the liposomes at the site of infection by virtue of locally enhanced vascular permeability. A long circulation time of the liposomes is therefore required for abscess uptake. Our

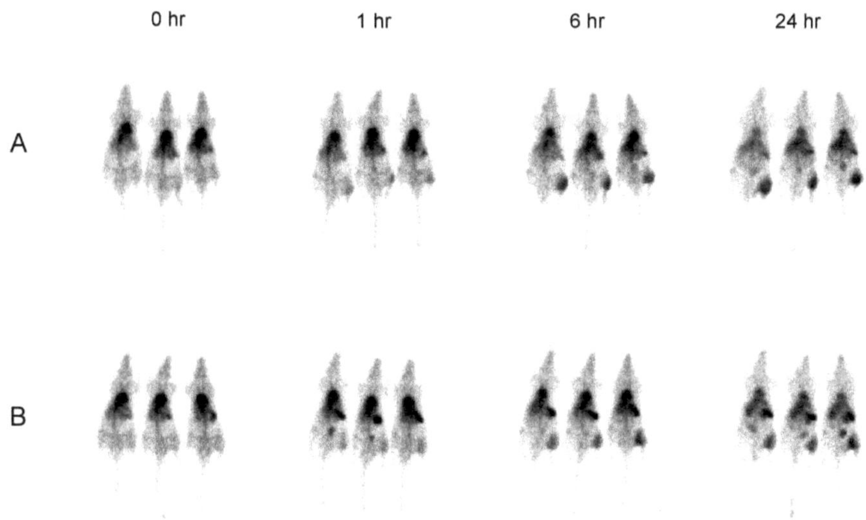

FIGURE 12.2. Images of rats with unilateral *S. aureus* abscess in calf muscle, recorded at 0 h (= 5 min), 1, 6, and 24 h post-injection of ^{99m}Tc-labeled HYNIC-liposomes (**A**) and HMPAO-liposomes (**B**).

microscopy studies revealed that retention of the PEG-liposomes in the abscess could be due to phagocytosis of the liposomes by macrophages. Although it is known that PEG-liposomes evade the MPS, they apparently finally end up in cells of the MPS, as located mainly in liver and spleen.

12.3.3. Sterile Inflammation

In a study comparing three different models of infection, we investigated the application of radiolabeled PEG-liposomes in model of sterile turpentine-induced muscular inflammation (Oyen et al., 1996). The uptake in such an inflammatory focus was at least as high as in the *S. aureus* infection model (1.18 %ID/g vs 0.98 %ID/g, 24 hr p.i.). In addition, high abscess-to-muscle ratios were obtained (44.5, 24 h p.i.). So, whether or not the inflammation was induced by a microorganism appeared irrelevant for the uptake of the liposomes. These data indicate that infection is not crucial and that microorganisms as such are not responsible for the uptake of the liposomes in abscesses. Moreover, since turpentine causes only a minor influx of white blood cells (van der Laken et al., 1998), accumulation seemed largely independent of infiltrating white blood cells.

12.3.4. Arthritis

Although the two previously described animal models are well established to investigate the potential of new radiopharmaceuticals for scintigraphic detection of infection and inflammation, clinically these models are less relevant since the inflammatory foci are already obvious by visual inspection. A model more relevant to clinical practice and clinical demand is a rat model of adjuvant arthritis (Boerman et al., 1997; Love et al., 1989a, 1990).

In 1987, Williams and colleagues showed the clinical application of ^{99m}Tc-labeled negatively charged unilamellar liposomes in patients with rheumatoid arthritis (Williams et al., 1987). To further optimize the liposomes for imaging arthritis, a study was performed with cholesterol-rich liposomes of 100 and 200 nm (Love et al., 1990). The use of cholesterol-rich liposomes did not improve the uptake in arthritic paws as compared to the cholesterol-poor liposomes. Additionally, activity in background organs like liver and spleen remained high. However, the 100 nm liposomes showed a threefold higher uptake in the affected paws than the 200 nm liposomes. It was shown that the use of sterically stabilized liposomes, in contrast to conventional liposomes, greatly improved the use of radiolabeled liposomes to image adjuvant arthritis in rats (Boerman et al., 1997). Whereas the uptake of the liposomes in the inflamed ankle was similar to the uptake reported by Love et al. (1989a, 1990), uptake in liver and spleen appeared significantly lower.

12.3.5. Lung Infections

Another clinically relevant model of infection has been developed by Bakker-Woudenberg and colleagues (Bakker-Woudenberg et al., 1992, 1993). They studied the localization of liposomes of different lipid composition in a rat model of unilateral pneumonia caused by *Klebsiella pneumoniae*. Although the studies were primarily carried out to study the localization of liposomes aiming to deliver therapeutical agents to the infected tissues, they also gave an impression of the utility of liposomes to image such lung infections. Equally sized liposomes of different lipid composition ($\sim$100 nm) were tested in this model. Liposomes composed of hydrogenated phosphatidyl inositol/hydrogenated phosphatidylcholine/cholesterol were compared with liposomes composed of egg phosphatidylglycerol/egg phosphatidylcholine/cholesterol. Both preparations were labeled with ^{67}Ga (Bakker-Woudenberg et al., 1992). The first formulation showed significantly higher uptake in the infected lung than the latter formulation: probably due to the relatively longer blood residence time of the first formulation. Furthermore, a positive relation was found between the uptake of the liposomes in the infected tissue and the severity of the infection. In a second study, the same group investigated the behavior of PEG-liposomes in the same model (Bakker-Woudenberg et al., 1993). It was shown that PEG-liposomes performed better than non-PEG-liposomes, apparently also due to their long-circulating characteristics. At 40 h p.i., the uptake in the infected lung was ten times higher than after administration of the non-PEG-liposomes. For imaging purposes these liposome should preferably be labeled with more suitable imaging radionuclide such as ^{99m}Tc or ^{111}In.

12.3.6. Acute colitis

Currently, the radiopharmaceutical of choice for imaging inflammatory bowel disease (IBD) is a preparation of radiolabeled leukocytes. Unfortunately, such preparation has several disadvantages limiting their application (laborious, handling of possibly contaminated blood). Therefore, research is directed toward the development of an agent that is at least as good as radiolabeled leukocytes, but being more easy to prepare. In a rabbit model of acute colitis, ^{111}In-labeled PEG-liposomes were compared with ^{111}In–IgG- and with ^{111}In-labeled leukocytes (Oyen et al., 1997). In comparison with the labeled leukocytes, radiolabeled PEG-liposomes (~90 nm) showed higher absolute uptake as well as higher diseased-to-normal colon ratios. The inflamed colon could be delineated well with the leukocytes, since uptake in background tissues was extremely low. The liposomes performed at least as good as the labeled leukocytes. However, they are more easy to prepare.

In general, a ^{99m}Tc label is preferred over an ^{111}In label. Therefore, we also studied the performance of ^{99m}Tc-labeled PEG-liposomes in the same rabbit model and compared their in vivo characteristics with ^{99m}Tc-labeled white blood cells from donor rabbits (Dams et al., 1998). Results were similar to those obtained after injection of ^{111}In-labeled liposomes.

12.3.7. Osteomyelitis

For imaging osteomyelitis, three-phase ^{99m}Tc-MDP bone scanning is considered a most useful tool, especially when combined with a ^{67}Ga-citrate scan (Crim, 1994). A single radiopharmaceutical, which can combine the two agents just mentioned, would imply a major step forward. Awasthi and colleagues (1998a) investigated the feasability of dual-labeled ^{99m}Tc/^{111}In-PEG-liposomes in a rabbit model of acute osteomyelitis. The lesions could be visualized well within 8 h p.i. in the ^{99m}Tc window. Highest target-to-non-target ratios were obtained at 48 h p.i. in the ^{111}In window. Liposomes and ^{67}Ga-citrate were positive in eight rabbits, but equivocal in one rabbit. The ^{99m}Tc-MDP scan was positive in only five rabbits. In a rabbit model of chronic osteomyelitis, ^{99m}Tc-PEG-liposomes performed equally well as ^{67}Ga-citrate and ^{111}In-granulocytes (Dams et al., 2000a). These results indicated that liposomes are at least as good as ^{67}Ga-citrate and ^{111}In-granulocytes in imaging osteomyelitis than the bone scan. Given the early localization and the better image qualities, radiolabeled PEG-liposomes could be preferred over ^{67}Ga-citrate and ^{111}In-granulocytes.

12.4. Clinical Studies

The first clinical studies with radiolabeled liposomes for imaging infection were described by Morgan et al. in 1985 (Morgan et al., 1985). They used ^{99m}Tc-labeled negatively charged phospholipid liposomes to detect deep-seated infection and rheumatoid arthritis. With this liposomal formulation, in both patients and in a volunteer, high uptake in liver, spleen, and bone-marrow was observed, while no uptake was observed in the bowel. Infected prosthetic joints, tubo-ovarian abscess, and osteomyelitis were clearly visualized at 20 h p.i. Furthermore, in a patient with active rheumatoid arthritis, high uptake of the radiolabel in the affected joints was observed. In a separate study, the same group compared the liposome scan with bone scanning and with conventional radiography in patients with rheumatoid arthritis (O'Sullivan et al., 1988). It was concluded that the ^{99m}Tc-MDP bone scan as well as the ^{99m}Tc liposome scan were more sensitive than radiography in detecting joint disease. However, the liposome scans were positive only in clinically active disease. Furthermore, liposome scintigraphy was able to discriminate between various grades of joint tenderness in patients with rheumatoid arthritis. In these studies, negatively charged liposomes were used which led to rapid MPS uptake of the liposomes.

The promising results of radiolabeled PEG-liposomes in various animal studies prompted us to initiate a clinical study with ^{99m}Tc-HMPAO liposomes (Dams et al., 2000b). In 35 patients suspected of infectious or inflammatory disease, ^{99m}Tc-PEG liposomes were directly compared to ^{111}In-IgG-scintigraphy. ^{99m}Tc-PEG liposomes scans were performed at 4 and 24 h after injection. Scintigraphic results were compared and verified by culture, biopsy, surgery, and follow-up of at least 6 months. Most patients had suspected infections of the locomotor system, 4 patients suffered from fever of unknown origin, four from bacteraemia, and two had suspected abdominal pathologies. Of the 16 proven infections and inflammations, 15 were detected. Both ^{99m}Tc-PEG liposomes and ^{111}In-IgG missed a case of endocarditis. Furthermore, in one patient the liposome scan was truely positive, while the ^{111}In-IgG scan was falsely negative (Figure 12.3). This first clinical evaluation of ^{99m}Tc-PEG liposomes showed that focal infection and inflammation can be adequately imgaged with this liposomal preparation. In a subsequent study, we investigated the potential role of ^{99m}Tc-PEG liposome to determine the extent and severity of active disease

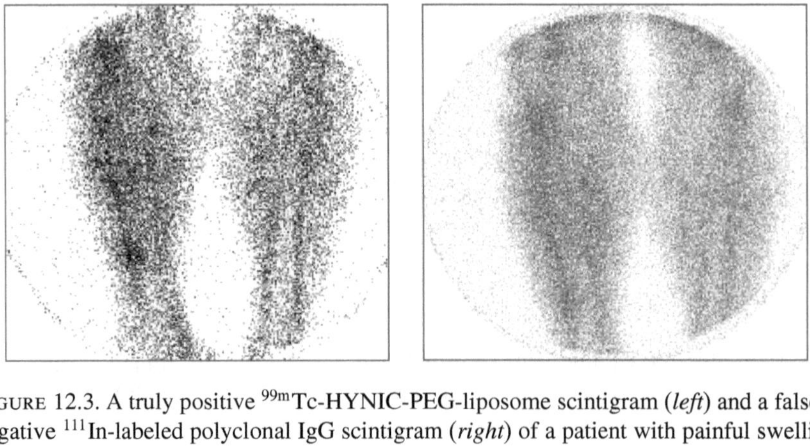

FIGURE 12.3. A truly positive ^{99m}Tc-HYNIC-PEG-liposome scintigram (*left*) and a falsely negative ^{111}In-labeled polyclonal IgG scintigram (*right*) of a patient with painful swelling and redness at the level of an old tibial fracture.

of Crohn's colitis. Patients suspected of having an exacerbation of Crohn's disease underwent a ^{99m}Tc-PEG liposome scan. A barium enema or endoscopy was performed as the standard verification procedure. Disease activity indices (Clinical Disease Activity Index and Van Hees Activity Index) were calculated. In 7 out of 9 patients, positive images of colon segments affected by Crohn's colitis were obtained using ^{99m}Tc-PEG liposomes. Only a moderate relation between ^{99m}Tc-liposome scan grading and verification procedures was found. In accordance with previous studies, no significant correlation was found between the clinical disease activity indices and the verification procedures.

12.4.1. Liposome Related Side-Effects

In many clinical studies with the commercial PEG-liposome preparation Caelyx® (Doxil® in the US), a preparation of PEG-liposomes containing doxorubicin, acute, as well as subacute, side effects were encountered during or after the infusion of the liposomal preparation (Goebel et al., 1996; Koukourakis et al., 1999; Uziely et al., 1995). Signs and symptoms of the acute reactions consisted of shortness of breath, widespread erythema, facial flushing, chest pain, back pain, and changes in blood pressure, all resolving within minutes after termination of the liposome infusions. These side effects were similar to those observed in clinical studies with non-PEGylated liposomes containing amphotericin (Laing et al., 1994; Ringden et al., 1994). Not surprisingly, these acute and subacute side effects were no reason for alarm as the events were minor compared to the usual toxicity of anticancer drugs, and seemed to balance the therapeutic benefit of the drug. However, similar acute side effects were encountered using ^{99m}Tc-labeled PEG-liposomes (without encapsulated drugs) for the scintigraphic detection of infection and inflammation (Brouwers et al., 2000; Dams et al., 2000b). Obviously, when using PEG-liposomes as a diagnostic agent, these adverse events are unacceptable. In our first clinical study (Dams et al., 2000b), 2 patients experienced mild flushing and tightness of the chest during the infusion of the labeled liposomes, which subsided when the infusion was temporarily stopped (Dams et al., 2000b). When the infusion was continued, no side effects were noted. Remarkably, on the scintigrams acquired 4 h later, strongly increased uptake in liver and spleen was observed, whereas the activity in the heart region was decreased, indicating enhanced clearance of the PEG-liposomes from the blood to the liver and spleen. In the second clinical study, acute adverse reactions were again observed (Brouwers et al., 2000). Although the incidence was much higher (3 out of 9 patients), the symptoms were similar to those observed in the first clinical study. The liposomal preparation used in the scintigraphic studies is comparable to the Doxil® formulation (however without doxorubicin-content), but subacute side effects reported in the clinical studies with Doxil®, such as PPE, mucositis, or stomatitis, were not observed (Brouwers et al., 2000; Dams et al., 2000b; Goebel et al., 1996; Koukourakis et al., 1999; Uziely et al., 1995).

In all reported cases, subsequent infusions could be given without problems regardless of whether these patients had side effects during the first infusion. However, Koukourakis et al. (1999) described one patient who experienced the same symptoms as during the first infusion ('heavy chest feeling'). The fact that even a minimal amount of liposomes could induce these side effects suggests that the observed reactions could be classified as hypersensitivity or pseudoallergic reaction (Szebeni et al., 1999).

12.5. Conclusion

Soon after their development it was recognized that liposomes labeled with gamma radiation–emitting radionuclides could be used as radiopharmaceuticals to visualize pathological sites such as tumors and infections. Liposomes as formulated in the past were rapidly taken up by cells of the mononuclear phagocyte system (MPS), primarily those located in liver and spleen. At the end of the 1980s, it was found that liposomes coated with hydrophilic molecules evaded recognition by the MPS. The development of these long-circulating liposomes greatly broadened the in vivo applicability of liposomes. Currently, the most generally applied method is to coat liposomes with polyethyleneglycol (PEG) chains: PEG-coated liposomes have long-circulating properties irrespective of lipid composition, lipid dose, and/or surface charge. In addition, various methods have been developed to label liposomes with various gamma emitters, providing perspective for the development of liposome-based radiopharmaceuticals. Preformed liposomes can now be labeled rapidly and efficiently with ^{67}Ga, ^{111}In, or ^{99m}Tc, using the so-called 'after-loading' procedures. With these methods radiolabeled liposomal preparations can be obtained with high in vivo stability.

The potential of radiolabeled liposomes to image infectious and inflammatory foci has been demonstrated in a series of preclinical studies. In various infection and inflammation models (*S. aureus* infection, *E. coli* infection, sterile inflammation, arthritis, colitis, etc.), radiolabeled PEG-liposomes performed equal to or even better than conventional radiopharmaceuticals for infection imaging (^{67}Ga-citrate, radiolabeled leukocytes, and radiolabeled non-specific polyclonal IgG). Clinical results exemplify the potential of ^{99m}Tc-PEG liposomes to image infection and inflammation. More extensive studies with radiolabeled liposomes in various patient populations are warranted to define the place of a liposome-based imaging agent in routine clinical practice. However, one has to consider that PEG-liposomes are not inert vehicles in vivo. Pharmacological effects—induced solely by using a liposomal formulation—can occur irrespective of the drug contents. Considering the encountered side effects, the PEG-liposomal preparation may have to be modified.

References

Abrams, M.J., Juweid, M., tenKate, C.I., Schwartz, D.A., Hauser, M.M., Gaul, F.E., Fuccello, A.J., Rubin, R.H., Strauss, H.W., Fischman, A.J., 1990. Technetium-99m-human polyclonal IgG radiolabeled via the hydrazino nicotinamide derivative for imaging focal sites of infection in rats. J Nucl Med 31, 2022–2028.

Ahkong, Q.F., Tilcock, C., 1992. Attachment of 99mTc to lipid vesicles containing the lipophilic chelate dipalmitoylphosphatidylethanolamine-DTTA. Int J Rad Appl Instrum B 19, 831–840.

Andreopoulos, D., Kasi, L.P., 1997. Tc-labelled diphytanoylphosphatidylcholine liposomes: In vitro and in vivo studies. J Microencapsul 14, 427–436.

Awasthi, V., Goins, B., Klipper, R., Loredo, R., Korvick, D., Phillips, W.T., 1998a. Imaging experimental osteomyelitis using radiolabeled liposomes. J Nucl Med 39, 1089–1094.

Awasthi, V.D., Goins, B., Klipper, R., Phillips, W.T., 1998b. Dual radiolabeled liposomes: Biodistribution studies and localization of focal sites of infection in rats. Nucl Med Biol 25, 155–160.

Bakker-Woudenberg, I.A., Lokerse, A.F., ten Kate, M.T., Mouton, J.W., Woodle, M.C., Storm, G., 1993. Liposomes with prolonged blood circulation and selective localization in Klebsiella pneumoniae-infected lung tissue. J Infect Dis 168, 164–171.

Bakker-Woudenberg, I.A., Lokerse, A.F., ten Kate, M.T., Storm, G., 1992. Enhanced localization of liposomes with prolonged blood circulation time in infected lung tissue. Biochim Biophys Acta 1138, 318–326.

Bangham, A.D., Horne, R.W., 1964. Negative staining of phospholipids and their structural modification by surface-active agents as observed in the electron microscope. J Mol Biol 12, 660–668.

Boerman, O.C., Oyen, W.J., Storm, G., Corvo, M.L., Van Bloois, L., van der Meer, J.W., Corstens, F.H., 1997. Technetium-99m labelled liposomes to image experimental arthritis. Ann Rheum Dis 56, 369–373.

Boerman, O.C., Storm, G., Oyen, W.J., van, B.L., van der Meer, J.W., Claessens, R.A., Crommelin, D.J., Corstens, F.H., 1995. Sterically stabilized liposomes labeled with indium-111 to image focal infection. J Nucl Med 36, 1639–1644.

Brouwers, A.H., de Jong, D.J., Dams, E.T., Oyen, W.J., Boerman, O.C., Laverman, P., Naber, T.H.J., Storm, G., Corstens, F.H., 2000. Tc-99m-PEG-liposomes for the evaluation of colitis in Crohn's disease. J Drug Target 8, 225–233.

Caride, V.J., 1990. Technical and biological considerations on the use of radiolabeled liposomes for diagnostic imaging. Int J Rad Appl Instrum B 17, 35–39.

Corvo, M.L., Boerman, O.C., Oyen, W.J., Van Bloois, L., Cruz, M.E., Crommelin, D.J., Storm, G., 1999. Intravenous administration of superoxide dismutase entrapped in long circulating liposomes. II. In vivo fate in a rat model of adjuvant arthritis. Biochim Biophys Acta 1419, 325–334.

Crim, JR., Seeger, LL., 1994. Imaging evaluation of osteomyelitis. Crit Rev Diagn Imaging 35, 201–256.

Dams, E.T., Nijhof, M.W., Boerman, O.C., Laverman, P., Storm, G., Buma, P., Lemmens, J.A., van der Meer, J.W., Corstens, F.H., Oyen, W.J., 2000a. Scintigraphic evaluation of experimental chronic osteomyelitis. J Nucl Med 41, 896–902.

Dams, E.T., Oyen, W.J., Boerman, O.C., Storm, G., Laverman, P., Koenders, E.B., van der Meer, J.W., Corstens, F.H., 1998. Technetium-99m-labeled liposomes to image experimental colitis in rabbits: comparison with technetium-99m-HMPAO-granulocytes and technetium-99m-HYNIC-IgG. J Nucl Med 39, 2172–2178.

Dams, E.T., Oyen, W.J., Boerman, O.C., Storm, G., Laverman, P., Kok, P.J., Buijs, W.C., Bakker, H., van der Meer, J.W., Corstens, F.H., 2000b. ^{99m}Tc-PEG liposomes for the scintigraphic detection of infection and inflammation: clinical evaluation. J Nucl Med 41, 622–630.

Gabizon, A., Huberty, J., Straubinger, R.M., Price, D.C., Papahadjopoulos, D., 1988. An improved method for in vivo tracing and imaging of liposomes using a gallium 67-deferoxamine complex. J Liposome Res 1, 123–135.

Goebel, F.D., Goldstein, D., Goos, M., Jablonowski, H., Stewart, J.S., 1996. Efficacy and safety of Stealth liposomal doxorubicin in AIDS-related Kaposi's sarcoma. The International SL-DOX Study Group. Br J Cancer 73, 989–994.

Goins, B., Klipper, R., Rudolph, A.S., Cliff, R.O., Blumhardt, R., Phillips, W.T., 1993. Biodistribution and imaging studies of technetium-99m-labeled liposomes in rats with focal infection. J Nucl Med 34, 2160–2168.

Goto, R., Kubo, H., Okada, S., 1989. Liposomes Prepared from Synthetic Amphiphiles. I. Their Technetium Labeling and Stability. Chem Pharm Bull 37, 1351–1354.

Hnatowich, D.J., Friedman, B., Clancy, B., Novak, M., 1981. Labeling of performed liposomes with Ga-67 and Tc-99m by chelation. J Nucl Med 22, 810–814.

Hwang, K.J., Merriam, J.E., Beaumier, P.L., Luk, K.F.S., 1982. Encapsulation, with high efficiency, of radioactive metal ions in liposomes. Biochim Biophys Acta 716, 101–109.

Koukourakis, M.I., Koukouraki, S., Giatromanolaki, A., Archimandritis, S.C., Skarlatos, J., Beroukas, K., Bizakis, J.G., Retalis, G., Karkavitsas, N., Helidonis, E.S., 1999. Liposomal doxorubicin and conventionally fractionated radiotherapy in the treatment of locally advanced non-small-cell lung cancer and head and neck cancer. J Clin Oncol 17, 3512–3521.

Laing, R.B., Milne, L.J., Leen, C.L., Malcolm, G.P., Steers, A.J., 1994. Anaphylactic reactions to liposomal amphotericin. Lancet 344, 682.

Laverman, P., Dams, E.T., Oyen, W.J., Storm, G., Koenders, E.B., Prevost, R., van der Meer, J.W., Corstens, F.H., Boerman, O.C., 1999. A novel method to label liposomes with ^{99m}Tc by the hydrazino nicotinyl derivative. J Nucl Med 40, 192–197.

Laverman, P., Dams, E.T., Storm, G., Hafmans, T.G., Croes, H.J., Oyen, W.J., Corstens, F.H., Boerman, O.C., 2001. Microscopic localization of PEG-liposomes in a rat model of focal infection. J Control Release 75, 347–355.

Love, W.G., Amos, N., Kellaway, I.W., Williams, B.D., 1989a. Specific accumulation of technetium-99m radiolabelled, negative liposomes in the inflamed paws of rats with adjuvant induced arthritis: effect of liposome size. Ann Rheum Dis 48, 143–148.

Love, W.G., Amos, N., Kellaway, I.W., Williams, B.D., 1990. Specific accumulation of cholesterol-rich liposomes in the inflammatory tissue of rats with adjuvant arthritis. Ann Rheum Dis 49, 611–614.

Love, W.G., Amos, N., Williams, B.D., Kellaway, I.W., 1989b. Effect of liposome surface charge on the stability of technetium (^{99m}Tc) radiolabelled liposomes. J Microencapsul 6, 105–113.

Morgan, J.R., Williams, K.E., Davies, R.L., Leach, K., Thomson, M., Williams, L.A., 1981. Localisation of experimental staphylococcal abscesses by ^{99m}Tc- technetium-labelled liposomes. J Med Microbiol 14, 213–217.

Morgan, J.R., Williams, L.A., Howard, C.B., 1985. Technetium-labelled liposome imaging for deep-seated infection. Br J Radiol 58, 35–39.

O'Sullivan, M.M., Powell, N., French, A.P., Williams, K.E., Morgan, J.R., Williams, B.D., 1988. Inflammatory joint disease: a comparison of liposome scanning, bone scanning, and radiography. Ann Rheum Dis 47, 485–491.

Oku, N., Namba, Y., Takeda, A., Okada, S., 1993. Tumor imaging with technetium-99m-DTPA encapsulated in RES-avoiding liposomes. Nucl Med Biol 20, 407–412.

Oyen, W.J., Boerman, O.C., Dams, E.T., Storm, G., van, B.L., Koenders, E.B., van Haelst, U.J., van der Meer, J.W., Corstens, F.H., 1997. Scintigraphic evaluation of experimental colitis in rabbits. J Nucl Med 38, 1596–1600.

Oyen, W.J., Boerman, O.C., Storm, G., van, B.L., Koenders, E.B., Claessens, R.A., Perenboom, R.M., Crommelin, D.J., van der Meer, J.W., Corstens, F.H., 1996. Detecting infection and inflammation with technetium-99m-labeled Stealth liposomes. J Nucl Med 37, 1392–1397.

Phillips, W.T., Rudolph, A.S., Goins, B., Timmons, J.H., Klipper, R., Blumhardt, R., 1992. A simple method for producing a technetium-99m-labeled liposome which is stable in vivo. Nucl Med Biol 19, 539–547.

Rennen, H.J., Boerman, O.C., Koenders, E.B., Oyen, W.J., Corstens, F.H., 2000. Labeling proteins with Tc-99m via hydrazinonicotinamide (HYNIC): optimization of the conjugation reaction. Nucl Med Biol 27, 599–604.

Rennen, H.J., Boerman, O.C., Oyen, W.J., Corstens, F.H., 2001. Imaging infection/inflammation in the new millennium. Eur J Nucl Med 28, 241–252.

Ringden, O., Andstrom, E., Remberger, M., Svahn, B.M., Tollemar, J., 1994. Allergic reactions and other rare side-effects of liposomal amphotericin. Lancet 344, 1156–1157.

Senior, J.H., 1987. Fate and behavior of liposomes in vivo: a review of controlling factors. Crit Rev Ther Drug Carrier Syst 3, 123–193.

Szebeni, J., Fontana, J.L., Wassef, N.M., Mongan, P.D., Morse, D.S., Dobbins, D.E., Stahl, G.L., Bunger, R., Alving, C.R., 1999. Hemodynamic changes induced by liposomes and liposome-encapsulated hemoglobin in pigs: a model for pseudoallergic cardiopulmonary reactions to liposomes. Role of complement and inhibition by soluble CR1 and anti-C5a antibody. Circulation 99, 2302–2309.

Uziely, B., Jeffers, S., Isacson, R., Kutsch, K., Wei-Tsao, D., Yehoshua, Z., Libson, E., Muggia, F.M., Gabizon, A., 1995. Liposomal doxorubicin: antitumor activity and unique toxicities during two complementary phase I studies. J Clin Oncol 13, 1777–1785.

van der Laken, C.J., Boerman, O.C., Oyen, W.J., Laverman, P., van de Ven, M.T., Corstens, F.H., van der Meer, J.W., 1998. In vivo expression of interleukin-1 receptors during various experimentally induced inflammatory conditions. J Infect Dis 177, 1398–1401.

Williams, B.D., O'Sullivan, M.M., Saggu, G.S., Williams, K.E., Williams, L.A., Morgan, J.R., 1987. Synovial accumulation of technetium labelled liposomes in rheumatoid arthritis. Ann Rheum Dis 46, 314–318.

Woodle, M.C., Lasic, D.D., 1992. Sterically stabilized liposomes. Biochim Biophys Acta 1113, 171–199.

13
Sulphur Colloid for Imaging Lymph Nodes and Bone Marrow

Frederick L. Moffat, Jr. and Seza A. Gulec

Abstract: ^{99m}Technetium sulphur colloid (TcSC) has become very widely used since the advent of lymphoscintigraphy (LS) in the late 1970s and, especially, sentinel lymph node biopsy (SLNB) in 1992. This colloid provides very good external imaging of lymph node drainage patterns in oncology patients. Moreover, this radiocolloid's in vivo kinetics and relative time-independence make it an excellent agent for localizing sentinel lymph nodes in the operating room. SLNB is well on its way to becoming the standard of care for surgical lymph node staging in melanoma and breast cancer and is being investigated in a variety of other solid tumors. Bone marrow imaging with this agent provides valuable diagnostic information in patients with blood dyscrasias and bone marrow neoplasms and those with osteomyelitis, infected orthopedic prostheses, and neuropathic pedal conditions.

13.1. Introduction

Radiolabeled colloidal sols were developed in the late 1940s and 1950s for interstitial injection with the goal of irradiating metastasizing tumor cells in their travel from the primary lesion to regional lymph nodes (Krag, 1998a). Lymphoscintigraphy (LS) was first reported in 1958 (Sage and Gozun, 1958) and was established as a useful method of determining lymphatic drainage patterns in melanoma patients in 1977 (Fee et al., 1978).

Many radiocolloids are now in use around the world. These vary widely in particle size and pharmacokinetics (Strand and Persson, 1979; Gulec et al., 1997). The particle size of most radiocolloids used for lymphatic mapping can be tailored by altering preparation techniques (Higashi et al., 2004). ^{99m}Tc is now the preferred radiolabel as it is a pure γ-emitter of low energy (photopeak 140 meV) and short half-life (6 hours).

Clinical experience with ^{99m}Tc-sulphur colloid (TcSC) in imaging lymph nodes and bone marrow is reviewed here with emphasis on LS and sentinel lymph node biopsy (SLNB), as these are by far the more prevalent clinical applications of TcSC in North America today.

13.2. TcSC and Regional Lymph Node Imaging and Targeting

Sentinel lymph nodes (SLNs) are the one or more lymph nodes which lymphatic vessels draining a malignant neoplasm reach first in their course to at-risk regional lymph node basin(s). Tumor cells should seed out first in SLNs before involving more distal nodes in affected basin(s). The presence or absence of nonpalpable metastatic disease in regional nodes can be determined very accurately with SLNB. In patients in whom all SLNs are free of tumor, no further regional nodal surgery is necessary. In those with positive SLNs, completion lymphadenectomy frequently yields additional nodal metastases. These patients are usually candidates for adjuvant systemic therapy regimens for which they would not otherwise be eligible.

13.2.1. In Vivo Kinetics of Radiocolloids Following Interstitial or Intravascular Injection

Functional lymphatic anatomy relevant to SLNB is reviewed elsewhere (Gulec et al., 1997). When injected interstitially, radiocolloids enter lymphatic capillaries (Leak 1970; Spiegel et al., 1992) and travel to the regional nodes where they are phagocytosed by lymph node macrophages (Nopajaroonsri and Simon, (1971a); Nopajaroonsri et al., 1971b; Sainte-Marie et al., 1982; Luk et al., 1982; Sainte-Marie et al., 1986). The rate of progression of radiocolloids through lymph nodes and the avidity with which they are sequestered and immobilized within individual nodes vary with radiocolloid particle size. Radiocolloids with a larger particle size tend to be more avidly retained by phagocytes in SLN(s). The larger the particle size, the smaller the fraction of injectate which migrates from the injection site and the slower the rate of migration through the first (or sentinel) node to second and third echelon lymph nodes within the affected nodal basin.

While larger radiocolloid particle size yields more restricted lymph node imaging, and therefore less elegant lymphoscintigraphic imaging, this lymph node labeling is also less time-dependent. These attributes are advantageous for SLNB. When injected intravenously, radiocolloids are rapidly taken up and sequestered by phagocytes within the liver, spleen, and bone marrow. This phenomenon provides the rationale for the use of radiocolloids to investigate disease processes in these organs by external scintigraphy.

13.2.2. Radiopharmaceuticals for Lymphoscintigraphy and Sentinel Lymph Node Biopsy

Radiopharmacological properties of different radioparticles and their kinetics have been described by Strand and Persson (1979). The earliest lymphoscintigraphic agent used was colloidal gold (^{198}Au). Colloidal ^{198}Au has a uniform particle size of 5 nm and has the greatest proportional and most rapid rate of uptake into the lymphatics of any of the available radiocolloids. However, this radioisotope being a ß-emitter, the radiation dose to tissues at the site of injection, at 50 to 100 rad/μCi, is unacceptably high.

For LS visualization of the maximum number of nodes in a given nodal basin, the optimal radioparticle size is in the range of 5–10 nm (Strand and Persson, 1979). Particles smaller than 5 nm are preferentially taken up by vascular capillaries and venules. LS with very large radioparticles (sulphur colloid) fails to image 50 % of the normal nodes in a draining basin. As already noted, the avidity with which radiocolloids are retained in sentinel nodes correlates directly with particle size. Small-sized colloids are better suited for external gamma camera imaging of lymph flow through nodal basins while larger-sized radiocolloids are more amenable to gamma detection probe (GDP)-directed SLNB.

Many particulate radioactive tracers have been studied including colloidal preparations, radiolabeled antibodies, sestamibi, dextrans, and several albumin-based preparations. There is as yet no consensus on what is the optimal radiocolloid for SLNB. Strand and Persson (1979) compared the kinetics of ^{99m}Tc-sulphur colloid (TcSC), tin colloid, antimony sulphur colloid, sodium tin phytate, colloidal ^{198}Au, and ^{99m}Tc-human serum albumin (^{99m}Tc-HSA) in a rabbit model. The time-activity curve pattern for regional nodal uptake was similar for all radiocolloids tested. Uptake in the nodes began immediately after injection, and a plateau was reached after about 2 hours. The highest proportion of injected radiocolloid taken up by the nodes was observed with colloidal ^{198}Au (9 %). Antimony sulphur colloid had the second highest fractional migration at 5 %. Whole-body LS imaging showed no extranodal activity at any time after injection of colloidal ^{198}Au. However, there was escape of some ^{99m}Tc-antimony sulphur colloid to the liver, spleen, kidneys, and bladder. The highest extranodal "escape activity" was seen with ^{99m}Tc-phytate, and the lowest with TcSC.

Among tested radiocolloids, the fraction of injectate which migrates through lymphatics to lymph nodes was lowest for TcSC in two series (Strand and Persson, 1979; Krag et al., 1998b).

TcSC is the only LS agent available for clinical use in the United States. Sulphur-based colloid formation depends on the tendency of sulphide ions to form bonds with proteins and other materials containing -SH groups. The biochemistry of colloid formation involves conversion of sulphur to soluble polysulphides (Frier et al., 1981). ^{99m}Tc is incorporated into the body of the particle and is not simply bound to its surface. The particle size of unfiltered TcSC (uTcSC) ranges from 50 to 1000 nm (mean size 200 nm). Filtering TcSC through a 220 nm filter (fTcSC) yields a mean particle size of about 100 nm. There are approximately 5.5×10^8 particles per ml of uTcSC solution, as estimated by autoradiography (Frier et al., 1981). This figure underestimates the actual number by about 14 % as autoradiography does not count particles of less than 200 nm in size. Sulphur colloid particles undergo changes over time, both when left standing in a container and after injection into body fluids (Frier et al., 1981).

Solutions of smaller particle size comparable to antimony-sulphur colloid can be obtained by various preparation techniques and filtration of TcSC. One such preparation technique, and the kinetic properties of the resultant TcSC filtrate, was recently described by Hung et al. (1995). Particle size ranged from 15 to 50 nm. The average radioactivity retained in the filter ranged from 10.1 % to 62.7 % of the total activity. Nineteen patients studied with this highly filtered preparation of TcSC demonstrated LS imaging characteristics similar to those seen on antimony–sulphur colloid images.

13.2.3. The Kinetics of TcSC in SLNB

13.2.3.1. Comparisons of uTcSC and fTcSC

Krag et al. (1998b) compared dextran-40, Microlite, Cardiolite, fTcSC, and uTcSC in their initial investigation of the radiocolloid/GDP technique for SLNB in melanoma and breast cancer. Among the radiopharmaceuticals assayed, TcSC provided the highest SLN radiolocalization, and uTcSC was superior to fTcSC (86 % vs 77 %).

Linehan et al. (1999a) reported a comparison of fTcSC (220 nm) and uTcSC in breast cancer patients (Table 13.1), all of whom were also injected with isosulfan blue. uTcSC gave superior radiolocalization of SLNs both by LS and by SLNB, and localization rates were higher with each preparation for SLNB as compared to LS, as would be predicted by the inverse square law. However, there was no difference in SLN localization rate between the two groups when nodes detected by blue dye were factored in.

Tafra et al. (2001) reported a prospective multicenter series of SLNB in 532 breast cancer patients in which the use of either uTcSC (321 patients) or fTcSC (211 patients) was permitted. Time intervals from injection of radiocolloid to SLNB ranged from less than 20 minutes to over 240 minutes. For patients studied with uTcSC, both the mean hottest SLN GDP count and mean GDP count of all SLNs increased steadily over time. With fTcSC, SLN GDP counts increased rapidly between 21 and 60 minutes, with maximum GDP counts being 7.4-fold (mean for hottest SLNs) and 6-fold (mean for all SLNs) higher than those for uTcSC over that time interval. Thereafter, GDP counts in patients studied with fTcSC decreased markedly. Between 61 and 240 minutes, mean count of hottest SLNs with fTcSC was half that of uTcSC and mean count of all SLNs was 31 %

TABLE 13.1. Successful SLN radiolocalization with filtered and unfiltered ^{99m} TcSC in external LS and SLNB in breast cancer patients Linehan 1999A.

	Filtered TcSC ($n = 57$ pts)	Unfiltered TcSC ($n = 77$ pts)	p
Lymphoscintigraphy	29 (52 %)	50 (65 %)	0.05
Sentinel node biopsy	42 (73 %)	68 (88 %)	0.03
SLN localized by either radiocolloid or blue dye	54 (95 %)	71 (92 %)	NS

lower. After 240 minutes, SLN GDP counts with fTcSC were only 22% and 26%, respectively, of those with uTcSC.

There was less variation in GDP counts with uTcSC than fTcSC at all time intervals. Nonetheless, localization and false-negative (FN) rates with uTcSC and fTcSC were comparable (18). Both uTcSC and fTcSC are used throughout North America with reproducibly excellent SLN radiolocalization rates and other measures of clinical outcome.

13.2.3.2. Kinetics of uTcSC

A failure analysis (Gulec et al., 1998; Moffat et al., 1999) was reported in 70 breast cancer patients undergoing LS/SLNB with intraparenchymal peritumoral uTcSC (1 mCi in 8 mls saline), with immediate confirmatory axillary node dissection (AND). AND specimens were interrogated ex vivo with the GDP at the operating room table, and radiolabeled nodes which had escaped detection during SLNB were removed and submitted separately labeled as "not under (cutaneous) hot spot" (NUHS) to Pathology. These NUHS nodes had to meet the minimum criterion for a SLN ($\geq$10% of the radioactivity of the hottest SLN removed); in essence, these were SLNs missed by the surgeon. At least one NUHS nodal specimen was identified in 20 patients, and in three of these no SLNs had been found.

Localization rate was 89% and two false-negative SLNBs occurred among 20 node-positive patients, one of which would have been clinically significant. Time elapsed from injection of radiocolloid to lymphatic mapping at surgery ranged from 0.75 to 6.25 h (median 2.5 h). Correlations between time elapsed and cutaneous hot spot counts, hot spot to adjacent skin (HS:AS) count ratio, number of SLNs resected, number of SLNs plus NUHS nodes identified, SLN radioactivity, and activity of SLNs plus NUHS nodes revealed no time dependence in any of these variables (Figures 13.1–13.3). Thus, over the 6-h time period studied there was no decrement in SLN localization with peritumoral uTcSC in early breast cancer patients.

These data suggest that SLNB with uTcSC is well suited to real-world surgical practice with its unpredictable operating room schedules and other temporal concerns. As hot spot and nodal activity neither decreased nor number of SLNs increased with time suggests that uTcSC migrates continuously from the peritumoral site, but does not migrate beyond the SLN(s). Median SLN radioactivity amounted to 0.09% of the injected dose of uTcSC, consistent with its large particle size.

The radioactive diffusion zone created by peritumoral intraparenchymal radiocolloid injection is a significant impediment to SLN localization. In breast cancer, SLNs are relatively close to the primary tumor. This challenge is only modestly susceptible to surgical experience. While NUHS nodes tended to be deeper and less radioactive than SLNs, in 8 of 17 patients in whom both SLNs and NUHS nodes were found, the NUHS nodes were as or more superficial than the SLNs (Moffat et al., 1999). Despite an exquisitely sensitive, well–shielded, and collimated GDP, they had been indiscernible from injection site background radiation

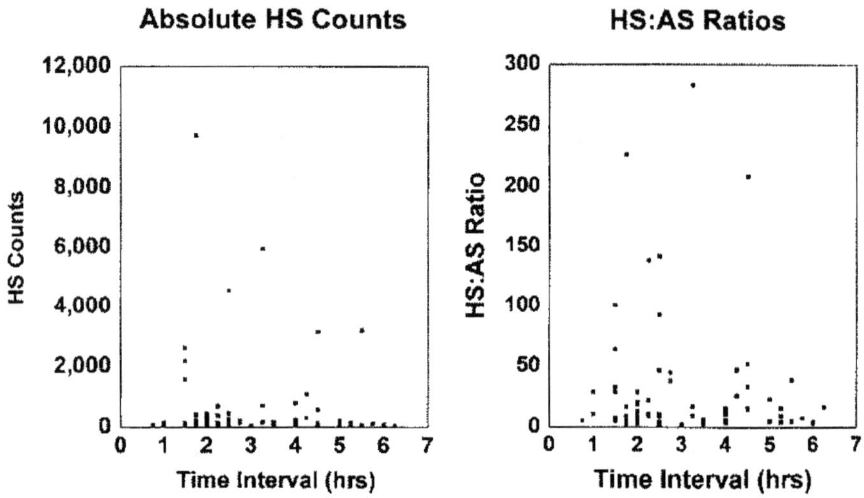

FIGURE 13.1. Scattergrams of axillary hot spot radioactivity, expressed as absolute hot spot (HS) 10-second GDP counts and as hot spot/adjacent skin (HS:AS) GDP count ratios, versus time elapsed from uTcSC injection to start of operative GDP mapping, by patient. There was no correlation between skin-level radioactivity and time (absolute HS counts: $r_s = 0.053$, $p = 0.667$ and HS:AS ratios: $r_s = 0.049$, $p = 0.688$). Reproduced with permission from Moffat et al. (1999)

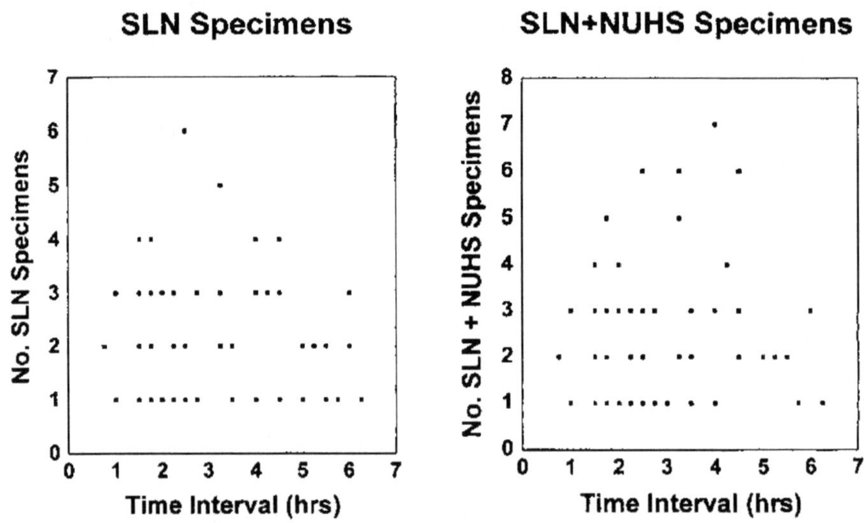

FIGURE 13.2. Scattergrams of number of radiolabeled nodal specimens (SLN specimens and SLN plus NUHS specimens) versus time elapsed from uTcSC injection to operative GDP mapping, by patient. There was no time dependence for number of specimens (SLN specimens: $r_s = 0.085$, $p = 0.510$ and SLN+NUHS specimens: $r_s = 0.130$, $p = 0.755$). Specimens could contain more than one node, and analysis by node number yielded similar findings (SLNs: $r_s = 0.130$, $p = 0.314$ and SLNs+NUHS: $r_s = 0.106$, $p = 0.399$) (data not shown). Reproduced with permission from Moffat et al. (1999).

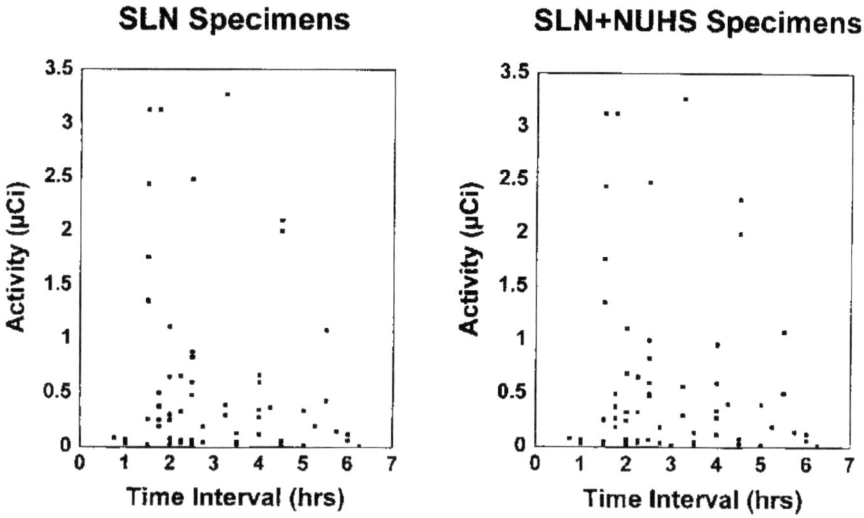

FIGURE 13.3. Scattergrams of SLN and SLN+NUHS radioactivity per patient versus time elapsed from uTcSC injection to operative GDP mapping, by patient. Again, no time dependence was demonstrated (SLNs: $r_s = 0.079$, $p = 0.538$ and SLNs+NUHS: $r^s = 0.111$, $p = 0.379$). Reproduced with permission from Moffat et al. (1999).

prior to AND. Concomitant use of blue dye and alternative routes of injection may offset this limitation to some degree.

McCarter et al. (2001) compared SLNB performed on the same day as (933 patients) or one day after (387 patients) intradermal uTcSC injection (0.1 mCi for the former and 0.5 mCi for the latter, to compensate for the time difference between them of four ^{99m}Tc half-lives). Slightly, more SLNs were localized in the day-after group (3.1 vs 2.7, $p < 0.05$). However, absolute GDP counts, SLN-to-background count ratios, and SLN localization rates with and without blue dye were essentially identical. Thus, uTcSC lymphatic flow rates do not change and migration through SLNs to more distal nodes does not occur over 18 to 24 hours after injection.

13.2.3.3. Kinetics of fTcSC

White et al. (1999) reported 46 patients in whom fTcSC was injected 18–24 hours pre-operatively. LS was performed right after injection and again on the day of surgery, with no change in findings between the two time points. SLNS were removed from all patients, 12 of whom proved to have metastatic disease.

Solorzano et al. (2001) studied the effects of day-before-surgery peritumoral intraparenchymal injection of 2.5 mCi fTcSC (five times the activity used for same-day injection) in 117 patients. Interval between injection and SLNB ranged from 15 to 20 hours. SLN localization was achieved in 97.5 % of patients, and mean number of SLNs per patient was 2.3. These results were comparable to those obtained with same-day injection.

13.2.4. Comparison of TcSC with Other Lymph Node Labeling Agents

13.2.4.1. TcSC and Vital Blue Dyes

Most North American surgeons who perform SLNB use vital blue dye (usually 1 % isosulfan blue) and TcSC/GDP in combination. SLN localization rates are superior when both methods are used, as compared to blue dye alone. The learning curve is shorter and wound complication rate lower with TcSC/GDP. The extent of surgical dissection tends to be more limited with TcSC/GDP, especially in the axillary and head and neck lymphatics. The cutaneous hot spot mapped by the GDP represents the point of closest approach to underlying SLNs from the skin surface. This minimizes the need to dissect skin flaps back toward the primary tumor to identify blue-stained lymphatics (Moffat et al., 2000).

Isosulfan blue and other colloidal blue dyes are much more time-sensitive than TcSC, migrating through SLNs very quickly. For this reason, blue dye is injected at the time of surgery, and SLNB commences within 5–10 minutes so that the dye does not diffuse beyond the SLNs. Re-injection of blue dye after 20–30 minutes is sometimes necessary in challenging cases.

Table 13.2 summarizes SLN localization experience in melanoma and breast cancer patients in whom blue dye and TcSC/GDP were used in combination. These techniques complement each other; each localizes SLNs missed by the other. In general, the TcSC/GDP method localizes more SLNs missed by blue dye than nonradiolabeled SLNs. Cody et al. (2001) documented factors contributing to the complementarity of the blue dye and TcSC/GDP SLNB techniques. In multivariate analysis, blue dye success correlated with upper outer quadrant tumor location (where radioactive background for the GDP is maximal because of proximity of the TcSC injection site to the axilla) while TcSC success correlated with intradermal as compared to intraparenchymal route of injection (smaller area of radioactivity at the injection site).

Derossis et al. (2001) reported a progressive reduction in contribution of blue dye to SLN localization with increasing surgical experience. SLN localization rate for TcSC increased from 86 % in the first 500 (of 2000) cases to 94 % in the last 500 ($p < 0.00005$), as compared to 82 % and 83 %, respectively, for blue dye ($p =$ NS). In the last 500 patients, the fractional localization benefit attributable to blue dye was reduced to 3 % for SLN localization and 2 % for positive SLN localization.

Woznick et al. (2006) reported 172 pathological node-positive breast cancer patients who underwent SLNB with uTcSC and isosulfan blue dye. Complementarity between the two tracers with respect to false-negative rate was demonstrated. Had only blue dye been used, 8.2 % of patients would have been understaged, whereas had SLNB been done using only the uTcSC/GDP method, the understaging error would have been 5.7 %. Each tracer contributed to the accuracy of nodal staging independently of the other.

While blue dye may account for less SLN localization complementarity than TcSC/GDP in some series, it is nonetheless a very useful visual aid to surgeons

TABLE 13.2. Sentinel lymph node biopsy in melanoma and breast cancer using TcSC and isosulfan blue

	No. of pts	Radiocolloid	No. of SLNs*	Blue only	Hot only	Blue+hot
Malignant melanoma						
White et al., 1999	46	fTcSC	122	4 (3.3 %)	38(31.2 %)	80(65.6 %)
Gershenwald et al., 1998a	374	fTcSC	787	4 (0.7 %)	176(22.4 %)	607(77.1 %)
Leong et al., 1997	163	fTcSC	303	2 (0.8 %)	83(27.3 %)	218(71.9 %)
Ariyan et al., 2004	183	fTcSC	467	0 (0.0 %)	229(49.0 %)	238(51.0 %)
Krag et al., 1995	44	uTcSC	44 pts	0 (0.0 %)	4 (9.1 %)	40(90.9 %)
Breast cancer						
Cox et al., 2000	1147	fTcSC	2395	472(19.7 %)	645(26.9 %)	1278(53.4 %)
Hill et al., 1999	500	uTcSC	458 pts	39(9.0 %)	65(14.0 %)	354(77.0 %)
Teal et al., 2005	101	fTcSC	220	33(15.0 %)	81(36.8 %)	106(48.2 %)
Teal et al., 2005	101	fTcSC	100 pts	2 (2.0 %)	18(18.0 %)	80(80.0 %)
Woznick et al., 2006	172†	uTcSC	170 pts	2 (1.2 %)	6(3.5 %)	162(95.3 %)

*Series marked "pts" in this column were analyzed by number of patients, not SLNs
†All patients in this series were node-positive

as they gain experience in SLNB. Experts in the blue dye method can achieve localization rates which approximate those attainable with radiocolloids (Morton et al. 2005).

A few patients in whom blue dye is used may have persistent, residual tattooing of the skin, an especially vexacious problem for female patients and those with head and neck tumors (Ariyan et al., 2004). While isosulfan blue dye carries a 0.7 % to 1.9 % risk of allergic and adverse reactions including anaphylaxis, the most severe reactions can be mitigated by pre-operative prophylaxis with hydrocortisone, diphenhydramine, and famotidine (Hirsch et al., 1982; Leong et al., 2000; Cimmino et al., 2001; Albo et al., 2001; Montgomery et al., 2002; Raut et al., 2004).

13.2.4.2. TcSC and Human Serum Albumin

Bostick et al. (1999) compared ^{99m}Tc-HSA and ^{99m}Tc-albumin colloid with fTcSC as LS/SLNB adjuncts to isosulfan blue-directed SLNB in 87 melanoma patients. Both albumin and sulphur colloids were filtered (200 nm). At surgery, 144 SLNs were localized and resected from 98 of 100 nodal basins identified on pre-operative LS (98 % localization rate). All three radiopharmaceuticals were equally sensitive on pre-operative LS. The isosulfan blue technique accounted for SLNs in 94 basins and the radiopharmaceuticals for the remaining four.

Tonakie et al. (2000) studied reproducibility of LS in 38 patients with truncal melanomas by dual sequential noncolloidal ^{99m}Tc-HSA and TcSC. This was solely a study of LS imaging, and ^{99m}Tc-HSA was chosen for its excellent external gamma camera imaging characteristics related to its nonparticulate nature. ^{99m}Tc-HSA LS was performed several days prior to SLNB, and TcSC LS was done on the day of surgery, as radiocolloids are less prone to pass through to nonSLNs. In 32 patients, LS imaging was concordant. In the other six, imaging findings were discordant, ^{99m}Tc-HSA LS suggesting drainage to two lymphatic basins and TcSC LS demonstrating only one. The six discordant TcSC-imaging studies showed less intensity, and all were done using only 1 mCi of activity as compared to TcSC studies done earlier in the series with 3 mCi activity, none of which were discordant. Nine patients had positive SLNs including two with discordant LS findings, but the positive SLNs were not in basins imaged only by ^{99m}Tc-HSA. The imperatives for LS imaging can be at cross-purposes with the surgical requirements for SLNB. The reported discordances are most likely a function of flow and particle dynamics in the lymphatic system; the extra drainage routes imaged by HSA may have been small, secondary channels which tumor cells, because of size and flow characteristics, would not preferentially traverse.

Essner et al. (2000) reported 247 patients with clinical Stage I and II melanoma who were studied with either ^{99m}Tc-HSA injected 24 hours prior to SLNB, fTcSC injected 4 hours prior to SLNB, or fTcSC injected 18 hours prior to SLNB. Isosulfan blue was used in all cases and patients in the first group underwent a repeat injection of ^{99m}Tc-HSA immediately prior to surgery. There was no difference between groups in LS-imaging accuracy. Same-day fTcSC gave better

SLN-to-background ratios than the other two, yielded the greatest difference between radiolabeled SLNs and the operative bed after removal, and localized the largest number of nodes which were both blue and hot.

13.2.4.3. TcSC, ^{99m}Tc-DTPA-Mannosyl-Dextran, and MIBI

The receptor agent ^{99m}Tc-DTPA-mannosyl-dextran is under investigation as a lymph node marker as it clears injection sites faster than conventional colloids, while binding avidly to a mannose-binding receptor on the surface of macrophages, with little pass through to distal lymph nodes (Vera et al., 2001). A Phase I radiopharmaceutical study in 12 breast cancer patients demonstrated an injection site clearance half-time of 2.7 hours as compared to 49.5 hours for fTcSC, 0.55 % SLN uptake as compared to 0.65 % for fTcSC, a lower number of SLNs and less pass through to distal nodes than fTcSC, and a higher concordance with isosulfan blue labeling of SLNs (46). No adverse effects were seen. Further studies are planned.

MIBI is being studied for SLNB based on its activity as a "tumor-seeking agent," the mechanism of which has not yet been elucidated. In a study of 72 breast cancer patients, 42 of whom were injected peritumorally with fTcSC 2–12 hours pre-operatively, 25 patients were injected with ^{9mm}Tc-MIBI 2–12 hours and an additional 5 patients injected 12–24 hours prior to surgery. All patients underwent injection of isosulfan blue at surgery. Performance of both radiolabeled marking agents was similar (Arican et al., 2005).

13.2.5. TcSC and Lymphatic Mapping in Cancer Patients

13.2.5.1. Cutaneous Malignant Melanoma

Surgery is the sole treatment modality with curative potential in patients with locoregional stage melanoma. It is vital that the primary tumor be resected with tumor-free margins, and the presence or absence of regional nodal metastases established at the time of initial treatment. Unlike lymph node metastases in breast cancer, untreated occult regional melanoma metastases almost invariably progress and prejudice the prospects for cure. In patients with clinically palpable regional nodal metastases, therapeutic radical lymphadenectomy is essential to long-term survival and cure.

The advent of effective adjuvant systemic therapy for node-positive melanoma has heightened the importance of accurate nodal staging. In 1996, an ECOG prospective randomized double-blind study of interferon α_{2b} (IFN α_{2b}) as adjuvant therapy in node-positive and high-risk primary melanoma demonstrated a 10 % increase in 5-year survival in favor of the immunotherapy arm in node-positive patients (Kirkwood et al., 1996). The findings of this and subsequent studies, and those of SLNB, hold forth real hope that significant improvements in treatment outcome for cutaneous melanoma may be realized in the near future.

Formal lymph node dissections are expensive, morbid, and can disfigure and disable to the point of unemployability. Such risks are justified in melanoma patients with known nodal metastases, not for nodal staging only. These operations can only be of benefit if at risk nodal basins harbor metastatic melanoma, and should only be performed for node-positive disease.

Prior to SLNB, surgical management of the regional lymphatics in clinical node-negative (cN-) melanoma was a major point of contention. The alternatives at issue were immediate "elective lymph node dissection" (ELND) and "watchful waiting," with delayed therapeutic lymphadenectomy only if clinical nodal metastases developed in follow-up. Analyses (Balch et al., 2001a) continue to suggest a possible survival advantage in patients undergoing lymphadenectomy for micrometastatic nodal disease as compared to those with palpable nodal metastases. Each strategy was energetically advocated in the absence of definitive evidence for either. None of the four published prospective randomized trials of ELND and observation alone have shown a difference in clinical outcome (Veronesi et al., 1977, 1982; Sim et al., 1986; Balch et al., 1996, 2000; Cascinelli et al., 1998).

The ELND controversy was effectively preempted by SLNB as this newer procedure identifies micrometastatic nodal disease in cN-melanoma patients. These are the only patients who can possibly benefit from formal lymphadenectomy.

13.2.5.1.1. External LS

Traditional concepts of cutaneous lymphatic drainage in the head and neck and truncal regions, the so-called "zones of ambiguity," were refuted by studies of LS in melanoma patients (Berman et al., 1992; Norman et al., 1991; O'Brien et al., 1995; Pijpers et al., 1995; Ross et al., 1993; Uren et al., 1994; Wells et al., 1994). A higher than expected incidence of bilateral drainage, unexpected "skip" drainage to more distal nodes in a basin, drainage to multiple nodal basins, and patterns of lymphatic drainage hitherto considered "unorthodox" have been documented. LS has proven that historical concepts of lymphatic anatomy and surgical experience do not reliably predict the distribution of nodal melanoma metastases. Pre-operative LS is essential for melanoma because of the unpredictability and variability of lymphatic drainage patterns, especially in truncal and head and neck sites.

13.2.5.1.2. Sentinel Node Biopsy for cN-Malignant Melanoma

In the United States, TcSC is injected into the dermis around the tumor and SLNB is directed with a GDP and, commonly, isosulfan blue (Albertini et al., 1996; Alex and Krag, (1993a); Alex et al., 1993b, Glass et al., 1996; Godellas et al., 1995; Gulec et al., 1997; Krag et al., 1993; North and Spellman, 1996; Pijpers et al., 1995; Uren et al., 1994; Reintgen et al., 1995; Thompson et al., 1995). The TcSC/GDP method, first reported by Krag and colleagues (Alex and Krag, (1993a); Alex et al., 1993b; Krag et al., 1993), was developed because the blue node technique has a long learning curve and is prone to wound complications

(Morton et al., 1992). In addition, the extent of dissection in search of blue-stained nodes can be quite significant, especially in the axilla, occasionally approaching that of therapeutic lymphadenectomy.

The use of TcSC and the GDP with or instead of blue dye allows placement of the biopsy incision directly over the radiolabeled sentinel node(s). The surgeon then uses the GDP to establish the "line of sight" to the underlying radiolabeled nodes, dissecting directly to these without disturbing surrounding tissues.

Sentinel node localization rates with blue dye as compared to radiocolloid and GDP, with or without blue dye, are shown in Table 13.3. SLNB using radiocolloid and a GDP alone or in combination with blue dye yields SLNs in 82 % to 100 % of cases, with a very modest false-negative rate (Glass et al., 1996; Krag et al., 1995). SLNB with radiocolloid and a GDP, with or without blue dye, has a shorter learning curve than the blue dye technique alone. That said, with intensive training, a highly experienced international group of participating surgeons achieved a very high localization rate with isosulfan blue alone in the Multicenter Selective Lymphadenectomy Trial I (Morton et al., 1999).

13.2.5.1.3. Selection of Patients for LS and SLNB

The probability of nodal metastases varies by tumor site (extremities, trunk, head and neck) and correlates directly with primary tumor microstage (Clark's levels and Breslow's depth, especially the latter), AJCC T category and AJCC stage I and II subcategory (Balch et al., 2001a; Doubrovsky et al., 2004; Rousseau et al., 2003; Slingluff et al., 1994). In general, the risk of occult nodal metastases in patients with melanomas of less than 1 mm in thickness ("thin melanomas") is very low. A consensus as to when LS and SLNB should be offered to patients with thin melanomas has yet to be reached. Risk factors for the presence of nodal disease in these patients include Clark's level III and IV lesions (Corsetti et al., 2003; Zapas et al., 2003), lesions of over 0.5 mm (Ranieri et al., 2004), 0.76 mm (Bleicher et al., 2003; Jacobs et al., 2003), or 0.8 mm (Morton et al., 1992; Vacquerano et al., 2004 in thickness, young age, high mitotic rate, substantial regression, and presence of vertical growth phase (Bedrosian et al., 2000).

For cN-melanomas of 1–4 mm in thickness, the incidence of occult nodal disease is sufficiently high, and that of occult systemic disease sufficiently low, to justify LS and SLNB.

TABLE 13.3. SLNB localization rates for melanoma and breast cancer: Blue dye alone versus TcSC/GDP ± blue dye.

	Blue dye (%)	TcSC/GDP ± blue dye (%)
Krag et al., 1995	91	100
Glass et al., 1996	75	97
Albertini et al., 1996	70	96
Leong et al., 1997	73	99.2
Morton et al., 1999	95.2	99.1
Ariyan et al., 2004	51	100

In patients with cN-melanomas of over 4 mm in thickness, SLNB has prognostic value. In a series of 131 such patients Gershenwald et al. (2000) reported that those with negative sentinel nodes had significantly higher disease-free and overall survival as compared to those who were node-positive.

13.2.5.1.4. Failure Rates and Complications of SLNB in Melanoma

Failure of SLNB to identify node-positive disease has been documented in three studies of a total of 383 patients in whom concurrent confirmatory ELND was performed (Table 13.4). The overall false-negative rate (the proportion of pN+ patients in whom SLNB yielded nodes which did not harbor metastatic disease) was 5.5 %. Follow-up of patients in whom all resected SLNs were negative have documented relapses in the affected nodal basin(s) at a rate of about 1 % per year within the first 5 years (Table 13.5). This figure is concordant with that reported in the studies in which confirmatory ELND was performed. These data attest to the precision of current SLNB techniques when used in melanoma patients and validate the clinical utility of the sentinel lymph node hypothesis.

The Sunbelt Melanoma Trial (Wrightson et al., 2003) showed that complications with SLNB alone, performed in patients whose SLNs prove to be negative, are neither as common nor severe as those in patients undergoing formal lymphadenectomy for positive sentinel nodes. For patients with cervical SLNs, the complication rate for SLNB alone was 2.4 % as compared to 10 % in those requiring neck dissections ($p = 0.008$). For patients with axillary SLNs, complication rates were 4.4 % and 19 %, respectively ($p < 0.0001$) and for inguinal SLNs, 8.1 % and 51.2 % ($p < 0.0001$). Lymphedema rates for axillary SLNB alone as compared to SLNB plus CLND were 0.3 % and 4.6 % ($p < 0.0001$) and the figures for inguinal procedures were 1.5 % and 31.5 %, respectively ($p < 0.0001$). SLNB is clearly less morbid than therapeutic lymphadenectomy, and one can reasonably infer a similar advantage over ELND.

13.2.5.2. Interval and Unexpected SLNs in Melanoma Patients

Drainage to unexpected "interval" lymph nodes outside any of the usual lymph node basins is observed in 19 % to 26 % of melanoma patients (O'Brien et al., 1995; Norman et al., 1993; Uren et al., 1993). For extremity melanomas, drainage to epitrochlear, popliteal, or supraclavicular nodes has been documented. Sentinel lymphatic drainage to occipital, retroauricular, facial artery, or even axillary

TABLE 13.4. False-negative rates with SLNB in cN-melanoma patients

	Total No. of patients	No. of pN+ patients	SLN localization rate (%)	False-negative rate (%)
Morton et al., 1992	223	40	82	5
Reintgen et al., 1994	42	8	100	0
Thompson et al., 1995	118	25	87	8
Total	383	73	85	5.5

TABLE 13.5. Nodal relapse after negative SLNB

	No. of pts	No. of Negative SLNs	Median follow-up (months)	Nodal relapse (%)	Isolated nodal relapse (%)
Gershenwald et al., 1998b	322	243	35	4.1 %	2.8
Morton et al., 1999	584	–	45	5.0 %	–
Essner et al., 1999	–	267	45	4.8 %	–
Cascinelli et al., 2000	710	569	29	5.6 %	–
Statius Muller et al., 2000	204	162	42	3.0 %	1.8
Ariyan et al., 2004	263	235	32	1.3 %	–

nodes has been observed in patients with head and neck melanoma. Truncal melanoma sites may drain directly to intermuscular, internal mammary, rectus sheath, mediastinal, para-aortic, or retroperitoneal lymph nodes.

In three recent studies (Table 13.6), interval SLNs (iSLNs) were resected in 3.2 % to 7.2 % of patients and iSLNs were as likely to harbor metastatic melanoma as SLNs removed from standard nodal basins. Moreover, a positive iSLN was the only positive node identified in 38 % to 85 % of patients from whom positive iSLNs were retrieved. Mediastinal, retroperitoneal, and para-aortic sites excepted, iSLNs should be resected whenever they occur.

13.2.5.2.1. SLNB After Definitive Resection of the Primary Lesion

Whether prior WLE of the melanoma primary increases the false-negative rate of SLNB because of perturbation of peritumoral lymphatics has been a point of contention. Wide excision with closure by skin graft or rotational flap probably does increase the probability of a false-negative SLNB. Wide excision with primary closure simply increases the number of sentinel nodes (and number of nodal basins) without adversely affecting the false-negative rate.

A meta-analysis compared seven series in which SLNB was performed at the time of WLE to four series in which SLNB was performed after WLE (Evans et al., 2003). SLN localization and false-negative rates were not adversely affected when SLNB was delayed, and the mean number of SLNs retrieved was only slightly increased.

TABLE 13.6. Interval sentinel nodes (iSLNs) in melanoma patients identified by TcSC

	No. of pts	No. of basins	No. of pts with iSLNS	No. of pts with + iSLNs	Only iSLN(s) +
Thelmo et al., 2001	557	–	30 (5 %)	6 (20 %)	3 (10 %)
Sumner et al., 2002	1145	–	54 (4.7 %)	7 (13 %)	4 (57 %)
McMasters et al., 2002	2000	2332	62 (3.2 %)	13 (21 %)	11 (18 %)

In 38% to 85% of patients with interval SLNs in these series, the positive interval SLN was the only tumor-infiltrated lymph mode. These Nodes are therefore clinically important and should be pursued whenever lymphatic mapping reveals their presence.

13.2.5.2.2. SLNB for Melanoma in the Head and Neck

The lymphatic system in the head and neck predisposes to multiple SLNs in ipsilateral, contralateral, or bilateral zones, and a higher likelihood of interval SLNs (O'Brien et al., 1995). Melanomas in this site have a higher risk of recurrence (Balch et al., 2001b), and a higher false-negative rate. Reports of SLNB in head and neck melanoma are summarized in Table 13.7. The false-negative rate for all seven series was 8.9 %. The false-negative rate with SLNB for head and neck melanoma was higher than that for truncal or extremity melanomas in the Sunbelt Melanoma Trial (Chao et al., 2003).

LS is critically important for identification of SLNs in this region. All SLNs must be resected; failure to do so portends subsequent clinical progression of disease (O'Brien et al., 1995).

13.2.5.3. Early Breast Cancer

Regional lymph node status is the most significant clinicopathological predictor of survival in patients with cN-breast cancer. Only 20 % to 30 % of such patients are pathologically node-positive (pN+). Routine axillary node dissection, the standard of care for surgical nodal staging in breast cancer since the 1890s, is inordinately expensive and puts pN-patients at risk of morbidity. SLNB has been investigated as an alternative since the mid-1990s. A validation study of SLNB with confirmatory AND in 443 breast cancer patients (Krag et al., 1998c) established the feasibility of this approach, and several prospective randomized studies are in progress.

Published localization and false negative rates for SLNB in breast cancer in single institution series are summarized elsewhere (Moffat et al., 2000;

TABLE 13.7. SLNB for head and neck melanoma

	No. of pts	SLN Localization rate (%)	SLNs in multiple zones (%)	Pts with + SLNs	F/U (mos)	LN recurrence	FN rate
Morton et al., 1993	70	90	10+	12	–	0	0 (0 %)
Bostick et al., 1997	117	92.2	10	14	48	0	0 (0 %)
Wells et al., 1997	58	95	36	6	13	0	0 (0 %)
Alex et al., 1998	23	96	–	3	17	1	1 (25 %)
Ollila et al., 1999	39	95	–	4	33	1	1 (20 %)
Jansen et al., 2000	30	90	–	8	23	3	2 (20 %)
Medina-Franco et al., 2001	38	92	40	4	18	2	1 (20 %)
Total	375			51		7	8.9 %

Moffat and Krag, 2003). As with melanoma, localization rates with radiocolloid are higher than with blue dye alone. However, SLNB false-negative rates in multicenter series in the US, at 8.3 % to 13 %, are significantly higher in breast cancer patients (Krag et al., 1998c; Moffat et al., 1999; Moffat, 2005; Tafra et al., 2001; Wong et al., 2001). The primary tumor and radioactive diffusion zone around the radiocolloid injection site is much closer to at-risk lymph nodes than is generally the case in melanoma. The resultant background radioactivity impedes GDP localization of SLNs (Moffat et al., 1999). The NSABP B-32 prospective randomized trial of SLNB versus SLNB plus confirmatory AND, with over 5600 patients accrued, has the statistical power to detect a 2 % difference in survival between the two arms, and will establish whether SLNB alone, despite the false-negative rate, is equivalent to SLNB+AND in terms of survival (Harlow et al., 2005).

13.2.5.4. Routes of TcSC Injection in SLNB for Breast Cancer

In most early studies of SLNB in breast cancer, radiocolloid and/or blue dye was injected into the peritumoral breast parenchyma to ensure that the node labeling agent is transported in the same lymphatic vessels which cancer cells would traverse in metastasizing to the regional nodes.

The density of breast tissue and its lymphatic vascular network is considerably less than the case in the dermis. By increasing radiocolloid injectate volume, interstitial tissue pressure at the breast cancer primary site is increased, ingress of radiocolloid into lymphatic lumina is augmented, and SLN localization is improved commensurately. Krag et al. (1998b) showed that for volumes of TcSC of less than 3 ml, SLN localization was achieved in only 56 % of cases as compared to 79 % for volumes of 3–7.9 ml and 100 % for 8 ml or more ($p < 0.01$). Using these volumes, SLN localization with TcSC alone (without blue dye) was achieved in over 90 % of cases in the University of Vermont validation trial (Krag et al., 1998c).

However, the resultant diffusion zone of radioactivity around the site of the primary tumor is considerably larger than that with intradermal injection. This hinders identification of cutaneous hot spots and SLN localization with the probe, especially for tumors in the upper outer quadrant. Borgstein et al. (1997) hypothesized that because both the breast parenchyma and skin parenchyma are of ectodermal origin, they share a common lymphatic pathway to the same sentinel node. In 30 consecutive patients, ^{99m}Tc-colloidal albumin was injected peritumorally and blue dye intradermally over the primary tumor. Blue dye and radiocolloid had labeled the same lymph node(s) in every patient. All 14 pN+ patients among the 25 who underwent concomitant AND were correctly identified by the status of the SLN (no false-negative SLNBs).

A number of studies have shown that SLN localization rates with intradermal, subdermal, subareolar, or periareolar TcSC (Bauer et al., 2002; Boobol et al., 2001; Klimberg et al., 1999; Lin et al., 2004; Linehan et al., 1999b; McMasters et al., 2001; Shen et al., 2001; Zogakis et al., 2005), or other radiocolloids (Borgstein et al., 2000; D'Eredita et al., 2006; De Cicco et al., 1997; Galimberti et al., 1998; Veronesi et al., 1997, 1999), with or without blue dye, are superior to those attained with peritumoral injection, and produce less background radiation interference, higher absolute SLN counts, and higher SLN-to-background count ratios.

Intradermal, subdermal, periareolar, or subareolar TcSC injection in small fluid volumes is conceptually appealing. There is less chance of technical failure as inadvertent injection of radiocolloid into the biopsy cavity or pectoralis muscle is avoided. Nuclear medicine physicians are more facile with superficial injection techniques than with intraparenchymal injection. Much of the radioactive diffusion zone may be eliminated, improving SLN localization and minimizing false-negative rates.

Superficial routes of injection may underestimate the frequency of sentinel lymphatic drainage to internal mammary lymph nodes, as compared to peritumoral administration (Krag, 1999, Shen et al., 2001; Tanis et al., 2005). There may be a separate deep lymphatic network in the breast, which drains preferentially to the internal mammary nodes (Tanis et al., 2005) (Figure 13.4). The prognostic impact of positive nodes in the internal mammary and axillary nodal metastases is similar, so it is important to identify and remove SLNs in this and other nonaxillary sites to avoid understaging errors (Alex and Krag, 1996; Gulec et al., 1998; Krag et al., 1993, 1998c,d; Moffat et al., 2000).

13.2.5.5. SLNB for Other Tumors

There are a substantial number of reports of SLNB in other tumor types. A brief survey of SLNB with TcSC in a variety of cancers is presented here.

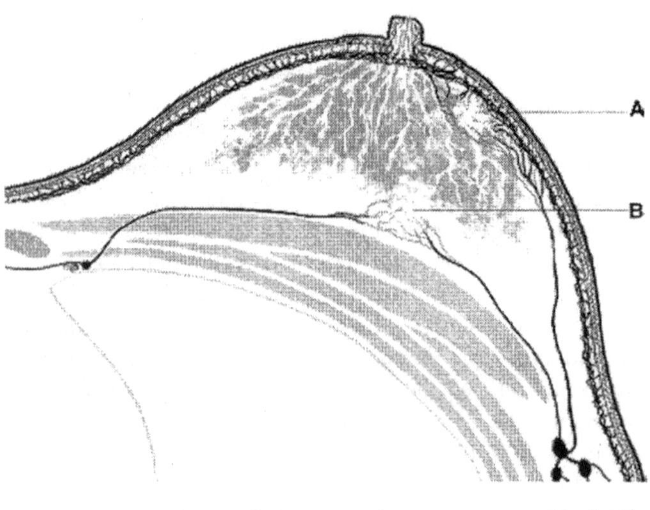

FIGURE 13.4. Lymphatic drainage of the breast, in transverse profile (122). Lymphatic channels draining to the internal mammary nodes are more deeply situated and have sparse, if any, interconnection with the superficial lymphatics which drain mainly to the axilla. Reproduced with permission from Tanis (2005).

13.2.5.5.1. Colorectal Adenocarcinoma

As reviewed by Stojadinovic et al. (2005), most studies of SLNB in colorectal cancer have been undertaken using vital blue dye without radiocolloid. Only recently has TcSC been assayed as an SLN-marking agent (Bilchik et al., 2003; Patten et al., 2004; Saha et al., 2004a, b). In a number of reports, dissection and mobilization of the tumor and surrounding large bowel was performed just prior to SLNB or at a prior operation, possibly disrupting important lymphatic pathways. False-negative rates were highest for rectal cancers below the peritoneal reflection, T3 and T4 tumors, incomplete or inadequate peritumoral subserosal/endoscopic injection, and injection of TcSC into the tumor or bowel lumen.

Saha et al. (2004a) reported an analysis of SLNB with TcSC and isosulfan blue in 57 patients. While lymphatic mapping was successful in more patients with blue dye than with TcSC, the difference was not significant. Of SLNs detected by blue dye alone, 10.7 % contained metastatic disease as compared to 19.8 % of SLNs which were both blue and hot ($p = 0.028$).

13.2.5.5.2. Mucosal Squamous Head and Neck Cancer

The management of the clinical N0 neck has been the subject of a decades-long debate not dissimilar to the ELND controversy in cutaneous melanoma. SLNB may provide a solution to this clinical dilemma as it has in melanoma. A prospective multicenter trial sponsored by the American College of Surgeons Oncology Group (ACoSOG) is currently accruing patients. Reports of SLNB for head and neck cancer in small to modest numbers of patients are currently in print (Alex et al., 2000; Altinyollar et al., 2002; Civantos et al., 2003, 2006; Payoux et al., 2005; Ross et al., 2002).

Civantos et al. (2006) compared SLNB with uTcSC in 63 patients with cutaneous cancers of the head and neck and 43 patients with mucosal squamous head and neck cancer. The latter group underwent confirmatory neck dissection. SLNs were identified outside the traditional lymphatic drainage zones in eight mucosal squamous cancer patients (18.6 %). In 2 of the 20 mucosal cancer patients with metastatic nodal disease, labeled SLNs were adjacent to nodes replaced with cancer but were themselves free of metastatic disease (10 % false-negative rate).

Payoux et al. (2005) performed SLNB with TcSC and confirmatory neck dissection in 30 patients. There was only one false-negative among a total of 7 patients with positive nodes (14.3 % false-negative rate), this occurring in the first patient operated upon in the series. In a review of 12 published series including their own in which a total of 383 patients were operated upon, the overall false-negative rate was 7.5 % (7 of 94 pN+ patients).

13.2.5.5.3. Gynecological Malignancies

SLNB has been successful in patients with carcinoma of the cervix uteri and vulvar carcinoma. Lin et al. (2005) published a series of 30 Stage IA2 and IIA cervical cancer patients undergoing SLNB following cervical injection of uTcSC with 100 % localization. All underwent confirmatory lymphadenectomy. There were no false-negatives among the seven pN+ patients. Five cited series of open or laparoscopic SLNB for cervical cancer had a cumulative false-negative rate of 2 %. Similarly, vulvar cancers, both squamous carcinomas and melanomas, are very amenable to this technique (Abramova et al., 2002).

13.2.5.5.4. Carcinoma of the Prostate

Silva et al. (2005) reported 23 patients with prostate cancer undergoing SLNB following ulrasouond-guided transrectal injection of fTcSC into each lobe of the gland. SLNs were localized with the GDP in 21 patients, three of whom were pN+, and in one the positive SLN was outside the obturator and external iliac lymphatic chains.

13.2.5.5.5. Adenocarcinoma of the Stomach

Zulfikaroglu et al. (2005) studied 32 patients with gastric cancer in whom fTcSC was injected gastroscopically prior to surgery. SLNs were localized with the GDP in 31 patients and no false-negatives were seen.

13.2.5.5.6. Primary and Metastatic Pulmonary Neoplasms

Faries et al. (2004) performed SLNB in 67 patients with lung neoplasms in whom isosulfan blue plus either 99mTc-HSA or TcSC was injected peritumorally in the operating room after exploration of the involved hemithorax. SLNs were found in all patients and positive SLNs were found in 19. Almost all patients with clinical stage I lung cancer had positive SLNs, and in 33 patients with metastatic melanoma, SLN positivity portended a 2-year survival of 0 % as compared to 48 % when SLNs were negative.

13.3. TcSC and Bone Marrow Imaging

TcSC was first reported as an imaging agent for reticuloendothelial system (RES) organs in the early 1960s. When injected intravenously, clearance of TcSC particles from the circulating blood volume is biphasic, the first phase with a half-time of 3 minutes. The half-time of the second phase lasts several minutes to a few hours, varying with preparation technique. About 90 % of the activity localizes in the liver, 5 % in the spleen, and 5 % in the bone marrow. Whereas imaging of liver and spleen are satisfactory with 3 to 5 mCi of TcSC, bone marrow imaging requires injectate activity in the range of 5–10 mCi because of the relatively sparse uptake of radioactivity in a much wider distribution of tissue (Desai et al., 1985). By comparison, TcSC injectate activity for SLNB is typically in the range of 350–500 μCi for melanoma and 0.5–1.0 mCi for breast cancer. The typical radiation dose to the bone marrow is 0.26–0.34 rad/mCi (Datz and Taylor, 1985).

Marrow radiolocalization is optimized when radiocolloid particle size is 100 nm or less (Desai et al., 1985; Kloiber et al., 1981). Thus, antimony sulphur colloid has very favorable bone marrow imaging characteristics. That said, TcSC is an excellent bone marrow imaging agent and remains one of the most commonly used radiocolloids worldwide for this purpose.

Bone marrow imaging has applications in hematology and oncology, and for diagnosis of musculoskeletal infections.

13.3.1. Hematology and Oncology

Normal TcSC distribution in bone marrow imaging studies is typically limited to the axial skeleton and the proximal one-third of humerus and femur (Henry et al., 1975). Perturbations of this normal pattern are observed in patients with various anemias (Henry et al., 1975; Kahn et al., 1988), bone marrow infiltrative or obliterative processes (Elster et al., 1992; Henry et al., 1975; Lorberboym et al., 1997; Siddiqui et al., 1979), hematological malignancies (Henry et al., 1975; Hoppin et al., 1979; Kniseley et al., 1966; Siegel et al., 1999), myelophthisic or aplastic/myelofibrotic processes (the latter often attended by extramedullary hematopoiesis which is imaged by TcSC) (Coates et al., 1994; Carloss et al., 1977; Tanner et al., 2003), bone marrow transplantation (Berlangieri et al., 1993; Tanner et al., 2003), or treatment with cytotoxic chemotherapy or ionizing radiations (Achong and Oates, 1995; Berlangieri et al., 1993; Roeske et al., 2005; Siddiqui et al., 1979; Tanner et al., 2003). This technique has also been used to avoid bone marrow sampling errors in patients with conditions in which such events are liable to occur (Datz and Taylor, 1985).

13.3.2. Musculoskeletal Infections and Related Conditions

Labeled leukocyte (WBC) imaging with ^{111}In or, in children, ^{99m}Tc-HMPAO provides a means of identifying inflammations and infections in the musculoskeletal system. However, the imaging efficacy of these scans is variable and subject to confounding nonspecific sequestration of labeled WBCs in the bone marrow. Osteomyelitis is imaged positively by labeled WBCs, and WBC localization which is asymmetrical and/or outside the normal distribution of bone marrow is considered pathognomonic. However, inter-individual variation in distribution of bone marrow is extreme, and marrow distribution is often markedly altered in patients with anemia, solid and hematological neoplasms, bone and marrow infarcts, and orthopedic prostheses (Palestro and Torres, 1997; Seabold et al., 1991).

TcSC marrow scintigraphy offers a solution to these challenges. Whereas labeled WBCs and TcSC both localize in marrow, the presence of infection stimulates localization of labeled WBCs while suppressing TcSC uptake. As osteomyelitis becomes established, the involved bone undergoes suppuration and necrosis with low local oxygen tension, extensive small vessel thrombosis, and increased interstitial pressure. Local marrow phagocytes are suppressed or destroyed, and TcSC localization is inhibited. Thus, when TcSC marrow scintigraphy and WBC imaging are done in sequence in patients with bone infection, the localization patterns of the two markers are spatially incongruent. WBC imaging in areas of osteomyelitis is intense whereas these areas either show no increase in radioactivity or are overtly photopenic on TcSC marrow studies. It is critical, however, that the TcSC be properly prepared and injected within the first 2 hours after preparation for optimal imaging quality (Palestro et al., 2006).

13.3.2.1. Osteomyelitis

Palestro et al. (1992) reported on 73 patients undergoing autologous WBC and TcSC marrow imaging for possible osteomyelitis. There were 22 confirmed cases of osteomyelitis in this cohort. The sensitivity, specificity, and accuracy of WBC imaging alone were 86 %, 12 %, and 34 %. When WBC scintigrams were read with the TcSC marrow studies, these measures improved to 100 %, 94 %, and 96 %, respectively. In other reports, sensitivity of WBC-TcSC marrow imaging was 87 % to 94 % and specificity 81 % to 91 % (Copping et al., 1992; 1996, 1996; Schauwecker, 1992).

For osteomyelitis of the skull, spine, or hips, WBC/TcSC marrow imaging with SPECT is necessary to overcome interference from adjacent bone, soft tissue, and background from liver and spleen. In the setting of possible spinal osteomyelitis in patients with ferromagnetic orthopedic hardware precluding MRI imaging, WBC/TcSC marrow SPECT is an excellent alternative (Seabold and Nepola, 1999).

13.3.2.2. Periprosthetic Infections in Orthopedics

Hematopoietically active bone marrow develops around joint prostheses in the absence of infection for reasons which are not well elucidated. Loosening of a joint prosthesis in the absence of infection is associated with florid periprosthetic inflammation. WBC/TcSC marrow imaging permits differentiation of infection from simple loosening. In this setting, osteomyelitis-related spatial incongruity between the two imaging studies is limited to the joint space and adjacent tissues (Palestro et al., 2006).

Palestro et al. (1991) in a study of infection in knee arthroplasties (9 of 41 prostheses) reported that accuracy of diagnosis of periprosthetic infection with WBC/TcSC marrow imaging was 95 % as compared to 78 % for WBC scintigraphy alone, 74 % for bone scintigraphy alone, and 75 % for WBC/bone scintigraphy in combination (Palestro et al., 1991). For total hip arthroplasties (23 of 92 infected), the same group reported sensitivity, specificity, and accuracy of WBC/TcSC marrow imaging as 100 %, 97 %, and 98 % (Palestro et al., 1990). Esper et al. (2004) reported similar results in 60 patients with suspected infection in hip or knee arthroplasties, noting markedly easier interpretation of WBC/marrow scintigraphy as compared to WBC/bone scintigraphy.

Love et al. (2004) compared ^{18}FDG PET to WBC/marrow imaging in 59 patients with failed hip or knee replacements, in 25 of whom periprosthetic infection was confirmed. Using several benchmark criteria and semiquantitative analysis, they showed that WBC/marrow imaging was clearly superior for detection of periprosthetic sepsis.

13.3.2.3. The Charcot (Neuropathic) Joint

The neuropathic foot is a major clinical problem, especially among diabetics (Palestro and Torres, 1997). Evaluation of the neuropathic foot and Charcot joint for infection is very challenging, and three-phase bone scintigraphy is only of modest value in detecting sepsis in this setting. Bone marrow generation and growth in the Charcot or neuropathic joint is quite prominent. Therefore, labeled WBCs readily accumulate at the site independent of whether infection is present.

Palestro et al. (1998) reported 17 patients with 20 Charcot joints, 4 of which had osteomyelitis (3) or adjacent soft tissue infection (1). All 4 were positive on WBC/marrow imaging, while the other 16 were negative. WBC/marrow imaging was superior to both three-phase bone scintigraphy and WBC/bone scintigraphy in this study.

13.4. Conclusion

Sulphur colloid is a safe, reliable, and valuable imaging and localization agent which facilitates management of the lymphatics in patients with malignant disease, and for bone marrow imaging in patients with hematological and musculoskeletal conditions. The major indication for TcSC at present is lymphatic mapping and sentinel node biopsy. These techniques have greatly improved the accuracy of nodal staging and spare node-negative cancer patients the expense and morbidity of formal staging lymphadenectomies. TcSC in LS and SLNB is changing practice patterns and will affect standards of care once these techniques have been widely adopted throughout the surgical workforce.

References

Abramova, L., Parekh, J., Irvin, W.P. Jr., et al., 2002. Sentinel node biopsy in vulvar and vaginal melanoma: presentation of six cases and a literature review. Ann Surg Oncol 9, 840–846.

Achong, D.M., Oates, E., 1995. Diffusely discordant In-111 WBC/Tc-99m SC bone marrow uptake. A possible chemotherapeutic effect. Clin Nucl Med 20, 599–600.

Albertini, J.J., Cruse, C.W., Rapaport, D., et al., 1996. Intraoperative radiolymphoscintigraphy improves sentinel node identification in patients with melanoma. Ann Surg 223, 217–224.

Albo, D., Wayne, J.D., Hunt, K.K., et al., 2001. Anaphylactic reactions to isosulfan blue during sentinel lymph node biopsy for breast cancer. Amer J Surg 182, 393–398.

Alex, J.C., Krag, D.N., 1993a. Gamma-probe guided localization of lymph nodes. Surg Oncol 2, 137–143.

Alex, J.C., Krag, D.N., 1996. The gamma probe-guided resection of radiolabeled primary lymph nodes. Surg Oncol Clin North Am 5, 33–41.

Alex, J.C., Krag, D.N., Harlow, S.P., Meijer, S., Loggie, B.W., Kuhn, J.A., Gadd, M., Weaver, D.L., 1998. Localization of regional lymph nodes in melanomas of the head and neck. Arch Otolaryngol Head Neck Surg 124, 135–140.

Alex, J.C., Sasaki, C.T., Krag, D.N., Wenig, B., Pyle, P.B., 2000. Sentinel lymph node radiolocalization in head and neck squamous cell carcinoma. Laryngoscope 110 (2 Pt 1), 198–203.

Alex, J.C., Weaver, D.L., Fairbanks, J.T., Rankin, B.S., Krag, D.N., 1993b. Gamma probe-guided lymph node localization in malignant melanoma. Surg Oncol 2, 303–308.

Altinyollar, H., Berberoglu, U., Celen, O., 2002. Lymphatic mapping and sentinel lymph node biopsy in squamous cell carcinoma of the lower lip. Eur J Surg Oncol 28, 72–74.

Arican, P., Íbis, E., Aras, G., Çam, R., Kücük, N.Ö., 2005. Identification of sentinel node in Stage I-II breast cancer with lymphoscintigraphy and surgical gamma probe: comparison of Tc-99m MIBI and Tc-99m sulphur colloid. Clin Nucl Med 30, 317–321.

Ariyan, S., Ariyan, C., Farber, L.R., Fischer, D.S., Flynn, S.D., Truini, C., 2004. Reliability of identification of 655 sentinel lymph nodes in 263 consecutive patients with malignant melanoma. J Am Coll Surg 198, 924–932.

Balch, C.M., Soong, S-J., Bartolucci, A., et al., 1996. Efficacy of an elective regional lymph node dissection of 1 to 4 mm thick melanomas for patients 60 years of age or younger. Ann Surg 224, 255–266.

Balch, C.M., Soong, S-J., Gershenwald, J.E., Thompson, J.F., Reintgen, D.S., Cascinelli, N., et al., 2001a. Prognostic factors analysis in 17,600 melanoma patients: validation of the American Joint Committee on Cancer melanoma staging system. J Clin Oncol 19, 3622–3634.

Balch, C.M., Soong, S-J., Ross, M.I., et al., 2000. Long-term results of a multi-institutional randomized trial comparing prognostic factors and surgical results for intermediate thickness melanomas (1.0 to 4.0 mm). Ann Surg Oncol 7, 87–97.

Balch, C.M., Soong, S-J., Smith, T., et al., 2001b. Long-term results of a prospective surgical trial comparing 2 cm vs. 4 cm excision margins for 740 patients with 1–4 mm melanomas. Ann Surg Oncol 8, 101–108.

Bauer, T.W., Spitz, F.R., Callans, L.S., et al., 2002. Subareolar and peritumoral injection identify similar sentinel nodes for breast cancer. Ann Surg Oncol 9, 169–172.

Bedrosian, I., Faries, M.B., Guerry DuPont, I.V., Elenitsas, R., Schuchter, L., Mick, R., et al., 2000. Incidence of sentinel node metastasis in patients with thin primary melanoma ($\leq$ 1 mm) with vertical growth phase. Ann Surg Oncol 7, 262–267.

Berlangieri, S.U., Peters, W.P., Coleman, R.E., 1993. Distribution of 99mTc-sulphur colloid during granulocytye colony-stimulating factor administration in autologous bone marrow transplantation. Nucl Med Commun 14, 896–901.

Berman, C.G., Norman, J., Cruse, C.W., et al., 1992. Lymphoscintigraphy in malignant melanoma. Ann Plast Surg 28, 29–32.

Bilchik, A.J., Nora, D.T., Sobin, L.H., et al., 2003. Effect of lymphatic mapping on the new tumor-node-metastasis classification for colorectal cancer. J Clin Oncol 21, 668–672.

Bleicher, R.J., Essner, R., Foshag, L.J., Wanek, L.A., Morton, D.L., 2003. Role of sentinel lymphadenectomy in tin invasive melanomas. J Clin Oncol 21, 1326–1331.

Boobol, S.K., Fey, J.V., Borgen, P.I., et al., 2001. Intradermal isotope injection: a highly accurate method of lymphatic mapping in breast carcinoma. Ann Surg Oncol 8, 20–24.

Borgstein, P.J., Meijer, S., Pijpers, R. 1997. Intradermal blue dye to identify sentinel lymph node in breast cancer. Lancet 349, 1668–1669.

Borgstein, P.J., Meijer, S., Pijpers, R.J., van Diest, P.J. 2000. Functional lymphatic anatomy for sentinel node biopsy in breast cancer. Echoes from the past and the periareolar blue method. Ann Surg 232, 81–89.

Bostick, P., Essner, R., Glass, E., Kelley, M., Sarantou, T., Foshag, L.J., et al., 1999. Comparison of blue dye and probe-assisted intraoperative lymphatic mapping in m melanoma to identify sentinel nodes in 100 lymphatic basins. Arch Surg 134, 43–49.

Bostick, P., Essner, R., Sarantou, T., Kelley, M., Glass, E., Foshag, L., Stern, S., Morton, D. 1997. Intraoperative lymphatic mapping for early-stage melanoma of the head and neck. Am J Surg 174, 536–43.

Carloss, H., Winslow, D., Kastan, L., Yam, L.T., 1977. Bone marrow necrosis. Diagnosis and assessment of extent of involvement by radioisotope studies. Arch Int Med 137, 863–866.

Cascinelli, N., Belli, F., Santinami, M., Fait, V., Testori, A., Ruka, W., et al., 2000. Sentinel lymph node biopsy in cutaneous melanoma: the WHO Melanoma Program experience. Ann Surg Oncol 7, 467–474.

Cascinelli, N., Morabito, A., Santinami, M., et al., 1998. Immediate or delayed dissection of regional nodes in patients with melanoma of the trunk: a randomized trial. Lancet 351, 793–796.

Chao, C., Wong, S., Edwards, M.J., et al., 2003. Sentinel lymph node biopsy for head and neck melanomas. Ann Surg Oncol 10, 21–26.

Cimmino, V.M., Brown, A.C., Szocik, J.F., et al., 2001. Allergic reactions to isosulfan blue during sentinel node biopsy: a common event. Surgery 130, 439–442.

Civantos, F.J., Moffat, F.L., Goodwin, W.J., 2006. Lymphatic mapping and sentinel lymphadenectomy for 106 head and neck lesions: contrasts between oral cavity and cutaneous malignancy. Laryngoscope 116(Suppl. 109), 1–15.

Civantos, F.J., Gomez, C., Duque, C., et al., 2003. Sentinel node biopsy in oral cavity cancer: correlation with PET scan and immunohistochemistry. Head Neck 25, 1–9.

Coates, G.G., Eisenberg, B., Dail, D.H., 1994. Tc-99m sulphur colloid demonstration of diffuse pulmonary interstitial extramedullary hematopoiesis in a patients with myelofibrosis. A case report and review of the literature. Clin Nucl Med 19, 1079–1084.

Cody, H.S. III, Fey, J., Akhurst, T., Fazzari, M., Mazumdar, M., Yeung, H., et al., 2001. Complementarity of blue dye and isotope in sentinel node localization for breast cancer: univariate and multivariate análisis of 966 procedures. Ann Surg Oncol 8, 13–19.

Copping, C., Dalgleish, S.M., Dudley, N.J., Griffiths, P.A., Harrington, M., Potter, R., et al., 1992. The role of ^{99m}Tc-HMPAO white cell imaging in suspected orthopaedic infection. Br J Radiol 65, 309–312.

Corsetti, R.L., Allen, H.M., Wanebo, H.J., 2003. Thin $\leq$ 1mm Level II and IV melanomas are higher risk lesions for regional failure and warrant sentinel lymph node biopsy. Ann Surg Oncol 7, 456–460.

Cox, C.E., Bass, S.S., McCann, C.R., et al., 2000. Lymphatic mapping and sentinel lymph node biopsy in patients with breast cancer. Ann Rev Med 51, 525–542.

Datz, F.L., Taylor, A. Jr., 1985. The clinical use of radionuclide bone marrow imaging. Semin Nucl Med 15, 239–259.

De Cicco, C., Cremonesi, M., Chinol, M., et al., 1997. Optimization of axillary lymphoscintigraphy to detect the sentinel node in breast cancer. Tumori 83, 539–541.

D'Eredita, G., Giardina, C., Guerreri, A.M., Berardi, T. 2006. A further validation of subareolar injection technique for breast sentinel lymph node biopsy. Ann Surg Oncol 13, 1–7.

Derossis, A.M., Fey, J., Yeung, H., Yeh, S.D.J., Heerdt, A.S., Petrek, J., et al., 2001. A trend analysis of the relative value of blue dye and isotope localization in 2,000 consecutive cases of sentinel node biopsy for breast cancer. J Am Coll Surg 193, 473–478.

Desai, A.G., Thakur, M.L., 1985. Radiopharmaceuticals for spleen and bone marrow studies. Semin Nucl Med 15, 229–238.

Doubrovsky, A., De Wilt, J.H., Scolyer, R.A., McCarthy, W.H., Thompson, J.F., 2004. Sentinel node biopsy is more accurate than elective lymph node dissection in determining the regional node status of patients with cutaneous melanoma. Ann Surg Oncol 11, S122–S123.

Elster, A.D., Theros, E.G., Key, L.L., Stanton, C. 1992. Autosomal recessive osteopetrosis: bone marrow imaging. Radiology 182, 507–514.

Esper, I.E., Blondet, C., Moullart, V., et al., 2004. The usefulness of ^{99m}Tc sulphur colloid bone marrow scintigraphy combined with ^{111}In leucocyte scintigraphy in prosthetic joint infection. Nucl Med Commun 25, 171–175.

Essner, R., Bostick, P.J., Glass, E.C., Foshag, L.J., Haigh, P.I., Wang, H-J., Morton, D.L., 2000. Standardized probe-directed sentinel node dissection in melanoma. Surgery 127, 26–31.

Essner, R., Conforti, A., Kelley, M.C., et al., 1999. Efficacy of lymphatic mapping, sentinel lymphadenectomy and selective complete node dissection as a therapeutic procedure for early-stage melanoma. Ann Surg Oncol 6, 442–449.

Evans, H.L., Krag, D.N., Teates, C.D., Patterson, J.W., Meijer, S., Harlow, S.P., et al., 2003. Lymphoscintigraphy and sentinel node biopsy accurately stage melanoma in patients presenting after wide local excision. Ann Surg Oncol 10, 416–425.

Faries, M.B., Bleicher, R., Ye, X., Essner, R., Morton, D.L., 2004. Lymphatic mapping and sentinel lymphadenectomy for primary and metastatic pulmonary malignant neoplasms. Arch Surg 139, 870–877.

Fee, J.H., Robinson, D.S., Sample, W.F., et al., 1978. The determination of lymph shed by colloidal gold scanning in patients with malignant melanoma: a preliminary study. Surgery 84, 626–632.

Frier, M., Griffiths, P., Ramsey, A., 1981. The physical and chemical characteristics of sulphur colloids. Eur J Nucl Med 6, 255–260.

Galimberti, V., Zurrida, S., Zucali, P., Luini, A., 1998. Can sentinel node biopsy avoid axillary dissection in clinically node-negative breast cancer patients? The Breast 7, 8–10.

Gershenwald, J.E., Colome, M.I., Lee, J.E., Mansfield, P.F., Tseng, C-h., Lee, J.J., Balch, C.M., Ross, M.I., 1998b. Patterns of recurrence following a negative sentinel lymph node biopsy in 243 patients with Stage I or II melanoma. J Clin Oncol 16, 2253–2260.

Gershenwald, J.E., Mansfield, P.F., Lee, J.E., Ross, M.I., 2000. Role for lymphatic mapping and sentinel lymph node biopsy in patients with thick ($\geq$ 4 mm) primary melanoma. Ann Surg Oncol 7, 160–165.

Gershenwald, J.E., Tseng, C-h., Thompson, W., Mansfield, P.E., Lee, J.E., Bouvet, M., Lee, J.J., Ross, M.I., 1998a. Improved sentinel lymph node localization in patients with primary melanoma with the use of radiolabeled colloid. Surgery 124, 303–310.

Glass, L.F., Messina, J.L., Cruse, W., et al., 1996. The use of intraoperative radiolymphoscintigraphy for sentinel node biopsy in patients with malignant melanoma. Dermatol Surg 22, 715–720.

Godellas, C.V., Berman, C.G., Lyman, G., et al., 1995. The identification and mapping of melanoma regional nodal metastases: minimally invasive surgery for the diagnosis of nodal metastases. Amer Surg 61, 97–101.

Gulec, S.A., Moffat, F.L., Carroll, R.G., 1997. The expanding clinical role for intraoperative gamma probes. In: Freeman, L.M. (Ed.). Nuclear Medicine Annual 1997. Lippincott-Raven, Philadelphia, pp. 209–237.

Gulec, S.A., Moffat, F.L., Carroll, R.G., et al., 1998. Sentinel lymph node localization in early breast cancer. J Nucl Med 39, 1388–1393.

Harlow, S.P., Krag, D.N., Julian, T.B., et al., 2005. Prerandomization surgical training for the National Surgical Adjuvant Breast and Bowel Project (NSABP) B-32 Trial. Ann Surg 241, 48–54.

Henry, R.E., Fletcher, J.W., George, E.A., Donati, R.M., 1975. Bone marrow ^{99m} Tc sulphur colloid distribution and marrow cellularity. Amer J Med Sci 270, 419–425.

Higashi, H., Natsugoe, S., Uenosono, Y., Ehi, K., Arigami, T., Nakabeppu, Y., Nakajo, M., Aikuo, T. 2004. Particle size of tin and phytate colloid in sentinel node identification. J Surg Res 121, 1–4.

Hill, A.D.K., Tran, K.N., Akhurst, T., et al., 1999. Lessons learned from 500 cases of lymphatic mapping for breast cancer. Ann Surg 229, 5258–5535.

Hirsch, J.I., Tirado, J., Cho, S.R., et al., 1982. Use of isosulfan blue for identification of lymphatic vessels: experimental and clinical evaluation. AJR Amer J Roentgenol 139, 1061–1064.

Hoppin, E.C., Lewis, J.P., DeNardo, S.J., 1979. Bone marrow scintigraphy in the evaluation of acute nonlymphocytic leukemia. Clin Nucl Med 4, 296–301.

Hung, J.C., Wiseman, G.A., Wahner, H.W., Mullan, B.P., Taggart, T.R., Dunn, W.L., 1995. Filtered technetium 99m sulfur colloid evaluated for lymphoscintigraphy. J Nucl Med 36, 1895–1901.

Jacobs, I.A., Chang, C.K., DasGupta, T.K., Salti, G.I., 2003. Role of sentinel lymph node biopsy in patients with thin (<1 mm) primary melanoma. Ann Surg Oncol 10, 558–561.

Jansen, L., Koops, H.S., Nieweg, O.E., Doting, M.H.E., Kapteijn, B.A.E., Balm, A.J.M., Vermey, A., et al., 2000. Sentinel node biopsy for melanoma in the head and neck region. Head Neck 22, 27–33.

Kahn, C.E., Jr., Ryan, J.W., Hatfield, M.K., Martin, W.B., 1988. Combined bone marrow and gallium imaging. Differentiation of osteomyelitis and infarction in sickle hemoglobinopathy. Clin Nucl Med 13, 443–449.

Kirkwood, J.M., Strawderman, M.H., Ernsthoff, M.S., et al., 1996. Interferon α_{2b} adjuvant therapy of high-risk resected cutaneous melanoma: the Eastern Cooperative Oncology Group Trial 1684. J Clin Oncol 14, 7–17.

Klimberg, V.S., Rubio, I.T., Henry, R., et al., 1999. Subareolar versus peritumoral injection for location of the sentinel lymph node. Ann Surg 229, 860–865.

Kloiber, R., Damtew, B., Rosenthall, L., 1981. A crossover study comparing the effect of particle size on the distribution of radiocolloid in patients. Clin Nucl Med 6, 204–206.

Kniseley, R.M., Andrews, G.A., Tanida, R., Edwards, C.L., Kyker, G.C., 1966. Delineation of active marrow by wholebody scanning with radioactive colloids. J Nucl Med 575, 582.

Krag, D.N., Weaver, D.L., Alex, J.C., Fairbank, J.T., 1993. Surgical resection and radiolocalization of the sentinel node in breast cancer using a gamma probe. Surg Oncol 2, 335–340.

Krag, D.N., 1998a. Minimal access surgery for staging regional lymph nodes: the sentinel node concept. Curr Probl Surg 35, 951–1018.

Krag, D.N., 1999. Current status of sentinel lymph node surgery for breast cancer. J Natl Cancer Inst 91, 302–304.

Krag, D.N., Ashikaga, T., Harlow, S.P., Weaver, D.L., 1998b. Development of sentinel node targeting technique in breast cancer patients. Breast J 4, 67–74.

Krag, D.N., Meijer, S.J., Weaver, D.L., Loggie, B.W., Harlow, S.P., Tanabe, K.K., et al., 1995. Minimal access surgery for staging of malignant melanoma. Arch Surg 130, 654–658.

Krag, D.N., Harlow, S.P., Weaver, D.L., Ashikaga, T. 1998d. Technique of selected resection of radiolabeled lymph nodes in breast cancer patients. Semin Breast Dis 1, 111–116.

Krag, D.N., Weaver, D., Ashikaga, T., et al., 1998c. The sentinel node in breast cancer. A multicenter validation study. New Engl J Med 339, 941–946.

Krznaric, E., De Roo, M., Verbruggen, A., Stuyck, J., Mortelmans, L. 1996. Chronic osteomyelitis: diagnosis with technetium-99m-d,l-hexamethylpropylene amine oxime labeled leukocytes. Eur J Nucl Med 23, 792–797.

Leak, L.V., 1970. Electron microscopic observations on lymphatic capillaries and the structural components of the connective tissue-lymph interface. Microvasc Res 2, 361–391.

Leong, S.P.L., Steinmetz, I., Habib, F., et al., 1997. Optimal selective sentinel lymph node dissection in primary malignant melanoma. Arch Surg 132, 666–673.

Leong, S.P.L., Donegan, E., Heffernan, W., et al., 2000. Adverse reactions to isosulfan blue during selective sentinel lymph node dissection in melanoma. Ann Surg Oncol 7, 361–366.

Lin, K.M., Patel, T.H., Ray, A., et al., 2004. Intradermal radioisotope is superior to peritumoral blue dye or radioisotope in identifying breast cancer sentinel nodes. J Am Coll Surg 199, 561–566.

Lin, Y.S., Tzeng, C.C., Huang, K.F., Kang, C.Y., Chia, C.C., Hsieh, J.F., 2005. Sentinel node detection with radiocolloid lymphatic mapping in early invasive cervical cancer. Int J Gynecol Cancer 15, 273–277.

Linehan, D.C., Hill, A.D.K., Tran, K.N., et al., 1999a. Sentinel lymph node localization in breast cancer: unfiltered isotope is superior to filtered. J Am Coll Surg 188, 377–381.

Linehan, D.C., Hill, A.D.K., Akhurst, T., et al., 1999b. Intradermal radiocolloid and intraparenchymal blue dye injection optimize sentinel node identification in breast cancer patients. Ann Surg Oncol 6, 450–454.

Lorberboym, M., Pastores, G.M., Kim, C.K., Hermann, G., Glajchen, N., Machac, J., 1997. J Nucl Med 38, 890–895.

Love, C., Marwin, S.E., Tomas, M.B., et al., 2004. Diagnosing infection in the failed joint replacement: a comparison of coincidence detection ^{18}F-FDG and ^{111}In-labeled leukocyte/^{99m}Tc-sulphur colloid marrow imaging. J Nucl Med 45, 1864–1871.

Luk, S.C., Nopajaroonsri, C., Simon, G.T., 1982. The architecture of the normal lymph node and hemolymph node. Amer J Anat 164, 275–309.

McCarter, M.D., Yeung, H., Yeh, S., Fey, J., Borgen, P.I., Cody, H.S. III., 2001. Localization of the sentinel node in breast cancer: identical results with same-day and day-before isotope injection. Ann Surg Oncol 8, 682–686.

McMasters, K.M., Chao, C., Wong, S.L., et al., 2002. Interval sentinel lymph nodes in melanoma. Arch Surg 137, 543–549.

McMasters, K.M., Wong, S.L., Martin, R.C.G. II, et al., 2001. Dermal injection of radioactive colloid is superior to peritumoral injection for breast cancer sentinel lymph node biopsy: results of a multi-institutional study. Ann Surg 233, 676–687.

Medina-Franco, H., Beenken, S.W., Heslin, M.J., Urist, M.M., 2001. Sentinel node biopsy for cutaneous melanoma in the head and neck. Ann Surg Oncol 8, 716–719.

Moffat, F.L., 2005. Lymph node staging surgery and breast cancer: potholes in the fast lane from more to less. J Surg Oncol 89, 53–60.

Moffat, F.L., Ashikaga, T., Krag, D.N., 2000. Sentinel node biopsy for breast cancer: showtime or dress rehearsal? In Vivo 14, 255–264.

Moffat, F.L., Gulec, S.A., Sittler, S.Y., et al., 1999. Unfiltered sulphur colloid and sentinel node biopsy for breast cancer: technical and kinetic considerations. Ann Surg Oncol 6, 746–755.

Moffat, F.L., Krag, D.N., 2003. Should surgeons abandon routine axillary dissection for sentinel node biopsy in early breast cancer? In: Nabholtz, J-M., Tonkin, K., Reese, D.M., Aapro, M.S., Buzdar, A.U. (Eds.). Breast Cancer Management. Application of Clinical and Translational Evidence to Patient Care (2nd Ed.). Lippincott Williams & Wilkins, Philadelphia, pp. 59–66.

Montgomery, L.L., Thorne, A.C., Van Zee, K.J., et al., 2002. Isosulfan blue dye reactions during sentinel lymph node mapping for breast cancer. Anesth Analg 95, 385–388.

Morton, D.L., Cochran, A.J., Thompson, J.F., Elashoff, R., Essner, R., Glass, E.C., et al., 2005. Sentinel node biopsy for early-stage melanoma. Accuracy and morbidity in MSLT-I, an international multicenter trial. Ann Surg 242, 302–313.

Morton, D.L., Thompson, J.F., Essner, R., et al., 1999. Validation of the accuracy of intraoperative lymphatic mapping and sentinel lymphadenectomy for early-stage melanoma. A multicenter trial. Ann Surg 230, 453–465.

Morton, D.L., Wen, D-R., Foshag, L.J., Essner, R., Cochran, A. 1993. Intraoperative lymphatic mapping and selective cervical lymphadenectomy for early-stage melanomas of the head and neck. J Clin Oncol 11, 1751–1756.

Morton, D.L., Wen, D-R., Wong, J.H., et al., 1992. Technical details of intraoperative lymphatic mapping for early stage melanoma. Arch Surg 27, 392–399.

Nopajaroonsri, C., Simon, G.T., 1971a. Phagocytosis of colloidal carbon in a lymph node. Amer J Pathol 65, 25–42.

Nopajaroonsri, C., Luk, S.C., Simon, G.T., 1971b. Ultrastructure of the normal lymph node. Amer J Pathol 65, 1–24.

Norman, J., Cruse, C.W., Espinosa, C., et al., 1991. Redefinition of cutaneous lymphatic drainage with the use of lymphoscintigraphy for malignant melanoma. Amer J Surg 162, 432–437.

Norman, J., Wells, K., Kearney, R., Cruse, C.W., Berman, C., Reintgen, D. 1993. Identification of lymphatic drainage basins in patients with cutaneous melanoma. Semin Surg Oncol 9, 224–227.

North, J.H. Jr., Spellman, J.E., 1996. Role of sentinel node biopsy in the management of malignant melanoma. Oncology 10, 1237–1242.

O'Brien, C.J., Uren, R.F., Thompson, J.F., 1995. Prediction of potential metastatic sites in cutaneous head and neck melanoma using lymphoscintigraphy. Amer J Surg 170, 461–466.

Ollila, D.W., Foshag, L.J., Essner, R., Stern, S.L., Morton, D.L., 1999. Parotid region lymphatic mapping and sentinel lymphadenectomy for cutaneous melanoma. Ann Surg Oncol 6, 150–154.

Palestro, C.J., Kim, C.K., Swyer, A.J., et al., 1990. Total hip arthroplasty: periprosthetic indium-111-labeled leukocyte activity and complementary technetium-99 m-sulphur colloid imaging in suspected infection. J Nucl Med 31, 1950–1955.

Palestro, C.J., Mehta, H.H., Patel, M., et al., 1998. Marrow versus infection in the Charcot joint: indium-111 leukocyte and technetium-99 m sulphur colloid scintigraphy. J Nucl Med 39, 346–350.

Palestro, C.J., Love, C., Tronco, G.G., Tomas, M.B., Rini, J.N., 2006. Combined labeled leukocyte and technetium 99m sulphur colloid bone marrow imaging for diagnosing musculoskeletal infection. RadioGraphics 26, 859–870.

Palestro, C.J., Roumanas, P., Swyer, A.J., Kim, C.K., Goldsmith, S.J., 1992. Diagnosis of musculoskeletal infection using combined In-111 labeled leukocyte and Tc-99 m marrow imaging. Clin Nucl Med 17, 269–273.

Palestro, C.J., Swyer, A.J., Kim, C.K., Goldsmith, S.J., 1991. Infected knee prosthesis: diagnosis with In-111 leukocyte, Tc-99m sulphur colloid and Tc-99 m MDP imaging. Radiology 179, 645–648.

Palestro, C.J., Torres, M.A., 1997. Radionuclide imaging in orthopedic infections. Semin Nucl Med 27, 334–345.

Patten, L.C., Berger, D.H., Rodriguez-Bigas, M., et al., 2004. A prospective evaluation of radiocolloid and immunohistochemical staining in colon carcinoma lymphatic mapping. Cancer 100, 2104–2109.

Payoux, P., Dekeister, C., Lopez, R., Lauwers, F., Esquerré, J.P., Paoli, J.R., 2005. Effectiveness of lymphoscintigraphic sentinel node detection for cervical staging of patients with squamous cell carcinoma of the head and neck. J Oral Maxillofac Surg 63, 1091–1095.

Pijpers, R., Collet, G.J., Meijer, S., Hoekstra, O.S., 1995. The impact of dynamic lymphoscintigraphy and gamma probe guidance on sentinel node biopsy in melanoma. Eur J Nucl Med 22, 1238–1241.

Ranieri, J.M., Wagner, J.D., Wenck, S.D., Calley, C., Coleman, J.J. III, 2004. The prognostic relevance of sentinel lymph node biopsy in thin melanoma patients. Ann Surg Oncol 11, S122.

Raut, C.P., Daley, D., Hunt, K.K., et al., 2004. Anaphylactoid reactions to isosulfan blue dye during breast cancer lymphatic mapping in patients given preoperative prophylaxis. J Clin Oncol 22, 567–568.

Reintgen, D., Albertini, J., Miliotes, G., et al., 1995. The accurate staging and modern day treatment of malignant melanoma. Cancer Res Ther Control 4, 183–197.

Reintgen, D.S., Cruse, C.W., Wells, K., et al., 1994. The orderly progression of melanoma nodal metastases. Ann Surg 220, 759–767.

Roeske, J.C., Lujan, A., Reba, R.C., Penney, B.C., Yamada, S.D., Mundt, A.J., 2005. Incorporation of SPECT bone marrow imaging into intensity modulated whole-pelvic radiation therapy treatment planning for gynecologic malignancies. Radiother Oncol 77, 11–17.

Ross, G.L., Shoaib, T., Soutar, D.S., et al., 2002. The use of sentinel node biopsy to upstage the clinically N0 neck in head and neck cancer. Arch Otolaryngol Head Neck Surg 128, 1287–1290.

Ross, M.I., Reintgen, D., Balch, C.M., 1993. Selective lymphadenectomy: emerging role for lymphatic mapping and sentinel node biopsy in the management of early stage melanoma. Sem Surg Oncol 9, 219–223.

Rousseau, D.L., Ross, M.I., Johnson, M.M., Prieto, V.G., Lee, J.E., Mansfield, P.F., Gershenwald, J.E., 2003. Revised American Joint Committee on Cancer Staging criteria accurately predict sentinel lymph node positivity in clinically node-negative melanoma patients. Ann Surg Oncol 10, 569–574.

Sage, H., Gozun, B.V., 1958. Lymphatic scintigrams: method for studying function of lymphatics and lymph nodes. Cancer 11, 200–203.

Saha, S., Dan, A.G., Berman, B., et al., 2004a. Lymphazurin 1 % versus ^{99m}Tc sulphur colloid for lymphatic mapping in colorectal tumors: a comparative analysis. Ann Surg Oncol 11, 21–26.

Saha, S., Dan, A.G., Beutler, T., et al. 2004b. Sentinel lymph node mapping technique in colon cancer. Semin Oncol 31, 374–381.

Sainte-Marie, G., Peng, F.S., 1986. Diffusion of a lymph-carried antigen in the fiber network of the lymph node of the rat. Cell Tissue Res 245, 481–486.

Sainte-Marie, G., Peng, F.S., Belisle, C., 1982. Overall architecture and pattern of lymph flow in the rat lymph node. Amer J Anat 275–279.

Schauwecker, D.S., 1992. The scintigraphic diagnosis of osteomyelitis. Amer J Roentgenol 158, 9–18.

Seabold, J.E., Nepola, J.V., 1999. Imaging techniques for evaluation of postoperative orthopedic infections. Q J Nucl Med 43, 21–28.

Seabold, J.E., Nepola, J.V., Marsh, J.L., et al., 1991. Postoperative bone marrow alterations: potential pitfalls in the diagnosis of osteomyelitis with In-111-labeled leukocyte scintigraphy. Radiology 180, 741–747.

Shen, P., Glass, E.C., DiFronzo, L.A., Giuliano, A.E., 2001. Dermal versus intraparenchymal lymphoscintigraphy of the breast. Ann Surg Oncol 8, 241–248.

Siddiqui, A.R., Oseas, R.S., Wellman, H.N., Doerr, D.R., Baehner, R.L., 1979. Evaluation of bone-marrow scanning with technetium-99m sulphur colloid in pediatric oncology. J Nucl Med 20, 379–386.

Siegel, R.S., Hakimian, D., Spies, W., et al., 1999. Technetium-99m sulphur colloid scanning and correlative magnetic resonance imaging in patients with hairy cell leukemia and hypocellular bone marrow biopsies after 2-chlorodeoxyadenosine. Leuk Lymphoma 35, 171–177.

Silva, N. Jr., Anselmi, C.E., Anselmi, O.E., et al., 2005. Use of the gamma probe in sentinel lymph node biopsy in patients with prostate cancer. Nucl Med Commun 26, 1081–1086.

Sim, F.H., Taylor, W.F., Pritchard, D.J., et al., 1986. Lymphadenectomy in the management of stage I malignant melanoma: a prospective randomized study. Mayo Clin Proc 61, 697–705.

Slingluff, C.L., Stidham, K.R., Ricci, W.M., Stanley, W.E., Seigler, H.F., 1994. Surgical management of regional nodes in patients with melanoma. Experience with 4682 patients. Ann Surg 219, 120–130.

Solorzano, C.C., Ross, M.I., Delpassand, E., Mirza, N., Akins, J.S., Kuerer, H.M., et al., 2001. Utility of breast sentinel lymph node biopsy using day-before-surgery injection of high-dose ^{99m}Tc-labeled sulphur colloid. Ann Surg Oncol 8, 821–827.

Spiegel, M., Vesti, B., Shore, A., Franzeck, U.K., Becker, F., Bollinger, A., 1992. Pressure of lymphatic capillaries in human skin. Amer J Physiol 262, 1208–1210.

Statius Muller, M.G., Borgstein, P.J., Pijpers, R., van Leeuwen, P.A.M., van Diest, P.J., Gupta, A., Meijer, S., 2000. Reliability of the sentinel node procedure in melanoma patients: analysis of failures after long-term follow-up. Ann Surg Oncol 7, 461–468.

Stojadinovic, A., Allen, P.J., Protic, M., Potter, J.F., Shriver, C.D., Nelson, J.M., Peoples, G.E., 2005. Colon sentinel lymph node mapping: practical surgical applications. J Am Coll Surg 201, 297–313.

Strand, S.E., Persson, B.R.R., 1979. Quantitative lymphoscintigraphy: basic concepts for optimal uptake of radiocolloids in the parasternal lymph nodes of rabbits. J Nucl Med 20, 1038–1046.

Sumner, W.E. III, Ross, M.I., Mansfield, P.F., Lee, J.E., Prieto, V.G., Schacherer, C.W., Gershenwald, J.E., 2002. Implications of lymphatic drainage to unusual sentinel lymph node sites in patients with primary cutaneous melanoma. Cancer 95, 354–360.

Tafra, L., Lannin, D.R., Swanson, M.S., et al., 2001. Multicenter trial of sentinel node biopsy for breast cancer using both technetium sulphur colloid and isosulfan blue dye. Ann Surg 233, 51–59.

Tanis, P.J., van Rijk, M.C., Nieweg, O.E., 2005. The posterior lymphatic network of the breast rediscovered. J Surg Oncol 91, 195–198.

Tanner, M.L., Hoh, C.K., Bashey, A., et al., 2003. FLAG chemotherapy followed by allogeneic stem cell transplant using nonmyeloablative conditioning induces regression of myelolfibrosis with myeloid metaplasia. Bone Marrow Transplantation 32, 581–585.

Teal, C.B., Slocum, J.P., Akin, E.A., 2005. Evaluation of the benefit is using blue dye in addition to radioisotope for sentinel lymph node biopsy in patients with breast cancer. Breast J 11, 391–393.

Thelmo, M.C., Morita, E.T., Treseler, P.A., Nguyen, L.H., Allen, R.E. Jr., Sagebiel, R.W., Kashani-Sabet, M., Leong, S.P.L., 2001. Micrometastasis to in-transit lymph ndoes from extremity and truncal malignant melanoma. Ann Surg Oncol 8, 444–448.

Thompson, J.F., McCarthy, W.H., Bosch, C.M.J., 1995. Sentinel node status as an indicator of the presence of metastatic melanoma in regional lymph nodes. Melanoma Res 5, 255–260.

Tonakie, A., Sondak, V., Yahanda, A., Wahl, R.L., 2000. Reproducibility of lymphoscintigraphic drainage patterns in sequential ^{99m}Tc-human serum albumin and ^{99m}Tc-sulphur colloid studies: implications for sentinel node identification in melanoma. Surgery 127, 955–962.

Uren, R.F., Howman-Giles, R., Thompson, J.F., 1994. Lymphoscintigraphy to identify sentinel lymph nodes in patients with melanoma. Melanoma Res 4, 395–399.

Uren, R.F., Howman-Giles, R.B., Shaw, H.M., Thompson, J.F., McCarthy, W.H., 1993. Lymphoscintigraphy in high-risk melanoma of the trunk: predicting draining node groups, defining lymphatic channels and locating the sentinel node. J Nucl Med 34, 1435–1440.

Vacquerano, J., Kraybill, W., Cheney, R., Kane, J.M., 2004. Indications for sentinel lymph node biopsy in thin melanomas. Ann Surg Oncol 11, S77–S78.

Vera, D.R., Wallace, A.M., Hoh, C.K., Mattrey, R.F., 2001. A synthetic macromolecule for sentinel node detection: ^{99m}Tc-DTPA-Mannosyl-Dextran. J Nucl Med 42, 951–959.

Veronesi, U., Adamus, J., Bandiera, D.C., et al., 1977. Inefficacy of immediate node dissection in stage I melanoma of the limb. New Engl J Med 297, 627–630.

Veronesi, U., Adamus, J., Bandiera, D.C., et al., 1982. Delayed regional lymph node dissection in stage I melanoma of the skin of the lower extremities. Cancer 49, 2420–2430.

Veronesi, U., Paganelli, G., Viale, G., et al., 1999. Sentinel lymph node biopsy and axillary dissection in breast cancer: results in a large series. J Natl Cancer Inst 91, 368–373.

Veronesi, U., Paganelli, G., Galimberti, V., et al., 1997. Sentinel node biopsy to avoid axillary dissection in breast cancer with clinically negative lymph nodes. Lancet 349, 1864–1867.

Wells, K.E., Cruse, C.W., Daniels, S., Berman, C., Norman, J., Reintgen, D.S., 1994. The use of lymphoscintigraphy in melanoma of the head and neck. Plast Reconstr Surg 93, 757–761.

Wells, K.E., Rapaport, D.P., Cruse, C.W., Payne, W., Albertini, J., Berman, C., Lyman, G.H., Reintgen, D.S., 1997. Sentinel lymph node biopsy in melanoma of the head and neck. Plast Reconstr Surg 100, 591–594.

White, D.C., Schuler, F.R., Pruitt, S.K., Culhane, D.K., Seigler, H.F., Coleman, R.E., Tyler, D., 1999. Timing of sentinel lymph node mapping after lymphoscintigraphy. Surgery 126, 156–161.

Wong, S.L., Edwards, M.J., Chao, C., et al., 2001. Sentinel lymph node biopsy for breast cancer: impact of the number of sentinel nodes removed on the false-negative rate. J Am Coll Surg 192, 684–691.

Woznick, A., Franco, M., Bendick, P., Benitez, P.R., 2006. Sentinel lymph node dissection for breast cancer: how many nodes are enough and which technique is optimal. Amer J Surg 191, 330–333.

Wrightson, W.R., Wong, S.L., Edwards, M.J., Chao, C., Reintgen, D.S., Ross, M.I., et al., 2003. Complications associated with sentinel lymph node biopsy for melanoma. Ann Surg Oncol 10, 676–680.

Zapas, J.L., Coley, H.C., Beam, S.L., Brown, S.D., Jablonski, K.A., Elias, E.G., 2003. The risk of regional lymph node metastases in patients with melanoma less than 1.0 mm thick: recommendations for sentinel lymph node biopsy. J Am Coll Surg 197, 403–407.

Zogakis, T.G., Wetherille, R.E., Christensen, R.D., et al., 2005. Intraoperative subareolar injection of 99mTc-labeled sulphur colloid results in consistent sentinel lymph node identification. Ann Surg Oncol 12, 167–172.

Zulfikaroglu, B., Koc, M., Kucuk, O., Ozalp, N., Aras, G., 2005. Intraoperative lymphatic mapping and sentinel lymph node biopsy using radioactive tracer in gastric cancer. Surgery 138, 899–904.

Part 3
Acoustically Reflective Nanoparticles: Application in Ultrasound Imaging

14
Use of Ultrasound Bubbles in Lymph Node Imaging

Erik R. Wisner and Susannah H. Bloch

Abstract: Contrast-enhanced lymph node sonography with intravascularly administered contrast microbubbles has been used to characterize node status in cancer patients. In addition to the morphologic criteria of size, shape, internal architecture, and node margins used in non-contrast studies, contrast-assisted harmonic, color Doppler, and power Doppler imaging improve recognition of metastatic deposits as well as revealing altered intranodal vascular flow. More recently, contrast lymph node imaging following interstitial contrast media delivery has been proposed for sentinel lymph node localization. Sentinel node techniques reveal the location of high-risk regional lymph nodes that can then be biopsied for definitive determination of node status resulting in more accurate staging of regional disease. The interstitial delivery of ultrasound contrast medium poses a unique set of conflicting challenges in terms of optimizing contrast formulations for initial peripheral lymphatic uptake and maximal signal response in targeted nodes. Contrast microbubble size and shell characteristics are discussed in this review as are destructive and non-destructive imaging modes for optimal contrast response in sentinel node imaging. Recent preclinical contrast-assisted sentinel node studies are reviewed.

Keywords: Ultrasound, contrast media, microbubbles, lymphatic system, sentinel node, neoplasm staging.

14.1. Introduction

Contrast-assisted diagnostic imaging has been used for many years for detection and characterization of lymph nodes status in cancer patients. Direct lymphography involves cannulation of peripheral lymphatic vessels followed by injection of iodinated contrast media for contrast enhancement of lymph nodes receiving lymphatic drainage from these vessels (Koehler, 1976). Radiographic images reveal the distribution and internal architectural detail of regional lymph nodes and lymphatic vessels (Cox et al., 1966; Vaughan and Varey, 1964). Because the technique is cumbersome and associated with patient morbidity, it has been supplanted by CT, MR (Brown et al., 2003; Kim et al., 2004; King et al., 2000; Suzuma et al., 2002; van den Brekel, 2000), ultrasound (Alvarez et al., 2006; Hsu et al., 2006; Nori et al., 2005; Topal et al., 2005; van Rijk et al., 2006), and PET (Kumar et al., 2006; Quon and Gambhir, 2005). In addition, lymph node targeted contrast-enhanced MR using systemically delivered iron oxide particles has also recently been used clinically for this purpose (Harisinghani et al., 2006; Heesakkers et al., 2006; Will et al., 2006).

These cross-sectional imaging modalities are now considered standard of care for clinical cancer staging with nodal linear dimensions, shape, internal architectural characteristics, extent and pattern of contrast enhancement, and extracapsular spread used for characterization of node status. Although these non-invasive approaches are important for staging, treatment planning, and monitoring disease, their accuracy is limited, particularly when nodes are within normal limits for size and metastatic deposits are small. Another approach, sentinel node localization, has recently become an increasingly important adjunct for cancer staging and intraoperative decision-making.

14.2. Lymphatic Anatomy and Physiology

The tissue interstitium is composed of a gel-like matrix of collagen fibers and proteoglycan filaments within which the majority of interstitial fluid is trapped. Because of the gel-like character of the interstitial fluid space, movement of interstitially injected particles is restricted resulting in a relatively small dispersion volume (Guyton, 1991). Also within the interstitial matrix are interstitial macrophages that can phagocytize injected particles and either retain them at the injection site or transport them intracellularly to regional lymph nodes by traversing the lymphatic endothelium. Because interstitially injected ultrasound microbubbles reach draining lymph nodes within just a few minutes of injection, it is unclear to what degree intracellular transport plays a role in the initial distribution of nodal microbubbles.

The initial microlymphatics consist of an endothelial lining without a surrounding smooth muscle layer. Endothelial junctions are formed by loosely overlapping endothelial cells, and macroscopic openings form when the lymphatics are distended. Fenestrations up to a few microns in width can provide both uninhibited fluid flow as well as an avenue for microparticle uptake (Schmid-Schonbein, 1990).

Lymph nodes are surrounded by a thick fibrous capsule and the underlying parenchyma is divided into cortical and medullary regions. A lymph node hilus is present on the inferior surface of the node through which the efferent lymphatic and the nodal artery and vein enter or exit, respectively. Immediately deep to the capsule is a subcapsular sinus into which lymph fluid from the afferent lymphatic vessels drains. Lymph percolates through a complex cortical and medullary sinus system exiting through the efferent lymphatic. It is in the most peripheral regions of the sinus system that metastases lodge and grow. Contrast-enhanced lymph node ultrasound using intravenously administered contrast microbubbles relies on enhancement of the vascular tree and on non-uniform contrast distribution within the node for detection of metastatic infiltrates which appear as contrast filling voids.

Lymph node size, number, and distribution can vary widely from one individual to the next. In addition, there may be significant variability in lymphatic distribution between nodes and from draining tissues. It is this anatomical ambiguity, paired with the current lack of a definitive non-invasive diagnostic test for determining lymph node status, that has resulted in the need for sentinel node detection techniques.

14.3. The Sentinel Node Concept

A sentinel lymph node is the first node in a lymphatic draining sequence and is therefore at highest risk for metastasis from solid tumors with regional metastatic potential. As such, techniques have been established for intraoperative localization of sentinel nodes followed by excisional biopsy and intraoperative frozen-section review. When the sentinel node is deemed disease free, the remaining regional nodes are also considered negative for metastasis, and surgery is limited to excision of the primary tumor. When the sentinel node is positive for metastasis, other regional nodes are also considered to be positive, the cancer is up-staged leading to more aggressive regional lymphadenectomy in addition to excision of the primary neoplasm. It is important to recognize that sentinel localization techniques are not used to diagnose metastasis directly, rather they serve as a means to localize a node for definitive histologic diagnosis following selective excisional biopsy. When performed correctly, sentinel node biopsy has been shown to be highly accurate and can result in significantly less patient morbidity since surgical intervention is often less aggressive. The sentinel node technique is becoming a standard of care for melanoma (Berk et al., 2005; Chakera et al., 2004; Morton et al., 2005; Scoggins et al., 2005; Vaquerano et al., 2006) and breast cancer (Kelley et al., 2004; Kim et al., 2006; Krag, 2001; Krag et al., 2004; Maggard et al., 2005; Silberman et al., 2004; Thompson et al., 2005) and has also been used for management of pelvic (Di et al. Stefano et al., 2005; Doekhie et al., 2005; Gien et al., 2005; Saha et al., 2005), penile (Izawa et al., 2005; Perdona et al., 2005), gastrointestinal (Aikou et al., 2006; Bertagnolli et al., 2004; Cox et al., 2002; Saha et al., 2005), head and neck (Civantos et al., 2006; Ross et al., 2004; Rubello et al., 2006), and thoracic neoplasms (Liptay, 2004; Pulte et al., 2005; Rzyman et al., 2006).

Current approaches to sentinel node localization include some combination of preoperative lymphoscintigraphy, intraoperative peritumoral blue dye injection to stain local lymphatic vessels and first-order lymph nodes, and the use of intraoperative gamma probes to trace lymphatic migration of peritumorally injected particulate radiopharmaceuticals to sentinel nodes (Figure 14.1). The accuracy of current sentinel node localization techniques varies depending on the specific institutional protocol, the experience of the surgeon, and patient characteristics (Anan et al.,

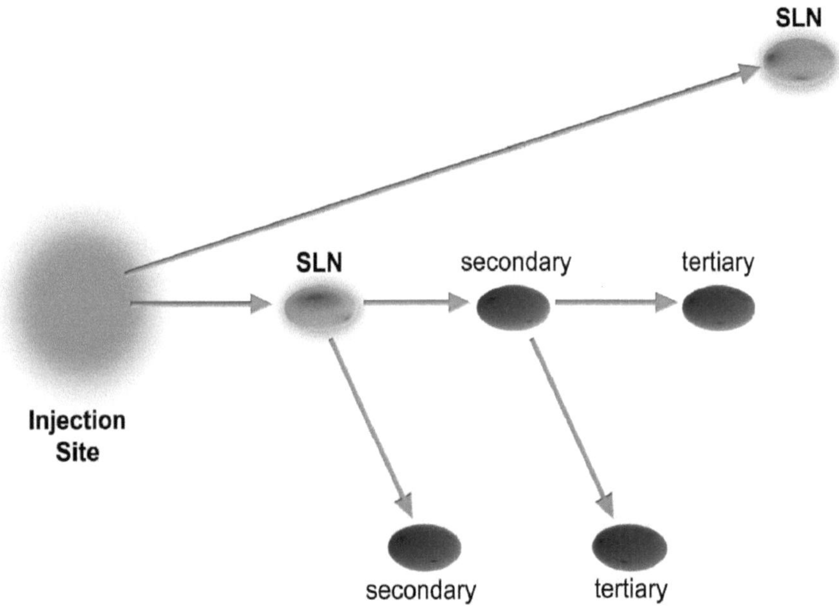

FIGURE 14.1. Diagrammatic representation of the sentinel node concept. A soluble or particulate tracer deposited at an injection site (or cells from a primary neoplasm) will migrate to the sentinel lymph nodes (SLNs) before continuing to the higher-order nodes in the chain.

2006; Kim et al., 2006; Posther et al., 2005; Radovanovic et al., 2004; Syme et al., 2005; Takei et al., 2006; Torrenga et al., 2004; Woznick et al., 2006). In a meta-analysis of sentinel node biopsy in early stage breast cancer, sentinel nodes were successfully mapped in 41%–100% of patients with a false negative rate ranging from 0% to 29% (average 7.3%) (Kim et al., 2006).

Recently, contrast-assisted ultrasound has been proposed both for characterization of lymph node status through sonographic assessment of nodal internal architecture following intravenous administration of ultrasound contrast microbubbles and for sentinel node localization following interstitial injection of ultrasound contrast media. These approaches show promise in improving accuracy of node status and sentinel node localization and could serve as viable adjuncts or alternatives to current techniques.

14.4. Ultrasound Contrast Agents Used in Lymphatic Imaging

14.4.1. General Considerations

Contrast agents used in intravascular studies of the lymph nodes are the same as those used in other intravascular applications. These are treated in other chapters of this book (see Chapters 16 and 17), so we will review them only briefly here.

Intravascular studies typically use microbubble-type contrast agents. These agents are typically 1–10 μm in diameter, and consist of a gaseous core covered by a stabilizing shell made of lipid, protein, or polymer. Many different microbubble agents are under development, and a few have achieved regulatory approval for clinical use (Becher and Burns, 2000; Blomley et al., 2001; Feinstein, 2004).

Fortuitously, the ultrasound frequencies used in clinical studies overlap the resonant frequencies of 1–10 μm microbubbles. Thus, microbubble-type agents create contrast in an ultrasound image not just because the gas inside them reflects ultrasound energy, but because they expand, collapse, and resonate in response to ultrasound pulses. This resonance effect can be exploited in several ways. Harmonic imaging modes detect the oscillations from resonating microbubbles at multiples or fractions of the transmitted ultrasound pulse frequency. Multi-pulse sequences, such as pulse-inversion mode, are designed to distinguish microbubble agents from tissue based on their characteristic echo signatures. High-power modes, such as power Doppler or "burst" modes, detect the strong signals from collapsing microbubbles, sensitively (though destructively) imaging the location of even small numbers of bubbles.

An agent may be chosen for a particular application based on its gas core, shell composition, size, or biological properties. In general, microbubbles containing perfluorocarbon gases generate a longer contrast effect because the higher molecular weight gases dissolve in blood more slowly than air or nitrogen. Lipid-shelled agents are more flexible than polymer- or protein-shelled agents, so the gas inside them is free to expand and contract with ultrasound waves without breaking the shell, making these agents more amenable to non-destructive imaging techniques. Microbubble agents with a narrow size distribution are in theory preferable because their resonant frequency is known, making customized contrast-specific imaging sequences possible; however, in practice, most agents have a wide size distribution. Finally, a significant area of current research is in targeted imaging with ultrasound contrast agents, where shell properties or biological ligands incorporated into the shell cause the agent to collect in an area of interest (e.g., in a tumor or lymph node) (Bloch et al., 2005; Klibanov, 2006).

14.4.2. Intra-Lymphatic Ultrasound Contrast Agents

The considerations for intra-lymphatic contrast agents are both similar and different from those used for intravascular studies. Size, desired imaging mode, and biological activity are each important additional areas of consideration.

14.4.2.1. Size

Particle size is a more important factor in intra-lymphatic studies than in intravascular studies. Lymphatic tracers used in other imaging modalities are usually at least an order of magnitude smaller than microbubbles although particles up to hundreds of nanometers in diameter have been shown to cross the lymphatic endothelium, the optimal inert particle diameter for lymphatic uptake is 10–50 nm (Bergqvist et al., 1983; Wolf, 1995).

Particles of less than a few nanometers in diameter are largely removed by blood capillaries at an interstitial injection site. Particles of up to a few tens of nanometers are absorbed rapidly into the lymph capillaries. Larger particles may remain trapped within the interstitium at the injection site for a much longer time (Bergqvist et al., 1983) and at least some stable particles may be phagocytized by mobile macrophages before entering the lymphatics (Ikomi et al., 1995). Optimal colloid particle size for scintigraphic agents has been reported to be between 1 and 10 nm (Strand and Persson, 1979) although a non-deformable particle size in the range of 10–1000 nm has been suggested as suitable for efficient transit across the interendothelial gaps (Wolf, 1995). While particle diameter is a major factor affecting the tendency to cross through the initial lymphatic capillary plexus, factors, such as shape, deformability, composition, charge, and other characteristics, may also influence uptake. Muscular activity, passive limb movement, and hyperthermia have also been shown to increase transport rate of interstitially injected nanoparticles (Elste et al., 1996; Ikomi et al., 1995; Wolf et al., 1994).

While the classic size criteria for optimal lymphatic endothelial transit may hold true for solid particles, ultrasound contrast microbubbles are highly deformable and may not be as limited. Although most microbubbles have a broad size distribution, the mean diameter of even the smallest formulations is many hundreds of nanometers in diameter; however, contrast formulations ranging in mean diameter from 0.5 to 4 microns empirically result in contrast enhancement of peripheral lymphatic vessels and regional lymph nodes following interstitial injection. One possible explanation is that deformability greatly increases the maximum particle size that can transit the peripheral lymphatic endothelium. A second possibility is that only the smallest subpopulation of microbubbles in a microbubble suspension is available to traverse the lymphatic endothelium. A third explanation is that interstitial injection of microbubble suspensions may transiently raise interstitial pressure establishing a gradient more conducive to lymphatic uptake. A fourth possibility is that interstitial injection causes trauma to peripheral lymphatics resulting in large fenestrations through which particles can transit.

At least two approaches have been taken to create submicron-sized intra-lymphatic ultrasound contrast agents. First, Omoto and Colleagues (2006, 2002) have found that a solution of 25% albumin injected into the skin near a tumor collects in the healthy sections of sentinel lymph nodes and can be detected with B-mode ultrasound. Second, Wisner et al. (2002, 2003), in cooperation with Point Biomedical, have investigated the use of bilayered albumin/polymer-shelled submicron contrast agents (Wisner et al., 2002, 2003). Microbubbles less than 1 μm in diameter are generally not stable in solution; however, dissolution of these hard-shelled agents is prevented by their thick shell, and they have been manufactured with mean sizes below 500 nm. Initial results indicated that smaller agents, with mean diameters less than or approximating 1 μm, might be more effective for sentinel lymph node imaging than larger formulations of the same type (Wisner et al., 2003). Subsequent investigation of these agents has revealed that they exhibit specific echo signatures which could be used for the development of an agent-specific imaging mode (Patel et al., 2002, 2004), although it is likely that echo response from these agents occurs only when the shell is disrupted, and that their response is related to both agent size and echo amplitude (Bloch et al., 2005) (Figure 14.2A). For these agents, there are two effective size thresholds for response: a lower limit, determined by the pressure amplitude of the acoustic pulse reaching the agent, and an upper limit, determined by the filtering properties of the lymphatic system. Therefore, the ideal agent for a particular application is the one containing the most bubbles within these two effective size thresholds (Figure 14.2B). It is probable that similar trade-offs exist for other types of agents; this is an area for future study.

14.4.2.2. Imaging Mode

A simplistic, but useful, distinction can be made between destructive and non-destructive contrast imaging modes:

Destructive imaging modes—such as power Doppler, color Doppler ("stimulated emission mode") (Blomley et al., 1999), and high-power B-mode or harmonic imaging—can be applied to a variety of microbubble contrast agents and are sensitive to even small numbers of bubbles. In addition, these modes are generally available on clinical ultrasound machines. However, these modes are destructive to contrast microbubbles: high-pressure ultrasound pulses disrupt the agents' shells, and the gas they contain dissolves into the body within milliseconds. In the extreme case, this means that the investigator has only one chance to scan a lymphatic vessel or node and detect contrast signal. Refill of lymphatic channels and lymph nodes is slow, so once the agents in a particular location have been imaged, the investigator must wait several minutes before the study can be repeated and any observations confirmed.

Non-destructive imaging modes, such has low-power harmonic grayscale modes, pulse-inversion mode, and other contrast-specific imaging modes, take advantage of the specific properties of the microbubble contrast agent under investigation. These modes have the distinct advantage that the contrast agents are imaged non-destructively—that is, one can scan the same area repeatedly to investigate lymphatic flow without losing signal strength.

However, in general, these modes are tailored to a specific agent and are less sensitive to individual agents than high-powered modes although their sensitivity is improving with continued research. They may also be less widely available to the general researcher or clinician.

In reality, no imaging mode is perfectly destructive or non-destructive to a specific agent, but it is useful to think about how these modes can be used together to optimize a specific investigation. Goldberg et al. (2005) used three different imaging modes in their preclinical investigations: conventional

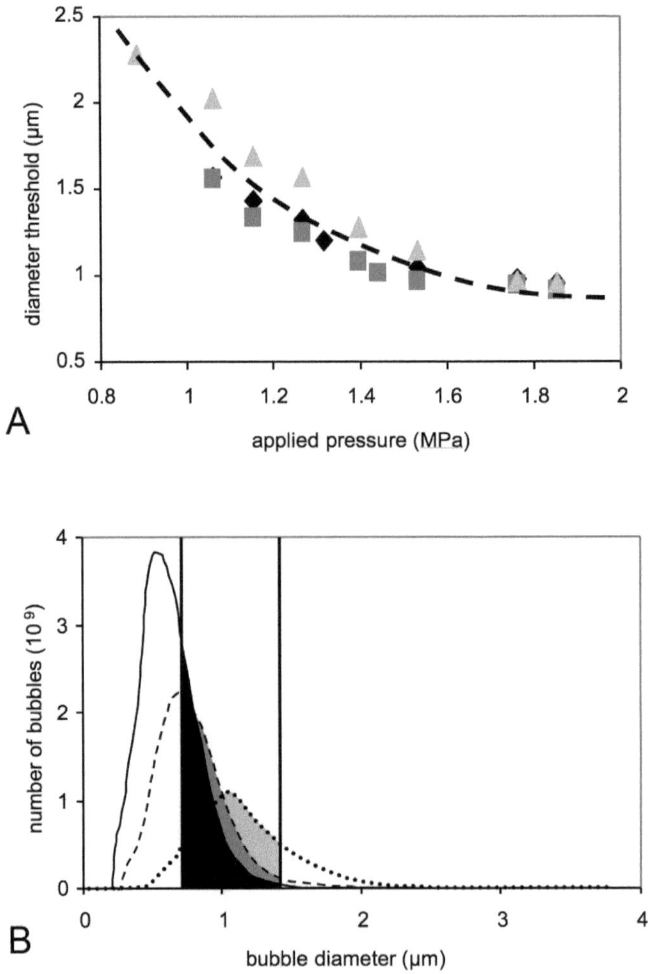

FIGURE 14.2. Illustration of the effect of agent size on likely signal intensity from polymer-shelled contrast agents. **A** Effective diameter threshold over which agents responded to acoustic interrogation in an in vitro investigation (Bloch et al., 2005). **B** The size distributions of three polymer-shelled agents with different mean diameters are shown. The shaded areas represent responding agents for a hypothetical experiment. The ideal agent for an application is the one containing the most microbubbles within the two effective-size thresholds determined by the lymphatic system and the acoustic pressure reaching the agents.

grayscale, pulse-inversion harmonic grayscale, and color flow. The conventional grayscale mode was used to image the tumor and surrounding anatomy before contrast injection and to identify potential sentinel lymph nodes (Figure 14.3A, *left*). Pulse-inversion harmonic grayscale is a non-destructive, contrast-specific mode: it was used after contrast injection to visualize contrast agents flowing

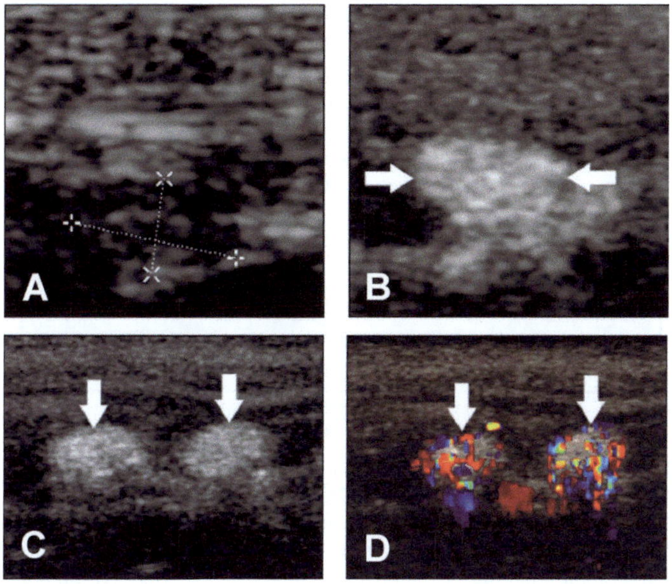

FIGURE 14.3. Different imaging modes are used to reveal a complete picture of sentinel lymph node location and anatomy. **A** Before contrast injection, grayscale mode shows the location of a lymph node and surrounding structures. **B** After contrast injection, contrast-specific grayscale mode demonstrates enhancement of this lymph node (right). **C** The enhancement demonstrated by two small sentinel lymph nodes in contrast-specific grayscale mode is confirmed in color flow mode (**D**) which shows contrast agent location with high sensitivity (Goldberg et al., 2005). Reprinted with permission from Goldberg et al., 2005. Contrast-enhanced sonographic imaging of lymphatic channels and sentinel lymph nodes. J Ultrasound Med 24, 953–965.

through lymphatic channels and toward sentinel lymph nodes (Figure 14.3A, *right*, and Figure 14.3B, *left*). Finally, once the likely sentinel nodes had been identified, high-power, destructive color flow mode was used to confirm the presence of contrast agents in each sentinel node, to distinguish between contrast-containing lymphatic channels and other bright structures, and to destroy agents within the nodes so that refill could be observed (Figure 14.3B, *right*).

14.4.2.3. Biological Activity

A final consideration for contrast agents is the biological activity of a given agent. "Targeted" or "molecular" imaging is an active area of ultrasound contrast agent research. The term *actively targeted* refers to a contrast agent that has one or more biologically active molecules attached such that the agent, when injected systemically, localizes in a specific area of interest; for example, in tumors, sites of inflammation, or specific organs. Hauff et al. (2004) have reported active targeting of the lymph nodes by an intravascular agent incorporating L-selectin-specific

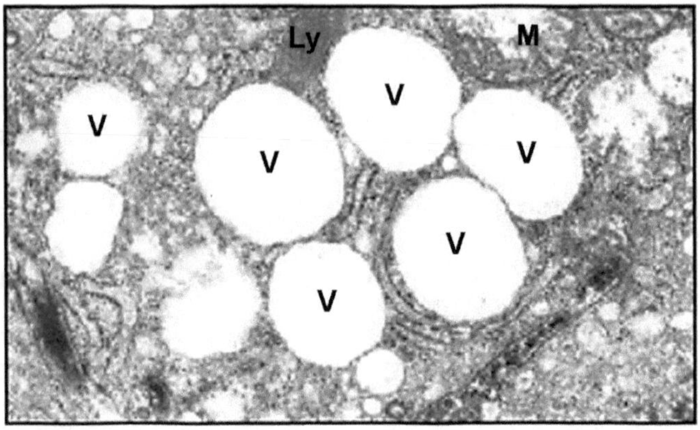

FIGURE 14.4. Electron microscopy of a sentinel lymph node. Scanning electron microscopy identified multiple spherical empty vacuoles (V) representing phagocytized intact contrast microbubbles. These vacuoles were not present in control lymph nodes. Ly indicates a primary lysosome and M indicates mitochondria with intact cristae (Goldberg et al., 2005). Reprinted with permission from Goldberg et al., 2005. Contrast-enhanced sonographic imaging of lymphatic channels and sentinel lymph nodes. J Ultrasound Med 24, 953–965.

ligands. To date, no studies have been published on active targeting of intralymphatic agents; however, several researchers have noted that these agents are *passively targeted*—that is, the pharmacokinetics of the agent as determined by its size and concentration serves to concentrate it in a particular region of interest. Kojima et al. (2003) injected contrast agents composed of carbon dioxide and soybean oil directly into swollen lymph nodes. They observed that heterogeneous enhancement and "filling defects" were observed in nodes containing metastatic lesions, but not in benign nodes. Goldberg et al. (2004) report a similar effect in their studies using Sonazoid, an agent known to be taken up by the reticuloendothelial system and have confirmed using histopathology that Sonazoid is taken up by macrophages in healthy lymph node tissue (Goldberg et al., 2005) (Figure 14.4). These researchers hypothesize that this cell uptake, as well as agent size, prevents microbubbles from traveling beyond the sentinel lymph nodes.

14.5. Contrast-Assisted Ultrasound for Diagnosis of Lymph Node Metastasis

Determination of lymph node status by non-contrast ultrasound is only moderately sensitive and specific when using size and morphology criteria (Alvarez et al., 2006) but has proved to be of benefit when combined with standard sentinel node biopsy techniques (Hsu et al., 2006; Nori et al., 2005). Contrast-enhanced harmonic lymph node imaging by intravascular administration of conventional ultrasound contrast agents increases accuracy of diagnosis and helps to discriminate between metastatic and benign reactive nodes when using linear dimensions, intranodal architecture, presence of discrete filling defects, margins, and location of associated vessels as criteria for evaluation (Rubaltelli et al., 2004). Others have shown contrast-enhanced color and power Doppler to be of additional value with reactive nodes revealing more pronounced branching vascular architecture and metastatic nodes having a non-uniform vascular pattern (Moritz et al., 2000; Schmid-Wendtner et al., 2004; Zenk et al., 2005). However, none of these techniques have proven to be definitive and some authors have found no additional benefit as compared to non-contrast imaging (Schulte-Altedorneburg et al., 2003).

Although contrast-enhanced lymph node imaging using intravenously administered ultrasound contrast media may improve anatomical detail, discrimination between malignant and benign reactive nodes, and the accuracy of cancer staging, it is not inherently different to contrast ultrasound of other organ systems using intravascular agents. This topic is addressed in greater detail in Chapter 16.

14.6. Preclinical Studies in Contrast-Assisted Ultrasound for Sentinel Lymph Node Localization

Mattrey et al. (2002) first proposed the use of interstitially delivered ultrasound contrast microbubbles for enhancement of normal and metastatic regional lymph nodes using a Rabbit VX2 carcinoma model. In this study, AF0150 (Imagent; Alliance Pharmaceutical Corp., San Diego, CA), a soft-shelled perflexane lipid microsphere suspension, was injected at specified sites in the extremities or in the tissues surrounding an implanted tumor. Ultrasound imaging of the regional lymphatics and lymph nodes was performed using real-time fundamental and harmonic imaging sequences. Intermittent imaging was also performed with frame delays ranging from 1 to 45 seconds. Contrast microbubbles could be consistently detected in afferent lymphatics and in regional lymph nodes with greater degrees of node enhancement occurring using intermittent imaging (Figure 14.5). Metastatic nodes enhanced non-uniformly with filling voids occurring in the region of the metastatic deposit. Enhancement degraded as the nodes were imaged, suggesting a destructive mechanism for contrast enhancement. The authors also reported seeing bubbles exiting the nodes via the efferent lymphatics documenting some degree of "passthrough," although the extent of this was not defined.

We have also explored the use of hard-shelled, minimally deformable contrast microbubbles for sentinel node detection (Wisner et al., 2002, 2003). Nitrogen containing bilayer shell microbubbles composed of either a polycaprolactone or polylactide inner layer and a cross-linked albumin outer layer were used for this study in a normal dog model and included mean bubble sizes of 0.6, 0.74, 1.52, and 3.91 microns. Of each suspension, 2 ml was injected into subcutaneous tissues of the front or hindlimb paw targeting the superficial cervical and popliteal lymph nodes, respectively. Bubble concentration in the contrast suspension

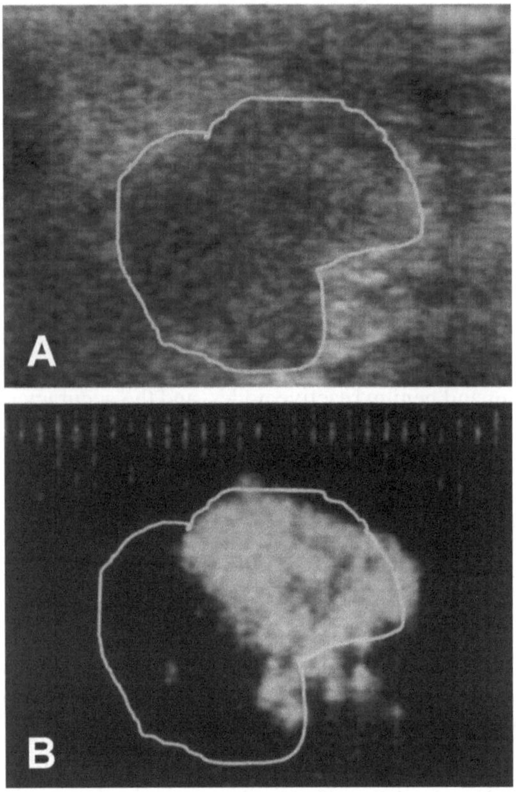

FIGURE 14.5. Metastatic popliteal lymph node in a rabbit implanted with VX2 carcinoma. **A** Standard grayscale image. **B** Burst image of the same node acquired following a 30 second delay. The portion to the right represents normal nodal tissue that accumulates contrast microbubbles. Insonation causes bubble rupture and an intense signal response. The segment on the *left* is replaced by tumor resulting in a contrast-filling void. Reproduced with permission from Mattrey (2002). Reprinted from Mattrey et al., 2002. Sentinel lymph node imaging with microbubble ultrasound contrast material. Acad Radiol 9 Suppl 1, S231–235 with permission of Association of University Radiologists.

containing the largest bubbles was diluted to produce a signal response to insonation equivalent to that of the smaller formulations in in vitro phantom studies. All lymph nodes and associated lymphatics were imaged intermittently in fundamental continuous power Doppler mode for up to 120 minutes following contrast media administration. Moderate to marked lymph node enhancement occurred within minutes in most subjects (94% positive response, 30/32 targeted nodes) when the submicron or near micron formulations were used and persisted for the imaging duration. Mild to moderate uptake occurred with the larger diameter formulation and the response diminished after the first hour. Node enhancement was less consistent when larger formulation was used with only 4/8 nodes responding. Contrast enhancement of afferent lymphatics was evident in most instances and contrast filling of the nodes was non-uniform with contrast material predominantly distributed in the periphery of the node which corresponds to the subcapsular space and sinuses. In all instances, contrast response diminished rapidly when scanning indicating a destructive imaging mechanism. In vitro studies of these formulations revealed that smaller particle diameters generated a better signal response at higher pulse repetition frequencies (PRF) while the 3.91 micron formulation performed better at low PRF. Subjectively, this also appeared to be true in vivo.

In a more recent study, we compared nodal uptake of the rigid shelled submicron and near-micron bubble formulations (0.75, 0.91, and 1.33 mean diameter formulations, POINT Biomedical Corporation, San Carlos, CA) to a commercially available soft-shelled formulation of larger diameter (Definity®, perflutren lipid microspheres, Bristol-myers Squibb Medical Imaging, North Andover, NJ) (Wisner et al., 2006). Because of their composition and the mechanism by which bubbles are formed and suspended, these larger microbubbles include a wide range of particle sizes but have a stated mean diameter of approximately 1.8 microns. Cervical and poplitieal lymph nodes

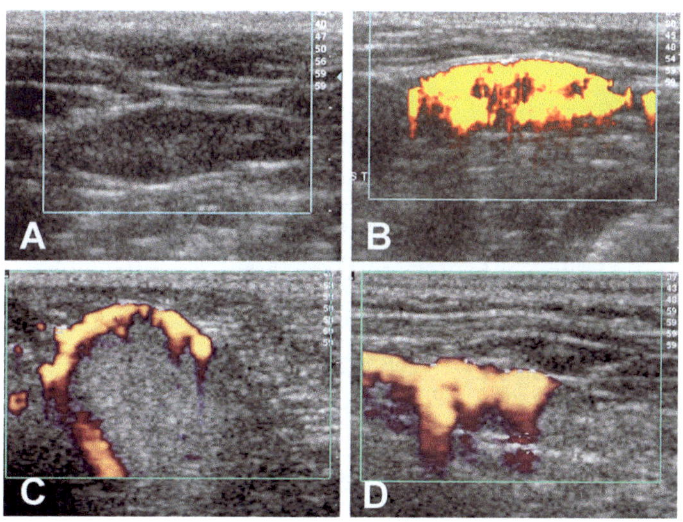

FIGURE 14.6. Power Doppler image of a normal canine cervical lymph node before (**A**) and 30 minutes after (**B**) subcutaneous injection of Definity (perflutren) into the metacarpus. Parenchymal contrast enhancement is virtually complete in this instance. The central unenhanced region represents the normal lymph node hilus. **C** Power Doppler image of a popliteal lymph node 15 minutes following injection of a hard-shelled contrast microbubble formulation (1.33 micron mean diameter). Early peripheral uptake corresponds to the region of the subcapsular sinus which receives lymphatic flow from afferent lymphatic channels. **D** Power Doppler image of a normal cervical lymph node 30 minutes following subcutaneous injection of Definity (perflutren) into the metacarpus. The linear enhancing structure on the *left side* of the image represents an afferent lymphatic channel.

were imaged using power Doppler at 15, 30, and 45 minutes following contrast media injection into the distal extremities of normal dogs. Nodal contrast uptake was evident at a minimum of one timepoint in 70%–85% of nodes in which the rigid shelled smaller formulations were used and in 95 % of nodes enhanced with the perflutren formulation (Figure 14.6). Regardless of formulation used, contrast enhancement occurred within 15 minutes in most instances. Given the anatomy and physiology of the peripheral lymphatic system, it is unclear why the larger particles appeared to be more efficacious in this study.

We have also used Definity for sentinel node localization in a study of ten dogs with spontaneously arising head and neck tumors (Lurie et al., 2006). In this investigation, the microbubble suspension was injected subcutaneously or submucosally around a primary tumor followed by power Doppler imaging of regional lymph nodes. Sentinel lymph nodes were successfully detected in eight of ten dogs and sentinel status of the mapped nodes was verified by lymphoscintigraphy performed contemporaneously (Figure 14.7). In the two dogs in which US mapping failed, it was thought that the contrast medium was inadvertently injected intralesionally rather than peritumorally since these were inflamed, indurated primary tumors with poorly defined margins.

Sentinel node localization using Definity has also been performed in 16 miniature swine with spontaneously arising melanoma. Regional draining lymph nodes

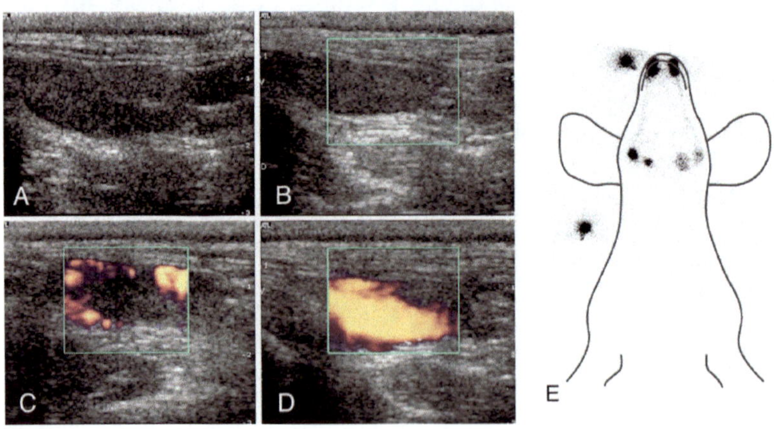

FIGURE 14.7. Ultrasound images of the right and left mandibular lymph nodes of a dog with nasal squamous cell carcinoma (**A** and **B**, respectively). Power Doppler images of the same two nodes (**C, D**) following perilesional injection of Definity (perflutren). The bilateral yet asymmetrical ultrasound microbubble uptake correlates with the similar pattern of ^{99m}Tc-sulfurcolloid uptake on a corresponding lymphoscintigram (**E**). The two foci of activity beyond the outline of the dog in (**E**) represent fiducial markers. Reproduced with permission from Lurie (2006). Reprinted with permission from Lurie et al., 2006. Contrast-assisted ultrasound for sentinel lymph node detection in spontaneously arising canine head and neck tumors. Invest Radiol 41, 415–421.

were imaged using power Doppler following peritumoral contrast suspension injection (wisner 2006). Nodal contrast enhancement occurred in all pigs with a high degree of correlation between contrast uptake on ultrasound versus that seen with comparable lymphoscintigraphy studies (Figure 14.8). It is important to note that contrast uptake was easily detected by ultrasound even in nodes in which normal architecture was predominantly replaced by tumor mass (Figure 14.9).

Goldberg et al. (2005) have used Sonazoid (GE Healthcare, Oslo, Norway) a lipid perflurocarbon microbubble suspension, for similar sentinel node and lymphatic localizing studies in dogs, rabbits, swine, and monkeys using grayscale pulse inversion harmonic imaging. In these studies, the investigators targeted a variety of superficial and deep lymphocenters by using a number of injection sites and found good agreement between lymphosonography results and findings based on comparable blue dye injection with lymph node dissection. Importantly, they also noted that second-order lymph nodes, those receiving lymphatic drainage from the primary or sentinel nodes, did not appear to accumulate contrast microbubbles. Lymphatic channels draining to the sentinel nodes also

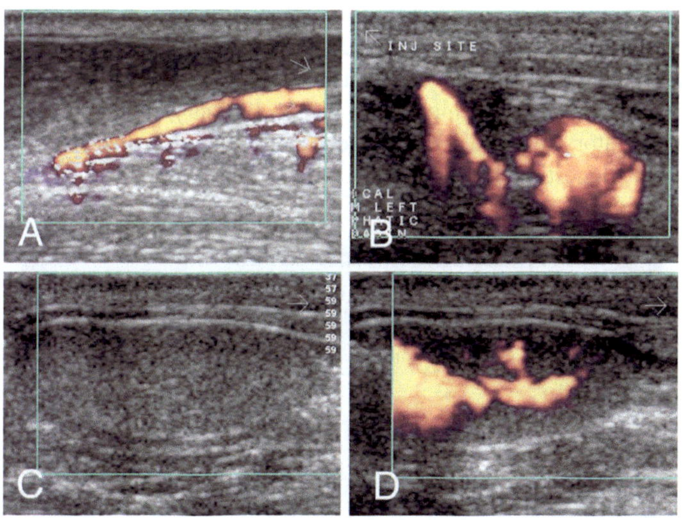

FIGURE 14.8. **A** Contrast enhanced afferent lymphatic vessel in the hindlimb of a melanoma-positive pig. The image was acquired 2 minutes following perilesional injection of Definity (perflutren) in the distal limb. **B** Contrast-enhanced cervical lymph node and associated afferent lymphatic channel 5 minutes following perilesional injection of Definity (perflutren) around a melanoma involving the external ear. **C** Pre-contrast and (**D**) 2 minute post-contrast images of a cervical lymph node following perilesional injection of Definity (perflutren) in a cervical melanoma. The porcine lymph node is anatomically different than that of other mammals with afferent lymphatics entering the node centrally and lymphatic flow progressing peripherally. As a result, only central filling is seen in this early post-contrast image.

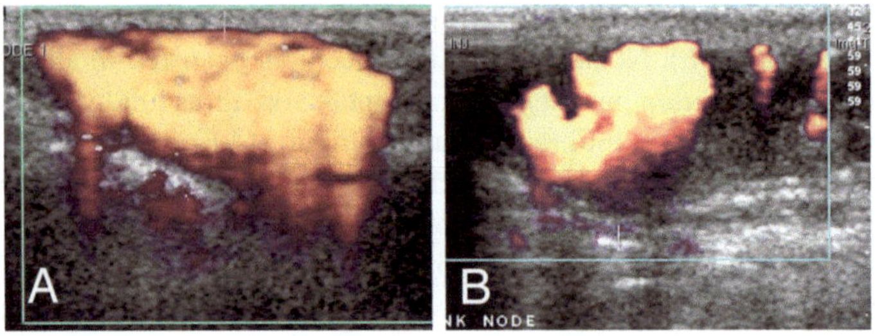

FIGURE 14.9. Complete (A) and incomplete (B) lymph node enhancement in a pig with melanoma. (B) represents a metastatic node with significant replacement of normal node architecture by melanoma cells. As a result, there is a contrast void in the right side of the node corresponding to the tumor deposit.

appeared to be easily identified (Figure 14.10). Another sentinel node mapping study by Goldberg et al. (2004) using Sonazoid in a swine melanoma model revealed very good correlation between contrast-assisted lymphosonography findings and comparable lymphoscinitgraphy studies.

Omoto and Colleagues (2006, 2002) have used a subcutaneously injected 25% albumen suspension as an ultrasound contrast medium to enhance regional lymph nodes in a normal pig model and in a preliminary clinical study in people. These investigators report a 1–5 mm echogenic region in the hilus of draining lymph nodes suggesting aggregation in the efferent lymphatics.

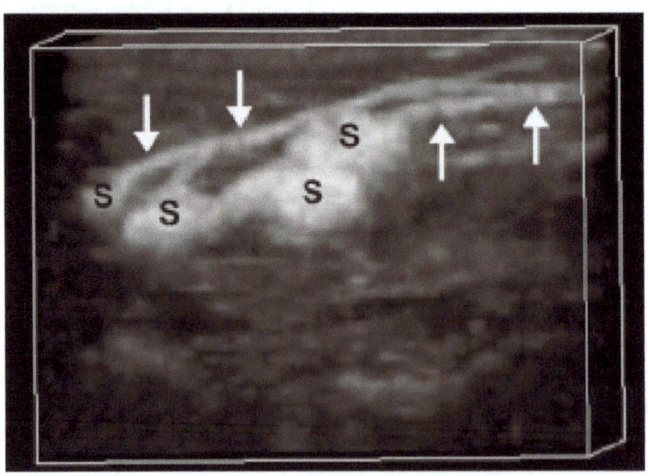

FIGURE 14.10. Three-dimensional pulse-inversion harmonic imaging was used to show multiple lymphatic channels (arrows) leading to multiple sentinel lymph nodes (S) in the pelvis of a swine after a single injection of Sonazoid into the rump (Goldberg et al., 2005). Reprinted with permission from Goldberg et al., 2005. Contrast-enhanced sonographic imaging of lymphatic channels and sentinel lymph nodes. J Ultrasound Med 24, 953–965.

14.7. Conclusions

Contrast-enhanced lymph node ultrasound using intravenously administered conventional contrast agents has been shown to improve detection of node metastasis and discriminate between malignant and benign nodes, although findings are not definitive using currently available ultrasound microbubble contrast formulations.

The unique characteristics of lymphatic anatomy and physiology pose a number of unique hurdles to optimal interstitial delivery and imaging of contrast microbubbles within the lymphatic system. Despite these challenges, initial preclinical studies using contrast-assisted ultrasound for sentinel node localization are promising and the technique could serve as an alternative or an adjunct to current approaches. Optimization of contrast microbubble formulations for interstitial administration and lymphatic uptake will likely improve the sensitivity and overall accuracy for sentinel node localization in the future.

References

Aikou, T., Kitagawa, Y., Kitajima, M., Uenosono, Y., Bilchik, A.J., Martinez, S.R., Saha, S., 2006. Sentinel lymph node mapping with GI cancer. Cancer Metastasis Rev 25, 269–277.

Alvarez, S., Anorbe, E., Alcorta, P., Lopez, F., Alonso, I., Cortes, J., 2006. Role of sonography in the diagnosis of axillary lymph node metastases in breast cancer: a systematic review. AJR Am J Roentgenol 186, 1342–1348.

Anan, K., Mitsuyama, S., Kuga, H., Saimura, M., Tanabe, Y., Suehara, N., Matsunaga, H., Nishihara, K., Abe, Y., Iwashita, T., Ihara, T., Tamae, K., Ono, M., Toyoshima, S., 2006. Double mapping with subareolar blue dye and peritumoral green dye injections decreases the false-negative rate of dye-only sentinel node biopsy for early breast cancer: 2-site injection is more accurate than 1-site injection. Surgery 139, 624–629.

Becher, H., Burns, P., 2000. Handbook of conteast echocardiography Springer Verlag, Frankfurt and New York.

Bergqvist, L., Strand, S.E., Persson, B.R., 1983. Particle sizing and biokinetics of interstitial lymphoscintigraphic agents. Semin Nucl Med 13, 9–19.

Berk, D.R., Johnson, D.L., Uzieblo, A., Kiernan, M., Swetter, S.M., 2005. Sentinel lymph node biopsy for cutaneous melanoma: the Stanford experience, 1997–2004. Arch Dermatol 141, 1016–1022.

Bertagnolli, M., Miedema, B., Redston, M., Dowell, J., Niedzwiecki, D., Fleshman, J., Bem, J., Mayer, R., Zinner, M., Compton, C., 2004. Sentinel node staging of resectable colon cancer: results of a multicenter study. Ann Surg 240, 624–630.

Bloch, S.H., Short, R.E., Ferrara, K.W., Wisner, E.R., 2005. The effect of size on the acoustic response of polymer-shelled contrast agents. Ultrasound Med Biol 31, 439–444.

Blomley, M.J., Albrecht, T., Cosgrove, D.O., Patel, N., Jayaram, V., Butler-Barnes, J., Eckersley, R.J., Bauer, A., Schlief, R., 1999. Improved imaging of liver metastases with stimulated acoustic emission in the late phase of enhancement with the US contrast agent SH U 508A: early experience. Radiology 210, 409–416.

Blomley, M.J., Cooke, J.C., Unger, E.C., Monaghan, M.J., Cosgrove, D.O., 2001. Microbubble contrast agents: a new era in ultrasound. BMJ 322, 1222–1225.

Brown, G., Richards, C.J., Bourne, M.W., Newcombe, R.G., Radcliffe, A.G., Dallimore, N.S., Williams, G.T., 2003. Morphologic predictors of lymph node status in rectal cancer with use of high-spatial-resolution MR imaging with histopathologic comparison. Radiology 227, 371–377.

Chakera, A.H., Drzewiecki, K.T., Eigtved, A., Juhl, B.R., 2004. Sentinel node biopsy for melanoma: a study of 241 patients. Melanoma Res 14, 521–526.

Civantos, F.J., Moffat, F.L., Goodwin, W.J., 2006. Lymphatic mapping and sentinel lymphadenectomy for 106 head and neck lesions: contrasts between oral cavity and cutaneous malignancy. Laryngoscope 112, 1–15.

Cox, E.D., Kellicut, D., Adair, C., Marley, K., Otchy, D.P., Peoples, G.E., 2002. Sentinel lymph node evaluation is technically feasible and may improve staging in colorectal cancer. Curr Surg 59, 301–306.

Cox, K.R., Hare, W.S., Bruce, P.T., 1966. Lymphography in melanoma. Correlation of radiology with pathology. Cancer 19, 637–647.

Di Stefano, A.B., Acquaviva, G., Garozzo, G., Barbic, M., Cvjeticanin, B., Meglic, L., Kobal, B., Rakar, S., 2005. Lymph node mapping and sentinel node detection in patients with cervical carcinoma: a 2-year experience. Gynecol Oncol 99, 671–679.

Doekhie, F.S., Peeters, K.C., Kuppen, P.J., Mesker, W.E., Tanke, H.J., Morreau, H., van de Velde, C.J., Tollenaar, R.A., 2005. The feasibility and reliability of sentinel node mapping in colorectal cancer. Eur J Surg Oncol 31, 854–862.

Elste, V., Wagner, S., Taupitz, M., Pfefferer, D., Kresse, M., Hamm, B., Berg, R., Wolf, K.J., Semmler, W., 1996. Magnetic resonance lymphography in rats: effects of muscular activity and hyperthermia on the lymph node uptake of intravenously injected superparamagnetic iron oxide particles. Acad Radiol 3, 660–666.

Feinstein, S.B., 2004. The powerful microbubble: from bench to bedside, from intravascular indicator to therapeutic delivery system, and beyond. Am J Physiol Heart Circ Physiol 287, H450–H457.

Gien, L.T., Kwon, J.S., Carey, M.S., 2005. Sentinel node mapping with isosulfan blue dye in endometrial cancer. J Obstet Gynaecol Can 27, 1107–1112.

Goldberg, B.B., Merton, D.A., Liu, J.B., Murphy, G., Forsberg, F., 2005. Contrast-enhanced sonographic imaging of lymphatic channels and sentinel lymph nodes. J Ultrasound Med 24, 953–965.

Goldberg, B.B., Merton, D.A., Liu, J.B., Thakur, M., Murphy, G.F., Needleman, L., Tornes, A., Forsberg, F., 2004. Sentinel lymph nodes in a swine model with melanoma: contrast-enhanced lymphatic US. Radiology 230, 727–734.

Guyton, A., 1991. Textbook of medical physiology. W.B. Saunders Company, Philadelphia.

Harisinghani, M.G., Saksena, M.A., Hahn, P.F., King, B., Kim, J., Torabi, M.T., Weissleder, R., 2006. Ferumoxtran-10-enhanced MR lymphangiography: does contrast-enhanced imaging alone suffice for accurate lymph node characterization? AJR Am J Roentgenol 186, 144–148.

Hauff, P., Reinhardt, M., Briel, A., Debus, N., Schirner, M., 2004. Molecular targeting of lymph nodes with L-selectin ligand-specific US contrast agent: a feasibility study in mice and dogs. Radiology 231, 667–673.

Heesakkers, R.A., Futterer, J.J., Hovels, A.M., van den Bosch, H.C., Scheenen, T.W., Hoogeveen, Y.L., Barentsz, J.O., 2006. Prostate cancer evaluated with ferumoxtran-10-enhanced T2*-weighted MR Imaging at 1.5 and 3.0 T: early experience. Radiology 239, 481–487.

Hsu, G.C., Ku, C.H., Yu, J.C., Hsieh, C.B., Yu, C.P., Chao, T.Y., 2006. Application of intraoperative ultrasound to nonsentinel node assessment in primary breast cancer. Clin Cancer Res 12, 3746–3753.

Ikomi, F., Hanna, G.K., Schmid-Schonbein, G.W., 1995. Mechanism of colloidal particle uptake into the lymphatic system: basic study with percutaneous lymphography. Radiology 196, 107–113.

Izawa, J., Kedar, D., Wong, F., Pettaway, C.A., 2005. Sentinel lymph node biopsy in penile cancer: evolution and insights. Can J Urol 12 Suppl 1, 24–29.

Kelley, M.C., Hansen, N., McMasters, K.M., 2004. Lymphatic mapping and sentinel lymphadenectomy for breast cancer. Am J Surg 188, 49–61.

Kim, J.H., Beets, G.L., Kim, M.J., Kessels, A.G., Beets-Tan, R.G., 2004. High-resolution MR imaging for nodal staging in rectal cancer: are there any criteria in addition to the size? Eur J Radiol 52, 78–83.

Kim, T., Giuliano, A.E., Lyman, G.H., 2006. Lymphatic mapping and sentinel lymph node biopsy in early-stage breast carcinoma: a metaanalysis. Cancer 106, 4–16.

King, A.D., Ahuja, A.T., Leung, S.F., Lam, W.W., Teo, P., Chan, Y.L., Metreweli, C., 2000. Neck node metastases from nasopharyngeal carcinoma: MR imaging of patterns of disease. Head Neck 22, 275–281.

Klibanov, A.L., 2006. Microbubble contrast agents: targeted ultrasound imaging and ultrasound-assisted drug-delivery applications. Invest Radiol 41, 354–362.

Koehler, P.R., 1976. Current status of lymphography in patients with cancer. Cancer 37, 503–516.

Kojima, S., Goto, H., Hirooka, Y., Itoh, A., Ishiguro, Y., Hashimoto, S., Hirai, T., Hayakawa, T., 2003. Differentiation of benign and malignant lymph nodes with contrast-enhanced echolymphography using endoscopic ultrasound-guided puncture. Hepatogastroenterology 50, 1285–1291.

Krag, D., 2001. Minimal invasive staging for breast cancer: clinical experience with sentinel lymph node biopsy. Semin Oncol 28, 229–235.

Krag, D.N., Julian, T.B., Harlow, S.P., Weaver, D.L., Ashikaga, T., Bryant, J., Single, R.M., Wolmark, N., 2004. NSABP-32: Phase III, randomized trial comparing axillary resection with sentinal lymph node dissection: a description of the trial. Ann Surg Oncol 11, 208S–210S.

Kumar, R., Zhuang, H., Schnall, M., Conant, E., Damia, S., Weinstein, S., Chandra, P., Czerniecki, B., Alavi, A., 2006. FDG PET positive lymph nodes are highly predictive of metastasis in breast cancer. Nucl Med Commun 27, 231–236.

Liptay, M.J., 2004. Sentinel node mapping in lung cancer. Ann Surg Oncol 11, 271S–274S.

Lurie, D.M., Seguin, B., Schneider, P.D., Verstraete, F.J., Wisner, E.R., 2006. Contrast-assisted ultrasound for sentinel lymph node detection in spontaneously arising canine head and neck tumors. Invest Radiol 41, 415–421.

Maggard, M.A., Lane, K.E., O'Connell, J.B., Nanyakkara, D.D., Ko, C.Y., 2005. Beyond the clinical trials: how often is sentinel lymph node dissection performed for breast cancer? Ann Surg Oncol 12, 41–47.

Mattrey, R.F., Kono, Y., Baker, K., Peterson, T., 2002. Sentinel lymph node imaging with microbubble ultrasound contrast material. Acad Radiol 9 Suppl 1, S231–S235.

Moritz, J.D., Ludwig, A., Oestmann, J.W., 2000. Contrast-enhanced color Doppler sonography for evaluation of enlarged cervical lymph nodes in head and neck tumors. AJR Am J Roentgenol 174, 1279–1284.

Morton, D.L., Cochran, A.J., Thompson, J.F., Elashoff, R., Essner, R., Glass, E.C., Mozzillo, N., Nieweg, O.E., Roses, D.F., Hoekstra, H.J., Karakousis, C.P., Reintgen, D.S., Coventry, B.J., Wang, H.J., 2005. Sentinel node biopsy for early-stage melanoma: accuracy and morbidity in MSLT-I, an international multicenter trial. Ann Surg 242, 302–311; discussion 311–313.

Nori, J., Bazzocchi, M., Boeri, C., Vanzi, E., Nori Bufalini, F., Mangialavori, G., Distante, V., Masi, A., Simoncini, R., Londero, V., 2005. Role of axillary lymph node ultrasound and large core biopsy in the preoperative assessment of patients selected for sentinel node biopsy. Radiol Med (Torino) 109, 330–344.

Omoto, K., Hozumi, Y., Omoto, Y., Taniguchi, N., Itoh, K., Fujii, Y., Mizunuma, H., Nagai, H., 2006. Sentinel node detection in breast cancer using contrast-enhanced sonography with 25 % albumin-Initial clinical experience. J Clin Ultrasound 34, 317–326.

Omoto, K., Mizunuma, H., Ogura, S., Hozumi, Y., Nagai, H., Taniguchi, N., Itoh, K., 2002. New method of sentinel node identification with ultrasonography using albumin as contrast agent: a study in pigs. Ultrasound Med Biol 28, 1115–1122.

Patel, D., Dayton, P., Gut, J., Wisner, E., Ferrara, K.W., 2002. Optical and acoustical interrogation of submicron contrast agents. IEEE Trans Ultrason Ferroelectr Freq Control 49, 1641–1651.

Patel, D.N., Bloch, S.H., Dayton, P.A., Ferrara, K.W., 2004. Acoustic signatures of submicron contrast agents. IEEE Trans Ultrason Ferroelectr Freq Control 51, 293–301.

Perdona, S., Autorino, R., De Sio, M., Di Lorenzo, G., Gallo, L., Damiano, R., D'Armiento, M., Gallo, A., 2005. Dynamic sentinel node biopsy in clinically node-negative penile cancer versus radical inguinal lymphadenectomy: a comparative study. Urology 66, 1282–1286.

Posther, K.E., McCall, L.M., Blumencranz, P.W., Burak, W.E., Jr., Beitsch, P.D., Hansen, N.M., Morrow, M., Wilke, L.G., Herndon, J.E., 2nd, Hunt, K.K., Giuliano, A.E., 2005. Sentinel node skills verification and surgeon performance: data from a multicenter clinical trial for early-stage breast cancer. Ann Surg 242, 593–602.

Pulte, D., Li, E., Crawford, B.K., Newman, E., Alexander, A., Mustalish, D.C., Jacobson, D.R., 2005. Sentinel lymph node mapping and molecular staging in nonsmall cell lung carcinoma. Cancer 104, 1453–1461.

Quon, A., Gambhir, S.S., 2005. FDG-PET and beyond: molecular breast cancer imaging. J Clin Oncol 23, 1664–1673.

Radovanovic, Z., Golubovic, A., Plzak, A., Stojiljkovic, B., Radovanovic, D., 2004. Blue dye versus combined blue dye-radioactive tracer technique in detection of sentinel lymph node in breast cancer. Eur J Surg Oncol 30, 913–917.

Ross, G.L., Soutar, D.S., Gordon MacDonald, D., Shoaib, T., Camilleri, I., Roberton, A.G., Sorensen, J.A., Thomsen, J., Grupe, P., Alvarez, J., Barbier, L., Santamaria, J., Poli, T., Massarelli, O., Sesenna, E., Kovacs, A.F., Grunwald, F., Barzan, L., Sulfaro, S., Alberti, F., 2004. Sentinel node biopsy in head and neck cancer: preliminary results of a multicenter trial. Ann Surg Oncol 11, 690–696.

Rubaltelli, L., Khadivi, Y., Tregnaghi, A., Stramare, R., Ferro, F., Borsato, S., Fiocco, U., Adami, F., Rossi, C.R., 2004. Evaluation of lymph node perfusion using continuous mode harmonic ultrasonography with a second-generation contrast agent. J Ultrasound Med 23, 829–836.

Rubello, D., Pelizzo, M.R., Al-Nahhas, A., Salvatori, M., O'Doherty M, J., Giuliano, A.E., Gross, M.D., Fanti, S., Sandrucci, S., Casara, D., Mariani, G., 2006. The role of sentinel lymph node biopsy in patients with differentiated thyroid carcinoma. Eur J Surg Oncol 32(9): 917–921.

Rzyman, W., Hagen, O.M., Dziadziuszko, R., Kobierska-Gulida, G., Karmolinski, A., Lothe, I.M., Babovic, A., Murawski, M., Paleczka, W., Jastrzebski, T., Kopacz, A., Jassem, J., Lass, P., Skokowski, J., 2006. Intraoperative, radio-guided sentinel lymph node mapping in 110 nonsmall cell lung cancer patients. Ann Thorac Surg 82, 237–242.

Saha, S., Dan, A.G., Viehl, C.T., Zuber, M., Wiese, D., 2005. Sentinel lymph node mapping in colon and rectal cancer: its impact on staging, limitations, and pitfalls. Cancer Treat Res 127, 105–122.

Schmid-Schonbein, G.W., 1990. Microlymphatics and lymph flow. Physiol Rev 70, 987–1028.

Schmid-Wendtner, M.H., Dill-Muller, D., Baumert, J., Wagner, A., Eberle, J., Tilgen, W., Plewig, G., 2004. Lymph node metastases in patients with cutaneous melanoma: improvements in diagnosis by signal-enhanced color Doppler sonography. Melanoma Res 14, 269–276.

Schulte-Altedorneburg, G., Demharter, J., Linne, R., Droste, D.W., Bohndorf, K., Bucklein, W., 2003. Does ultrasound contrast agent improve the diagnostic value of colour and power Doppler sonography in superficial lymph node enlargement? Eur J Radiol 48, 252–257.

Scoggins, C.R., Chagpar, A.B., Martin, R.C., McMasters, K.M., 2005. Should sentinel lymph-node biopsy be used routinely for staging melanoma and breast cancers? Nat Clin Pract Oncol 2, 448–455.

Silberman, A.W., McVay, C., Cohen, J.S., Altura, J.F., Brackert, S., Sarna, G.P., Palmer, D., Ko, A., Memsic, L., 2004. Comparative morbidity of axillary lymph node dissection and the sentinel lymph node technique: implications for patients with breast cancer. Ann Surg 240, 1–6.

Strand, S.E., Persson, B.R., 1979. Quantitative lymphoscintigraphy I: Basic concepts for optimal uptake of radiocolloids in the parasternal lymph nodes of rabbits. J Nucl Med 20, 1038–1046.

Suzuma, T., Sakurai, T., Yoshimura, G., Umemura, T., Tamaki, T., Yang, Q.F., Oura, S., Naito, Y., 2002. MR-axillography oriented surgical sampling for assessment of nodal status in the selection of patients with breast cancer for axillary lymph nodes dissection. Breast Cancer 9, 69–74.

Syme, D.B., Collins, J.P., Mann, G.B., 2005. Comparison of blue dye and isotope with blue dye alone in breast sentinel node biopsy. ANZ J Surg 75, 817–821.

Takei, H., Suemasu, K., Kurosumi, M., Horii, Y., Ninomiya, J., Kamimura, M., Naganuma, R., Uchida, K., Igarashi, K., Inoue, K., Tabei, T., 2006. Added value of the presence of blue nodes or hot nodes in sentinel lymph node biopsy of breast cancer. Breast Cancer 13, 179–185.

Thompson, J.F., Uren, R.F., Scolyer, R.A., Stretch, J.R., 2005. Selective sentinel lymphadenectomy: progress to date and prospects for the future. Cancer Treat Res 127, 269–287.

Topal, U., Punar, S., Tasdelen, I., Adim, S.B., 2005. Role of ultrasound-guided core needle biopsy of axillary lymph nodes in the initial staging of breast carcinoma. Eur J Radiol 56, 382–385.

Torrenga, H., Meijer, S., Fabry, H., van der Sijp, J., 2004. Sentinel node biopsy in breast cancer patients: triple technique as a routine procedure. Ann Surg Oncol 11, 231S–235S.

van den Brekel, M.W., 2000. Lymph node metastases: CT and MRI. Eur J Radiol 33, 230–238.

Vaquerano, J., Kraybill, W.G., Driscoll, D.L., Cheney, R., Kane, J.M., 3rd, 2006. American Joint Committee on Cancer clinical stage as a selection criterion for sentinel lymph node biopsy in thin melanoma. Ann Surg Oncol 13, 198–204.

van Rijk, M.C., Deurloo, E.E., Nieweg, O.E., Gilhuijs, K.G., Peterse, J.L., Rutgers, E.J., Kroger, R., Kroon, B.B., 2006. Ultrasonography and fine-needle aspiration cytology can spare breast cancer patients unnecessary sentinel lymph node biopsy. Ann Surg Oncol 13, 31–35.

Vaughan, B.F., Varey, J.F., 1964. The lymphographic appearances of malignant pelvic lymph node involvement. Clin Radiol 15, 329–340.

Will, O., Purkayastha, S., Chan, C., Athanasiou, T., Darzi, A.W., Gedroyc, W., Tekkis, P.P., 2006. Diagnostic precision of nanoparticle-enhanced MRI for lymph-node metastases: a meta-analysis. Lancet Oncol 7, 52–60.

Wisner, E., 2006. Contrast-assisted ultrasound for sentinel lymph node detection in normal nodes and in a melanoma model. European Radiol 16, 173–174.

Wisner, E., Bloch, S., Schneider, P., Short, R., 2006. Contrast-assisted ultrasound for sentinel node detection; the effect of particle characteristics and node status on imaging efficacy. Contrast Media Mol Imaging 1, 75.

Wisner, E.R., Ferrara, K., Gabe, J.D., Patel, D., Nyland, T.G., Short, R.E., Ottoboni, T.B., 2002. Contrast enhanced intermittent power Doppler ultrasound with sub-micron bubbles for sentinel node detection. Acad Radiol 9 Suppl 2, S389–S391.

Wisner, E.R., Ferrara, K.W., Short, R.E., Ottoboni, T.B., Gabe, J.D., Patel, D., 2003. Sentinel node detection using contrast-enhanced power Doppler ultrasound lymphography. Invest Radiol 38, 358–365.

Wolf, G., 1995. Specific imaging agents for lymph nodes. In: Torchilin, V. (Ed.), Handbook of targeted delivery of imaging agents. CRC Press, Boca Raton, pp. 365–384.

Wolf, G.L., Na, G.C., Gazelle, G.S., McIntire, G.L., Cannillo, J., Bacon, E.R., Halpern, E., 1994. Time-lapse quantitative computed tomography lymphography: assessing lymphatic function in vivo. Acad Radiol 1, 358–363.

Woznick, A., Franco, M., Bendick, P., Benitez, P.R., 2006. Sentinel lymph node dissection for breast cancer: how many nodes are enough and which technique is optimal? Am J Surg 191, 330–333.

Zenk, J., Bozzato, A., Steinhart, H., Greess, H., Iro, H., 2005. Metastatic and inflammatory cervical lymph nodes as analyzed by contrast-enhanced color-coded Doppler ultrasonography: quantitative dynamic perfusion patterns and histopathologic correlation. Ann Otol Rhinol Laryngol 114, 43–47.

15
Use of Ultrasound Microbubbles for Vascular Imaging

Yuko Kono, Thilo Hölscher, and Robert F Mattrey

Abstract: Recent advances in both ultrasound contrast agents and contrast-specific imaging techniques have significantly impacted the field of contrast sonography. Improvement of contrast-specific imaging techniques has made all bubble-based agents visible on gray-scale displays with high spatial resolution. Low mechanical index (MI) imaging now allows good parenchymal enhancement of various organs in real time, although intermittent imaging is still necessary in some occasions.

In this chapter, advantages of using microbubble-based contrast and its important clinical applications in vascular imaging are discussed. Clinical applications in vascular imaging include diagnosis of carotid artery stenosis, transcranial brain imaging, assessment of transjugular intrahepatic portocaval shunt (TIPS) patency, and others. Grayscale contrast-specific imaging enables visualization of the interior and exterior margins of the vessel wall, allowing accurate assessment of plaque, atherosclerotic disease, or any other wall abnormality, as well as accurate measurements of stenoses. Very slow flow or small-volume flow, such as channels in recanalized clots or very tight stenoses that are frequently invisible on Doppler imaging techniques, can be depicted with gray-scale contrast-enhanced sonography.

Keywords: Ultrasound, contrast agent, vascular imaging

15.1. Introduction

Clinical testing of the current ultrasound contrast agents began in the mid-1990s. Several contrast agents are now approved worldwide for a variety of indications except in the US, where they are limited to cardiac imaging. These agents are all microbubble-based and, except for one agent, are filled with fluorocarbon vapor. Their average diameter is from 1 to 3 μm and each milliliter of microbubble suspension contains 10^8–10^9 microbubbles. Their unique properties have been studied in detail resulting in several "contrast-specific imaging" techniques. Although microbubbles are visible with standard imaging (Girard et al., 1999), contrast-specific imaging techniques provided ultrasound with extreme sensitivity allowing the detection of a single moving microbubble (Klibanov et al., 2002). Their size is sufficiently small to allow them to traverse the microcirculation and recirculate for several minutes, but is too large to leak across capillaries limiting their distribution to the intravascular space. Therefore, unlike CT/MRI agents that rapidly equilibrate with the extracellular fluid, microbubbles are true blood pool agents. Depending on the ultrasound transmit power and frame-rate used, microbubbles enhance the entire vascular compartment from large vessels down to the capillaries. While a capillary cannot be resolved, capillary filling results in diffuse enhancement of perfused tissues. The other unique property of microbubbles in an ultrasound field is that the transmit power and frequency can be adjusted to allow microbubbles to simply reflect the transmitted sound, i.e. becoming an ultrasound transmitter themselves, or be destroyed to produce a broad-band sound. When microbubbles become transmitters or are destroyed, they generate sound at frequencies other than the transmitted frequency that is either broad band or at half, double, or triple the transmitted frequency. Contrast-specific imaging relies upon the generation of sound at frequencies other than the transmitted frequency to uniquely recognize microbubbles from tissues. The most popular techniques such as second harmonic imaging, pulse (or phase) inversion imaging, and amplitude modulation are now available on all high-end instruments and are discussed elsewhere in this book.

The applications of contrast agents in radiology are broad and varied. Essentially, contrast agents can be grouped into vascular (large vessels) and solid organ (capillary) imaging. This chapter will focus on vascular imaging.

15.2. Vascular Applications and Gray-Scale Imaging

Although color and power Doppler have improved significantly in the past few years, they still suffer from blooming artifacts and insufficient sensitivity. Gray-scale contrast-specific imaging with microbubbles overcomes these limitations to provide high spatial and temporal resolution of the luminal and advantitial surfaces to better characterize clots, plaques, and stenosis.

Earlier generation agents filled with air were not visible with standard gray-scale techniques. Enhanced Doppler signals are needed to increase the visualization of smaller and deeper vessels (Fan et al., 1993) and allow velocity assessment (Hilpert et al., 1989). Agents filled with perfluorocarbon (PFC) vapor are longer lived in the vascular space and are more resistant to insonation than their air-filled counterparts. Although the air and PFC-filled microbubbles are visible on gray-scale contrast-specific imaging techniques, only some of the PFC-filled agents are visible with standard ultrasound imaging (Girard et al., 1999). Gray-scale contrast-specific imaging allows the assessment of tissue enhancement caused by capillary filling that is not possible with Doppler techniques. When Doppler imaging that relies on the presence of flow to overlay color on the gray-scale image is used to define luminal stenosis or to assess plaque ulceration, errors occur when flow and structure no longer overlap in cases where flow is disturbed by plaque, aneurysm, clot, or tortuosity. Furthermore, regions of slow flow or small-volume flow, such as channels in recanalized clots or very tight stenoses, may be invisible with Doppler, but are depicted with gray-scale imaging (Coley et al., 1994). The interior and exterior arterial walls are clearly depicted, allowing accurate assessment of plaque and atherosclerotic disease. In fact, this technique may become the standard ultrasound method to assess carotid disease, rather than measuring velocity as a surrogate for stenosis.

15.3. Clinical Applications

15.3.1. Carotid Artery Imaging

Stroke is one of the leading causes of morbidity and mortality in the US. Atherosclerotic disease at the carotid bifurcation is a major cause of stroke, affecting 88% of patients with amaurosis fugax or hemispheric transient ischemic attacks (Eisenberg et al., 1977). Carotid endarterectomy has been shown to be effective for preventing stroke in patients with significant carotid stenosis.

Duplex ultrasound is the primary examination in patients considered for endarterectomy or who are suspected of having carotid artery disease. Duplex ultrasound is readily accessible, non-invasive, and relatively inexpensive. It is highly effective at detecting flow limiting disease (Moneta et al., 1993), but has been limited in detecting echolucent plaques and ulcerations (Widder et al., 1990; Young et al., 1992), precisely determining the degree of stenosis (Grant et al., 2000; Hood et al., 1996; van Everdingen et al., 1998) and differentiating critical stenosis from complete occlusion (AbuRahma et al., 1997; Bornstein et al., 1988).

Spectral Doppler has been used to measure flow velocity to evaluate stenosis. Velocity is affected by the cross-sectional area and increases as the cross sectional area decreases. However, velocity can be affected by multiple factors, such as systemic pressure, the presence or absence of one or more upstream or downstream stenoses, or the integrity of the contralateral carotid system and the ipsilateral external carotid artery. Therefore, velocity measurement remains a rough estimate of stenosis.

We evaluated 20 patients with abnormal duplex US scans, who would probably be advised to undergo conventional angiography or who were scheduled for endoarterectomy, with gray-scale phase inversion contrast-enhanced ultrasound (CEUS) with Optison (Kono et al., 2004). Longitudinal, as well as transverse cross-sectional, images were obtained with phase inversion gray-scale imaging on a Siemens Elegra within the 1- to 2-minute diagnostic window following the bolus injection of 0.5 to 1ml of Optison. The contrast-enhanced ultrasound clearly displayed the outer and inner luminal margins of the vessel (Figure 15.1) allowing the precise assessment of the arterial wall including intimal thickening, plaques, and ulcerations. When stenoses were measured directly as a percent diameter narrowing at the point of stenosis compared to the downstream normal

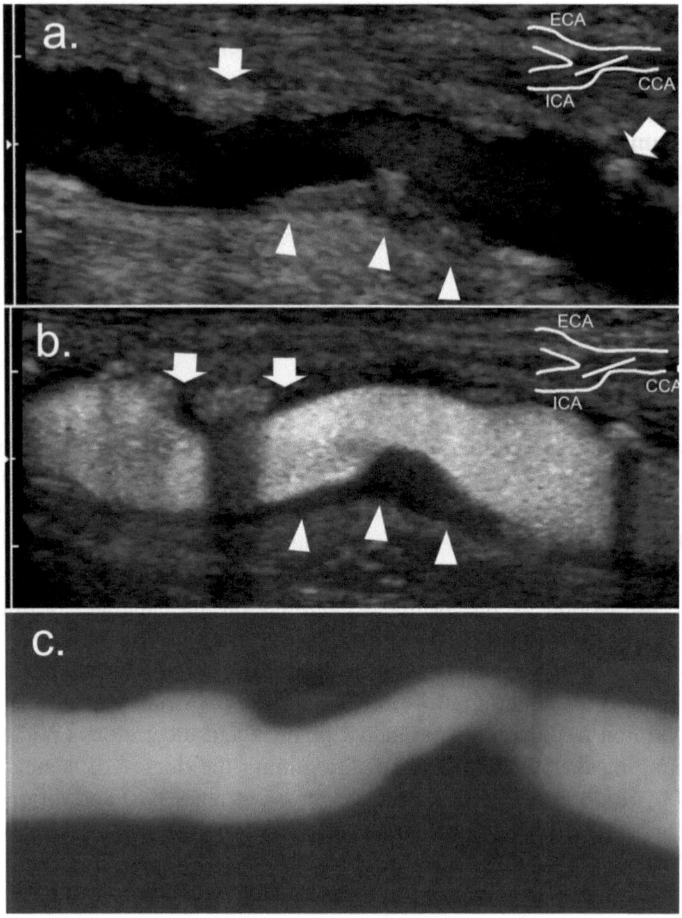

FIGURE 15.1. Images of a 69-year-old male patient with a history of transischemic attack. **a** Pre-contrast US angiogram in right common carotid artery shows calcified (arrows) and non-calcified (arrowheads) plaques. **b** Post-contrast US angiogram obtained after injection of 0.5 mL of Optison shows complete luminal enhancement. Soft plaque (arrowheads) along the posterior wall is well visualized. Also, note the non-calcified component (arrows) of the calcified plaque along the anterior wall that was not detected in **a**. CCA = common carotid artery, ECA = external carotid artery, ICA = internal carotid artery. **c** Conventional angiogram depicts the right common carotid artery (superior artery is on the left side and inferior artery is on the right side). Gray scale is inverted to emphasize the similarity of this image and the contrast-enhanced US angiogram. This image was interpreted as showing focal kinking, which was actually intimal thickening. Reproduced with permission from Kono et al. (2004).

diameter using CEUS and conventional angiography the measurements were highly correlated ($r = 0.988$; $p < 0.001$). Table 15.1 shows the ability of CEUS and duplex Doppler to properly classify patients.

TABLE 15.1. Accuracy of percentage diameter measurements at Duplex US, US angiography, and conventional angiography. Adapted from Kono et al. (2004) with permission

	Percentage stenosis at conventional angiography*				
Modality and percentage stenosis	0–49	50–69	70–89	≥90	100
US angiography†					
0–49	4	1	0	0	0
50–69	0	0	0	0	0
70–89	0	1	3	0	0
≥90	0	0	1	4	0
100	0	0	0	0	5
Duplex US‡					
0–49	2	1	0	0	0
50–69	2	0	0	0	0
70–89	0	1	0	0	0
≥90	0	0	0	3	0
100	0	0	1	0	4

Note: Data are the number of segments.
* Reference standard.
† Findings included 3 (16%) of 19 misclassification errors (95% CI: 3%, 40%)
‡ Findings included 5 (36%) of 14 misclassification errors (95% CI: 13%, 65%)

Direct measurement of percent cross-sectional area narrowing at the point of maximal stenosis is a more accurate predictor of the effect of stenosis on flow than percent diameter narrowing, since it is not affected by irregular or eccentric narrowing. To prove that CEUS allows the accurate measurement of percent area narrowing, ex-vivo, high spatial resolution, three-dimensional, gradient-echo MR imaging of carotid plaques was performed when patients studied with CEUS had an endarterectomy. The cross-sectional percentage area narrowing was measured at the point of maximal stenosis as the inner cross-sectional area relative to the outer cross-sectional area. The measurements were not only highly correlated ($r = 0.979$, $P < 0.001$) but also, plaques seen on CEUS, similar to those seen on ex-vivo MR images (Figure 15.2). Eccentric plaques depicted on the CEUS and ex-vivo MR were not recognized on conventional angiography. The degree of stenosis of one of two patients that had CEUS, conventional angiography, and ex vivo MR imaging was overestimated on angiography because it was eccentric (Figure 15.2).

Severe calcifications hindered our ability to fully visualize the lumen in two lesions. However, the presence of contrast did not preclude the ability to measure flow velocity. In fact, the presence of contrast enhances Doppler signal, as has been shown by others (Gutberlet et al., 1998).

In summary, CEUS of the carotid arteries with contrast-specific imaging is comparable to X-ray angiography and ex vivo MRI, providing an accurate depiction of carotid stenosis, plaques, and ulcerations and allows the calculation of percent area narrowing. The high correlation between this technique and angiography, its consistent display of intraluminal detail, and the lack of added risk when combined with the advantages of sonography warrant its further

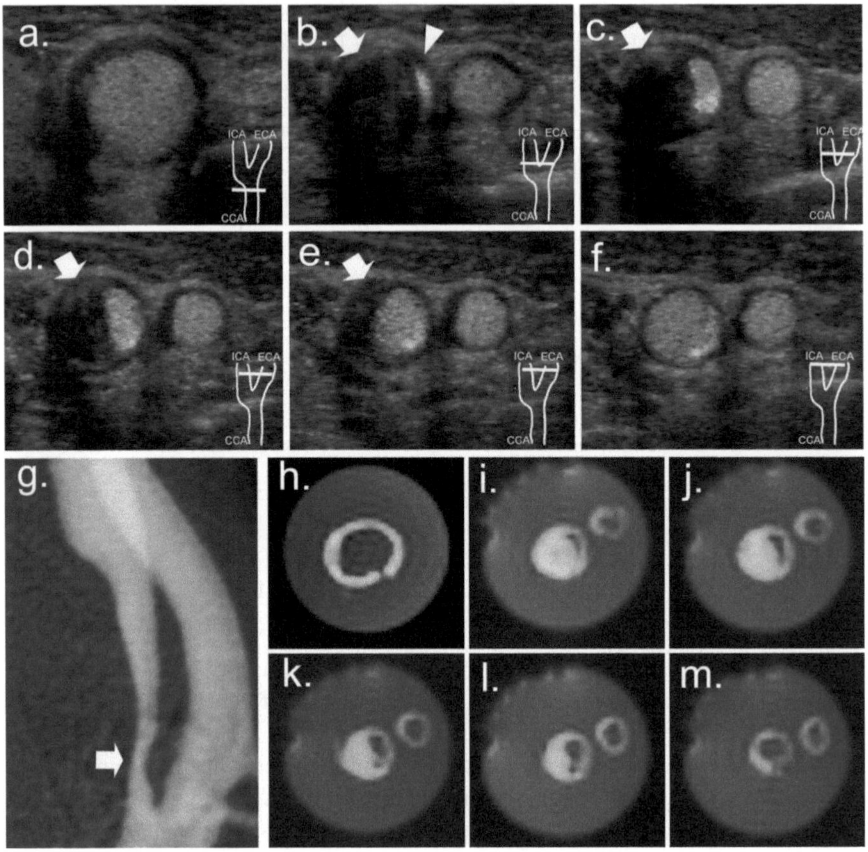

FIGURE 15.2. **a–f** Transverse images of carotid artery displaying a long non-calcified plaque (arrows) causing eccentric narrowing of the ICA (arrow head). Serial transverse images of the carotid artery from the common carotid (**a**) to the normal portion of the distal carotid (**f**). **g** X-ray angiogram showed a severe stenosis in the ICA (arrow). Since this is a 2D projection that happened to be along the long axis of the eccentric stenosis, the degree of the stenosis is overestimated. **h-m** Ex vivo high resolution MRI of the plaque shown in a-f confirmed the plaque configuration on display on the ultrasound angiogram. Reproduced with permission from Kono et al. (2004). Plane of view is shown by the diagrams right side of images **a–f**.

clinical development. It is one of the important and promising applications of contrast-enhanced ultrasound with microbubbles.

15.3.2. Transcranial Contrast-Enhanced Ultrasound: Advances in Diagnosis and Therapy

Innovative ultrasound techniques enabled by current systems and microbubble-based contrast agents allow real-time visualization of the intracranial macro- as well as micro-circulation. Unlike CT or MRI, transcranial CEUS can be performed at the patient's bedside and repeatedly not only to detect vascular occlusion in the acute setting, but also to monitor therapeutic intervention aimed at re-establishing normal intracranial blood flow. Further, it has been suggested that ultrasound has a therapeutic advantage as it appears to accelerate thrombolysis (sonothrombolysis) when patients are being treated with rtPA.

15.3.2.1. Macrovascular Imaging

The visualization of intracranial arteries (circle of Willis) is easily accomplished by transcranial color-coded sonography (TCCS). As in conventional transcranial Doppler (TCD), the insonation occurs through the temporal bone window. Acoustic absorption, phase aberration of the incident, and the backscattered sound degrade image quality. Since the introduction of ultrasound contrast agents, the assessment of intracranial vessels improved significantly (Bogdahn et al., 1993; Gerriets et al., 2000; Hölscher et al., 2001). The development of contrast-specific imaging techniques enabled the visualization of the circle of Willis in an angiography-like mode (Hölscher et al., 2005) (Figure 15.3).

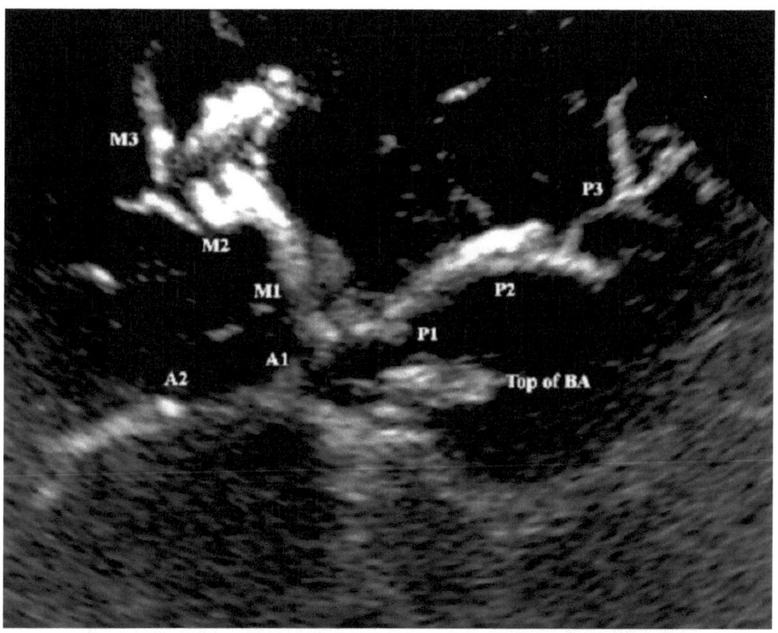

FIGURE 15.3. Transcranial CEUS image of the circle of Willis and its branches after 1 ml of perfluorocarbon-based microbubble contrast agent (Imagent®) injection. M1, M2, M3: middle cerebral artery segments, A1, A2: anterior cerebral artery segments, P1, P2, P3, P4: posterior cerebral artery segments, BS: brainstem, Top of BA: hyperechoic distal part of the basilar artery. Reproduced with permission from Hölscher et al. (2004).

Currently, the main clinical application for TCCS and/or CEUS is the assessment of stroke patients (Gerriets et al., 2002; Postert et al., 1998a, 1999). In 1995, Seidel et al. and coworkers (1995) discussed the potential and limitation of TCCS for stroke diagnosis in a case series. Forty-eight stroke patients were studied within 48 hours of onset of symptoms. Intracranial vessel occlusion was detected in 15 of 48 cases with TCCS. Further, TCCS allowed the monitoring of recanalization in 12 out of these 15 patients to detect continued occlusion, reperfusion, or hyperperfusion when the vessel was recanalized.

15.3.2.2. Microvascular Imaging

Similar to PET, SPECT, MRI, or CT, innovative contrast-enhanced transcranial ultrasound techniques aim to visualize and semiquantitatively measure brain tissue perfusion. The basic principles of microvascular imaging with ultrasound are based on the specific acoustic behavior of microbubble-based contrast agents. Current ultrasound perfusion techniques take advantage of specific acoustic properties of microbubbles to separate backscattered signals generated from microbubbles from those generated from tissue. Kinetic modeling enables off-line generation of time intensity curves, describing wash in and wash out rates in a region of interest. Parameters, like arrival time, time to peak intensity or peak intensity, are commonly used to semiquantify perfusion patterns (Figure 15.4). Further, because microbubbles can be destroyed by ultrasound, calculating the replenishment rate using kinetic models can also estimate tissue perfusion. Current limitations of brain perfusion parametric imaging with ultrasound are the quality of the temporal bone window, the assessment of a single imaging plane, and the need for post-processing.

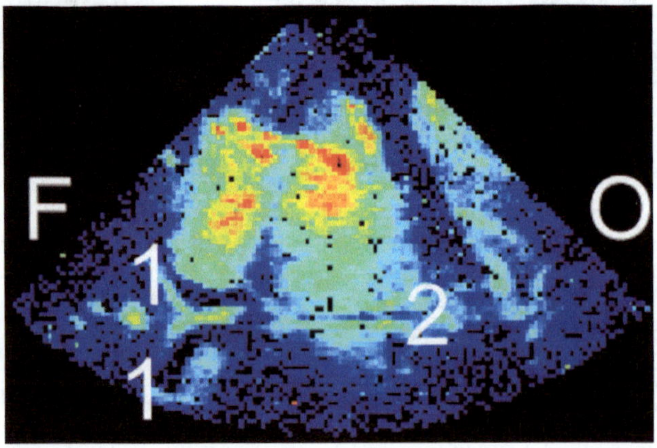

FIGURE 15.4. Parametric image, generated after ipsilateral transtemporal axial ultrasound images were acquired as a 1 ml bolus of contrast agent (Optison™) through the intracranial circulation. F: frontal, O: occipital, 1: anterior lateral ventricles, 2: third ventricle. Reproduced with permission from Hölscher et al. (2002).

In non-stroke patients, the feasibility and reproducibility of contrast-enhanced brain perfusion assessment in the supply area of the middle cerebral artery could be demonstrated (Postert et al., 1998b, 2000; Seidel et al., 2000, 1999). Early experience in stroke patients have been recently published showing abnormal perfusion patterns in the center of infracted areas (Federlein et al., 2000). These early results were confirmed by other groups (Meyer et al., 2003; Seidel et al., 2003). The potential to differentiate the infracted region from the tissue at risk, based on perfusion assessment, was demonstrated in a study by Eyding et al. (2006). However, stroke patient data, correlating ultrasound perfusion to other established perfusion techniques are still pending.

15.3.2.3. Contrast-Enhanced Sonothrombolysis

In the 1990s, the first experimental studies showed the positive effect of ultrasound on thrombolysis (Kornowski et al., 1994; Lauer et al., 1992; Olsson et al., 1994). This effect seems to be potentiated when ultrasound is used in combination with a thrombolytic agent or an ultrasound contrast agent (Behrens et al., 1999; Francis et al., 1995). This effect is due to the possibly improved penetration of lytic agents into the clot because of the temporary disaggregation of the fibrin mesh by ultrasound (Braaten et al., 1997; Devcic-Kuhar et al., 2004). Based on these and other experimental results, first clinical trials have been performed within the last few years. The further improvement of the positive effect of ultrasound in combination with rt-PA treatment has been shown in a randomized and prospective trial, in which the use of an ultrasound contrast agent, in combination with rt-PA and diagnostic ultrasound, could be shown to accelerate thrombolysis (Molina et al., 2006).

15.4. TIPS and Other Liver Vessels

Assessment of TIPS (transjugular intrahepatic porto-systemic shunt) patency is another important vascular application of CEUS. TIPS placement is a widely accepted treatment for the complication of portal hypertension, including variceal bleeding and refractory ascites that is much less invasive than portosystemic shunt surgery. However, TIPS stent dysfunction is common and the 1-year primary patency rate is reported to be between 25% and 66%. Duplex ultrasound is used to screen for TIPS dysfunction particularly when patient's symptoms recur to distinguish stent stenosis from other causes. The Duplex study is difficult due to the typically cirrhotic and shrunken liver requiring imaging through the intercostal space and limiting the angle of incidence to optimally evaluate the stent. Furthermore, the complex curvature of the stent makes angle correction difficult, decreasing the accuracy of velocity measurements. Because of these difficulties, results from retrospective studies have shown sensitivity that varied from 86% to 100% and specificity from 54% to 98% (Chong et al., 1993; Dodd et al., 1995; Feldstein et al., 1996; Foshager et al., 1995; Kanterman et al.,

1997). Venography is required for diagnosis and retreatment when ultrasound is positive, equivocal, or inconclusive.

Since ultrasound clearly shows the wall of the stent and CEUS shows complete filling of the stent lumen, assessment of intimal thickening is easily accomplished (Figure 15.5). In addition, the entire tract from the portal vein to the inferior vena cava is well visualized allowing the complete assessment of the portal vein, the seating of the stent within the vein, as well as the intrahepatic vein between the stent and the cava that can be frequently narrowed from increased intrahepatic capsular pressure. The filling of the lumen with microbubbles on gray-scale contrast-specific imaging is superior to Doppler, because it is complete and independent of the angle of incidence and flow velocity. As expected, a thrombosed TIPS stent does not fill with contrast and remains unenhanced.

Our study included 17 patients that had an abnormal or equivocal duplex study that were scheduled to have or had a recent X-ray venogram with pressure measurements from the portal vein to the right atrium. The CEUS had a 91%

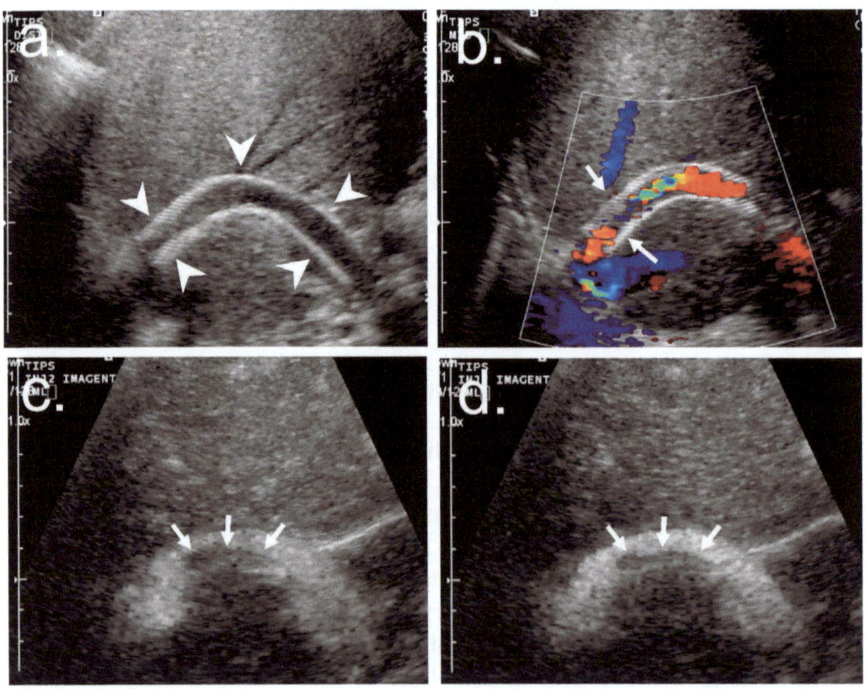

FIGURE 15.5. TIPS stent imaged with CEUS. **a** Conventional grayscale imaging showing the stent wall (arrowheads). Isoechoic intimal thickening cannot be visualized and stenoses are difficult to detect. **b** Color Doppler ultrasound image suggests the presence of narrowing (arrows). Assessment of stenoses is done by velocity measurement using duplex Doppler. **c, d** Grayscale contrast-enhanced ultrasound following the bolus injection of 1 ml of contrast clearly delineates the region of intimal thickening (arrows), location, and degree of stenosis.

positive predictive value for the presence of stenosis as compared to 69% for the duplex study. Using the pressure gradient measurements as gold standard, the sensitivity of CEUS to detect focal stenosis was 100%, while the morphological venogram was only 70% (Kono et al., 2001).

Hepatic arterial thrombosis (HAT) is a common complication of orthotopic liver transplantation (OLT) occurring in as high as 42% of children and 7% to 12% of adults (Nishida et al., 2002; Wozney et al., 1986). Doppler US can detect early HAT that when promptly treated can significantly reduce the incidence of biliary complication and graft loss (Nishida et al., 2002). As with other vascular beds, our experience with a small number of cases suggests that CEUS is more accurate in the detection of HAT, as well as recognizing diminutive hepatic arterial flow post-transplantation from occlusion compared to Doppler techniques.

Portal venous thrombosis (PVT) is a common complication of liver cirrhosis and hepatocellular carcinoma (HCC). Although Doppler imaging can detect and suggest that a thrombus is malignant, the ability to distinguish a benign from a malignant thrombus can be difficult on ultrasound. Since this distinction is important for patient management, and since both benign and malignant thrombi can co-exist even in the presence of an HCC in a cirrhotic liver, we aimed to determine whether CEUS can allow this distinction. Because of the extreme sensitivity of ultrasound to microbubbles and because a tumor thrombus is supplied by the hepatic artery that delivers contrast to the thrombus before the portal vein is enhanced, this distinction was fairly straightforward with CEUS (Kono et al., 2003). Malignant thrombi enhanced in a similar fashion as the primary tumor displaying similar washout characteristics, while benign thrombi appeared as filling defects within the enhanced portal vein or the entire portal vein failed to enhance (Figure 15.6).

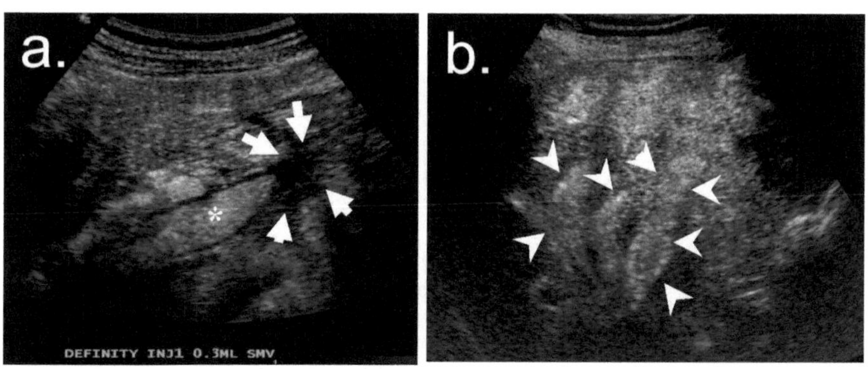

FIGURE 15.6. Thrombi in portal veins. **a** A bland thrombus in the superior mesenteric vein (SMV) (arrows) that appears as a filling defect compared to the enhanced normal lumen of the main portal vein (*) following the injection of 0.5 ml of contrast. **b** Tumor thrombus in the right portal vein and its branches (arrowheads) that enhance avidly during the arterial phase after contrast injection.

15.5. Conclusions

There are many other applications not mentioned here. Any vessel, disease, or equivocal ultrasound study can benefit from a contrast-enhanced study, particularly if slow or small volume flow needs to be detected, such as vessels with stenoses, thromboses, or aneurysms, or to assess for aortic or arterial stent stenoses or leaks. Contrast ultrasound will provide better quality images with higher spatial and temporal resolution than color or power Doppler imaging. Further, if duplex data is needed, contrast will also enhance Doppler signals to improve assessment.

Microbubbles have unique property that they are easily destroyed by ultrasound. By controlling frame rate and ultrasound transmit power, the destruction of the bubbles is easily controlled. If we minimize destruction by very low power and/or low frame rate, bubbles will reach capillary level and therefore will provide perfusion image. On the other hand, if we increase the transmit power that is still relatively low compared to conventional ultrasound imaging, with higher frame rate, the microbubbles are easily destroyed and therefore capillary or small vessels with slower blood flow will not be filled by new microbubbles before next frame and only larger, faster flow vessels will be imaged, providing angiography-like images. This allows the operator to image from feeding arteries to perfused capillaries.

While any body region that is imagable with ultrasound can benefit from CEUS, challenges in imaging the intracranial vessels and brain remain. With the advent of focused ultrasound therapy and the possibility to treat tumors and dissolve clots non-invasively, CEUS should not only potentiate the therapeutic effect, but prove to be essential for guiding the focused beam to the region of interest.

References

AbuRahma, A.F., Pollack, J.A., Robinson, P.A., Mullins, D., 1997. The reliability of color duplex ultrasound in diagnosing total carotid artery occlusion. Am J Surg 174,2, 185–187.

Behrens, S., Daffertshofer, M., Spiegel, D., Hennerici, M., 1999. Low-frequency, low-intensity ultrasound accelerates thrombolysis through the skull. Ultrasound Med Biol 25,2, 269–273.

Bogdahn, U., Becker, G., Schlief, R., Reddig, J., Hassel, W., 1993. Contrast-enhanced transcranial color-coded real-time sonography. Results of a phase-two study. Stroke, 24,5, 676–684.

Bornstein, N.M., Beloev, Z.G., Norris, J.W., 1988. The limitations of diagnosis of carotid occlusion by Doppler ultrasound. Ann Surg 207,3, 315–317.

Braaten, J.V., Goss, R.A., Francis, C.W., 1997. Ultrasound reversibly disaggregates fibrin fibers. Thromb Haemost, 78,3, 1063–1068.

Chong, W.K., Malisch, T.A., Mazer, M.J., Lind, C.D., Worrell, J.A., Richards, W.O., 1993. Transjugular intrahepatic portosystemic shunt: US assessment with maximum flow velocity. Radiology 189,3, 789–793.

Coley, B.D., Trambert, M.A., Mattrey, R.F., 1994. Perfluorocarbon-enhanced sonography: value in detecting acute venous thrombosis in rabbits. AJR Am J Roentgenol, 163,4, 961–964.

Devcic-Kuhar, B., Pfaffenberger, S., Gherardini, L., Mayer, C., Groschl, M., Kaun, C., Benes, E., Tschachler, E., Huber, K., Maurer, G., Wojta, J., Gottsauner-Wolf, M., 2004. Ultrasound affects distribution of plasminogen and tissue-type plasminogen activator in whole blood clots in vitro. Thromb Haemost, 92,5, 980–985.

Dodd, G.D., 3rd, Zajko, A.B., Orons, P.D., Martin, M.S., Eichner, L.S., Santaguida, L.A., 1995. Detection of transjugular intrahepatic portosystemic shunt dysfunction: value of duplex Doppler sonography. AJR Am J Roentgenol 164,5, 1119–1124.

Eisenberg, R.L., Nemzek, W.R., Moore, W.S., Mani, R.L., 1977. Relationship of transient ischemic attacks and angiographically demonstrable lesions of carotid artery. Stroke 8,4, 483–486.

Eyding, J., Krogias, C., Schollhammer, M., Eyding, D., Wilkening, W., Meves, S., Schroder, A., Przuntek, H., Postert, T., 2006. Contrast-enhanced ultrasonic parametric perfusion imaging detects dysfunctional tissue at risk in acute MCA stroke. J Cereb Blood Flow Metab 26,4, 576–582.

Fan, P., Czuwala, P.J., Nanda, N.C., Rosenthal, S.M., Yoganathan, A., 1993. Comparison of various agents in contrast enhancement of color Doppler flow images: an in vitro study. Ultrasound Med Biol 19,1, 45–57.

Federlein, J., Postert, T., Meves, S., Weber, S., Przuntek, H., Buttner, T., 2000. Ultrasonic evaluation of pathological brain perfusion in acute stroke using second harmonic imaging. J Neurol Neurosurg Psychiatry 69,5, 616–622.

Feldstein, V.A., Patel, M.D., LaBerge, J.M., 1996. Transjugular intrahepatic portosystemic shunts: accuracy of Doppler US in determination of patency and detection of stenoses. Radiology 201,1, 141–147.

Foshager, M.C., Ferral, H., Nazarian, G.K., Castaneda-Zuniga, W.R., Letourneau, J.G., 1995. Duplex sonography after transjugular intrahepatic portosystemic shunts (TIPS): normal hemodynamic findings and efficacy in predicting shunt patency and stenosis. AJR Am J Roentgenol 165,1, 1–7.

Francis, C.W., Blinc, A., Lee, S., Cox, C., 1995. Ultrasound accelerates transport of recombinant tissue plasminogen activator into clots. Ultrasound Med Biol 21,3, 419–424.

Gerriets, T., Goertler, M., Stolz, E., Postert, T., Sliwka, U., Schlachetzki, F., Seidel, G., Weber, S., Kaps, M., 2002. Feasibility and validity of transcranial duplex sonography in patients with acute stroke. J Neurol Neurosurg Psychiatry 73,1, 17–20.

Gerriets, T., Postert, T., Goertler, M., Stolz, E., Schlachetzki, F., Sliwka, U., Seidel, G., Weber, S., Kaps, M., 2000. DIAS I: duplex-sonographic assessment of the cerebrovascular status in acute stroke. A useful tool for future stroke trials. Stroke 31,10, 2342–2345.

Girard, M.S., Baker, K.G., Steinbach, G.C., Deiranieh, L.H., Peterson, T.M., Mattrey, R.F., 1999. Assessment of liver and kidney enhancement with a perfluorocarbon vapor-stabilized US contrast agent. Acad Radiol 6,5, 273–281.

Grant, E.G., Duerinckx, A.J., El Saden, S.M., Melany, M.L., Hathout, G.M., Zimmerman, P.T., Marumoto, A.K., Cohen, S.N., Baker, J.D., 2000. Ability to use duplex US to quantify internal carotid arterial stenoses: fact or fiction? Radiology 214,1, 247–252.

Gutberlet, M., Venz, S., Zendel, W., Hosten, N., Felix, R., 1998. Do ultrasonic contrast agents artificially increase maximum Doppler shift? In vivo study of human common carotid arteries. J Ultrasound Med 17,2, 97–102.

Hilpert, P.L., Mattrey, R.F., Mitten, R.M., Peterson, T., 1989. IV injection of air-filled human albumin microspheres to enhance arterial Doppler signal: a preliminary study in rabbits. AJR Am J Roentgenol 153,3, 613–616.

Hölscher, T., Schlachetzki, F., Bauer, A., Rosin, L., Gotz, B., Bogdahn, U., 2001. Echo-enhanced transcranial color-coded US: clinical usefulness of intravenous infusion versus bolus injection of SH U 508A. Radiology 219,3, 823–827.

Hölscher, T., Postert, T., Meves, S., Thies, T., Ermert, M., Bogdahn, U., 2002. Assessment of brain perfusion with echo contrast specific imaging modes and Optison™. Acad Radiol Suppl 2, S386–S388.

Hölscher, T., Wilkening, W.G., Lyden, P.D., Mattrey, R.F., 2005. Transcranial ultrasound angiography (T USA): a new approach for contrast specific imaging of intracranial arteries. Ultrasound Med Biol 31,8, 1001–1006.

Hood, D.B., Mattos, M.A., Mansour, A., Ramsey, D.E., Hodgson, K.J., Barkmeier, L.D., Sumner, D.S., 1996. Prospective evaluation of new duplex criteria to identify 70% internal carotid artery stenosis. J Vasc Surg 23,2, 254–262.

Kanterman, R. Y., Darcy, M. D., Middleton, W. D., Sterling, K. M., Teefey, S. A., and Pilgram, T. K. 1997. Doppler sonography findings associated with transjugular intrahepatic portosystemic shunt malfunction. AJR Am J Roentgenol, 168,2, 467–472.

Klibanov, A.L., Rasche, P.T., Hughes, M.S., Wojdyla, J.K., Galen, K.P., Wible, J.H.J., Brandenburger, G.H., 2002. Detection of individual microbubbles of an ultrasound contrast agent: fundamental and pulse inversion imaging. Acad Radiol Suppl 9, 279–281.

Kono, Y., Alton, K., Rose, S., Hassanein, T., Choi, S., Mattrey, R., 2003. The usefulness of contrast-enhanced ultrasound in the diagnosis of benign and malignant venous thrombosis. Ultrasound Med Biol 29,5, S210.

Kono, Y., Mattrey, R., Pinnell, S., Rose, S., Hassanein, T., 2001. Ultrasound Contrast Venography of TIPS - Gray-scale Ultrasound vs X-ray Venography. J Ultrasound Med 20,3, S82.

Kono, Y., Pinnell, S.P., Sirlin, C.B., Sparks, S.R., Georgy, B., Wong, W., Mattrey, R.F., 2004. Carotid arteries: contrast-enhanced US angiography–preliminary clinical experience. Radiology 230,2, 561–568.

Kornowski, R., Meltzer, R.S., Chernine, A., Vered, Z., Battler, A., 1994. Does external ultrasound accelerate thrombolysis? Results from a rabbit model. Circulation 89,1, 339–344.

Lauer, C.G., Burge, R., Tang, D.B., Bass, B.G., Gomez, E.R., Alving, B.M., 1992. Effect of ultrasound on tissue-type plasminogen activator-induced thrombolysis. Circulation 86,4, 1257–1264.

Meyer, K., Wiesmann, M., Albers, T., Seidel, G., 2003. Harmonic imaging in acute stroke: detection of a cerebral perfusion deficit with ultrasound and perfusion MRI. J Neuroimaging 13,2, 166–168.

Molina, C.A., Ribo, M., Rubiera, M., Montaner, J., Santamarina, E., Delgado-Mederos, R., Arenillas, J.F., Huertas, R., Purroy, F., Delgado, P., Alvarez-Sabin, J., 2006. Microbubble administration accelerates clot lysis during continuous 2-MHz ultrasound monitoring in stroke patients treated with intravenous tissue plasminogen activator. Stroke 37,2, 425–429.

Moneta, G.L., Edwards, J.M., Chitwood, R.W., Taylor, L.M., Jr., Lee, R.W., Cummings, C.A., Porter, J.M., 1993. Correlation of North American Symptomatic Carotid Endarterectomy Trial (NASCET) angiographic definition of 70% to 99% internal carotid artery stenosis with duplex scanning. J Vasc Surg 17,1, 152–159.

Nishida, S., Kato, T., Levi, D., Naveen, M., Thierry, B., Vianna, R., Selvaggi, G., Buitorago, E., Al-Niami, A., Nakamura, N., Vaidya, A., Nery, J., Tzakis, A., 2002. Effect of protocol Doppler ultrasonography and urgent revascularization on early hepatic artery thrombosis after pediatric liver transplantation. Arch Surg 137,11, 1279–1283.

Olsson, S.B., Johansson, B., Nilsson, A.M., Olsson, C., Roijer, A., 1994. Enhancement of thrombolysis by ultrasound. Ultrasound Med Biol 20,4, 375–382.

Postert, T., Braun, B., Meves, S., Koster, O., Przuntek, H., Weber, S., Buttner, T., 1999. Contrast-enhanced transcranial color-coded sonography in acute hemispheric brain infarction. Stroke 30,9, 1819–1826.

Postert, T., Federlein, J., Braun, B., Koster, O., Bornke, C., Przuntek, H., Buttner, T., 1998a. Contrast-enhanced transcranial color-coded real-time sonography: a reliable tool for the diagnosis of middle cerebral artery trunk occlusion in patients with insufficient temporal bone window. Stroke 29,5, 1070–1073.

Postert, T., Hoppe, P., Federlein, J., Helbeck, S., Ermert, H., Przuntek, H., Buttner, T., Wilkening, W., 2000. Contrast agent specific imaging modes for the ultrasonic assessment of parenchymal cerebral echo contrast enhancement. J Cereb Blood Flow Metab 20,12, 1709–1716.

Postert, T., Muhs, A., Meves, S., Federlein, J., Przuntek, H., Buttner, T., 1998b. Transient response harmonic imaging: an ultrasound technique related to brain perfusion. Stroke 29,9, 1901–1907.

Seidel, G., Albers, T., Meyer, K., Wiesmann, M., 2003. Perfusion harmonic imaging in acute middle cerebral artery infarction. Ultrasound Med Biol 29,9, 1245–1251.

Seidel, G., Algermissen, C., Christoph, A., Katzer, T., Kaps, M., 2000. Visualization of brain perfusion with harmonic gray scale and power doppler technology: an animal pilot study. Stroke 31,7, 1728–1734.

Seidel, G., Greis, C., Sonne, J., Kaps, M., 1999. Harmonic grey scale imaging of the human brain. J Neuroimaging 9,3, 171–174.

Seidel, G., Kaps, M., Gerriets, T., 1995. Potential and limitations of transcranial color-coded sonography in stroke patients. Stroke 26,11, 2061–2066.

van Everdingen, K.J., van der Grond, J., Kappelle, L.J., 1998. Overestimation of a stenosis in the internal carotid artery by duplex sonography caused by an increase in volume flow. J Vasc Surg 27,3, 479–485.

Widder, B., Paulat, K., Hackspacher, J., Hamann, H., Hutschenreiter, S., Kreutzer, C., Ott, F., Vollmar, J., 1990. Morphological characterization of carotid artery stenoses by ultrasound duplex scanning. Ultrasound Med Biol 16,4, 349–354.

Wozney, P., Zajko, A.B., Bron, K.M., Point, S., Starzl, T.E., 1986. Vascular complications after liver transplantation: a 5-year experience. AJR Am J Roentgenol 147,4, 657–663.

Young, N., Soo, Y.S., Fischer, P., 1992. Comparison of duplex ultrasound with angiography in assessment of carotid bifurcation disease. Australas Radiol 36,1, 54–58.

16
Targeted Microbubbles: Ultrasound Contrast Agents for Molecular Imaging

Alexander L. Klibanov

Abstract: Targeted ultrasound contrast microbubble materials are designed as hollow micron-size spheres with a biocompatible shell stabilizer. Excellent detection sensitivity of these particles (down to individual particles of a picogram mass) is achieved routinely with clinical ultrasound. To improve stability and in vivo circulation time, poorly soluble perfluorinated gases can be used as the particle core. Microbubble shell (from several to several hundred nanometers thick) can be made of protein, polymer, lipid, surfactant, or combinations of these. Targeting ligands (antibodies, peptides and mimetics, carbohydrates and combinations) can be attached to the shell, either directly or via a protein/polymer spacer arm. To improve targeting in fast flow conditions that are characterized by high wall shear stress, microbubbles can be outfitted with surface folds/microvilli. Alternatively, bio-inspired fast-binding ligands, such as peptides from P-selectin Glycoprotein Ligand-1 or sialyl Lewis carbohydrates can be applied.

Selective targeting of the microbubbles to the vascular markers of diseases was successful in multiple animal model studies. Selectins, integrins, or addressins on vascular endothelium surface that are upregulated in response to inflammation or ischemia-reperfusion injury have been targeted successfully, ultrasound contrast enhancement of target tissues achieved. Angiogenic endothelium in the tumor vasculature has been imaged by targeting microbubbles to $\alpha_v\beta_3$ and similar molecules. Overall, the convenience, low cost, and real-time capabilities of ultrasound imaging, combined with molecular imaging capabilities of targeted microbubble contrast agents, may offer significant improvement in patient diagnostics and treatment progress monitoring.

Keywords: ultrasound contrast, echo contrast, targeting, selectin, inflammation, microbubbles.

16.1. Introduction: Targeted Ultrasound Contrast

The concept of targeted ultrasound contrast imaging has been proposed decades ago. The general idea is the same as for targeted molecular imaging with other contrast detection modalities. Labeled contrast is administered in the patient body (most likely in the bloodstream) and accumulates in the disease area. Free-circulating contrast material clears from the bloodstream and non-target tissues. Imaging with contrast detection is then performed and the area of disease is detected and delineated. The substantial difference between ultrasound and other modalities is in the nature of contrast agent. All of the described ultrasound contrast agents are of particulate nature; molecules of targeting ligands selective for the disease markers are coupled to the particle surface. Therefore, hydrodynamic considerations become critical for success of target binding. Many of the described ultrasound contrast particles are characterized by a large size (micrometers) and contain an internal gas core, which provides improved detection capability for ultrasound clinical imaging systems. The contrast particle shell is usually a thin membrane (from several nanometers to submicron thickness); ligands are attached to the membrane directly or via a spacer molecule (Figure 16.1). Nanoengineering (with a strong self-assembly component) becomes critical for the design and manufacturing of these contrast agent particles.

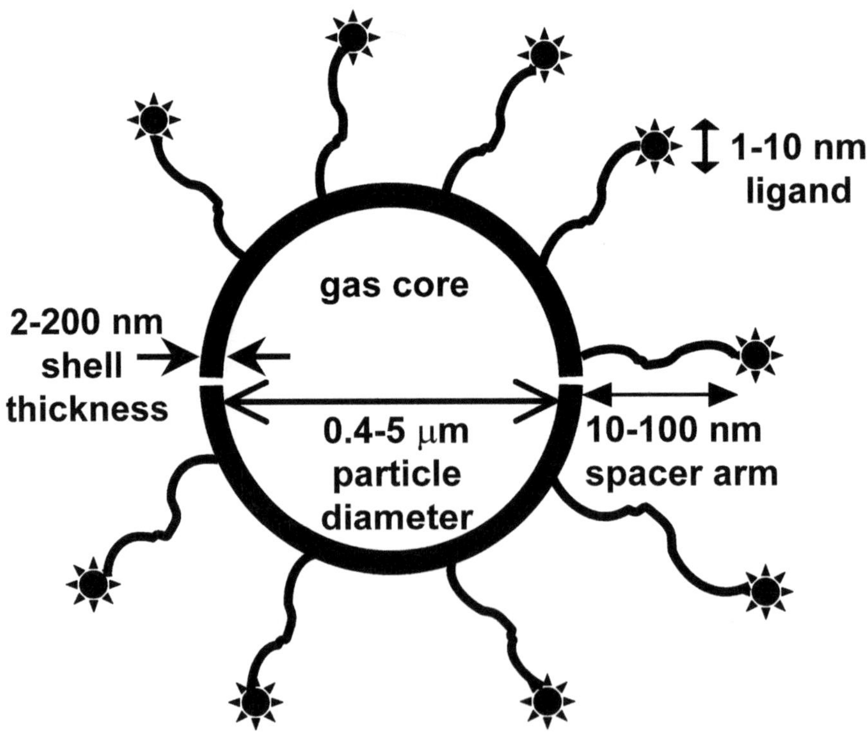

FIGURE 16.1. General design of a targeted microbubble ultrasound contrast agent. Microbubble shell (central circle) surrounds and stabilizes the gas core. Target-specific ligands (stars) are attached to the shell via an extended polymer spacer arm.

In order for the successful targeted ultrasound imaging to occur, several key requirements need to be addressed.

(1) The ultrasound contrast agent should be non-toxic, biocompatible/biodegradable and harmless to the patient.
(2) The ultrasound contrast agent needs to accumulate in the target area rapidly, selectively, and firmly, with minimal accumulation in the non-target tissues.
(3) The clearance of background ultrasound contrast agent from the circulation should be efficient, so that high target-to-blood contrast ratio could be achieved rapidly.
(4) The targeted contrast agent should be easy to prepare and administer; it should be stable on storage.
(5) The contrast particles accumulated in the target tissue should be sufficiently stable to perform an ultrasound-imaging exam.
(6) The sensitivity of contrast ultrasound imaging should be sufficient for the detection of contrast particles, with minimal interference signal from background tissues.

This set of requirements possesses a number of built-in inconsistencies; for instance, contrast is required to be stable on the target, yet to be removed from the bloodstream rapidly. However, reasonable compromises can be achieved. A number of targeted ultrasound contrast agents are already in preclinical testing, with some of them hopefully reaching clinical application in the next several years. In the following sections of this chapter, we will discuss the considerations for successful design and implementation of targeted microbubble contrast materials.

16.2. Microbubble Contrast Particle Design

Ultrasound imaging operates by sending pulses of ultrasound pressure waves (in the MHz frequency range) from the transducer into the tissue and listening to the echo response. Time delay of the echo is used to deduce the depth of the structures that produce echo, such as organ boundaries and tissue irregularities (Kremkau, 2002). Because gas is much more compressible than biological tissue (which is mostly water), a gas bubble subjected to the pressure field reduces size as the compression wave passes through its location; its size is increased during the rarefaction phase of the pressure wave. Such a cycle of movement of gas–liquid interface creates a secondary sound wave, which can be detected by the ultrasound probe as it listens to the reflected ultrasound signal (Leighton, 1994). Typical peak acoustic pressure values in diagnostic ultrasound imaging range from tens to hundreds of KPa (positive and negative), so the change in the volume of gas particle from its equilibrium value upon its interaction with ultrasound wave may be significant (Chomas et al., 2000). Most efficient scattering of ultrasound is achieved by unimpeded vibration of gas bubbles that completely lack the shell. However, a shell needs to be present—otherwise, bubbles would fuse with each other on storage and potentially in the bloodstream (which could make their practical application difficult and even unsafe). Shell-free bubbles are captured and destroyed in the lung and only make it into the right heart after intravenous injection (Vonbibra et al., 1989). Obviously, shell characteristics, such as material composition, structure, and thickness determine the efficacy of ultrasound scattering and stability of the bubble in the ultrasound field. Generally, microbubbles made with a lipid monolayer shell (a couple of nanometer thickness) possess excellent acoustic backscatter response (Albrecht et al., 2000). A polymer shell, such as polylactide (Narayan and Wheatley, 1999), is significantly thicker and much more difficult to compress and expand. Therefore, polymer bubbles may require significant ultrasound energy applied to "break the ice", and become acoustically reflective only when the acoustic pressure is above certain threshold level; sometimes hundreds of KPa in acoustic pressure is necessary (Bouakaz et al., 2005), which means that non-destructive imaging of polymeric targeted bubbles in the situations of low acoustic irradiation may not be possible. The destruction threshold pressure is dependent on the particle diameter (Bloch et al., 2005).

On a per-particle basis, sensitivity of detection of gas-filled microbubble contrast agents by ultrasound imaging exceeds significantly the ability to detect ultrasound contrast materials of other design types, such as nanoemulsion of liquid fluorocarbon (Lanza et al., 2000), or multilamellar echogenic liposomes (AlkanOnyuksel et al., 1996). Individual microbubbles with the size of one or several micrometers and picogram mass can be observed on the screen of clinical ultrasound imaging systems (Klibanov et al., 2002, 2004). The current generation of contrast-detection features in these systems provides the ability to suppress signal from the tissues and enhances the signal-to-noise ratio for microbubble detection in the clinical imaging situations. Essentially, all major ultrasound equipment manufacturers have contrast detection software packages available; these are based on the differential response of highly compressible bubbles and biological tissues to the compression waves of varying intensity and polarity, as well as frequency response differences. The high sensitivity of ultrasound contrast and its ability to detect individual contrast particles is the basis for ultrasound contrast molecular imaging. Of the 10^7–10^9 contrast particles injected intravenously, most circulate through non-target tissues and gradually lose entrapped gas and echogenicity. However, even several thousand particles that successfully target to the areas of interest would be sufficient for contrast-enhanced delineation of the target. The total mass of administered ultrasound contrast would still be negligible when compared with the MRI contrast agents.

The choice of gas core for microbubble preparation may determine the stability of the contrast particles and their circulation time. If the microbubble shell is thin ($\sim$2 nm lipid monolayer or $\sim$10–15 nm protein coat), the use of fluorinated gas with low solubility in water improves bubble circulation lifetime from seconds to minutes (Schneider, 1999). Thicker (submicron) polymer shells are much better barriers for gas diffusion, so even air-filled polymeric bubbles are quite stable and allow reasonable bloodstream circulation time (Narayan and Wheatley, 1999).

16.3. Manufacturing of Microbubble Contrast

For the manufacturing of ultrasound contrast microbubble agents, self-assembly, emulsion, and colloid chemistry approaches (which can be qualified as nanotechnology) are applied routinely. Usually, gas is dispersed in the liquid aqueous medium that contains shell components in the form of solution, liposomes, or micelles. Shell material from the aqueous phase is immediately deposited onto the newly formed gas–liquid interface and self-assembles into a membrane coat that stabilizes the gas bubble and prevents its dissolution in the surrounding medium. Hydrophobic interaction plays a crucial role in this process. Fusion of neighboring bubbles is also prevented after shell formation is completed. Denatured albumin protein (Grinstaff and Suslick, 1991; Keller et al., 1989; Porter and Xie, 1995) or lipid-based (Fritz et al., 1997; Klibanov et al., 1999; Schneider, 1999) microbubbles can be prepared this way. In some instances, the protein shell may be cross-linked (e.g., by disulfide bonds (Grinstaff and Suslick, 1991)), which provides further stabilization. Depending on their composition, some types of contrast particles prepared this way can be quite stable (up to several years) in the aqueous phase in refrigerated storage.

Microbubbles stabilized with the biodegradable polymeric shells (usually polylactide, lactide-co-glycolide, or cyanoacrylate) have been known in the drug delivery technology for many years. Some preparations are in clinical trials as blood pool ultrasound contrast agents (Wei et al., 2003). These materials are often manufactured by emulsion techniques followed by spray-drying and lyophilization. Generally, an oil–water microemulsion with the hydrophobic polymer is prepared by sonication or high-shear mixing. Volatile solvents are removed under reduced pressure, spray-drying, and subsequent lyophilization. Non-volatile polymer forms the shell that may surround a single void; alternatively, a collection of voids may exist in a single particle, creating a porous structure (Narayan et al., 2001). As vacuum is released after lyophilization, particles can be filled with a gas of choice, such as perfluorocarbon, for enhanced stability (Straub et al., 2005). Obviously, this technique is more complicated than the lipid monolayer self-assembly or stabilization by denatured protein. Polymer particles cannot be stored in the aqueous phase for extended period of time due to hydrolytic degradation of polylactide, so they are stored in the dry lyophilized form and reconstituted prior to use.

16.4. Targeting Ligands on the Microbubble Surface

Attachment of targeting ligands to the surface of gas-filled microbubbles can be accomplished by the standard protocols which are applied in biotechnology and bioconjugate chemistry for the coupling of proteins or polymers to microparticles and nanoparticles, such as liposomes (Hermanson, 1996). The most popular simple laboratory technique is the use of a biotin–streptavidin–biotin sandwich (Lindner et al., 2001). Briefly, a biotinylated lipid, such as a polymer or protein, is first incorporated in the shell of microbubbles. Next, excess of biotinylated material is removed from the particle dispersion by flotation and streptavidin is added. Because streptavidin has four binding pockets for biotin, it can serve as a bridge between the bubble and the biotinylated targeting ligand, which is added at the third step to top the sandwich. The main advantage of this approach is its "cassette" nature. Any biotinylated ligand or ligand combination can be added to streptavidin bubbles in a simple and straightforward procedure. The main disadvantage is in the use of streptavidin, a foreign protein that could be immunogenic. Therefore, while this approach is excellent for short-term feasibility studies, it will not be applicable in the clinical setting. A more clinically relevant approach would be to attach the targeting ligands to the bubble surface by covalent (Villanueva et al., 1998) or non-covalent, such as electrostatic (Toublan et al., 2006), interaction. In the first case, the bubble shell would carry a chemically reactive group, activated carboxyl, aldehyde, or maleimide, and the ligand would have an unprotected amine or thiol counterpart. In the second case, a targeting ligand is placed on a polyelectrolyte chain and is attached to the bubble shell via a charge interaction. An alternative would be to first attach the ligand covalently to the bubble-making material (polymer or lipid), and then add that to the component mixture prior to the bubble manufacturing. The advantage of this approach is in its ability to produce a purified and fully characterized covalent conjugate between targeting ligand and shell anchor. The disadvantage is in the limited range of suitable ligands (antibodies and other proteins may be inactivated during the bubble-making procedure).

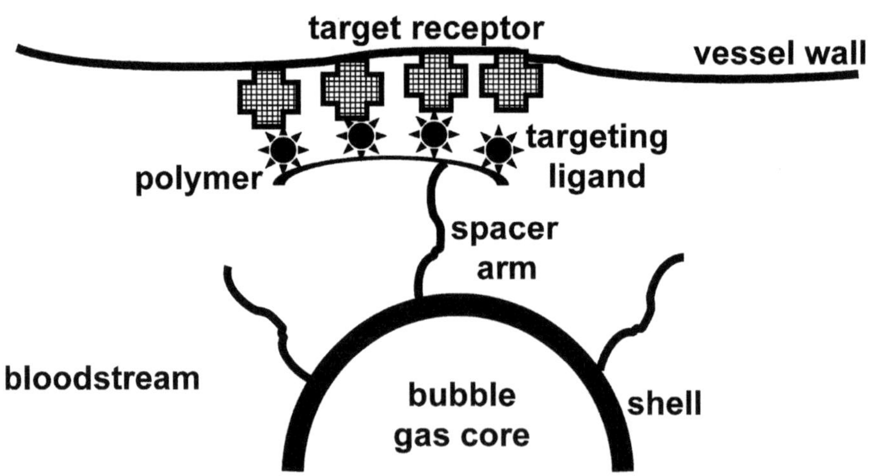

FIGURE 16.2. Umbrella-shaped polymer ligand supramolecular assembly on the surface of targeted microbubble. To improve targeting efficacy, a polymeric cluster carrying an assembly of targeting ligands (stars) is applied. Ligand-carrying polymer is coupled to the microbubble surface by an extended/flexible polymer spacer arm. Cooperative multi-point binding of ligand to the target receptor on the surface of the vessel wall (crosses) is achieved.

Additional benefits to improve the targeting efficacy may be achieved if the coupling between the microbubble shell and targeting ligand is performed indirectly via an extended spacer arm. This approach mimics the interaction between selectins and their ligands and has been in general use for particle-based drug delivery systems. Most beneficial could be the use of a flexible linear polymer spacer arm, such as poly(ethylene glycol), which for a typical molecular mass of several KDa may extend the ligand into the surrounding aqueous medium by tens of nanometers (Jeppesen et al., 2001), creating a nanoscale "fishing line" (Lee and Low, 1995) (Figures 16.1 and 16.2). Targeting ligands coupled to a PEG tip extended from the microbubble surface into the medium are attaching to the model target more efficiently than in case when they are connected to the bubble shell directly (Kim et al., 2000). A multi-micron contrast agent particle can have a significant contact area with the target surface. Therefore, it can be hooked to the target by multiples of the selective ligands that can better reach their target receptors on the endothelium and accommodate the necessary spatial conformation for firm binding. As a cooperative multipoint interaction is achieved, it should be difficult to break (as with Velcro tape).

16.5. Imaging Applications:Suitable Biological Targets and Current State of the Art

Initial studies with targeting monolayer-coated microbubbles were performed in a biotin–avidin model system. Biotinylated PEG-lipid was incorporated in the lipid monolayer membrane of decafluorobutane bubbles and a successful targeting to avidin-coated surfaces was achieved in vitro in a static environment in a Petri dish. Targeted bubbles on the dish surface were observed by microscopy and detected by ultrasound with a clinical imaging system (Klibanov et al., 1997, 1998, 1999).

One of the initial realistic targets tested for microbubble contrast imaging is the detection of a thrombus. Lipid microbubble preparations carrying RGD-peptide on the tip of a PEG spacer arm were successfully attaching to clots in vitro (Unger et al., 1998) and in vivo (Christiansen et al., 2001). Further development of this strategy is not to only image the clot, but to use ultrasound treatment in therapeutic interventions in combination with a plasminogen activator enzyme to enhance the rate of clot dissolution (Wu et al., 1998).

The presence of the clot manifests late critical-stage events in the bloodstream. It may be more beneficial to target earlier markers of disease, e.g. on the surface of the vascular endothelium. Such markers include P- and E-selectins, ICAM-1 or VCAM-1, that are known markers of inflammatory events that are presented on vascular endothelium (including its luminal surface). Targeting molecules to these receptors may include specific antibodies, peptides, oligosaccharides and polysaccharides, and various synthetic mimetic compounds.

An initial study with microbubble targeting to the receptors was performed in vitro in a parallel plate flow chamber in a cell culture setting (Villanueva et al., 1998). Microbubbles carrying the anti-ICAM-1 antibody were attaching selectively to endothelial cells that were carrying ICAM-1 at high surface density after stimulation by interleukin-1β. A nearly 40-fold targeting enhancement was observed over binding to control unstimulated cells.

This study was subsequently extended to the animal model setting, to monitor rejection in a heterotopic heart transplant in rats. Ultrasound contrast backscatter intensity as estimated by image pixel intensity over myocardium was significantly higher in the rejecting hearts (Weller et al., 2003).

Selectins (P- and E-selectin) are expressed on the surface of vascular endothelium as an early response to the inflammatory events in the underlying tissues, to attract leukocytes into the area (Ley, 1992). Selectins are also upregulated in response to ischemia-reperfusion injury (Jordan et al., 1999). Therefore, placing selectin-specific ligands on the surface of microbubbles should allow targeted contrast imaging of these regions of interest. Indeed, targeting of microbubbles to the areas of inflammation via the anti-P-selectin antibody was achieved in a cremaster muscle mouse model of TNF-α-induced inflammation, as well as ischemia-reperfusion injury in a kidney model (Lindner et al., 2001). In nature, P-selectin glycoprotein ligand-1 (PSGL-1) on the surface of leukocytes is used to achieve their deceleration and rolling on activated endothelium (McEver, 1997). Truncated versions of PSGL-1 with proper sulfation of Tyr-46,48,51 and modification of Thr-57 with sialyl Lewis^x (Rychak et al., 2004) or even simpler molecules, such as sialyl Lewis^x itself (Weller et al., 2005a), or its polymeric version (Klibanov et al., 2006) can be placed on the microbubble surface for use as targeted ultrasound contrast. The advantage of these simplified molecules over antibodies is in their ability to bind to selectins regardless of the species of animal model. More importantly, the kinetic rate of association between selectin and its ligand is fast, which brings an important advantage for binding in fast flow conditions. When a particle of contrast agent is moving by the target surface at ~0.1 m/s, and target molecule size is, e.g., ~10 nm, binding has to occur within ~100 ns timeframe, otherwise the ligand–receptor bond would not have a chance to form. While some of the available monoclonal antibodies can provide faster binding, generally, antibody–antigen interaction is firm, but slow. Monoclonal antibody is usually selected for production by a solid phase immunoassay testing, i.e. due to the ability to hold the antigen tightly and release it slowly. In some instances, a combination of a fast ligand (such as sialyl Lewis^x) and a slow but firm ligand, such as a monoclonal antibody, may ensure firm binding of targeted ultrasound contrast microbubbles (Weller et al., 2005a) and drug carrier particles in general (Eniola et al., 2003).

As an alternative to a combination of the fast-binding and slow/firm-binding ligands, a nanoengineering approach can be applied. In nature, leukocytes apply externally pulled microvillous tethers to decelerate and attain slow rolling in the fast-flow conditions (Park et al., 2002). Microbubble contrast agent can be produced not to possess a standard spherical shape and smooth shell, but with the surface irregularities, such as ridges and folds. As a result, the bubbles become more deformable. The contact area between the bubble and the target is increased, which makes firm binding to the target surface more likely (Rychak et al., 2006).

Instead of (or in addition to) surface features on the microbubble shell, specialized supramolecular patterns could be applied. A clustered polymeric version of a selectin ligand sialyl Lewis^x has been available for nearly a decade (WeitzSchmidt et al., 1996). This polymer carries approximately a dozen sialyl Lewis^x residues per chain. Such a polymer has a significantly higher avidity to a selectin-coated surface as compared with monomeric sialyl Lewis^x (~1 mM Kd for the latter is not suitable for targeting use on its own). Placing such polymer ligand on the surface of microbubbles, preferably via a PEG spacer arm, forming "umbrella" supramolecular structures on the surface of the bubbles (Figure 16.2), results in the successful targeting to selectin-expressing surfaces in vitro and in vivo. Targeting is efficient even if the wall shear stress exceeds 4 dyn/cm^2 (Klibanov et al., 2006).

Antibodies provide a high degree of selectivity and will be quite useful as microbubble targeting agents in the conditions of moderate flow. Several important examples of antibody use for microbubble targeting include imaging of lymph nodes by intravenous administration of ultrasound contrast particles carrying antibodies against L-selectin (Hauff et al., 2004), targeted ultrasound imaging of ICAM-1 expression in the lesion-associated vasculature in the experimental encephalomyelitis in rats (Linker et al., 2005), and targeting of inflamed ileum with an antibody against MAdCAM-1 addressin in a murine model (Bachmann et al., 2006). The latter study achieved selective ultrasound contrast enhancement of the areas of ileal inflammation in the conditions of experimental Crohn's disease in mice, but anti-MAdCAM-1-targeted ultrasound contrast did not accumulate in normal, non-inflamed tissue (Figure 16.3). In this study, ultrasound imaging data correlated well with immunohistological findings.

Many tumors have a tortuous neovasculature, where blood flow can be extremely slow. Therefore, antibodies should be perfectly capable for targeting the angiogenic endothelium in these conditions. Specific surface receptors on the endothelium in the tumor/angiogenic setting have been widely described in the literature. Potential targets of interest include $\alpha_V\beta_3$ (Leong-Poi et al., 2003), VCAM-1 (Dienst et al., 2005), and receptors to VEGF, such as KDR (Lu et al., 2003). All these molecules can be useful as targets to delineate angiogenic or apoptotic tumor endothelium location and molecular surface properties by ultrasound molecular imaging. Using antibodies or other ligands, such as targeted ultrasound contrast imaging, has been achieved in animal models of angiogenesis in matrigel plugs in mice (Leong-Poi et al., 2003) and rat tumors (Ellegala et al., 2003). Specific accumulation of targeted microbubbles in the areas that overexpressed $\alpha_V\beta_3$ was observed, with histological correlation, by targeting either with a monoclonal antibody against $\alpha_V\beta_3$ or with a disintegrin peptide echistatin that has a high affinity toward $\alpha_V\beta_3$ or $\alpha_5\beta_1$. An alternative to echistatin on the bubble surface could be a positively charged RRL-carrying peptide, a lipid monolayer decafluorobutane microbubbles carrying a cyclic peptide with RRL

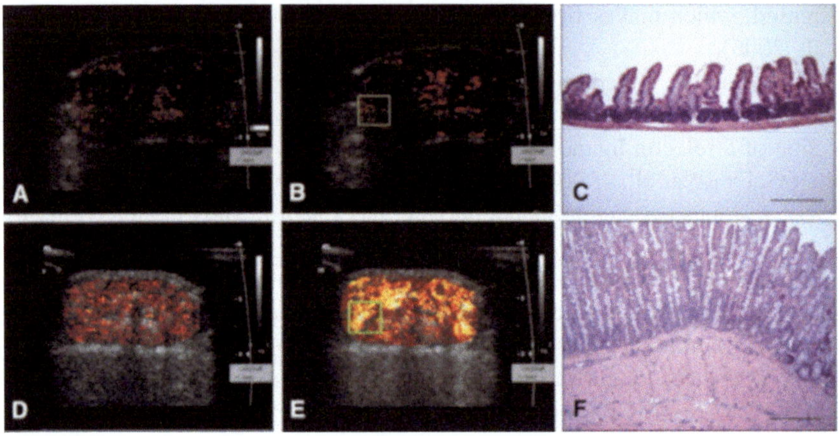

FIGURE 16.3. MBM-targeted, contrast-enhanced transabdominal US in SAMP mice. **A** Healthy AKR mice showed weak background signal following either MBI or (**B**) MBM infusion, whereas (**E**) accumulations of MBM produced strong acoustic echoes that appear as bright regions in color-coded images. **D** Background subtracted color-coded images of MBI in SAMP mice produced little baseline signal. **C** Histologic analysis of ileal tissues isolated from AKR control (**B**, green box) showed normal bowel architecture with no evidence of inflammation, whereas (**F**) ileal tissues obtained from bright regions of SAMP mice showed gross morphologic changes, including acute and chronic inflammation (scale bar = 250 mm), indicative of severe disease (scale bar = 250 μm). Copyright 2006, American Gastroenterological Association. Reprinted from Bachmann et al. (2006), with permission of Elsevier B.V.

sequence, which selectively accumulates in the tumor vasculature (but not in the myocardium) in a mouse model (Weller et al., 2005b).

The efficacy of therapeutic angiogenesis (such as in the ischemic hindlimb model in rodents treated with FGF-2) can be evaluated in detail by contrast ultrasound imaging (Leong-Poi et al., 2005): blood pool ultrasound contrast is used to estimate blood volume and velocity and the presence of $\alpha_V\beta_3$ or $\alpha_5\beta_1$ on the endothelium is evaluated by targeted bubbles decorated with echistatin.

Accumulation of targeted bubbles may be dependent not only on the presence of target receptors on the endothelium, but also on the influx of contrast material coming into the particular areas of the target with the flow of blood. Therefore, it may be necessary to normalize the targeted bubble ultrasound backscatter signal by the perfusion flux data that can be obtained from the non-targeted ultrasound contrast imaging with a blood pool agent.

A targeted ultrasound contrast agent that is in clinical trials and might be first to reach clinical use is based on a simpler principle than having a targeted ligand attached to the microbubble shell. In its composition, the targeting ligand constitutes the shell itself (Hvattum et al., 2006). Phospholipid phosphatidylserine (PS), the external marker of apoptosis on cell membranes, is used to attract phagocytic cells to capture and degrade apoptotic cells and cell vesicles in an organized fashion (Fadok and Chimini, 2001). Model lipid membrane particles that contain PS (e.g., liposomes) are rapidly taken up by the cells of the mononuclear phagocytic system after intravenous administration (Klibanov et al., 1991). Therefore, one can expect that PS-coated decafluorobutane gas bubbles (e.g., Sonazoid) should be targeted to and internalized by Kupffer cells in the liver. This is applied for negative contrast imaging of tumor nodes in the liver tissue (Watanabe et al., 2005), as was done earlier with Levovist contrast (Albrecht et al., 2000). Sonazoid may provide enhanced accumulation efficacy in the normal liver tissue and extended in vivo stability, as compared with the first-generation Levovist contrast agent, which uses air instead of insoluble fluorocarbon gas and palmitic acid coating instead of PS membrane. In addition to selectively targeting Kupffer cells, PS-containing bubbles can be applied for targeted imaging of inflammatory events. An activated leukocyte attached to activated endothelium in the region of inflammation or ischemia-reperfusion injury is capable to capture and internalize microbubbles, especially PS-containing bubbles (Lindner et al., 2000; Christiansen et al., 2002). Despite internalization, microbubbles maintain the ability to scatter ultrasound in the conditions of viscoelastic damping the microenvironment inside the cells (Dayton et al., 2001). After bubble uptake by leukocyte, the loss of gas due to diffusion and dissolution in the surrounding medium should be inhibited and contrast lifetime on the target increased.

The practical importance of clinical applications of targeted molecular imaging with ultrasound contrast may be not just in diagnostic imaging, but also in image-guided therapy to monitor and improve the efficacy of therapeutic interventions. There are procedures (such as tumor radiofrequency ablation) that are already performed under real-time ultrasound guidance. Addition of targeted contrast to the imaging protocol will achieve comprehensive treatment of a tumor mass, while minimizing undesired effects on the surrounding normal tissue.

16.6. Conclusions

Ultrasound contrast materials, stabilized biocompatible gas-filled microspheres, with a particle diameter of several micrometers and stabilizer shell thickness from several to several hundred nanometers have been described. These microbubble particles are decorated with antibodies or other selective ligands (e.g., peptides and carbohydrates). Ligand-carrying microbubbles selectively attach to the surface receptors on vascular endothelium in the areas of inflammation or angiogenesis. Targeted microbubbles are detected by ultrasound imaging with excellent sensitivity: single particles with picogram mass are visualized by clinical ultrasound imaging systems.

Acknowledgment. The author is grateful to the UVA Cardiovascular Division and Cardiovascular Research Center for help and support. Valuable discussions with

Klaus Ley, Jonathan Lindner, Sanjiv Kaul, John Hossack, Michael Lawrence, Kathy Ferrara, Richard Price, and Joshua Rychak are appreciated. Generous support and research collaboration with Philips Research (Christopher Hall, Chien Ting Chin and Marcel Bohmer) is gratefully acknowledged. Generous equipment donation to UVA by Mallinckrodt Inc (Hazelwood, MO) is appreciated. This work is supported in part via NIH BRP EB002185.

References

Albrecht, T., Blomley, M.J.K., Heckemann, R.A., Cosgrove, D.O., Jayaram, V., Butler-Barnes, J., Eckersley, R.J., Hoffmann, C.W., Bauer, A., 2000. Stimulated acoustic emission with the ultrasound contrast agent levovist: a clinically useful contrast effect with liver-specific properties. Rofo-Fortschritte Auf Dem Gebiet Der Rontgenstrahlen Und Der Bildgebenden Verfahren 172, 61–67.

AlkanOnyuksel, H., Demos, S.M., Lanza, G.M., Vonesh, M.J., Klegerman, M.E., Kane, B.J., Kuszak, J., McPherson, D.D., 1996. Development of inherently echogenic liposomes as an ultrasonic contrast agent. J Pharm Sci 85, 486–490.

Bachmann, C., Klibanov, A.L., Olson, T.S., Sonnenschein, J.R., Rivera-Nieves, J., Cominelli, F., Ley, K.F., Lindner, J.R., Pizarro, T.T., 2006. Targeting mucosal addressin cellular adhesion molecule (MAdCAM)-1 to noninvasively image experimental Crohn's disease. Gastroenterology 130, 8–16.

Bloch, S.H., Short, R.E., Ferrara, K.W., Wisner, E.R., 2005. The effect of size on the acoustic response of polymer-shelled contrast agents. Ultrasound in Medicine and Biology 31, 439–444.

Bouakaz, A., Versluis, M., de Jong, N., 2005. High-speed optical observations of contrast agent destruction. Ultrasound in Medicine and Biology 31, 391–399.

Chomas, J.E., Dayton, P.A., May, D., Allen, J., Klibanov, A., Ferrara, K., 2000. Optical observation of contrast agent destruction. Applied Physics Letters 77, 1056–1058.

Christiansen, J.P., Leong-Poi, H., Klibanov, A.L., Kaul, S., Lindner, J.R., 2002. Noninvasive imaging of myocardial reperfusion injury using leukocyte-targeted contrast echocardiography. Circulation 105, 1764–1767.

Christiansen, J.P., Song, J., Matsunaga, T., Lindner, J.R., 2001. Microbubbles targeted to the platelet IIb/IIIa integrin adhere to microvascular thrombi in-vivo. Circulation 104, 589–589.

Dayton, P.A., Chomas, J.E., Lum, A.F.H., Allen, J.S., Lindner, J.R., Simon, S.I., Ferrara, K.W., 2001. Optical and acoustical dynamics of microbubble contrast agents inside neutrophils. Biophys J 80, 1547–1556.

Dienst, A., Grunow, A., Unruh, M., Rabausch, B., Nor, J.E., Fries, J.W.U., Gottstein, C., 2005. Specific occlusion of murine and human tumor vasculature by VCAM-1-targeted recombinant fusion proteins. J Natl Cancer Inst 97, 733–747.

Ellegala, D.B., Poi, H.L., Carpenter, J.E., Klibanov, A.L., Kaul, S., Shaffrey, M.E., Sklenar, J., Lindner, J.R., 2003. Imaging tumor angiogenesis with contrast ultrasound and microbubbles targeted to alpha(v)beta(3). Circulation 108, 336–341.

Eniola, A.O., Willcox, P.J., Hammer, D.A., 2003. Interplay between rolling and firm adhesion elucidated with a cell-free system engineered with two distinct receptor-ligand pairs. Biophys J 85, 2720–2731.

Fadok, V.A., Chimini, G., 2001. The phagocytosis of apoptotic cells. Semin Immunol 13, 365–372.

Fritz, T.A., Unger, E.C., Sutherland, G., Sahn, D., 1997. Phase I clinical trials of MRX-115. A new ultrasound contrast agent. Invest Radiol 32, 735–740.

Grinstaff, M.W., Suslick, K.S., 1991. Air-Filled Proteinaceous Microbubbles – Synthesis of an Echo-Contrast Agent. Proc Natl Acad Sci U S A 88, 7708–7710.

Hauff, P., Reinhardt, M., Briel, A., Debus, N., Schirner, M., 2004. Molecular targeting of lymph nodes with L-selectin ligand-specific US contrast agent: A feasibility study in mice and dogs. Radiology 231, 667–673.

Hermanson, G.T., 1996. Bioconjugate techniques. Academic Press, San Diego.

Hvattum, E., Uran, S., Sandbaek, A.G., Karlsson, A.A., Skotland, T., 2006. Quantification of phosphatidylserine, phosphatidic acid and free fatty acids in an ultrasound contrast agent by normal-phase high-performance liquid chromatography with evaporative light scattering detection. J Pharm Biomed Anal 42, 506–512.

Jeppesen, C., Wong, J.Y., Kuhl, T.L., Israelachvili, J.N., Mullah, N., Zalipsky, S., Marques, C.M., 2001. Impact of polymer tether length on multiple ligand-receptor bond formation. Science 293, 465–468.

Jordan, J.E., Zhao, Z.Q., Vinten-Johansen, J., 1999. The role of neutrophils in myocardial ischemia-reperfusion injury. Cardiovasc Res 43, 860–878.

Keller, M.W., Glasheen, W., Kaul, S., 1989. Albunex: a safe and effective commercially produced agent for myocardial contrast echocardiography. J Am Soc Echocardiogr 2, 48–52.

Kim, D.H., Klibanov, A.L., Needham, D., 2000. The influence of tiered layers of surface-grafted poly(ethylene glycol) on receptor-ligand-mediated adhesion between phospholipid monolayer-stabilized microbubbles and coated class beads. Langmuir 16, 2808–2817.

Klibanov, A.L., Gu, H., J.K. Wojdyla, Jr., J.H.W., Kim, D.H., Needham, D., Villanueva, F.S., Brandenburger, G.H., 1999. Attachment of ligands onto the gas-filled microbubbles via PEG spacer arm and lipid residues anchored at the interface., Proceedings of 26th International Symposium on Controlled Release of Bioactive Materials, Boston. Controlled Release Society, pp. 124–125.

Klibanov, A.L., Hughes, M.S., Marsh, J.N., Hall, C.S., Miller, J.G., Wible, J.H., Brandenburger, G.H., 1997. Targeting of ultrasound contrast material - An in vitro feasibility study. Acta Radiologica 38, 113–120.

Klibanov, A.L., Hughes, M.S., Wojdyla, J.K., Marsh, J.N., Hall, C.S., Miller, J.G., Wible, J.H., Brandenburger, G.H., 1998. Targeting of ultrasound contrast material: Selective imaging of microbubbles in vitro. Acad Radiol 5, S243–S246.

Klibanov, A.L., Maruyama, K., Beckerleg, A.M., Torchilin, V.P., Huang, L., 1991. Activity of Amphipathic Poly(Ethylene Glycol)-5000 to prolong the circulation time of liposomes depends on the liposome size and is unfavorable for immunoliposome binding to target. Biochimica Et Biophysica Acta 1062, 142–148.

Klibanov, A.L., Rasche, P.T., Hughes, M.S., Wojdyla, J.K., Galen, K.P., Wible, J.H., Brandenburger, G.H., 2002. Detection of individual microbubbles of an ultrasound contrast agent: Fundamental and pulse inversion imaging. Acad Radiol 9, S279–S281.

Klibanov, A.L., Rasche, P.T., Hughes, M.S., Wojdyla, J.K., Galen, K.P., Wible, J.H., Brandenburger, G.H., 2004. Detection of individual microbubbles of ultrasound contrast agents – Imaging of free-floating and targeted bubbles. Invest Radiol 39, 187–195.

Klibanov, A.L., Rychak, J.J., Yang, W.C., Li, B., Acton, S., Lindner, J.R., Ley, K.F., Kaul, S., 2006. Targeted ultrasound contrast agents for molecular imaging in high-shear flow. Contrast Media & Molecular Imaging 1, 259–266.

Kremkau, F.W., 2002. Diagnostic Ultrasound: Principles and Instruments, 7th ed. Elsevier Health Sciences, NY.

Lanza, G.M., Abendschein, D.R., Hall, C.S., Scott, M.J., Scherrer, D.E., Houseman, A., Miller, J.G., Wickline, S.A., 2000. In vivo molecular imaging of stretch-induced tissue factor in carotid arteries with ligand-targeted nanoparticles. J Am Soc Echocardiogr 13, 608–614.

Lee, R.J., Low, P.S., 1995. Folate-mediated tumor-cell targeting of liposome-entrapped doxorubicin in-vitro. Biochimica Et Biophysica Acta-Biomembranes 1233, 134–144.

Leighton, T., 1994. The acoustic bubble. Academic Press, NY.

Leong-Poi, H., Christiansen, J., Heppner, P., Lewis, C.W., Klibanov, A.L., Kaul, S., Lindner, J.R., 2005. Assessment of endogenous and therapeutic arteriogenesis by contrast ultrasound molecular Imaging of integrin expression. Circulation 111, 3248–3254.

Leong-Poi, H., Christiansen, J., Klibanov, A.L., Kaul, S., Lindner, J.R., 2003. Noninvasive assessment of angiogenesis by ultrasound and microbubbles targeted to alpha(v)-integrins. Circulation 107, 455–460.

Ley, K., 1992. Leukocyte Adhesion to Vascular Endothelium. J Reconstr Microsurg 8, 495–503.

Lindner, J.R., Song, J., Christiansen, J., Klibanov, A.L., Xu, F., Ley, K., 2001. Ultrasound assessment of inflammation and renal tissue injury with microbubbles targeted to P-selectin. Circulation 104, 2107–2112.

Lindner, J.R., Song, J., Xu, F., Klibanov, A.L., Singbartl, K., Ley, K., Kaul, S., 2000. Noninvasive ultrasound imaging of inflammation using microbubbles targeted to activated leukocytes. Circulation 102, 2745–2750.

Linker, R.A., Reinhardt, M., Bendszus, M., Ladewig, G., Briel, A., Schirner, M., Maurer, M., Hauff, P., 2005. In vivo molecular imaging of adhesion molecules in experimental autoimmune encephalomyelitis (EAE). J Autoimmun 25, 199–205.

Lu, E.X., Wagner, W.R., Schellenberger, U., Abraham, J.A., Klibanov, A.L., Woulfe, S.R., Csikari, M.M., Fischer, D., Schreiner, G.F., Brandenburger, G.H., Villanueva, F.S., 2003. Targeted in vivo labeling of receptors for vascular endothelial growth factor – Approach to identification of ischemic tissue. Circulation 108, 97–103.

McEver, R.P., 1997. Selectin-carbohydrate interactions during inflammation and metastasis. Glycoconj J 14, 585–591.

Narayan, P.M., Marchant, D., Wheatley, M.A., 2001. Optimization of spray drying by factorial design for production of hollow microspheres for ultrasound imaging. J Biomed Mater Res 56, 333–341.

Narayan, P., Wheatley, M.A., 1999. Preparation and characterization of hollow microcapsules for use as ultrasound contrast agents. Polym Eng Sci 39, 2242–2255.

Park, E.Y.H., Smith, M.J., Stropp, E.S., Snapp, K.R., DiVietro, J.A., Walker, W.F., Schmidtke, D.W., Diamond, S.L., Lawrence, M.B., 2002. Comparison of PSGL-1 microbead and neutrophil rolling: microvillus elongation stabilizes P-selectin bond clusters. Biophys J 82, 1835–1847.

Porter, T.R., Xie, F., 1995. Transient Myocardial Contrast after Initial Exposure to Diagnostic Ultrasound Pressures with Minute Doses of Intravenously Injected Microbubbles – Demonstration and Potential Mechanisms. Circulation 92, 2391–2395.

Rychak, J.J., Klibanov, A.L., Leppanen, A., Cummings, R.D., Ley, K., 2004. Enhanced binding of ultrasound contrast microbubbles targeted to P-selectin using a physiological capture ligand. Faseb J 18, A446–A446.

Rychak, J.J., Lindner, J.R., Ley, K., Klibanov, A.L., 2006. Deformable gas-filled microbubbles targeted to P-selectin. J Control Release 114, 288–299.
Schneider, M., 1999. Characteristics of SonoVue (TM). Echocardiography-a J Cardiovasc Ultrasound and Allied Tech 16, 743–746.
Straub, J.A., Chickering, D.E., Church, C.C., Shah, B., Hanlon, T., Bernstein, H., 2005. Porous PLGA microparticles: AI-700, an intravenously administered ultrasound contrast agent for use in echocardiography. J Control Release 108, 21–32.
Toublan, F.J.J., Boppart, S., Suslick, K.S., 2006. Tumor targeting by surface-modified protein microspheres. J Am Chem Soc 128, 3472–3473.
Unger, E.C., McCreery, T.P., Sweitzer, R.H., Shen, D.K., Wu, G.L., 1998. In vitro studies of a new thrombus-specific ultrasound contrast agent. Am J Cardiol 81, 58g–61g.
Villanueva, F.S., Jankowski, R.J., Klibanov, S., Pina, M.L., Alber, S.M., Watkins, S.C., Brandenburger, G.H., Wagner, W.R., 1998. Microbubbles targeted to intercellular adhesion molecule-1 bind to activated coronary artery endothelial cells. Circulation 98, 1–5.
Vonbibra, H., Hartmann, F., Petrik, M., Schlief, R., Renner, U., Blomer, H., 1989. Contrast Color-Coded Doppler Flow Imaging - Improved Diagnosis of Right Heart-Disease after Intravenous-Injection of Echovist. Zeitschrift Fur Kardiologie 78, 101–108.
Watanabe, R., Matsumura, M., Chen, C.J., Kaneda, Y., Fujimaki, M., 2005. Characterization of tumor imaging with microbubble-based ultrasound contrast agent, Sonazoid, in rabbit liver. Biol Pharm Bull 28, 972–977.
Wei, K., Crouse, L., Weiss, J., Villanueva, F., Schiller, N.B., Naqvi, T.Z., Siegel, R., Monaghan, M., Goldman, J., Aggarwal, P., Feigenbaum, H., DeMaria, A., 2003. Comparison of usefulness of dipyridamole stress myocardial contrast echocardiography to technetium-99m Sestamibi single-photon emission computed tomography for detection of coronary artery disease (PB127 multicenter phase 2 trial results). Am J Cardiol 91, 1293–1298.
WeitzSchmidt, G., Stokmaier, D., Scheel, G., Nifantev, N.E., Tuzikov, A.B., Bovin, N.V., 1996. An E-selectin binding assay based on a polyacrylamide-type glycoconjugate. Anal Biochem 238, 184–190.
Weller, G.E.R., Lu, E., Csikari, M.M., Klibanov, A.L., Fischer, D., Wagner, W.R., Villanueva, F.S., 2003. Ultrasound Imaging of acute cardiac transplant rejection with microbubbles targeted to intercellular adhesion molecule-1. Circulation 108, 218–224.
Weller, G.E.R., Villanueva, F.S., Tom, E.M., Wagner, W.R., 2005a. Targeted ultrasound contrast agents: In vitro assessment of endothelial dysfunction and multi-targeting to ICAM-1 and sialyl Lewis. Biotechnol Bioeng 92, 780–788.
Weller, G.E.R., Wong, M.K.K., Modzelewski, R.A., Lu, E.X., Klibanov, A.L., Wagner, W.R., Villanueva, F.S., 2005b. Ultrasonic imaging of tumor angiogenesis using contrast microbubbles targeted via the tumor-binding peptide arginine-arginine-leucine. Cancer Res 65, 533–539.
Wu, Y.Q., Unger, E.C., McCreery, T.P., Sweitzer, R.H., Shen, D.K., Wu, G.L., Vielhauer, M.D., 1998. Binding and lysing of blood clots using MRX-408. Invest Radiol 33, 880–885.

17
Use of Acoustically Active Contrast Agents in Imaging of Inflammation and Atherosclerosis

Patrick H. Kee and David D. McPherson

Abstract: Inflammation plays an important role in the development of atherosclerosis. The endothelium is an active organ that forms a barrier between the circulation and the arterial wall. In response to pro-atherogenic factors, the endothelium is induced to become an adhesive and pro-thrombotic surface. A range of molecular markers associated with early and late changes in atherogenesis have been identified in the endothelium. Those pathological changes in the endothelium are potential targets for early detection of atherosclerosis and may precede advanced changes that can be detected by conventional imaging modalities, such as coronary angiography. Acoustically active contrast agents have been widely used for clinical applications such as enhancing cardiac chamber definition and measuring myocardial perfusion in diagnostic ultrasound imaging. In the context of molecular imaging, those agents are pure intravascular tracers and are ideally suited for interrogating the expression of molecular markers on the endothelium. Studies have demonstrated how microbubbles can detect inflammation by means of the interactions between their lipid shell components and leukocytes that co-localize on the surface of inflamed endothelium. More sophisticated acoustically active targeting agents, however, involve the incorporation of high-affinity peptides or antibodies into their lipid shell that highlight inflammatory markers, thrombosis, and neovascularization in the arterial wall in atherosclerotic animal models. Before those agents can be widely used in clinical practice, they will require further refinements to reduce immunogenicity of targeting ligands, minimize toxicity of lipid shell components, and improve acoustic stability after intravenous administration. The most challenging aspect of this research is, however, the identification of clinically relevant markers that can accurately predict the presence and progression of atherosclerosis.

17.1. Introduction

Currently available imaging modalities are suitable for detecting advanced stages of atherosclerosis (ATH) in which luminal narrowing and flow disruption in macrovasculature have occurred (O'Rourke et al., 2004). It is becoming more evident that a number of pathological changes occur at very early stages of atheroma development. By using knowledge of inflammation and tailoring techniques to highlight these earlier markers of inflammation, we can identify and direct treatment to atheroma at earlier or more amenable stages.

The endothelium forms a barrier between the circulation and the arterial wall. Inflammatory changes in the endothelium are expressed early in ATH. Inflammation also occurs in the dense adventitia. Both areas of inflammation are easily accessible via the circulation.

Acoustically active contrast agents are pure intravascular tracers. Apart from their clinical applications in contrast-enhanced ultrasound and myocardial perfusion imaging, acoustically active contrast agents are being developed as molecular targeting agents. If successful, these acoustically active contrast agents will provide expanded clinical application for highlighting pathological inflammatory changes at a molecular level. In this chapter, we will explore potential applications of these agents as intravascular tracers for detecting inflammatory processes.

17.2. Role of Inflammation in Atherosclerosis

Normal arterial endothelium is an active organ with vasodilatory, anti-thrombotic, and anti-inflammatory properties. It provides a "non-adhesive" surface for the unimpeded flow of erythrocytes and other cellular components in the circulation. When exposed to pro-atherogenic factors, early inflammatory changes occur, which renders the endothelium "adhesive" to circulatory components in blood. A number of pro-atherogenic factors have been identified and some of them are potential targets for targeted imaging (Table 17.1). This inflammatory cascade has been divided into a number of stages: capture, rolling, adhesion, and transmigration (Figure 17.1). The initial leukocyte capture is mediated by the expression of P-selectin in the endothelium. Once captured,

TABLE 17.1. Possible pro-atherogenic factors

Class	Molecules
Cytokines	Interleukin-1β (IL-1β) Tumor necrosis factor-α (TNF-α) Interleukin-6 (IL-6)
Lipids	Oxidized LDL
Acute phase proteins	Platelet-activating factor CRP (produced by the liver in response to IL-6)
Others	Protease-activated receptor signaling CD40/ CD40 ligand interactions

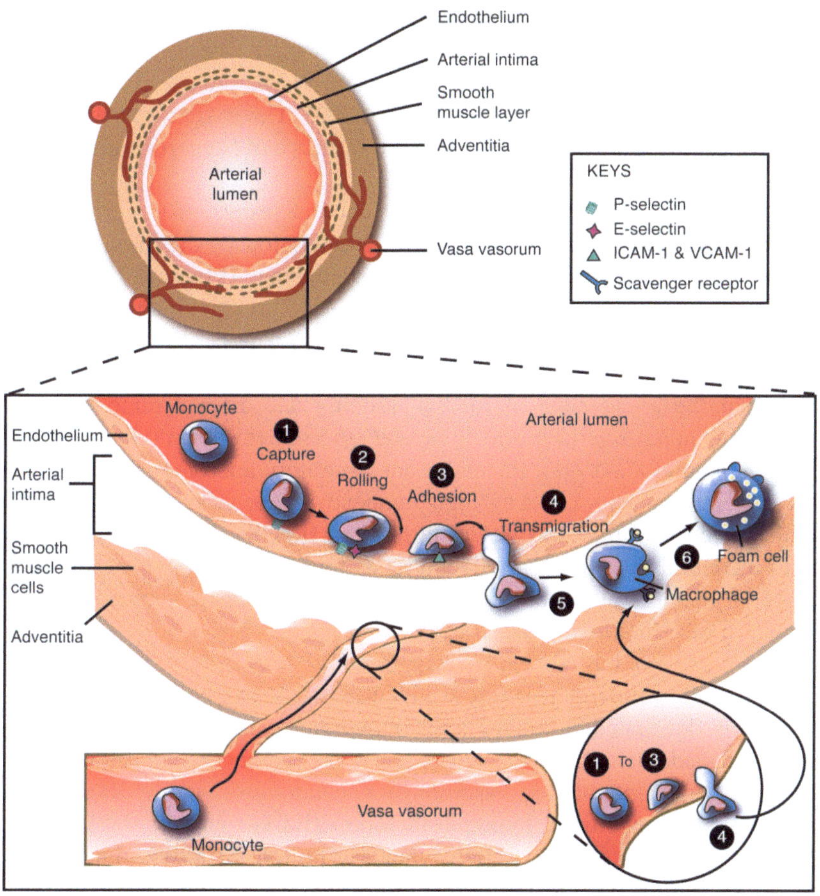

FIGURE 17.1. Possible pathways describing the infiltration of inflammatory cells into the arterial wall in the development of atherosclerosis. One pathway suggests the monocytes originate from the arterial lumen and transmigrate through the endothelium into the intima and smooth muscle cells. The inflammatory cascade is divided into (1) capture, (2) rolling, (3) adhesion, and (4) transmigration. P- and E-selectins are responsible for the initial capture and rolling of monocytes while ICAM-1 and VCAM-1 for enhanced attachment of monocytes to the endothelium. Once transmigrated through the endothelium into the intima, monocytes differentiate into macrophages (5), which engulf extracellular lipids to form lipid-laden macrophages or foam cells (6). In the other pathway, vasa vasorum in the adventitia provides a conduit for the infiltration of monocytes through the adventitia into the intima (ICAM-1, intercellular adhesion molecule-1; VCAM-1, vascular cell adhesion molecule-1).

leukocytes begin to roll on the endothelial surface under the influence of a family of selectins, which includes L-selectin on leukocyte and P- and E-selectins on the endothelium. Leukocyte adherence is strengthened under the effect of intercellular adhesion molecule-1 (ICAM-1) and vascular cell adhesion molecule-1 (VCAM-1) (Cybulsky and Gimbrone, 1991). Once firmly attached to the endothelium, leukocytes transmigrate through the endothelium into the intima. These leukocytes differentiate into macrophages and express scavenger receptors such as SR-A, CD36, and LOX-1 (Chen et al., 2002), which enable them to ingest modified low-density lipoprotein (LDL) to form lipid-laden macrophages or foam cells (Kunjathoor et al., 2002). These foam cells continue to synthesize chemotactic factors that promote further leukocyte recruitment and macrophage replication.

Early pathophysiologic studies focused on intimal insults as the originating factor. More recent studies have demonstrated alteration of the vasa vasorum and adventitial inflammation in atheroma development (Figure 17.1). Thus, atherosclerosis development seems to occur both outward from the intima to the media and inward from the adventitia to the media (Kwon et al., 1998; Moreno et al., 2002; van der Loo and Martin, 1997). This theory is important and would better explain the phenomena of medial thinning and arterial remodeling in early advancing atherosclerosis, which we and others have shown (Glagov et al., 1987; McPherson et al., 1991). Recently, our laboratory and others have shown increased arterial wall thickness by high-frequency echocardiography, which is not demonstrated well by intravascular ultrasound (IVUS). This is presumed to be due to increased inflammatory adventitia, as traditional non-enhanced IVUS cannot identify arterial layers beyond the media/external elastic lamina (Gradus-Pizio et al., 2002; Hamilton et al., 2002a).

The vasa vasorum has also been suggested to play a role in the initial pathogenesis and intimal changes in atherosclerosis. It has been observed that atherosclerosis developed in rabbits after disruption of the arterial vasculature with an intact endothelium (Booth et al., 1989). It has been demonstrated that occlusion of the adventitial vasa vasorum in Yucatan miniswine resulted in the production of a focal intimal hyperplasia lesion (Barker et al., 1993). These phenomena were attributed to arterial hypoxia. The role of adventitial inflammation in the pathogenesis of unstable coronary syndrome has also led some investigators to examine the role of adventitial inflammation in the early pathogenesis of atherosclerosis. Inflammatory infiltrates in the media, consisting of macrophages and T lymphocytes, localized predominantly around adventitial vasa vasorum. In advanced atherosclerosis they spread more diffusely (van der Wal et al., 1993). Using micro computed tomography, it was demonstrated in pigs that neovascularization of coronary vasa vasorum occurred within the first weeks of experimental hypercholesterolemia and prior to the development of endothelial dysfunction of the host vessel, suggesting a role for vasa vasorum neovascularization in the initial stages of atherosclerotic vascular disease (Herrmann et al., 2001). It was also demonstrated that ICAM-1 expression decreased on the endothelium of the main vessel with increasing stages of atherosclerosis, but ICAM-1 expression remained relatively constant in the adventitial vasa vasorum during growing atherosclerosis (Kasper et al., 1996). This suggests a prominent role of the vasa in the progression of atherosclerosis.

Inflammation also plays an important role in plaque disruption and thrombotic complications in the late phase of ATH. Advanced ATH is characterized by the presence of a necrotic lipid core and a thin overlying fibrous cap. Inflammatory processes weaken the integrity of the fibrous cap and lead to plaque disruption, thrombus formation, and vascular occlusion. Inflammatory cells and smooth muscle cells produce proteolytic enzymes, such as matrix metalloproteinases and myeloperoxidase, that weaken the fibrous cap (Fu et al., 2001; Galis and Khatri, 2002). Inflammatory cells produce cytokines, such as IFN-γ and Fas ligand, that induce apoptosis in smooth muscle cells (Geng et al., 1997). Death of foam cells and release of their intracellular lipid in the necrotic core promote further inflammation. Rupture of the fibrous cap and the resultant release of tissue factor and necrotic lipid material are not only pro-thrombotic, but can induce further inflammation. Thrombin is a potent inducer of interleukin-6 (IL-6) expression by vascular smooth muscle cells, which can contribute to elevation in C-reactive protein (CRP) and fibrinogen levels (Kranzhofer et al., 1996). Neovessel formation in the vasa vasorum and adventitial inflammation promote atheroma formation and destabilization and intraplaque hemorrhage.

Although angiography has been regarded as the gold standard in detecting obstructive arterial lesions, it is well documented that arterial remodeling progresses to an advanced stage before encroachment of the arterial lumen occurs. Effective therapeutic interventions rely on early detection of ATH. The endothelium and adventitia are clearly important organs that participate in the initiation and progression of inflammation in atherosclerosis and may provide a useful and accessible early warning system.

As intravascular tracers demonstrate similar rheology as erythrocytes, acoustically active contrast agents are well suited for interrogating and monitoring the upregulation of inflammatory markers in the endothelium as well as in the adventitia via the vasa vasorum.

17.3. Acoustically Active Contrast Agents for Ultrasound Imaging

Without the aid of acoustically active contrast agents, ultrasound imaging of small vessels is technically challenging. Coronary arteries and other smaller blood vessels are too small to be resolved by conventional ultrasound. High-resolution intravascular or transvascular ultrasound is required. Although high-resolution techniques provide good data, atheroma components and early inflammatory components are often poorly resolved due to the physics of ultrasound imaging. Color Doppler and spectral Doppler are generally not sensitive enough to detect low flow velocities. Red blood cells are weak acoustic reflectors due to their similar density and compressibility as surrounding plasma. Thus, imaging with ultrasound cannot use red blood cells as contrast agent to differentiate between the arterial lumen and the arterial wall. Conventional Doppler techniques rely upon the relative velocity difference between blood and tissue to detect blood vessels within tissues. In a moving organ like the heart, signals from blood become undetectable when blood and myocardium travel at similar velocities relative to the transducer.

Acoustically active contrast agents enhance the detection of small vessels by generating high backscatter signal. Entrapped air or gas within the contrast agents are better acoustic reflectors than red blood cells due to the large difference in impedance between air/gas and the surrounding plasma. Acoustically active contrast agents also oscillate non-linearly in response to an incident acoustic pressure and generate significant scattered echoes at harmonic multiples of the transmitted pressure wave. Thus, the signal-to-noise ratio generated by acoustically active contrast agents is high and echoes from contrast agents can be distinguished from tissue echoes. Current ultrasound imaging technology is sensitive enough to detect individual bubbles even if the density of tracer particles in the tissue is low (Klibanov et al., 2002, 2004).

A number of competing formulations have been investigated as acoustically active contrast agents (Table 17.2). They include liquid-core microemulsions,

TABLE 17.2. Ultrasound contrast microbubbles currently available for clinical research

Agent name (manufacturer)	Bubble size mean (range)	Gas core	Shell composition
Levovist (Schering AG)	2–3 μm (2–8 μm)	Air	Palmitic acid
Albunex (MBI/ Mallinckrodt)	4.5 μm (1–10 μm)	Air	Albumin
Optison (Mallinckrodt/ Amersham)	4.7 μm (1–10 μm)	Perfluoropentane	Albumin
Definity (Bristol Myers Squibb Medical Imaging)	1.5 μm (1–10 μm)	Perfluoropentane	Phospholipids (DPPC, MPEG-DPPE, DPPA)
Imagent (Alliance/ Photogen)	5 μm	N2/perfluorohexane vapor	Phospholipid (DMPC)
SonoVue (Bracco)	2.5 μm (1–10 μm)	Sulfur hexafluoride	Phospholipid (DSPC, DPPG, palmitic acid)
EchoGen (SONUS pharmaceuticals)	2–5 μm (1–30 μm)	Perfluoropentane	Stabilized surfactant
Sonogen (SONUS pharmaceuticals)	2–5 μm (1–30 μm)	Perfluoropentane	Anionically charged surfactant
Cardiosphere (Point biomedical)	4.0 μm (3–5 μm)	Nitrogen	Biodegradable polymer bilayer
NC100100	3.4 μm	Unspecified Perfluorocarbon	Unknown
AI-700 (Acusphere)	2 μm	Perfluorocarbon	Synthetic polymer (poly (D, L-lactide coglycolide) or PLGA
Echogenic liposomes	0.06–0.1 μm before lyophilization; 0.5–1.0 μm after lyophilization	Air	Phospholipids (Egg PC, DPPE, DPPG)

gas-filled microbubbles, and liposome-based agents. Liquid-core microemulsions are particles consisting of a liquid gas, such as perfluorocarbon, in the core that is stabilized by a phospholipid shell. When administered into the circulation, body temperature converts the liquid perfluorocarbon into a gas and renders the particle echogenic. Formulations of gas-filled microbubbles range from phospholipid-stabilized microbubbles containing air as the gas phase to formulations consisting of fluorine-containing inert insoluble gases, such as decafluorobutane, octafluoropropane, or sulfur hexafluoride. Lipids in the shell typically comprise a mixture of saturated phospholipids with high transition temperature (e.g., DPPC, 42°C, or DSPC, 56°C). To prevent coalescence and aggregation between microbubbles, negatively charged phospholipids or polyethylene glycol (PEG)–containing phospholipids are incorporated into the lipid shell (Fisher et al., 2002; Klibanov, 1999).

First generation echogenic contrast agents use air as the gaseous phase and are stabilized by palmitic acid or albumin. They include Levovist (Schering, AG) and Albunex (Mallinckrodt). Echogenicity of those agents is short-lived and they are not widely used these days.

Second- and third-generation echogenic contrast agents use perfluorocarbon as the gaseous phase and are stabilized by phospholipids or other polymers or surfactants. Agents such as SonoVue, Optison, and Definity are now commercially available.

ImaRx Corporation has produced a liposomal-based microbubble formulation, originally called MRX-115 or Aerosomes (ImaRx, Tucson, AZ) and now called Definity (Bristol-Myers Squibb) (Unger et al., 1992). The original formulation was a gas-filled microbubble consisting of nitrogen as the gaseous phase, but later was replaced by a liquid-core containing perfluorbutane, stabilized by a phospholipid shell. This formulation currently provides clinical applications in endocardial border enhancement and chamber detection, as well as myocardial perfusion imaging in cardiac ultrasound imaging. The original formulation has also been modified and investigated as tracers for detecting thrombus formation (Takeuchi et al., 1999; Wu et al., 1998) and vehicles for local delivery of dexamethasone (Unger et al., 1998a) and paclitaxel (Unger et al., 1998b).

Mallinckrodt Corporation has developed a protein-encapsulated microbubble formulation, now called Optison (GE Healthcare), which shares similar clinical indications as Definity. Optison is comprised of gaseous microbubbles containing octafluoropropane in the gaseous core that is stabilized by human serum albumin. It has been studied as a passive targeting agent for inflammation (Lindner et al. 1998, 2000a,b 2001). In the second-generation microbubble formulation, mixtures of phospholipids containing phosphatidylcholine and peglated phospholipids have been used to encapsulate the gaseous core. They are currently being investigated as agents for detecting thrombus formation (Schumann et al., 2002), adhesion molecule expression (Villanueva et al., 1998), and neovessel formation (Leong-Poi et al., 2003).

Lanza et al. (1996) have developed a perfluorocarbon emulsion nanoparticle containing a non-gaseous perfluorocarbon emulsion that is stabilized by a mixture of phospholipids. Those nanoparticles have poor inherent echogenicity but become echogenic when they aggregate and reach a high local concentration. Such a design may provide a better signal-to-noise ratio that is suitable for target imaging. Those nanoparticles have been studied as in vivo targeting agents for thrombus formation using a stepwise avidin–biotin interaction technique (Lanza et al., 1996) (see Section 17.6.2).

Our group pioneered the development of liposome-based acoustically active contrast agents (Huang et al., 2001). Liposomes, in general, are non-echogenic. We hypothesized that multilamellar liposomes with separate bilayers might produce sufficient backscatter to achieve enhancement of acoustic reflectivity.

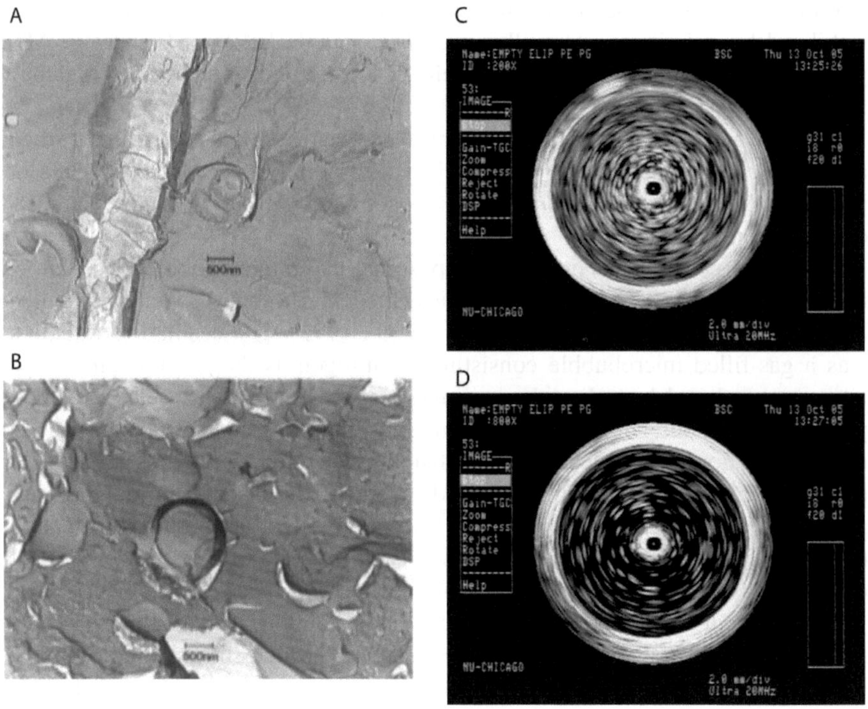

FIGURE 17.2. Freeze fracture electro microscopy and intravascular ultrasound images of echogenic immunoliposomes. Freeze fracture micrographs showing the structure of liposomes (×12000, reproduced at 50% of original size). (**A**) Liposomes with high echogenicity composed of a 60:8:2:30 molar mixture of PC:PE:PG:CH (10 mg of lipid/mL of buffer). (**B**) Liposomes with low echogenicity composed of a 58:8:4:30 molar mixture of PC:PE:PG:CH (10 mg of lipid/mL of buffer). Reproduced with permission from Alkan-Onyuksel et al. (1996). Ultrasound images of liposomes: intravascular ultrasound catheter in 10 mL glass vial containing liposomes with high echogenicity composed of a 69:8:8:15 molar mixture of PC:PE:PG:CH (10 mg of lipid/mL of buffer) at 200-fold dilution (**C**) and 800-fold dilution (**D**).

In our initial studies, we prepared liposomes containing phosphatidylcholine (PC), cholesterol (CH), and phosphatidylethanolamine (PE) by a dehydration–rehydration method (Alkan-Onyuksel et al., 1996). Liposomes prepared from PC and CH were not markedly echogenic. Addition of PE, which is required for antibody conjugation, gave enhanced echogenicity, probably due to the effect of the small polar head group of PE on the liposome structure (Figure 17.2). Subsequently, we added a negatively charged phospholipid, phosphatidylglycerol (PG), to our formulation to improve separation of internal bilayers, enhance acoustic reflectivity, prevent liposomal aggregation, and prolong the shelf life. Lyophilization of liposomes in the presence of mannitol is an important process for the entrapment of air inside the liposomes. During lyophilization, mannitol forms ice crystals and creates defects in the lipid shell. When lyophilized liposomes are rehydrated, air is entrapped in the defects of the lipid bilayers. The in vitro acoustic levels obtained in our echogenic liposomes exceeded levels obtained for Albunex (Alkan-Onyuksel et al., 1996). We demonstrated that the acoustic properties of echogenic liposomes did not change under simulated in vivo conditions, indicating that blood components and elevation of temperature to 37°C do not alter liposomal morphology (Alkan-Onyuksel et al., 1996). When echogenic liposomes were stored as dispersion in buffer in a refrigerator, 67 % of the original acoustic properties were retained at the end of the two-week storage. These results indicate that our inherently echogenic liposomes are more stable and acoustically reflective than other air-containing contrast agents. Furthermore, liposomes can be modified for antibody or peptide conjugation in molecular imaging. Unlike gas-filled microbubbles, liposomal-based acoustically active contrast agents can act as vehicles for drug or gene loading for therapeutic drug delivery.

17.4. Design of Acoustically Active Contrast Agents as Molecular Targeting Agents

A number of strategies have been investigated to engineer acoustically active contrast agents into molecular targeting agents. All of them involve methods of incorporating or conjugating high-affinity targeting ligands to the lipid shell of the tracer while maintaining acoustic reflectivity in the tracer agent (Figure 17.3).

One strategy involves the incorporation of small organic ligands into the lipid shell. Ligands are first coupled to the shell-forming material such as phospholipids or PEG-containing phospholipids. The ligand–shell conjugate is then mixed with other shell components to form a shell around gas bubbles. This strategy has been successfully employed for ligands such as peptides, peptide mimetics, carbohydrates, hormones, and vitamins (such as bitoin) (Klibanov et al., 1997, 1998). The method has the advantage of producing molecular targeting agents of high purity and high yield. Some biochemically labile molecular ligands such

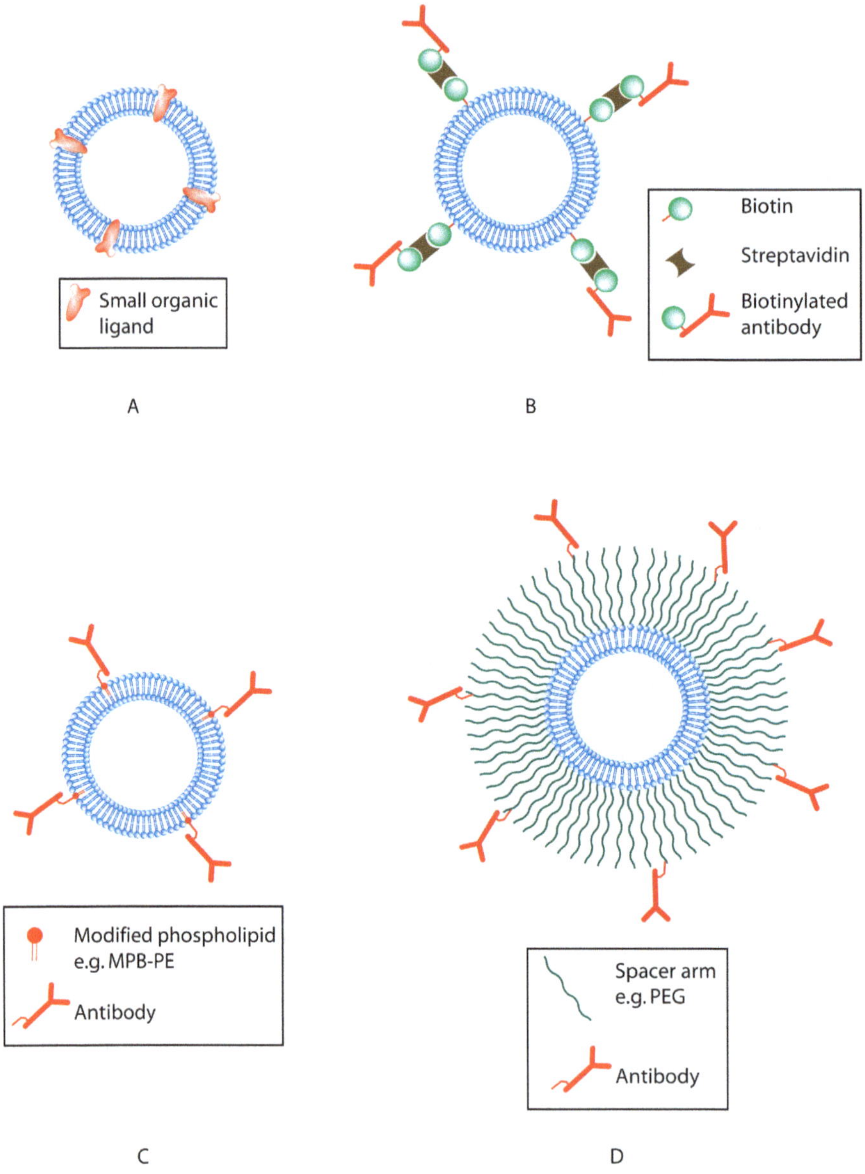

FIGURE 17.3. Attachment of ligands or antibodies to acoustically active contrast agent. **A** Incorporation of small organic ligands into the lipid shell. **B** Non-covalent attachment of antibodies to lipid shell by biotin–streptavidin conjugation. **C** Covalent attachment of antibodies to lipid shell. **D** Covalent attachment of antibodies to flexible polymer space arm such as polyethylene glycol (PEG).

as antibodies may not be able to withstand the preparation conditions; therefore, other processes may be necessary.

Another strategy involves the conjugation of ligands to pre-formed acoustically active contrast agents by covalent or non-covalent attachment. Non-covalent methods using biotin–streptavidin conjugation are well established (Klibanov, 1999). The attractiveness of this technique is the wide availability of biotinylated antibodies and the ability to process chemically labile ligands such as antibodies and peptides. In this stepwise procedure, microbubbles are first prepared and purified, followed by the attachment of biotin to the spacer arm in pre-formed microbubbles. Excess biotin is separated from biotynlated microbubbles by low-speed centrifugation floatation method. Excess streptavidin is added to biotynlated microbubbles to prevent cross-linkage of microbubbles. Unconjugated streptavidin is again separated by low-speed centrifugation. Finally, a biotinylated antibody is then added to complete the conjugation process. Other investigators have simplified this 3-step process into a 2-step one by incorporating avidin into the lipid shell of microbubbles as an anchor for biotinylated antibodies, thus eliminating the need for incubating biotinylated microbubbles with avidin (Korpanty et al., 2005). This strategy is ideal for initial proof of concept studies. Possible immunogenicity of streptavidin may limit its clinical utility.

Covalent conjugation relies on the presence of anchors in the shell of pre-formed microbubbles. In albumin-coated microbubbles, the amino group of albumin can serve as an anchor. More sophisticated methods involve the incorporation of derivatives into the shell in the pre-formed microbubbles. One of the covalent attachment strategies involves the coupling of amino-containing ligand to carboxyl-carrying microbubbles via water-soluble carbodiimide and *N*-hydroxysulfosuccinimide. In this method, using lipid-based microbubbles, the carboxylated lipid derivative is first incorporated in the lipid shell of the microbubbles. Carboxylate can be activated by a water-soluble carbodiimide, such as 1-ethyl-3-(3'-dimethylaminopropyl) carbodiimide in the presence of *N*-hydroxysulfosuccinimide, and the resulting active ester can react with an amino group of an antibody or other protein ligand. Anti-ICAM-1–conjugated microbubbles have been prepared by this method and successfully targeted activated endothelial cells in vitro (Villanueva et al., 1998). The disadvantage of this technique is the large antibody wastage to achieve a desirable level of coating of antibody on the microbubbles.

Using our echogenic liposomes, we prefer another covalent attachment technique that directly conjugates the antibody to microbubbles. Investigators have successfully conjugated thiolated monoclonal antibodies to liposomes containing *N*-[4-(p-maleimidophenyl)-butyryl] phosphatidylethanolamine (MPB-PE) (Heath et al., 1983; Martin et al., 1990). The amount of MPB-PE in liposomes is important in determining liposomal integrity and stability. The conjugation of "Fab" fragments to liposomes containing 5 mol% MPB-PE resulted in a concentration-dependent increase in size and polydispersity of liposomes. Reduction of MPB-PE concentration to 1 mol% improved liposome stability and entrapment efficiency (Bredehorst et al., 1986). Antibody ligand was thiolated with *N*-succinimidyl 3-(2-pyridyldithio)-propionate (SPDP) and then reacted with MPB-ELIP. Bound antibody was separated from unbound antibody by chromatography on a Sepharose 4B column. We have demonstrated that this technique has been successfully used for conjugating monoclonal antibodies against fibrinogen, ICAM-1 and VCAM-1 to echogenic liposomes (Demos et al., 1999, 1998, 1997, 1996; Hamilton et al., 2004).

Some investigators have advocated the use of a flexible polymer spacer arm (as in PEG) as a linker between contrast agent and ligand. This strategy may improve ligand mobility along the surface of the contrast agent, enhance molecular targeting, and prevent dislodgement of contrast agent by shear force in circulation. This may also be important in lipid-based microbubbles if PEG-containing phospholipid is used as a component in the lipid shell.

Clinically useful tracers should provide a high signal-to-noise ratio shortly after administration. This is achievable if the tracer bound to the marker of interest has a longer persistence at the site of interest than the unbound tracer. In comparison with tracers and nanoparticles used for SPECT or MRI, which can persist for days in the circulation, the clearance of unbound acoustically active agents is quite rapid. A high target-to-background signal is generally achievable within a few minutes of intravenous administration. Unbound acoustically active agent is then filtered by repeat pulmonary recirculation and loses its gaseous content and the lipid shell is cleared by the reticuloendothelial system. Thus, the combination of acoustically active contrast agents and ultrasound imaging has a special role in targeted imaging.

17.5. Potential Markers of Inflammation

Acoustically active contrast agents are pure intravascular tracers and are ideally suited for the detection of inflammatory markers on the endothelium. While the understanding of atherobiology continues to evolve, certain endothelial markers emerge at different stages of atherosclerosis and may serve as useful markers in ascertaining and staging the progress of atherosclerosis (Table 17.3).

In the early stage of atheroma formation, there are a number of integrins and selectins, which lead to the recruitment of leukocytes to the intima and adventitia. In the more advanced stage of atherosclerosis, erosion of atheroma and apoptosis of endothelial and smooth muscle cells expose the subendothelial atheroma matrix and negative glycocalyx for passive interaction with albumin and negatively charged phospholipids. The damaged endothelial surface also releases tissue factors that can lead to activation of the coagulation cascade and platelet aggregation. Macrophages undergoing apoptosis results in loss of phospholipid asymmetry and exposes phosphatidylserine on the surface (Fadok et al., 1992).

TABLE 17.3. Potential endothelial markers of atherosclerosis

Class	Molecules
Early endothelial markers	
Selectins	P-selectin E-selectin
Integrins	LFA-1 (CD11a/CD18) Mac-1 (CD11b/CD18) VLA ($\alpha_4\beta_1$ integrin)
Immunoglobulin superfamily	ICAM-1 VCAM-1
Late endothelial markers	
Nonspecific markers	Subendothelial matrix/negative glycocalyx
Platelet receptors	Glycoprotein IIb/IIIa
Enzyme	Tissue factors
Cellular components	Oxidized LDL Scavenger receptors on foam cells Activated leukocytes
Apoptosis markers	Phosphatidylserine
Angiogenesis factors	Integrin $\alpha_v\beta_3$
Other novel markers	Endothelial markers identified by high throughput technologies such as in vivo phage display

Later in atheroma formation, other markers are expressed. Naturally occurring protein, annexin V, has a high affinity toward phosphatidylserine. This protein has been tested as a potential target for detecting advanced atherosclerosis using nuclear imaging (Kolodgie et al., 2003). The accumulation of foam cells and oxidized LDL in advanced atheroma may also be potential imaging targets (Shaw et al., 2001).

As the vasa vasorum is also lined by endothelium, it too expresses inflammatory markers, such as integrins and selectins, during inflammation. During the course of atherosclerosis, the expression of integrins and selectins remains relatively constant in the adventitia at progressive stages, but their expression becomes less important in the arterial endothelium (Kasper et al., 1996).

Neovessel proliferation is promoted by angiogenesis factors, such as integrin $\alpha_v\beta_3$, which may serve as another useful marker of advanced atherosclerosis. A number of soluble proteins and proteolytic enzymes have been identified as important components responsible for the progression and destabilization of atheroma. High-throughput screening technologies, such as in vivo phage display, have identified some endothelially bound ligands that might serve as useful markers of atherosclerosis (Houston et al., 2001; Liu et al., 2003). The functional role of these ligands, although uncertain, is of particular practical interest as some ligands are expressed only in certain vascular beds and may be useful for localizing inflammatory processes in certain organs, such as coronary arteries.

17.6. Acoustically Active Contrast Agents for Detection of Endothelial Inflammation

17.6.1. Passive Targeting

Passive targeting relies on the inherent characteristics of protein or phospholipids in the shell of unmodified microbubbles for tissue targeting. The hypothesis stemmed from the observation that acoustically active contrast agents without surface modifications appear to accumulate in inflamed vasculature and co-localize with inflammatory cells. Several lines of evidence suggest that interactions between microbubble shells and inflamed tissues may be mediated via electrostatic and chemical mechanisms. In inflammation, the disruption of endothelium exposes negatively charged glycocalyx in the subendothelial matrix. This denuded endothelium provides a field for electrostatic interactions with components, such as albumin in the microbubble shell (Figure 17.4) (Basalyga et al., 1998; Lindner et al., 1998). The recruitment of leukocytes to the surface of inflamed endothelium provides another target for interactions with microbubbles. Depending on the shell composition in microbubbles, protein microbubbles (Optison; Mallinckrodt Medical, Inc) and lipid microbubbles

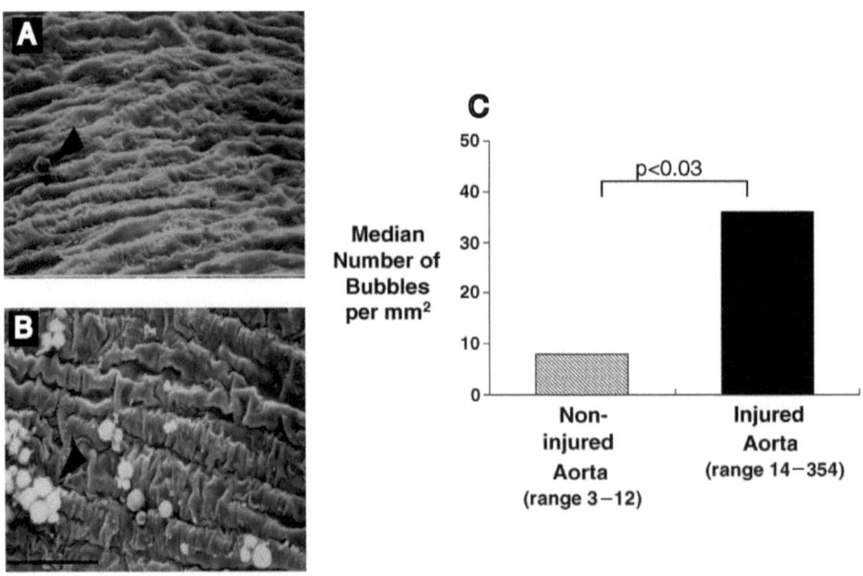

FIGURE 17.4. Binding of albumin in lipid shell to damaged subendothelial matrix. Scanning electron microscopy (original magnification ×350) of rabbit aortic segments perfused *ex vivo* with culture medium microbubbles (**A**) before and (**B**) after endothelial denudation by balloon injury. Microbubbles adhered to denuded endothelium (*black arrow*, **B**) in significantly greater numbers than to non-denuded segments (**C**). Arrow in (**A**) indicates an artifact. Shear rate tested is 100 s^{-1}. The black bar indicates 100 μm. Reproduced with permission from Basalyga et al. (1998).

(MP1950; Mallinckrodt Medical, Inc) interact with leukocytes via different mechanisms (Lindner et al., 2000a). The protein shell of Optison appears to adhere to leukocytes via β_2 integrin Mac-1, which plays an important role in neutrophil and monocyte adhesion. On the other hand, the phospholipid shell of MP1950 appears to become opsonized by complement in the circulation and becomes adherent to leukocytes on the inflamed endothelium. A similar study with albumin microspheres containing perfluoropropane also demonstrated enhanced adherence to leukocytes localized in inflamed venules, but the retention of micropheres was reduced by infusing blocking antibodies against ICAM-1, P-selectin, and common leukocyte antigen (Yasu et al., 2005). These studies suggest that the adherence of leukocytes to inflamed endothelium may be a prerequisite for passive targeting of microbubbles. Protein (albumin) or lipid microbubbles were subsequently phagocytosed intact by activated neutrophils and monocytes and remained detectable by ultrasound (Lindner et al., 2000b). By taking advantage of the non-specific affinity of ultrasound contrast agent to damaged endothelium and the natural targeting mechanism of leukocytes to inflamed endothelium, both protein and lipid microbubbles may be useful for detection and monitoring of inflammation.

Negatively charged phospholipid species, such as phosphatidylglycerol and phosphatidylserine, have been used in the preparation of acoustically active contrast agents to prevent aggregation. The activation of the complement and immune system by those negatively charged phospholipids might pose safety concerns. Others and we have found that the activation of complement could trigger anaphylaxis in animals and adversely limit the amount of contrast agents that can be safely administered. The presence of the polyethylene glycol on the surface of the contrast agents could not reduce complement activation by negatively charged phospholipids (Fisher et al., 2002). Furthermore, the capture and subsequent removal of these agents by leukocytes, macrophages, and Kupffer cells may enhance the clearance of the contrast agent from the circulation and reduce the number of tracers for imaging. Thus, for image enhancement the use of mainly negatively charged phospholipids in microbubble preparations should be avoided.

17.6.2. Active Targeting

In active targeting, the surface components of acoustically active contrast agents are modified to interact with and highlight the surface markers of inflammation.

Villanueva et al. have demonstrated the feasibility of targeting activated coronary artery endothelium by microbubbles in vitro (Villanueva et al., 1998). Anti-human anti-ICAM-1-conjugated microbubbles were able to adhere to activated endothelium and remained relatively adherent at wall shear rates of 100 s^{-1} and 1000 s^{-1}, which are typical for the shear rate in larger arteries and capillaries (Figure 17.5).

In an ischemia-reperfusion kidney injury model in mice, P-selectin-targeted microbubbles markedly enhanced the signal in inflamed kidneys

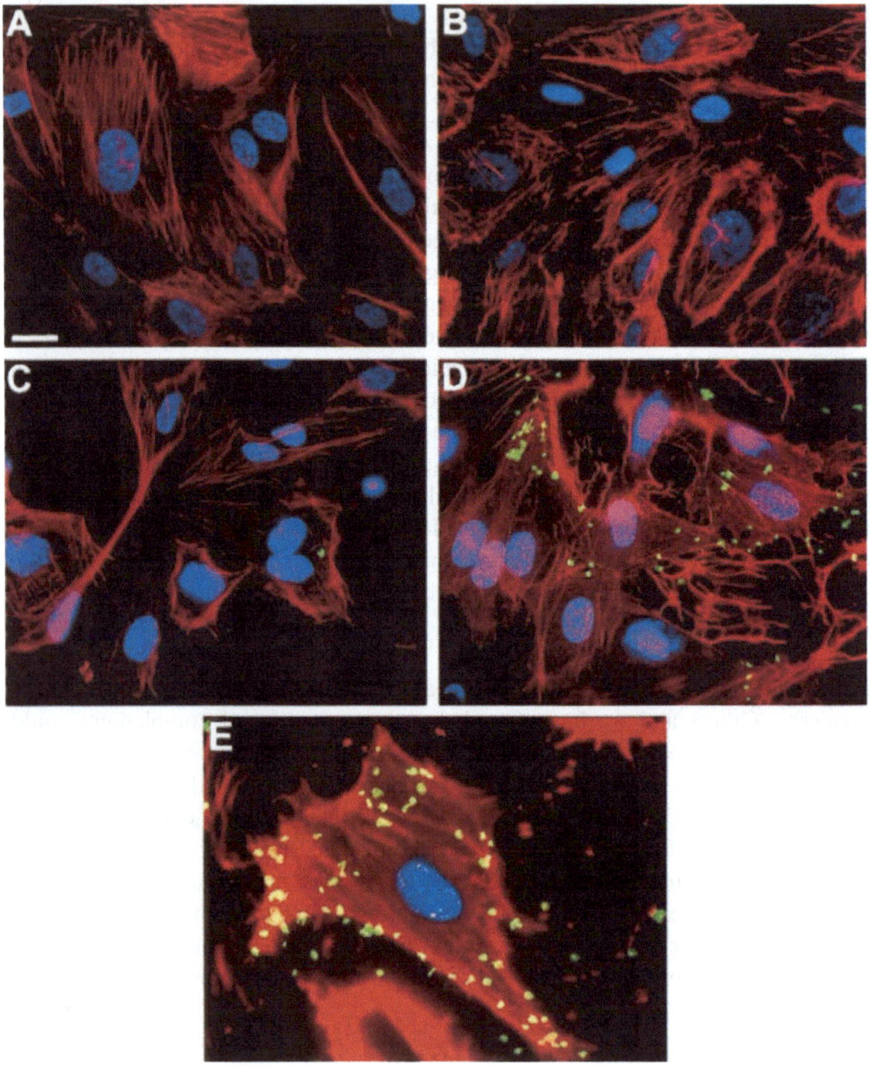

FIGURE 17.5. In vitro experiment demonstrating the attachment of anti-ICAM-1-conjugated microbubbles to activated coronary artery endothelial cells. Fluorescent micrographs of endothelial cells (ECs) after exposure to either non-specific IgG-conjugated bubbles (**A** and **B**) or anti-ICAM-1-conjugated bubbles (**C** and **D**). Cytosolic F-actin is stained with rhodamine-phalloidin (red), nuclei are stained with Hoechst dye (blue), and microbubbles are labeled with fluorescein (green). Microbubbles have lost their gas and spherical shape during preparation for microscopy. **A** Scale bar = 10 μm. There was no adherence of non-specific IgG-labeled bubbles to normal (**A**) or activated (**B**) cells. Although small numbers of anti-ICAM-1 bubbles adhered to normal ECs (**C**), numerous bubbles attached to activated ECs (**D**). Note intense rhodamine-phalloidin staining of F-actin indicative of activation (**B** and **D**) in cells exposed to IL-1β. **E** Enlargement of a single activated cell exposed to anti-ICAM-1 bubbles). Reproduced with permission from Villanueva et al. (1998).

(Lindner et al., 2001). In contrast, enhancement was reduced in transgenic mice with P-selectin deficiency. In a transplant rejection model in rats, ICAM-1-targeted microbubbles were able to highlight the inflamed allografts in strain-mismatched donors relative to matched donors (Weller et al., 2003).

Our laboratory was the first to demonstrate the feasibility of a single-step active targeting of inflammation using antibody-conjugated echogenic immunoliposomes. We have demonstrated the stability of ELIP targeting under physiologic conditions while retaining echogenic characteristics. To determine duration of liposomal attachment to fibrin, I^{125}-labeled ELIP were applied to fibrin-coated paper. More than 84 % of the ELIP remained attached to the fibrin after being washed five times with saline (Demos et al., 1997) and under physiologic shear/flow (Demos et al., 1998; Hamilton et al., 2002b). We also demonstrated that our ELIP were able to highlight components of atheroma in vivo. We prepared anti-ICAM-1–conjugated echogenic immunoliposomes and injected them into Yucatan miniswine that had been induced to develop atherosclerosis (Demos et al., 1999). Anti-ICAM-1–conjugated liposomes were attached to atherosclerotic endothelium (Figure 17.6). Although speculative, liposomal bridging, aggregation, and the multidimensional binding surface of atheroma probably enhance the ability of these agents to be acoustically reflective. Subsequently, we extended our work by preparing and injecting ELIP conjugated to different antibodies into the Yucatan miniswine model of atherosclerosis. We demonstrated ELIP binding to endothelial or atheroma targets by intravascular ultrasound and confirmed their binding specificity with immunohistochemical analysis (Figure 17.7) (Hamilton et al., 2004).

In the advanced stage of atherosclerosis, angiogenesis develops as an important feature and is seen as proliferation of the vasa vasorum in the adventitia and even intramural hemorrhage in the atherosclerotic plaque. This neovascularization phenomenon provides another potential target for staging inflammation in atherosclerosis. Biotinylated microbubbles coupled to biotinylated antibody against $\alpha_v\beta_3$ or to a biotinylated 5.4 KD RGD disintegrin peptide echistatin have successfully targeted angiogenesis in a matrigel plug mouse model (Leong-Poi et al., 2003). In our previous study, high-resolution imaging with IVUS had the capability of differentiating the different layers of the arterial wall and provided direct in vivo evidence of enhancement of the adventitial layer when echogenic liposomes conjugated with anti-ICAM-1 or anti-VCAM-1 were injected into a Yucatan miniswine model of advanced atherosclerosis.

Thrombus formation and arterial occlusion are important features of inflammation and atherosclerosis. Non-invasive detection of thrombus formation has been studied with thrombus-targeted microbubbles and echogenic liposomes. The delivery of large amounts of acoustically active agents at the site of thrombus may provide added opportunities for mechanical disruption of thrombus by ultrasound and chemical clot lysis by high local concentration of thrombolytics. Lanza et al. first demonstrated the feasibility of a multi-step active targeting of thrombi in vivo (Lanza et al., 1996). In his studies, thrombi were pre-targeted with antifibrin antibodies and avidin in situ, followed by administration of biotinylated

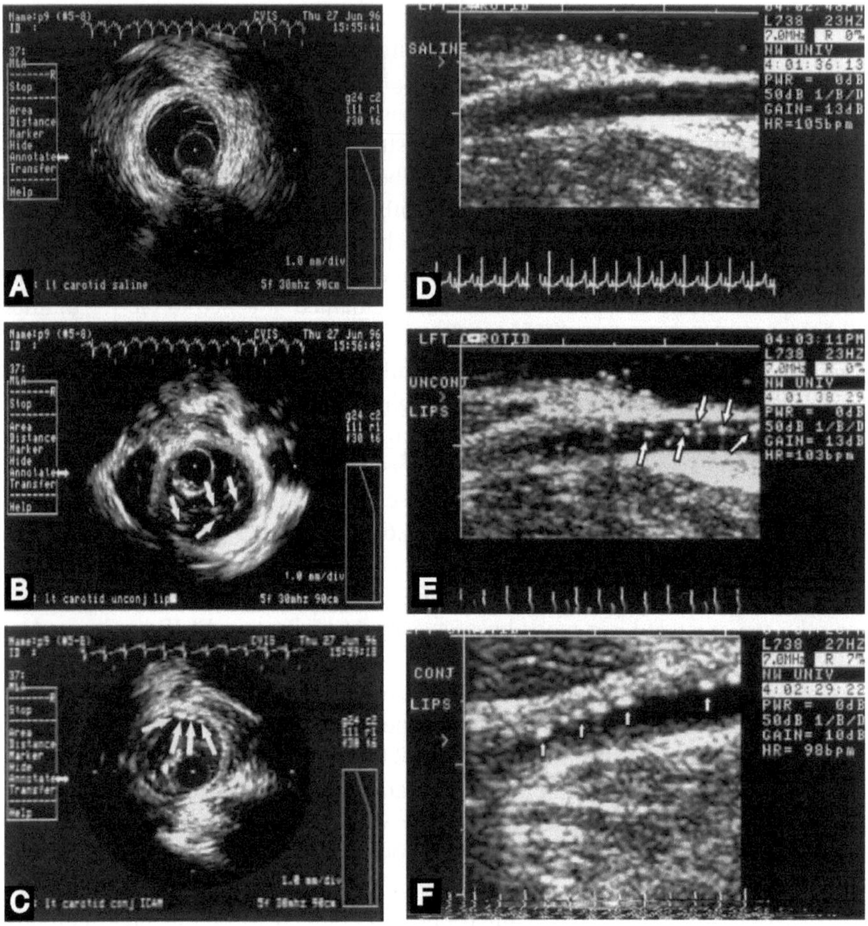

FIGURE 17.6. In vivo experiment demonstrating the attachment of anti-ICAM-1-conjugated echogenic immunoliposomes to atherosclerotic arterial wall. Intravascular ultrasound imaging of the atherosclerotic left carotid artery of a Yucatan miniswine (notice the circular atheroma deposition). **A** After injection of saline (arrows point to early atheroma). **B** After injection of unconjugated liposomes (arrows point to liposomes within lumen). **C** After injection of anti-ICAM-1-labeled liposomes (arrows point to liposomes attached to the atherosclerotic plaque). Transvascular ultrasound images of the atherosclerotic left carotid artery of a Yucatan miniswine. **D** After injection of saline. **E** After injection of unconjugated liposomes (arrows point to liposomes within the lumen). **F** After injection of anti-ICAM-1-labeled liposomes (arrows point to liposomes attached to the atherosclerotic plaque). Reproduced with permission from Demos et al. (1999).

perfluorocarbon nanoemulsion (Figure 17.8A). That technique was successful in highlighting arterial thrombi in vivo (Figure 17.8B). By targeting platelet glycoprotein IIb/IIIa receptor, platelet aggregation can also be detected. One potential candidate is the linear hexapeptide (lysine-glutamine-alanine-glycine-asparatate-

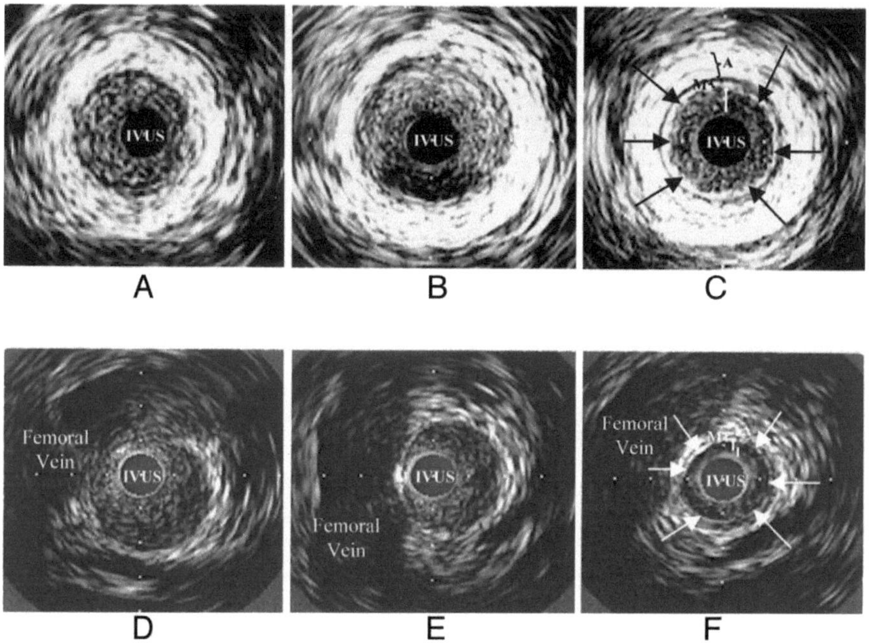

FIGURE 17.7. In vivo experiment demonstrating the attachment of anti-ICAM-1– and anti-VCAM-1-conjugated echogenic immunoliposomes to atherosclerotic arterial wall. Intravascular ultrasound imaging of the atherosclerotic left carotid artery of a Yucatan miniswine. **A** After injection of saline. **B** Five minutes after injection of unconjugated liposomes. **C** After injection of anti-ICAM-1-labeled liposomes (arrows point to enhanced atheroma). Intravascular ultrasound imaging of the atherosclerotic femoral artery of a Yucatan miniswine. **D** After injection of saline. **E** Five minutes after injection of unconjugated liposomes. **F** After injection of anti-VCAM-1-labeled liposomes (arrows point to enhanced atheroma). Reproduced with permission from Hamilton et al. (2004).

valine) that is coupled to the lipid shell via a PEG spacer in MRX-408 (ImaRX Pharmaceutical Corp, Tucson, Arizona) (Unger et al., 1998c). Preliminary studies demonstrated MRX-408 could specifically bind and enhance the detection of blood clot by ultrasound in vitro and in vivo, and may have the potential to enhance thrombolysis when activated by ultrasound (Schumann et al., 2002; Unger et al., 1998c; Wu et al., 1998) (Figure 17.9). Our laboratory has studied the in vitro binding characteristics and in vivo targeting of anti-fibrinogen–conjugated ELIP to thrombus formation. At a physiological shear stress of 15 dyn/cm^2, over 70 % of the liposomes remained attached to fibrin after 2 hours (Demos et al., 1998). Thus, the retention time of ELIP targeted to fibrin should be long enough for imaging after injection. We also demonstrated the binding of anti-fibrinogen conjugated ELIP to thrombotic components of atheroma and provided acoustic enhancement for intravascular and transvascular ultrasound imaging in a Yucatan miniswine model of atherosclerosis (Figure 17.10) (Demos et al., 1999; Hamilton et al., 2004).

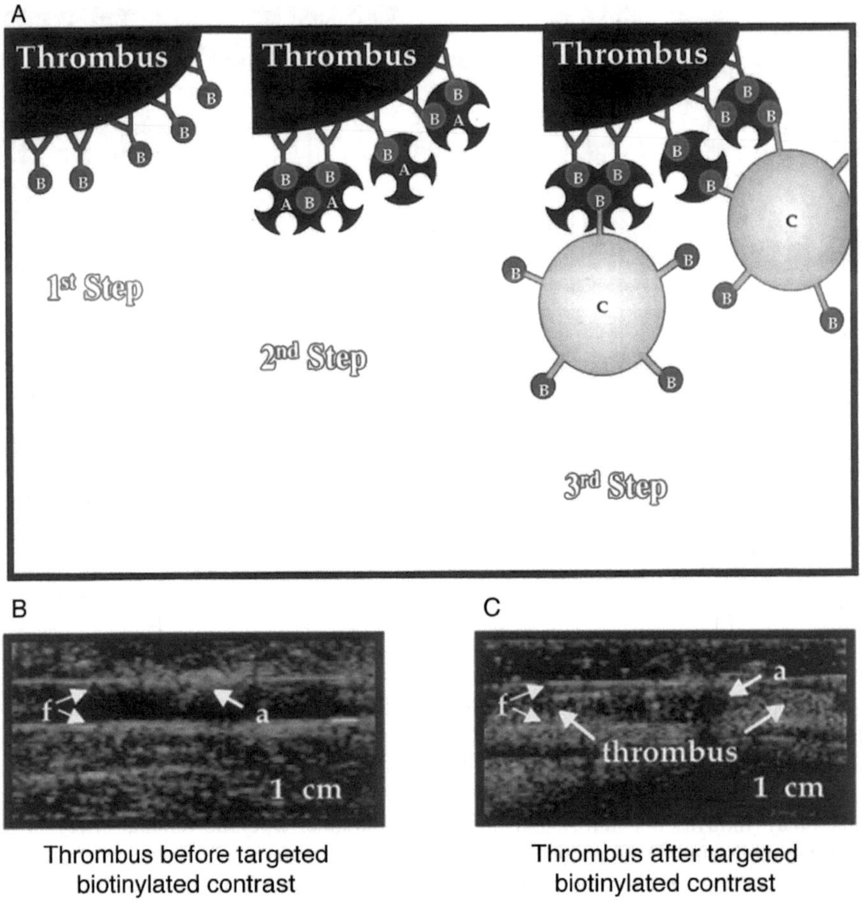

FIGURE 17.8. (**A**) Use of avidin–biotin interaction to target specific molecular epitopes through a triphasic (three step), pretargeting approach. In phase 1, a biotinylated ligand, specific for molecular epitopes of interest, e.g. a fibrin domain, is administered, binded, and equilibrated with the target. In phase 2, avidin is given, which conjugates and cross-links the biotinylated ligand, increasing the "avidity" of the complex for the tissue surface. In phase 3, the biotinylated perfluorocarbon emulsion particle is administered and attaches to the bound antibody–avidin complexes through unoccupied biotin receptor sites. The circulating emulsion particles have poor acoustic reflectivity, but the aggregated particles markedly increase the acoustic reflectivity at the targeted site. Although ligands are typically envisioned to be monoclonal antibodies or Fab fragments, any biotinylated macro-molecule with receptor specificity (e.g., nucleic acids, lectins, peptides, drugs, or viruses) could be used. Acoustic enhancement of canine femoral artery thrombus, targeted with biotinylated antifibrin antibody, before (**B**) and after (**C**) exposure to biotinylated perfluorocarbon emulsion. The acute arterial thrombus is poorly visualized with a 7.5-MHz linear-array, focused transducer. The transmural electrode (a) and the wall boundaries of the femoral artery (f) are clearly delineated (a). After exposure to the biotinylated emulsion, the thrombus is easily visualized. The anode (a) produces an ultrasonic shadowing effect in the midportion of the contrast-enhanced thrombus. Reproduced with permission from Lanza et al. (1996).

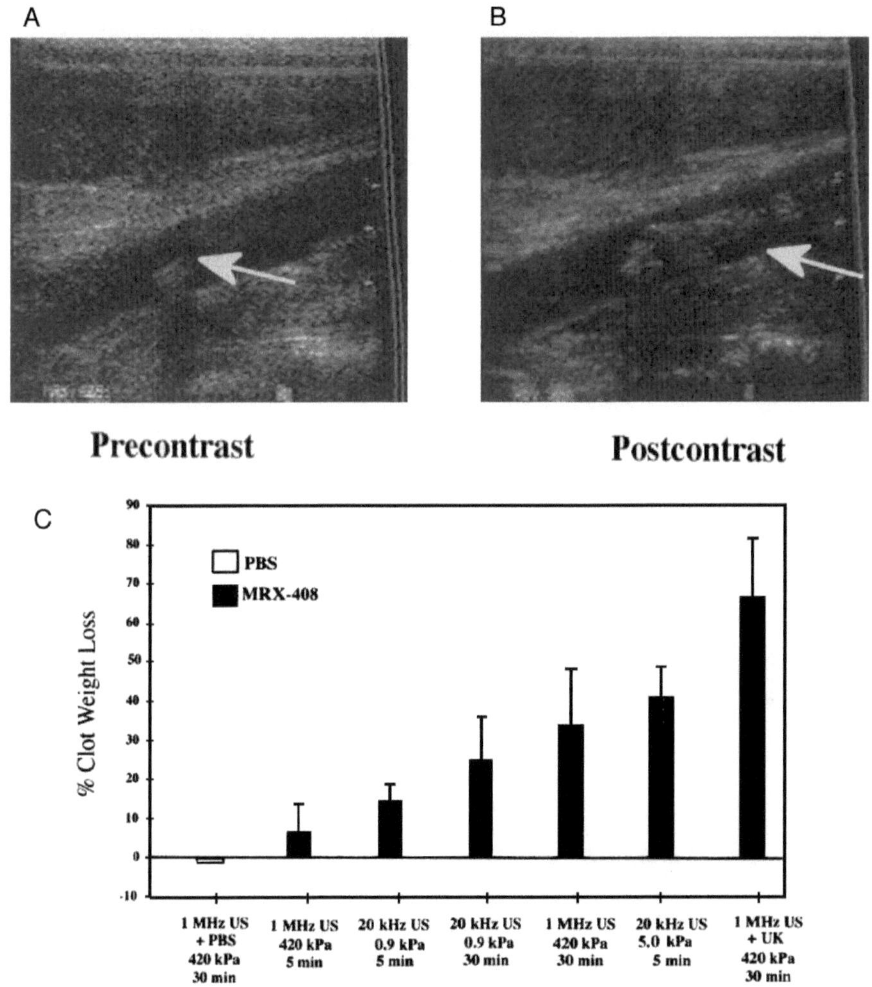

FIGURE 17.9. In vivo enhancement of thrombus by MRX-408. Ultrasound images of a blood clot in a dog's femoral vein (**A**) before and (**B**) after infusion with MRX-408. (**C**) Sonothrombolysis efficiency of MRX-408 under either 20-kHz or 1-MHz ultrasound. Reproduced with permission from Wu et al. (1998).

Antibodies (especially of murine origin or against murine antigens) are generally more widely available and more suitable for the initial evaluation of potential markers of inflammation in the endothelium. However, the use of full-size antibody molecules as targeting ligands raises the concern about the development of immune response to targeted particles and the ability to use these targeted agents for repeated administration. The use of humanized/human antibodies may partly overcome this issue. Ultimately, the better solution may be the use of small antibody fragments or short peptides with high affinity to the

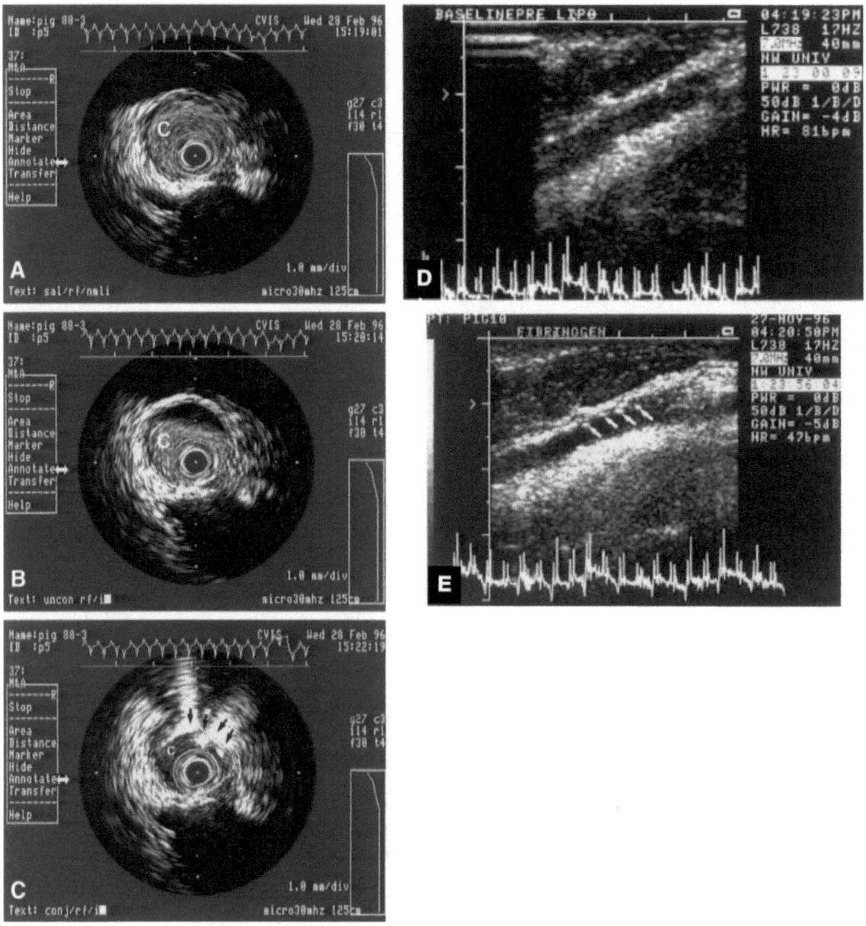

FIGURE 17.10. In vivo enhancement of complex atheroma by anti-fibrinogen–conjugated echogenic immunoliposomes. Intravascular ultrasound images of the right femoral artery of a Yucatan miniswine with a large central thrombus. **A** After injection of saline. **B** After injection of unconjugated liposomes. **C** After injection of antifibrinogen-labeled liposomes. (The dark central area is catheter imaging artifact. C = intraluminal thrombus/clot; *arrows* point to liposomes attached to thrombus.) Transvacular ultrasound images of an atherosclerotic left carotid artery of a Yucatan miniswine (**D**) after injection of saline and (**E**) after injection of anti-fibrinogen-labeled liposomes (*arrows* point to liposomal attachment to fibrinogen in the atherosclerotic plaque). Reproduced with permission from Demos et al. (1999).

endothelial markers of interest. Recent advances in library screening methods and in vivo phage display technique have led to the development of high-affinity short peptide sequences suitable for identifying markers of atherosclerosis (Houston et al., 2001; Liu et al., 2003).

17.7. Conclusion

Due to a lack of accurate techniques for detecting early stages of atherosclerosis, current management of cardiovascular disease focuses on the detection of advanced stages of atherosclerosis and palliation of obstructive arterial lesions. Better understanding of atherogenesis and the expression of pathological markers in the arterial wall have generated potential targets for imaging and therapy. While other imaging modalities have their own merits, the combination of diagnostic ultrasound and acoustically active tracers with targeting ability plays a special role in interrogating and highlighting markers of inflammation and atherosclerosis. Traditional vascular risk factors, such as the presence of hypertension and hypercholesterolemia, are too generalized and do not accurately identify the vascular risk of a subject at a particular time point. Atherosclerosis is a systemic condition and the measurement of the intensity of inflammation in arteries may potentially provide an assessment of the global vascular risk in a subject. Potential applications of diagnostic ultrasound and targeted acoustically active tracers may include the detection of inflammation or thrombus formation in superficial vessels, such as the carotid and femoral arteries, by transvascular ultrasound, or the detection of vulnerable plaque in coronary arteries with intravascular ultrasound. Effective therapies that alter the progression of atherosclerosis also rely upon an accurate monitoring tool that detects changes in vascular inflammation. If targeted acoustically active tracers can be administered repeatedly, this technique can provide a means to follow progress of treatment.

Before these promises of targeted imaging can be realized, the design of acoustically active targeting agents must be optimized. Ideal properties of the tracers should include improved longevity of echogenicity in the circulation and reduced immunogenicity of the shell components and targeting ligands. The use of monoclonal antibodies on the surface of microbubbles may provide a useful means to detect potential markers of inflammation, but their immunogenicity may hamper repeated use. Ultimately, short peptides against potential endothelial markers may provide the solution. However, the reliability of this technique rests upon the identification of markers that can accurately detect early changes of atherosclerosis and follow its progression and regression.

References

Alkan-Onyuksel, H., Demos, S. M., et al., 1996. Development of inherently echogenic liposomes as an ultrasonic contrast agent. J Pharm Sci 85,5, 486–490.

Barker, S.G., Talbert, A., et al., 1993. Arterial intimal hyperplasia after occlusion of the adventitial vasa vasorum in the pig. Arterioscler Thromb 13,1, 70–77.

Basalyga, D.M., Wagner, W.R., et al., 1998. Albumin microbubbles adhere to exposed extracellular matrix of perfused whole vessels. Circulation 98, I-290.

Booth, R.F., Martin, J.F., et al., 1989. Rapid development of atherosclerotic lesions in the rabbit carotid artery induced by perivascular manipulation. Atherosclerosis 76,2–3, 257–268.

Bredehorst, R., Ligler, F.S., et al., 1986. Effect of covalent attachment of immunoglobulin fragments on liposomal integrity. Biochemistry 25,19, 5693–5698.

Chen, M., Masaki, T., et al., 2002. LOX-1, the receptor for oxidized low-density lipoprotein identified from endothelial cells: implications in endothelial dysfunction and atherosclerosis. Pharmacol Ther 95,1, 89–100.

Cybulsky, M.I., Gimbrone, M.A., Jr., 1991. Endothelial expression of a mononuclear leukocyte adhesion molecule during atherogenesis. Science 251,4995, 788–791.

Demos, S.M., Alkan-Onyuksel, H., et al., 1999. In vivo targeting of acoustically reflective liposomes for intravascular and transvascular ultrasonic enhancement. J Am Coll Cardiol 33,3, 867–875.

Demos, S.M., Dagar, S., et al., 1998. In vitro targeting of acoustically reflective immunoliposomes to fibrin under various flow conditions. J Drug Target 5,6, 507–518.

Demos, S.M., Onyuksel, H., et al., 1997. In vitro targeting of antibody-conjugated echogenic liposomes for site-specific ultrasonic image enhancement. J Pharm Sci 86,2, 167–171.

Demos, S.M., Ramani, K., et al., 1996. Targeted acoustic liposomes for atherosclerotic enhancement during intravascular and transvascular ultrasonic imaging. Circulation 94, I-209.

Fadok, V.A., Voelker, D.R., et al., 1992. Exposure of phosphatidylserine on the surface of apoptotic lymphocytes triggers specific recognition and removal by macrophages. J Immunol 148,7, 2207–2216.

Fisher, N.G., Christiansen, J.P., et al., 2002. Influence of microbubble surface charge on capillary transit and myocardial contrast enhancement. J Am Coll Cardiol 40,4, 811–819.

Fu, X., Kassim, S.Y., et al., 2001. Hypochlorous acid oxygenates the cysteine switch domain of pro-matrilysin (MMP-7). A mechanism for matrix metalloproteinase activation and atherosclerotic plaque rupture by myeloperoxidase. J Biol Chem 276,44, 41279–41287.

Galis, Z.S., Khatri, J.J. 2002. Matrix metalloproteinases in vascular remodeling and atherogenesis: the good, the bad, and the ugly. Circ Res 90,3, 251–262.

Geng, Y.J., Henderson, L.E., et al., 1997. Fas is expressed in human atherosclerotic intima and promotes apoptosis of cytokine-primed human vascular smooth muscle cells. Arterioscler Thromb Vasc Biol 17,10, 2200–2208.

Glagov, S., Weisenberg, E., et al., 1987. Compensatory enlargement of human atherosclerotic coronary arteries. N Engl J Med 316,22, 1371–1375.

Gradus-Pizio, I., Bigelow, B., et al., 2002. The role of adventitia in coronary atherosclerosis: results of echocardiographic imaging of the left anterior descending coronary artery. J Am Coll Cardiol 39, 246A.

Hamilton, A., Benzuly, K., et al., 2002a. Adventitial thickening in non-occlusive atherosclerosis determined by high resolution echocardiographic imaging of the left anterior descending coronary artery. J Invest Med 50, 235A.

Hamilton, A., Rabbat, M., et al., 2002b. A physiologic flow chamber model to define intravascular ultrasound enhancement of fibrin using echogenic liposomes. Invest Radiol 37,4, 215–221.

Hamilton, A.J., Huang, S.L., et al., 2004. Intravascular ultrasound molecular imaging of atheroma components in vivo. J Am Coll Cardiol 43,3, 453–460.

Heath, T.D., Montgomery, J.A., et al., 1983. Antibody-targeted liposomes: increase in specific toxicity of methotrexate-gamma-aspartate. Proc Natl Acad Sci U S A 80,5, 1377–1381.

Herrmann, J., Lerman, L.O., et al., 2001. Coronary vasa vasorum neovascularization precedes epicardial endothelial dysfunction in experimental hypercholesterolemia. Cardiovasc Res 51,4, 762–766.

Houston, P., Goodman, J., et al., 2001. Homing markers for atherosclerosis: applications for drug delivery, gene delivery and vascular imaging. FEBS Lett 492,1–2, 73–77.

Huang, S.L., Hamilton, A.J., et al., 2001. Improving ultrasound reflectivity and stability of echogenic liposomal dispersions for use as targeted ultrasound contrast agents. J Pharm Sci 90,12, 1917–1926.

Kasper, H.U., Schmidt, A., et al., 1996. Expression of the adhesion molecules ICAM, VCAM, and ELAM in the arteriosclerotic plaque. Gen Diagn Pathol 141,5–6, 289–294.

Klibanov, A., Gu, H., et al., 1999. Attachment of ligands to gas-filled microbubbles via PEG spacer and lipid residues anchored at the interface. Controlled Release Society, Boston.

Klibanov, A.L., 1999. Targeted delivery of gas-filled microspheres, contrast agents for ultrasound imaging. Adv Drug Deliv Rev 37,1–3, 139–157.

Klibanov, A.L., Hughes, M.S., et al., 1997. Targeting of ultrasound contrast material. An in vitro feasibility study. Acta Radiol Suppl 412, 113–120.

Klibanov, A.L., Hughes, M.S., et al., 1998. Targeting of ultrasound contrast material: selective imaging of microbubbles in vitro. Acad Radiol 5,Suppl 1, S243–S246.

Klibanov, A.L., Rasche, P.T., et al., 2002. Detection of individual microbubbles of an ultrasound contrast agent: fundamental and pulse inversion imaging. Acad Radiol 9,Suppl 2, S279–S281.

Klibanov, A.L., Rasche, P.T., et al., 2004. Detection of individual microbubbles of ultrasound contrast agents: imaging of free-floating and targeted bubbles. Invest Radiol 39,3 187–195.

Kolodgie, F.D., Petrov, A., et al., 2003. Targeting of apoptotic macrophages and experimental atheroma with radiolabeled annexin V: a technique with potential for noninvasive imaging of vulnerable plaque. Circulation 108,25, 3134–3139.

Korpanty, G., Grayburn, P.A., et al., 2005. Targeting vascular endothelium with avidin microbubbles. Ultrasound Med Biol 31,9, 1279–1283.

Kranzhofer, R., Clinton, S.K., et al., 1996. Thrombin potently stimulates cytokine production in human vascular smooth muscle cells but not in mononuclear phagocytes. Circ Res 79,2, 286–294.

Kunjathoor, V.V., Febbraio, M., et al., 2002. Scavenger receptors class A-I/II and CD36 are the principal receptors responsible for the uptake of modified low density lipoprotein leading to lipid loading in macrophages. J Biol Chem 277,51, 49982–49988.

Kwon, H.M., Sangiorgi, G., et al., 1998. Enhanced coronary vasa vasorum neovascularization in experimental hypercholesterolemia. J Clin Invest 101,8, 1551–1556.

Lanza, G.M., Wallace, K.D., et al., 1996. A novel site-targeted ultrasonic contrast agent with broad biomedical application. Circulation 94,12, 3334–3340.

Leong-Poi, H., Christiansen, J., et al., 2003. Noninvasive assessment of angiogenesis by ultrasound and microbubbles targeted to alpha(v)-integrins. Circulation 107,3, 455–460.

Lindner, J.R., Coggins, M.P., et al., 2000a. Microbubble persistence in the microcirculation during ischemia/reperfusion and inflammation is caused by integrin- and complement-mediated adherence to activated leukocytes. Circulation 101,6, 668–675.

Lindner, J.R., Dayton, P.A., et al., 2000b. Noninvasive imaging of inflammation by ultrasound detection of phagocytosed microbubbles. Circulation 102,5, 531–538.

Lindner, J.R., Ismail, S., et al., 1998. Albumin microbubble persistence during myocardial contrast echocardiography is associated with microvascular endothelial glycocalyx damage. Circulation 98,20, 2187–2194.

Lindner, J.R., Song, J., et al., 2001. Ultrasound assessment of inflammation and renal tissue injury with microbubbles targeted to P-selectin. Circulation 104,17, 2107–2112.

Liu, C., Bhattacharjee, G., et al., 2003. In vivo interrogation of the molecular display of atherosclerotic lesion surfaces. Am J Pathol 163,5, 1859–1871.

Martin, F.J., Heath, T.D., et al., 1990. Covalent attachment of proteins to liposomes. New York, IRL Press.

McPherson, D.D., Sirna, S.J., et al., 1991. Coronary arterial remodeling studied by high-frequency epicardial echocardiography: an early compensatory mechanism in patients with obstructive coronary atherosclerosis. J Am Coll Cardiol 17,1, 79–86.

Moreno, P.R., Purushothaman, K.R., et al., 2002. Intimomedial interface damage and adventitial inflammation is increased beneath disrupted atherosclerosis in the aorta: implications for plaque vulnerability. Circulation 105,21, 2504–2511.

O'Rourke, R., O'Gara, P., et al., 2004. Diagnosis and management of patients with chronic ischemic heart disease. Hurst's the heart. V. Fuster, R. Alexander and R. O'Rourke. New York, McGraw-Hill, 1465–1494.

Schumann, P.A., Christiansen, J.P., et al., 2002. Targeted-microbubble binding selectively to GPIIb IIIa receptors of platelet thrombi. Invest Radiol 37,11, 587–593.

Shaw, P.X., Horkko, S., et al., 2001. Human-derived anti-oxidized LDL autoantibody blocks uptake of oxidized LDL by macrophages and localizes to atherosclerotic lesions in vivo. Arterioscler Thromb Vasc Biol 21,8, 1333–1339.

Takeuchi, M., Ogunyankin, K., et al., 1999. Enhanced visualization of intravascular and left atrial appendage thrombus with the use of a thrombus-targeting ultrasonographic contrast agent (MRX-408A1): In vivo experimental echocardiographic studies. J Am Soc Echocardiogr 12,12, 1015–1021.

Unger, E.C., Lund, P.J., et al., 1992. Nitrogen-filled liposomes as a vascular US contrast agent: preliminary evaluation. Radiology 185,2, 453–456.

Unger, E.C., McCreery, T., et al., 1998a. MRX 501: a novel ultrasound contrast agent with therapeutic properties. Acad Radiol 5,Suppl 1, S247–S249.

Unger, E.C., McCreery, T.P., et al., 1998b. Acoustically active lipospheres containing paclitaxel: a new therapeutic ultrasound contrast agent. Invest Radiol 33,12, 886–892.

Unger, E.C., McCreery, T.P., et al., 1998c. In vitro studies of a new thrombus-specific ultrasound contrast agent. Am J Cardiol 81,12A, 58G-61G.

van der Loo, B., Martin J.F., 1997. The adventitia, endothelium and atherosclerosis. Int J Microcirc Clin Exp 17,5, 280–288.

van der Wal, A.C., Becker, A.E., et al., 1993. Medial thinning and atherosclerosis–evidence for involvement of a local inflammatory effect. Atherosclerosis 103,1, 55–64.

Villanueva, F.S., Jankowski, R.J., et al., 1998. Microbubbles targeted to intercellular adhesion molecule-1 bind to activated coronary artery endothelial cells. Circulation 98,1, 1–5.

Weller, G.E., Lu, E., et al., 2003. Ultrasound imaging of acute cardiac transplant rejection with microbubbles targeted to intercellular adhesion molecule-1. Circulation 108,2, 218–224.

Wu, Y., Unger, E.C., et al., 1998. Binding and lysing of blood clots using MRX-408. Invest Radiol 33,12, 880–885.

Yasu, T., Greener, Y., et al., 2005. Activated leukocytes and endothelial cells enhance retention of ultrasound contrast microspheres containing perfluoropropane in inflamed venules. Int J Cardiol 98,2, 245–252.

Part 4
Iodinated Nanoparticles: Applications in Computed Tomography

18
Iodinated Liposomes as Contrast Agents

Andreas Sachse

Abstract: The use of liposomal contrast agents for computed tomography (CT) imaging applications is considered. The properties of the various liposome types as well as methods for their preparation are discussed. Specific challenges for the preparation and use of CT contrast-carrying liposomes are outlined. Available experience on the use of CT liposomes for liver and spleen imaging in preclinical animal models and humans is reviewed. The use of surface-modified and conventional liposomes for CT blood-pool imaging is discussed in light of the published animal data. Finally, new trends in the use of CT liposomes are considered based on the limited data available to date.

Keywords: Liposomes, preparation methods, computed tomography (CT), iodinated contrast agents, surface modification, liver imaging, blood-pool imaging, animal experience, human experience, liver tumors.

18.1. Introduction

The potential usefulness of liposomal contrast agents for liver and spleen imaging has already been demonstrated in the early 1980s (e.g., Havron et al., 1981). Based on the encouraging liver and spleen imaging results in animals, clinical studies (phase II) were contemplated as early as 1984 (Ryan et al., 1984). However, due to technological hurdles associated with the reproducible large-scale production of pharmaceutically acceptable liposomes with high encapsulation (Tilcock, 1999), it took until the second half of the 1990s to finally study the computed tomography (CT) imaging properties of such liposomes in humans (Leander et al., 1998).

Initial development centered on passive targeting of liposomes to the Kupffer cells in the liver and macrophages in the spleen (=reticuloendothelial system (RES)). Since phagocytic activity is not displayed by tumor tissue, the contrast agent can selectively be concentrated in the healthy tissue resulting in an increased tissue to tumor density difference (expressed in Hounsfield Units (HU)).

By the middle of the 1990s, new methods for liposome production (e.g., extrusion) were available which allowed reproducible large-scale production of stable liposomes with encapsulation efficiencies up to 50% and corresponding iodine/lipid ratios (mg/mg) well above 1. The resulting formulations showed acceptable tolerability, elimination, and imaging properties in various animal models (i.e., rat, rabbit, and non-human primates) so that eventually clinical testing (phase I) in humans could be performed by various groups.

Due to the introduction of methods for modification of liposome surfaces, it became possible to produce liposomes which could avoid the RES for prolonged timeperiods (Allen, 1994). Since CT liposomes with extended circulation times may be useful as blood-pool (vascular) imaging agent, various groups evaluated the imaging properties of unmodified as well as PEG-liposomes, in rabbits and non-human primates (e.g., Sachse et al., 1997; Schmiedl et al., 1999).

Recently, CT contrast agent-carrying immunoliposomes which carry an antibody (or their fragments) to a specific cell-surface protein to target the vesicles to the tissue of interest have been contemplated. In the present chapter, the available data on CT liposomes is reviewed and discussed.

18.2. Liposome Types

Liposomes are self-closed spherical or elliptical lipid vesicles composed of one or more lipid bilayers, which entrap an aqueous phase (Figure 18.1). Depending on their size, one can differentiate small unilamellar vesicles (SUVs) with mean diameters up to 100 nm and large unilamellar vesicles (LUVs) with diameters above 100 nm up to the micrometer range. Vesicles with several bilayers (lamellae) are called multilamellar vesicles (MLVs).

For the production of liposomes, mainly phospholipids like phosphatidylcholine (PC) or phosphatidylethanolamine (PE), which are main constituents of biological membranes, are employed. Additionally cholesterol (CH) is often used for membrane stabilization. The physico-chemical properties of the liposomes are dependent on the lipid composition (qualitative and quantitative), size, temperature, pH, and type and strength of the employed buffer.

Liposomes can be loaded with various polar (hydrophilic) or non-polar substances which can be entrapped in the aqueous liposome interior, dissolved in the hydrophobic bilayer, or adsorbed on the liposome surface. For encapsulation of hydrophilic compounds like X-ray (CT) contrast agents, unilamellar liposomes are most effective since the volume of the aqueous compartment in relation to a given lipid concentration is larger compared to multilamellar vesicles (MLVs). Most preferable are large unilamellar vesicles which have the highest captured volume, i.e. liter aqueous phase per mole of lipid compared to the other liposome types. In Table 18.1, the properties of the various liposome types are listed including encapsulation efficiencies (i.e., percent of total employed drug encapsulated) which were obtained with various hydrophilic drugs.

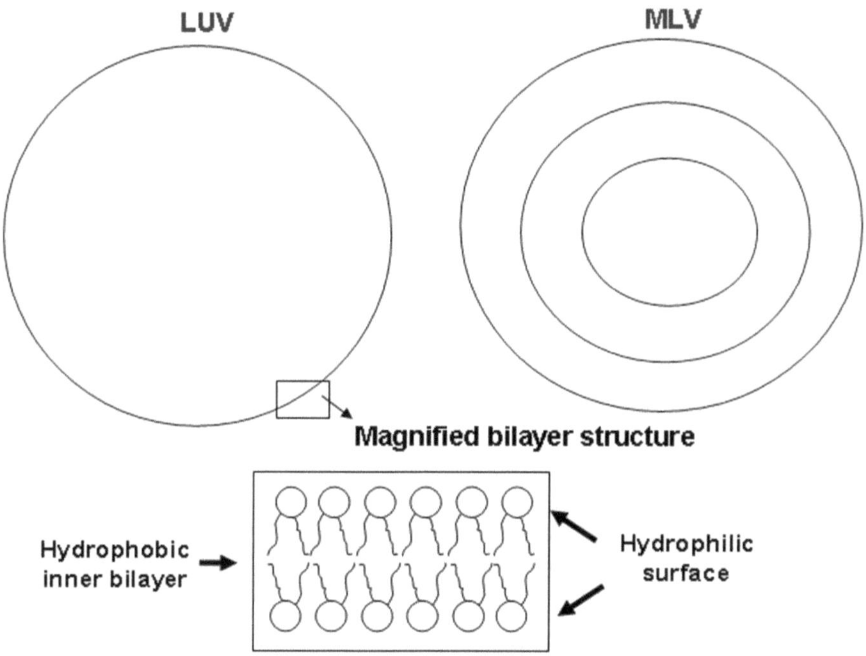

FIGURE 18.1. Schematic diagram of the various liposome types and bilayer structure.

TABLE 18.1. Properties of the different liposome types prepared by various methods

Liposome type	Liposome size (nm)	Captured volume (l H_2O/mol Lipid)	Encapsulation efficiency (%)
SUV	20–60	0.2–1.5	≤30
LUV	50–12000	≤22.5	≤75
MLV	100–10000	1–10	≤89

18.3. Challenges for Liposome Preparation and Use

For contrast-carrying liposomes for CT imaging the goal has been to produce liposomes with the following properties (Seltzer, 1988b, Seltzer et al., 1995):

1. High iodine entrapment, either expressed as encapsulation efficiency (see above) or as iodine to lipid ratio (i.e., amount of iodine that is delivered by a unit weight of lipids to the target organ expressed as mg/mg (Cheng et al., 1987)).
2. Sufficient plasma stability.
3. Satisfactory organ opacification (e.g., at least 50 ΔHU for liver and spleen imaging) at reasonable doses (i.e., iodine and lipid). For CT 1 mg iodine/g tissue is needed to result in an attenuation of approximately 25–30 HU.

4. Narrow size range and a maximum size below 5 μm (for i.v. injection).
5. Acceptable storage stability (especially absence of drug leakage and liposome size increases).
6. Sterile and pyrogen-free product.
7. Acceptable toxicity.

Overall, the respective liposome production method needs to be easy to perform also on a large-scale, reproducible, and reasonable in cost.

It has to be noted that because much higher concentrations of contrast media are needed for detection with computed tomography, organ-specific contrast media that are safe and effective are much harder to develop for this modality compared to MR or scintigraphy (Desser et al., 1999).

18.4. Preparation Methods

According to New (1990) the methods for the preparation of liposomes can basically be classified into three categories:

1. Two-phase dispersion
2. Detergent solubilization
3. Mechanical dispersion

The main disadvantage of the "two-phase dispersion" method is the use of organic solvents (e.g., diethyl ether, ethanol, etc.) which have to be completely removed during the preparation process. However, some of these methods potentially allow production of unilamellar liposomes (SUVs and LUVs) with high encapsulation over a wide size range (New, 1990; Martin, 1990). The most prominent method from this category is the reverse-phase evaporation method (REV) developed by Szoka et al. (1978) which was also successfully used for preparation of CT liposomes with high encapsulation efficiency by various groups (Table 18.2). Another method belonging to this category is the ethanol evaporation method used for the preparation of iopromide liposomes, which is also listed in Table 18.2.

Depending on the method of detergent removal, production of SUVs and LUVs in a size range of approximately 30–180 nm has been reported for "detergent solubilization" method. The main shortcomings of this method are leakage and dilution of the drug during liposome formation, as well as the difficulty of completely removing detergent from the final liposome suspension (Martin, 1990). This may also be the reason for the absence of any publications on the use of such methods for the production of CT liposomes.

Among the "mechanical dispersion" methods, high pressure homogenization or microfluidization (MF) are most often used for fast and reproducible large-scale production of SUVs (<100 nm) at high lipid concentrations (Lasic, 1993). The disadvantages of these methods include problems with heat regulation, as well as

TABLE 18.2. Properties of liposomes for spleen and liver CT imaging

Contrast agent(s)	Lipid composition (molar ratio)	Lipid concentration (mg/ml)	Liposome size (mean diameter) (nm)	Liposome type	Total initial iodine concentration (mg/ml)	Encapsulation efficiency (%)	Iodine/Lipid ratio	Animal model	CT imaging results	Reference
Diatrizoate	SPC/CH/SA (8:1:1)	583	1400 ± 130 (EM)	MLV	55.5* (final concentration)	11	Approximately 0.1	Sprague-Dawley rats	Liver: 13 ΔHU after 30 min Spleen: 97 ΔHU after 30 min (133.2 mg iodine/kg b.w.)	Havron et al., 1981
Diatrizoate	PC/CH (1:1)	90 μmol/ml	200–500 (EM)	MLV (REV)	300*	22–25	≤ 1.5	Various species including rabbit and dog	No values given, X-ray images shown after doses of 5 ml/kg	Zherbin, 1983 (German)
		150 μmol/ml				40				
Diatrizoate	EPC/CH (7.7:3.3)	50	30–600 (EM)	LUV (REV)	300*	9		Sprague-Dawley rats	Liver: 15–25 ΔHU at 60 min	Ryan et al., 1983
					150	35	Approximately 1		Spleen: 130 ΔHU at 60 min	
Metrizamide					307	18			(30 mg iodine (diatrizoate) /kg b.w.)	
Diatrizoate	EPC/CH/SA (4:1:1)	1320	100–2000 (EM + light microscope)	MLV	76 * (final concentration)	N/A	0.06	Sprague-Dawley rats (normal + carcinoma)	Liver: 70 ΔHU (rat, 5 min), 71 ΔHU (dog, 60–90 min), 124 ΔHU (baboon, 60 min)	Seltzer, 1984a
								Dogs	Spleen: 159 ΔHU (rat, 30 min), 122 ΔHU (dog, 60–90 mii), 257 ΔHU (baboon, 60 min)	
								Baboon	Abdominal aorta: 99 ΔHU (rat, at 1 h), 35 ΔHU (dog, 60–90 min), 39 ΔHU (baboon, 60 min) Tumor rim enhancement of 52 ΔHU within 20 min (190 (rat), 564 (dog) and 500 (baboon) mg I/kg b.w.)	

(Continued)

TABLE 18.2. (Continued)

Contrast agent(s)	Lipid composition (molar ratio)	Lipid concentration (mg/ml)	Liposome size (mean diameter) (nm)	Liposome type	Total initial iodine concentration (mg/ml)	Encapsulation efficiency (%)	Iodine/Lipid ratio	Animal model	CT imaging results	Reference
Na-Ioxitalamate	SPC/CH/SA or DCP (various molar ratios)	20	135–145	MLV (sonication)	100*	≤ 17	Approximately 1	None	None	Benita et al., 1984
Diatrizoate and other contrast agents	EPC/CH (2:1)	22.9	N/A	LUV (REV)	9.2* (final concentration)	16–23	0.4	Sprague-Dawley rats	Liver: 7.5 ΔHU peak enhancement Spleen: 156 ΔHU at 60 min (30 mg I (diatrizoate) /kg b.w.)	Ryan et al., 1984
Iosefamate (biliary contrast agent)	PC/CH/SA (4:1:1)	N/A	<2000 (light microscope)	MLV	18–35* (final concentration)	N/A	N/A	Dogs	Liver: 94 ΔHU at 120 min Spleen: 106 ΔHU at 240 min (300 mg I/kg b.w.)	Seltzer, 1984
Diatrizoate	EPC/CH (7:3)	80.7	2950 ± 1930	MLV (Proliposomes)	148	28	N/A	Sprague-Dawley rats	Liver: <20 ΔHU peak enhancement	Payne and Whitehouse, 1987
Iopamidol			4400 ± 1530			38			Spleen: 43 ΔHU peak enhancement (240 mg I (diatrizoate)/kg b.w.)	
Diatrizoate	EPC/CH/EPG (4:5:1)	60 μmol/ml	100–3000 (bimodal distribution)	LUV (MF)	300*	0.75	0.06	New Zealand white rabbits	Liver: 60 ± 3 ΔHU within 30 min	Cheng et al., 1987
Iotrolan and other contrast agents						3.38	0.26		Spleen: 306 ± 32 ΔHU within 30 min (250 mg I (iohexol)/kg b.w.)	

Diatrizoate iotrolan	SPC/CH/SA (8:1:1) DSPC/SM (5:2)	80 μ mol/mlØ	1400 1000–4000	MLV (film method)	N/A *	N/A	N/A	CD rats	None, only biodistribution studied using radiolabel	Zalutsky et al., 1987
Diatrizoate	EPC/CH (1:1)	65 μ mol/ml	570 ± 330	MLV	300*	8.53 ± 1.13	0.68 ± 0.09	Sprague-Dawley rats	Liver: 15 ΔHU (rat), 23 ΔHU (rabbit)	Seltzer et al., 1988
iotrolan				(DRV)		15.07 ± 4.70		Spleen: 109 ΔHU (rat), 150 ΔHU (rabbit)		
(+ iohexol)				(MF-DRV)		6.72 ± 4.61	*0.54 ± 0.36	New Zealand white rabbits	(500 (rat) or 250 (rabbit) mg I (diatrizoate) /kg b.w.)	
Diatrizoate	EPC/CH (7:3)	130	1000–1800 (EM)	LUV/MLV (REV)	300–320*	N/A	1.77	Wistar rats, nitrosamine-induced liver tumors	Liver: 40 ΔHU at 60 min p.i. for up to 5 h (260 mg I/kg b.w.), tumor lesions of 2–3 mm were detectable on enhanced images	Henze et al., 1989
iotrolan and other contrast agents	SPC/CH (4:6)						2.51			

*Unencapsulated iodine removed by dialysis, centrifugation, or gel filtration

N/A = not available

formation of metal abrasion. Furthermore, homogeneous (i.e., unimodal) preparations of larger liposomes (>100 nm) cannot be produced (Martin, 1990). As shown in Table 18.2, microfluidization has been used with some success for the production of contrast-carrying liposomes.

Another method belonging to this category is filter extrusion which allows the production of unilamellar liposomes in the size range of 40–150 nm originally developed by Olson et al. (1979). Due to its ease, reproducibility, and flexibility, extrusion has become the method of choice for the production of well-defined liposomes on the laboratory scale (Lasic, 1993). Recently, the usefulness of the extrusion method was considerably extended by the introduction of the continuous high-pressure extrusion method which allowed reproducible production of liposomes with very high encapsulation of CT or MR contrast agents. Thus, Schneider et al. (1995) obtained encapsulation efficiencies above 50% for iopromide for vesicles with mean diameters around 100 nm at high lipid concentrations. Encapsulation efficiency decreased with contrast agent concentration and increased with lipid concentration; however, at the cost of decreasing iodine/lipid ratios. At medium iodine and lipid concentrations (100 mg/g and 150 mg/g, respectively), encapsulation efficiencies as high as 40% could still be achieved for iopromide (batch size $\leq$ 200 ml).

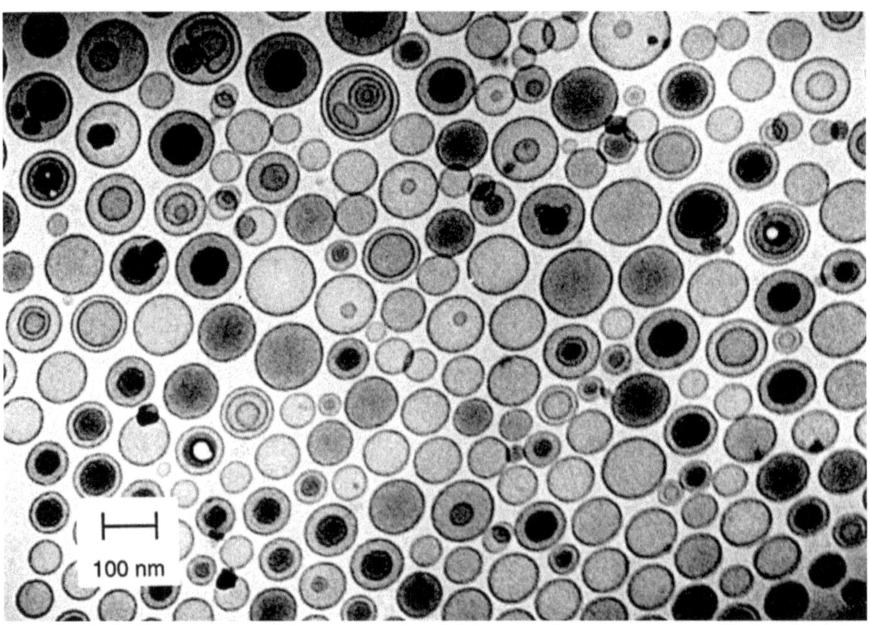

FIGURE 18.2. Cryo-electron micrograph of CHHS-polyethylene glycol liposomes prepared by continuous high-pressure extrusion. Reproduced from Sachse et al. (1997), with kind permission of Lippincott Williams & Wilkins.

Subsequently, the usefulness of continuous high-pressure extrusion for the rapid and reproducible production of large batches of storage-stable contrast-carrying liposomes with very high encapsulation could be confirmed (Sachse, 1998). In Figure 18.2, a typical electron micrograph of liposomes prepared by this method is shown. It is worth noting that the majority of liposomal CT preparations tested after 1995 were produced by filter extrusion, as shown in Tables 18.3 and 18.4.

18.5. Employed Contrast Agents

Most studies with contrast-carrying liposomes for CT employed tri-iodinated derivatives of benzoic acid which are the standard compounds for intravascular X-ray imaging today. They are small (molecular weights from approximately 800 to 1600Da) water-soluble molecules with iodine as the contrast-generating component. For liposome production, either monomeric ionic compounds, such as diatrizoate, or monomeric non-ionic compounds, such as iopromide, or iopamidol were used. Alternatively, dimeric non-ionic compounds, like iotrolan and ioxaglate, which exhibit a very high X-ray density due to their high iodine content were encapsulated (Desser et al., 1999). In Figure 18.3, the structures of diatrizoate (ionic monomer), iopromide (non-ionic monomer), and ioxaglate (non-ionic dimer) are shown as examples. The main difference between the various contrast agent classes (i.e., ionic and non-ionic, monomeric and dimeric) is osmolality. While ionic monomers exhibit very high osmolalities (1,500–2,300 mOsm/kg) at clinically relevant concentrations, non-ionic monomers and ionic dimers are slightly hypertonic (500–800 mOsm/kg) and non-ionic dimers are isotonic to blood (approximately 300 mOsm/kg). Osmolality does not only impact the tolerability of the unencapsulated contrast agents, but also plays an important role regarding their applicability for the entrapment in liposomes (Krause et al., 1995).

After intravenous injection, these extracellular contrast agents briefly stay in the bloodstream and then rapidly diffuse to the extracellular spaces of all non-neural tissues. They are excreted rapidly by the kidneys (Seltzer et al., 1988).

All, but two, studies with CT liposomes were performed using the above-mentioned water-soluble contrast agents encapsulated in the aqueous liposome phase. In the first study, Caride et al. (1982) studied the hepatosplenic CT-imaging properties of brominated radiopaque liposomes which were obtained by simple bromination of phosphatidylcholine (PC) in chloroform. According to the authors, this approach would also be suitable for iodination of liposomes. In the second study, an iodinated oil (Ethiodol, iodinated ester of poppy seed oil) has been encapsulated in the liposome membrane and been used for tumor imaging in VX2-carcinoma carrying rabbits (Jendrasiak et al., 1985).

TABLE 18.3. Physico-chemical properties of liver liposomes

Contrast agent	Product designation (originator)	Lipid composition (molar ratio)	Lipid concentration (mg/ml)	Liposome size (mean diameter) (nm)	Liposome type	Total iodine concentration (mg/ml)	Encapsulation efficiency (%)	Iodine/lipid ratio	Reference
Iopromide	NN (Schering AG)	PC/CH/SS (4:5:1)	99.3	439 ± 87	LUV /MLV (ethanol evaporation)	97.3	42.5	Approximately 0.4	Sachse et al., 1993
Iomeprol	Lipiom (Bracco)	HSPC or DSPC/ DPPA-Na (9.5:0.5)	15	1200	LUV/MLV (2 μm filter extrusion)	260	27	4.7	Fouillet et al., 1995
Iotrolan	NN (The Liposome Company)	DSPC	N/A	2900	LUV (interdigitation-fusion)	100*	N/A	4.6	Seltzer et al., 1995
Iopromide	SH U 561 A (Schering AG)	PC/CH/SS (5:4:1)	Approximately 100	467 ± 66	LUV /MLV (ethanol evaporation)	102.7	30–40	Approximately 0.4	Schmiedl et al., 1995
Iodixanol	NN (Nycomed)	N/A	N/A	550	LUV (1 μm filter extrusion)	100 (after dilution)	50	N/A	Leander, 1996 Dick et al., 1996
Iopromide	None	SPC/CH/SA (7:4:2)	N/A	1200	LUV (filter extrusion)	15.2 (after removal)	Approximately 10	N/A	Erdogan et al., 1998
Iomeprol	BR2 (Bracco)	PC-DPPA/CH (2:1)	40	400	LUV (N/A)	260	N/A	N/A	Petersein et al., 1999
Iomeprol	BR21 (Bracco)	PC-DPPA/CH (2:1)	20	400	LUV (N/A)	320	12.5 (Katayama et al. 2001)	N/A	Petersein et al., 1999
Iodixanol	CTP-10 (Nycomed Amershan)	N/A	N/A	550	LUV (1 μm filter extrusion)	200	40	N/A	Desser et al., 1999
Iomeprol	BR21 (Bracco)	HSPC/CH/DPPA (6.3:3.2:5)	20	400	LUV (N/A)	320 (40 encapsulated)	12.5	2	Spinazzi et al., 2000
Iodixanol	CTP (Nycomed Amersham)	PC/PS (N/A)	N/A	340	MLV (N/A)	200 (80 encapsulated)	40	1.5	Leander et al., 2001
Iomeprol	BR24 (Bracco)	Phospholipids/CH	Approximately 20	400	LUV (N/A)	300–320 (approximately 40 encapsulated)	Approximately 13	Approximately 2	Rummeny et al., 2002

N/A = not available

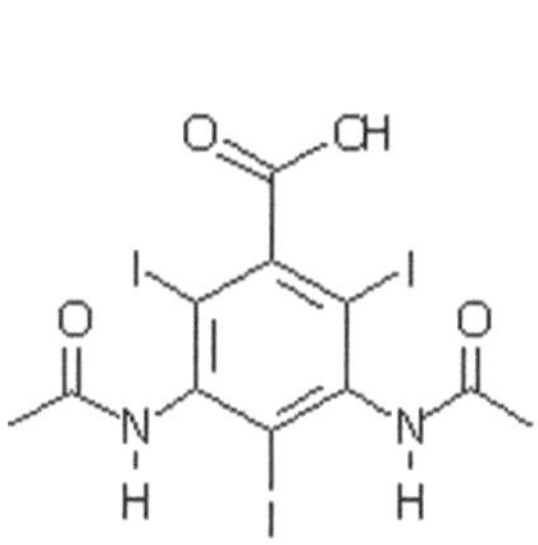

Diatrizoate

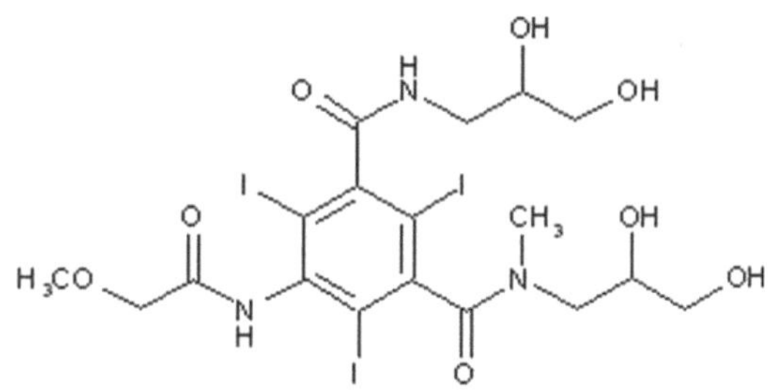

Iopromide

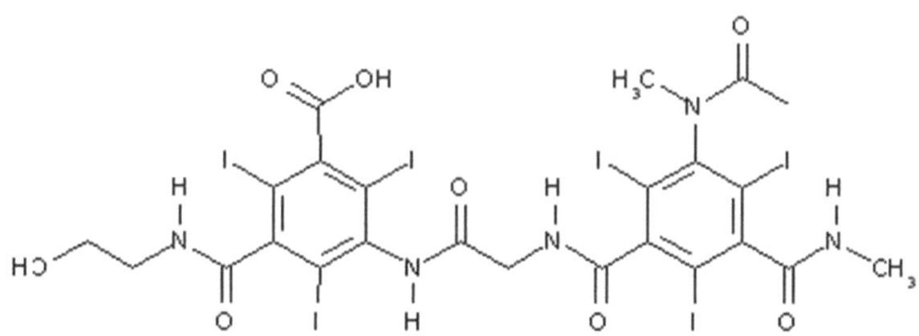

Ioxaglate

FIGURE 18.3. Structural formulas of various CT contrast agents.

18.6. Liver and Spleen Imaging

After i.v. injection, liposomes are mainly taken up by the phagocytic cells of the reticuloendothelial system (RES) which is also referred to as mononuclear phagocytic system (MPS). The RES comprises circulating blood macrophages, as well as the localized macrophages in liver (Kupffer cells), spleen, and bone marrow. This naturally occurring biodistribution pattern can be used for passive targeting of diagnostics to liver and spleen as the main sites of liposome uptake (Poste, 1983).

Larger liposomes (MLVs, LUVs) are preferentially taken up by the Kupffer cells in the liver and smaller liposomes (SUVs) are more rapidly taken up. Liposome uptake can be increased by incorporation of negatively charged lipids (e.g., phosphatidic acid (PA), phosphatidyl glycerol (PG) or phosphatidyl serine (PS)), the effect being more pronounced for smaller liposomes (Senior, 1987). Incorporation of cholesterol (CH) into the liposome membrane results in an increased plasma stability, prolonged blood circulation, and decreased intracellular degradation. Additionally, CH-containing liposomes also show increased spleen uptake (Roerdink et al., 1989). The organ distribution of liposomes is a function of liposome dose which is reflected by a reduced liver uptake at higher lipid doses. This seems to be due to a saturation of the endocytotic uptake mechanism into the Kupffer cells (Senior, 1987).

Since most malignant liver tumors including metastases do not contain any RES cells, a liposomal CT contrast agent can selectively be concentrated in healthy tissue. Thus, tumorous tissue will appear hypointense to normal liver tissue post-injection (Rummeny et al., 2002).

Liver imaging with various modalities (MR, CT, and ultrasound (US)) is frequently performed to detect and characterize focal liver lesions. From a clinical viewpoint, detection of primary liver tumors (i.e., hepatocellular carcinoma, cholangiocarcinomas) is important to facilitate surgical treatment. Detection of secondary malignant liver lesions (i.e., metastases of tumors arising from another site, such as colon, lung, pancreas) is important for pretherapeutic staging. Ideally, detection of focal lesions well below 0.5 cm should be possible (Fouillet et al., 1995). To date, computed tomography remains the primary imaging modality (screening and staging) for most patients with suspected liver disease. Therefore, development of liver-specific CT liposomes, which are both safe and highly sensitive, is of great clinical importance for non-invasive staging in patients with known or suspected cancers (Rummeny et al., 2002).

Already in 1980, a patent application on the preparation and use of X-ray contrast agent containing liposomes (MLV) for liver imaging was published (Daeniker, 1980). The use of tri-iodinated derivatives of benzoic acid was claimed. Intravenous application of such liposomes in rabbits and dogs resulted in marked opacification in liver and spleen already 15 minutes p.i.

Throughout the 1980s various investigators have published evidence for the ability of iodinated-liposomal contrast agents to opacify the liver and spleen in computed tomography as summarized in Table 18.2. The early experience with CT liposomes was reviewed by Seltzer in 1988. The author concluded that the liposomes available at that time showed insufficient encapsulation efficiencies requiring very high lipid doses for efficient organ opacification which result in at least temporary saturation of the RES. Furthermore, most available production methods did not allow reproducible production of pharmaceutically acceptable liposomes on a large scale.

Thus, there was a need for the development of new preparation methods for iodinated liposomes which would address these shortcomings and eventually allow to perform preclinical testing of such preparations in larger animals and ultimately clinical testing in humans. As shown in Table 18.3, eventually three new preparation methods were introduced to meet these criteria, namely (1) ethanol evaporation, (2) interdigitation-fusion, and (3) filter extrusion. The preclinical and clinical experience with the respective liver liposomes is described below:

18.6.1. Animal Experience

18.6.1.1. Rats and Rabbits

Sachse et al. (1993) studied the use of freeze-dried iopromide liposomes for CT tumor imaging in a VX-2 rabbit model. Prior to application, the lyophilized liposomes were reconstituted with 135 mM mannitol solution to result in a liposome suspension with a total iodine content of 97.3 mg/ml of which 42.5% were encapsulated (Table 18.3). The rabbits (Hare-rabbits) were injected manually via an ear vein at a rate of approximately 6 ml/min at doses between 50 and 200 mg total iodine/kg b.w. Within a few minutes after injection, liver attenuation reached a plateau which remained unchanged over the whole study period (up to 130 minutes). Mean liver attenuation was linearly correlated to the dose with a maximum density difference of 33, 40, and 54 HU at doses of 100, 150, and 200 mg iodine/kg b.w., respectively. As shown in Figure 18.4, the 150 mg dose allowed delineation of a very small tumor (<0.5 cm) in the post-contrast image (6 minutes p.i., verified by necropsy). In the same study, elimination of liposomal iopromide was investigated in tumor-free rabbits. Twenty-four hours after administration of 250 mg total iodine/kg b.w., 76% of the administered dose were eliminated via the urine while liver, spleen and blood concentrations amounted to 7.7%, 2.5%, and 6.3%, respectively. Seven days after injection complete urinary excretion was found.

Furthermore, biodistribution of the iopromide liposomes was studied in rats (Wistar strain) at doses of 100, 250, and 1000 mg total iodine/kg b.w. (total iodine (unencapsulated + encapsulated) to lipid ratio = 1). The respective biodistribution in the major organs of interest is shown in Figure 18.5. Within the 24-hour study period, the maximum iodine concentration in the liver was 23.6%, 16.1%, and 8.8%, respectively. The corresponding maximum spleen concentrations amounted to 9.7%, 8.9%, and 8.4%, respectively. On a per-gram organ weight basis, the spleen displayed an approximately 10 times higher capacity for liver uptake than the liver in this study. Increasing the liposome dose led to a decreased liver uptake (as percent of initial dose) which was accompanied by a marked increase in blood iodine concentration. At 1000 mg total iodine (or lipid)/kg b.w., 29.5% of dose were found in the blood at 15 minutes and 21.6% at 1 hour. In kidney and lung, the highest iodine concentration was consistently measured 15 minutes after injection. Except for the highest dose, lung accumulation was found to be negligible demonstrating absence of physical entrapment of large liposomes.

The authors concluded that the ethanol evaporation method proved to be suitable for the reproducible large-scale (up to 1 L) production of iopromide liposomes with favorable elimination and imaging characteristics in rabbits.

Fouillet et al. (1995) studied iomeprol-containing liposomes (Lipiom) for computed tomography liver scanning in an animal model of chemically induced hepatocellular carcinomas and other liver tumors in Sprague-Dawley rats. The sterile and pyrogen-free liposome suspension contained 260 mg iodine/ml of which 70 mg (27%) were encapsulated (see Table 18.3). Rats were injected intravenously via the tail vein at a rate of about 1 ml/min with doses between 33 and 330 mg entrapped iodine/kg b.w.

The initial enhancement observed in the blood (cardiac cavities) which peaked during and immediately after the bolus injection disappeared rapidly (no enhancement after 20 minutes) due to elimination of the non-entrapped iomeprol

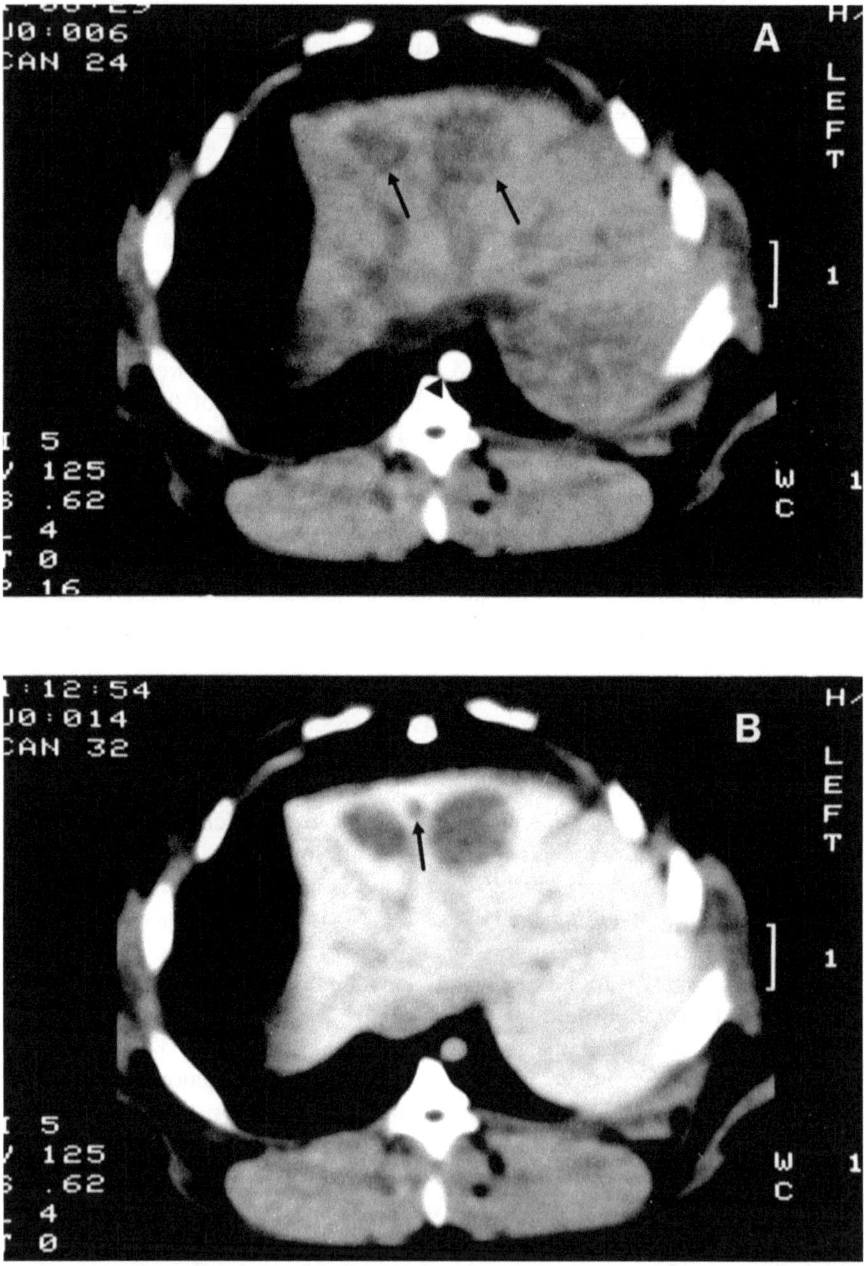

FIGURE 18.4. Demonstration of VX-2 carcinomas implanted in the liver of a rabbit. **A** Immediately after start of injection (150 mg total iodine/kg b.w.) there was a significant increase in density in the aorta (arrow head). The tumors (arrows) were slightly hypodense and barely visible. **B** Six minutes after injection, the tumors were well delineated and a very small tumor (arrow) could be detected. Reproduced from Sachse et al. (1993), with kind permission of Lippincott Williams & Wilkins.

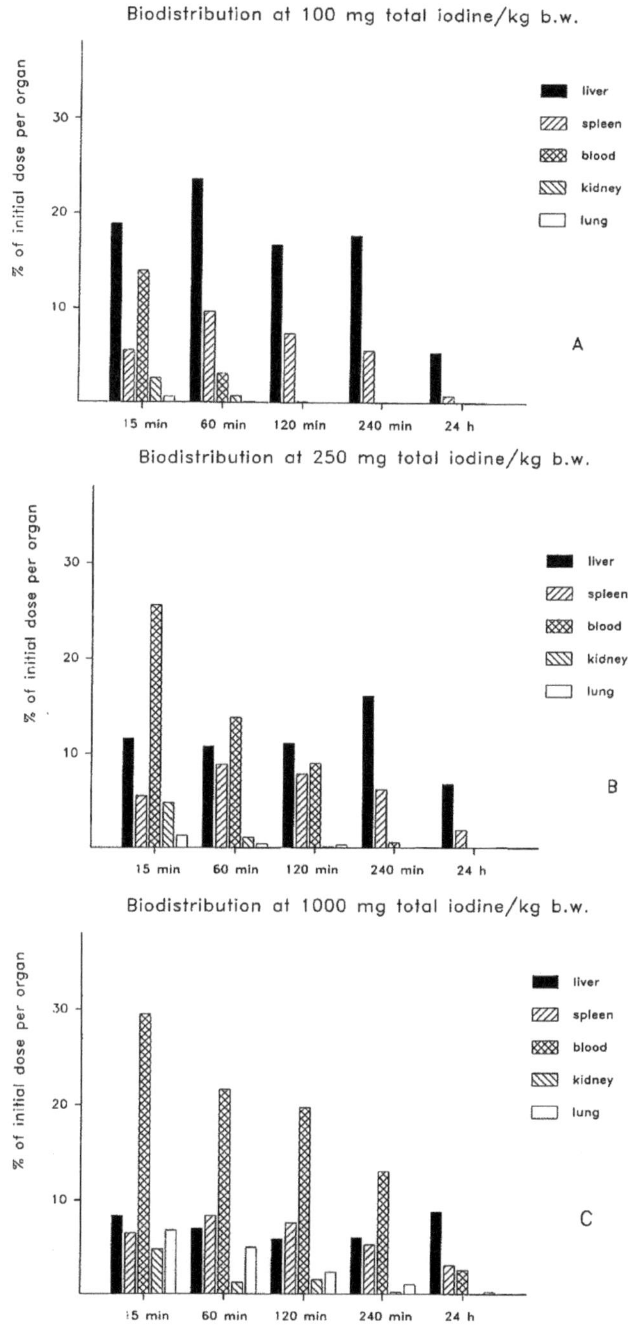

FIGURE 18.5. Biodistribution of iopromide liposomes at different doses.Total iodine-to-lipid ratio is approximately equal to 1. Reproduced from Sachse et al. (1993), with kind permission of Lippincott Williams & Wilkins.

via the kidneys and phagocytosis of the encapsulated contrast agent by the Kupffer cells in liver and spleen. Enhancement in the liver parenchyma rapidly increased within 5–10 minutes in a dose-dependent manner (approximately between 315 and 400 HU (90 minutes p.i.) at 165 and 330 mg iodine/kg, respectively) and thereafter remained constant for several hours. Spleen enhancement was remarkably higher (650 and 1338 HU (90 minutes p.i.) at 165 and 330 mg iodine/kg, respectively) and showed an overproportional increase with dose compared to the liver, which could have been due to saturation of the liver Kupffer cells. The contrast enhancement of liver and spleen at doses >150 mg encapsulated iodine was higher than necessary for lesion detection. A dose of 70 mg entrapped iodine, which resulted in an enhancement of 45 ΔHU in the liver (65% increase above post-contrast), appeared to be the adequate scanning dose (imaging window: 20 minutes–4 hours). The smallest detectable lesions were 3–4 mm in diameter. Microscopic observation after autopsy showed that the Kupffer cells were restricted to the normal liver parenchyma and absent or rare in the tumoral areas. The induced tumors had a histology similar to that of the most common primary liver tumors in humans.

The authors concluded that the tested iomeprol liposomes (Lipiom) show some promise as organ-specific contrast agent for liver and spleen imaging if the ratio between administered dose and organ enhancement proves to be the same in humans as it is in animals.

The biodistribution, clearance, and imaging properties of iotrolan-carrying liposomes produced by the interdigitation-fusion (IF) method were studied by Seltzer et al. (1995) in rats. The respective DSPC liposomes which contained only encapsulated drug (see Table 18.3) displayed very high iodine/lipid ratios which in certain cases were as high as 9.

At doses between 10 and 1,000 mg iodine/kg b.w., the liver contained 60%–95% of the injected dose after 1 hour demonstrating rapid liver uptake. Increasing doses led to increasing absolute hepatic uptake while the relative uptake as percent of dose decreased. Liposome uptake in the spleen ranged from 3% to 12% while uptake in other organs was mostly below 3%. There was only a slight increase in lung and blood uptake at the highest dose. IF liposomes showed a remarkably slower clearance compared to other liposomes with liver half-lives of 8.5, 10.8, and 12.1 days at 25, 100, and 250 mg iodine/kg b.w., respectively. By 8 weeks, the liver still contained 0.8%–2.6% of the injected dose, while in the spleen less than 0.1% were found. According to the authors, the prolonged half-live may be due to the physico-chemical properties of iotrolan which may show slower clearance from the intracellular compartment due to its hydrophilicity and higher molecular weight (dimeric compound).

In rats, CT liver enhancement of 46, 96, and 311 HU was obtained for doses of 100, 250, and 1,000 mg iodine/kg. The spleen showed the largest enhancement with 1,050 HU at the highest dose. In addition, CT imaging was performed in 2 dogs at a dose of 100 and 250 mg iodine/kg ($n = 1$ each). In contrast to rats, higher liver-to-spleen enhancement was found here resulting in 116 HU of hepatic and 65 HU splenic enhancement at 250 mg iodine/kg.

Thus, 65 mg of liposomal iodine would be required in dogs to obtain 50 HU hepatic enhancement.

The authors concluded that iotrolan IF vesicles have superior physico-chemical properties, as well as favorable biodistribution and imaging characteristics which warrant evaluation in clinical studies.

Leander (1996) studied the tumor imaging properties of iodixanol liposomes in a VX2-carcinoma rabbit model. Fourteen days after tumor cell inoculation, seven Swedish loop rabbits received injections of 200 mg liposomal iodine/kg b.w. (50% encapsulation, see Table 18.3), given as manual injection of 2 ml/kg during 30 seconds. Compared to precontrast images, the enhancement (ΔHU) of normal liver tissue to tumorous tissue was significantly higher at all timepoints during the observation period (up to 120 minutes). Ten minutes p.i., a slightly higher attenuation in tumorous tissue compared to precontrast was observed which may be explained by the presence of non-encapsulated iodixanol. At the same timepoint, enhancement in the abdominal aorta and inferior vena cava (IVC) amounted to approximately 11 ΔHU, corresponding to a CM concentration of about 0.4 mg iodine/ml. The highest liver-to-tumor contrast was reached 10 minutes post-injection at 68 ΔHU reaching a plateau around 55 ΔHU at 30 minutes throughout the observation period. According to the author, this attenuation difference should be fully satisfactory to allow tumor differentiation.

Dick et al. (1996) studied the CT-imaging properties of the iodixanol liposomes employed by Leander (1996), described above, in a pyrogenic liver abscess model in 17 rabbits. The imaging properties after application of 200 mg total liposomal iodine/kg (50% encapsulated) were compared to 600 mg iodine/kg applied as unencapsulated iopentol (1.7 ml/kg) which equals the amount given to humans in this diagnostic setting. Liposome application resulted in a clear abscess wall enhancement in six of seven rabbits (86%) between 10 and 120 minutes p.i. due to avid uptake by macrophages abundantly present in the abscess wall. The contrast between liver and abscess center, abscess wall and center, and between liver and portal vein was significantly higher 5 to 75 minutes p.i. for the iodixanol liposomes compared to iopentol. Spleen attenuation after liposome application increased asymptotically in the form of a sigmoid curve toward a maximum of 500 HU (3.2 times higher than maximum liver enhancement). For the liposomes, the overall abscess contrast was best approximately 40 minutes p.i., when maximum abscess wall–liver contrast (112 ΔHU) fell together with a near maximum abscess wall–center and abscess center–liver contrast. All abscesses were visualized with confidence between 30 and 120 minutes post liposome injection, while after iopentol the best contrast was obtained 1–2 minutes p.i. as verified by histology.

The authors concluded that liposomal iodixanol is superior to iopentol regarding abscess imaging in terms of both overall contrast enhancement and the length of the imaging window. To date, this is the only report on the use of CT liposomes for liver abscess imaging.

Erdogan et al. (1998) studied the CT-imaging properties of extruded iopromide liposomes which did not contain any unencapsulated contrast agent (Table 18.3) in rats. Liposomes (E2 formulation) were injected manually at a dose of 0.5 ml resulting in moderate liver and remarkable kidney opacification at 14.7 ΔHU and 63.3 ΔHU, respectively.

The CT liver-imaging properties of two liposomal formulations (BR2 + BR21, Bracco) was studied in 31 healthy male chinchilla bastard rabbits (Petersein et al. 1999). The respective iomeprol-containing liposomes contained encapsulated and unencapsulated contrast agent (encapsulation efficiency not given, see Table 18.3).

The two liposomal formulations, as well as free iomeprol (Imeron® – 300), were administered manually at a flow rate of 0.5 ml/s at doses between 0.5 and 2.5 ml/kg (0.5 ml increments). Doses below 1.5 ml/kg did not result in adequate liver enhancement for any of the tested agents at any time. At doses ≥1.5 ml/kg, the two liposomal agents produced superior enhancement compared to iomeprol. For doses ≥2 ml/kg, these agents resulted in a density increase of approximately 40 HU for up to 60 minutes. The two liposomal agents showed an almost linear dose–response (i.e., liver enhancement) relationship over the entire dose range, suggesting the absence of a saturation of RES uptake. At the highest dose (2.5 ml/kg), peak liver enhancement amounted to 102 ΔHU (BR2) and 84 ΔHU (BR21) compared to 87 ΔHU for free iomeprol. After 5 minutes, liver enhancement amounted to 61 ΔHU, 45 ΔHU, and 38 ΔHU, respectively. Delayed images on three animals 6 hours after injection showed liver densities close to precontrast values. After 2 weeks, no residual contrast agent was detected in the liver.

Due to the limited cross-section of the chinchilla rabbit spleen on axial scans (3 × 2 mm), which does not allow accurate positioning of the ROI without including partial volume effects, spleen densities could not be studied here. The authors concluded that at doses above 1.5 ml/kg BR2 and BR21 are likely to improve liver–tumor contrast. The prolonged retention in the RES provides a longer imaging window which would avoid the need for examination of the entire liver within a few minutes as needed for extracellular contrast agents.

18.6.1.2. Non-human Primates

Iopromide-containing liposomes were investigated as a RES-specific contrast agent for CT liver imaging in ten normal adult (five male + five female) baboons (Schmiedl et al. 1995). The respective freeze-dried liposomal formulation which was developed by Schering AG (SH U 561 A) had to be reconstituted with 135 mM mannitol prior to use. This resulted in a liposome suspension with a total iodine content of approximately 100 mg/ml of which 30%–40% were encapsulated (see Table 18.3).

Liposomes were injected at doses of 200 or 400 mg iodine/kg b.w. with a power injector at 0.1 ml/s or 0.2 ml/s, respectively, to standardize injection rates. No changes in vital signs, heart rhythm, or EKG waveform were recorded at any time during the procedure or subsequent observation period. No clinically significant changes were found during blood testing done on seven animals.

Persistent enhancement of the liver was found reaching a plateau 10–60 minutes after injection with maximum enhancement of 36 ΔHU (60% over baseline, 200 mg dose) and 61 ΔHU (100% over baseline, 400 mg dose). The mean time to plateau enhancement was 20 minutes for the low dose and 10 minutes for the high dose.

The authors described the following distribution phases for the liposomal contrast agent:

- Vascular phase: High intravascular concentration of free and encapsulated iodine during and immediately after the injection. Liver vasculature enhanced relative to parenchyma.
- Equilibrium phase: Approximately 8 minutes after start of the injection, vessels and liver parenchyma became isointense, indicating equal iodine concentrations. It was speculated that this may be due to (1) equilibration of free iopromide between the intravascular and interstitial spaces and (2) equilibration of encapsulated iopromide between vasculature and parenchyma (RES).
- Parenchymal phase: Starting 10 minutes after initiation of the injection, the liver parenchyma is enhanced compared to the vasculature due to RES uptake of the liposomal iopromide while the free iopromide is undergoing rapid renal excretion.

Spleen enhancement continuously increased post-injection with maximum values of 181 ΔHU for the low and 301 ΔHU for the high dose. The much higher spleen enhancement is likely due to the greater number of macrophages per gram of tissue in the spleen compared to the number of Kupffer cells in the liver.

The authors concluded that the tested iopromide liposomes are suitable as a liver CT contrast agent in non-human primates and that further studies are needed to maximize the lesion-to-liver contrast. However, they point out that the total amount of iodine that can be delivered to the hepatic RES is limited by toxicity concerns and possible saturation of the RES.

The liver- and spleen-imaging properties of liposomes containing the non-ionic, water-soluble contrast agent iomeprol were studied in 15 cynomolgus monkeys in comparison to non-liposomal iomeprol (Imeron® – 300) (Rummeny et al., 2002). The respective liposomal contrast agents, which were developed by Bracco (BR 23 and BR 24), consisted of unilamellar liposomes with a diameter of 400 nm which contained encapsulated as well as unencapsulated iomeprol (Table 18.3).

All contrast agents were injected as a bolus at a dose of 2 ml/kg (300–320 mg total iodine/ml) which was well tolerated by all animals without any adverse events or changes in vital signs. In all cases, significant enhancement in the aorta and hepatic arteries was found immediately after the injection which peaked at 10 seconds p.i. Both liposomal products showed a significantly prolonged opacification in the abdominal aorta compared to iomeprol with a plateau for several minutes after the injection.

Enhancement of the liver reached its initial maximum at 2 minutes p.i., with peak values between 124 ± 12 HU for iomeprol and 140.5 ± 17 HU for BR 23. Subsequently, a small dip in liver enhancement of about 20 HU was observed for the liposomal preparations reaching a plateau phase with values well above 100 HU between 5 and 30 minutes p.i. After 2 hours, liver enhancement was still persistent showing a somewhat higher enhancement compared to the plateau phase.

In the spleen, both liposomal formulations showed an increase in enhancement over the 2-hour observation period with maximum values of 283 ± 27 HU (360%) for BR 23 and 176 ± 14 HU (207%) for BR 24.

The authors concluded that BR 23 and BR 24 may be used to improve detection of small lesions (≤1 cm) in liver and spleen. Hypervascular lesions which may be isodense to normal enhanced liver early after injection will be detectable on delayed images following washout of the unencapsulated contrast agent.

18.6.1.3. Human Experience

There is only limited experience with liposomal X-ray/CT contrast agents in humans. To date, only three studies have been published.

Rozenberg et al. (1993, Russian paper) studied the imaging properties of liposomal triombrast (sodium amidotrizoate, 90 mg iodine/ml) in 23 patients with Hodgkin's disease, 9 with liver cirrhosis, and 6 with malignant tumors. On plane radiography, distinct liver and spleen images appeared 2–2.5 hours after i.v. administration lasting 4–6 hours. Lesions down to 0.8–1.0 cm could be detected. Administration of liposomal doses above 0.5 ml/kg was associated with fever and hyperthermia in 30% of the patients, which could be avoided by premedication with prednisolone and pipolphen.

Leander et al. (1998, 2001) first published clinical experience with a liposomal CT contrast agent (CT particles, CTP). These CTP were manufactured by Nycomed and consisted of a ready-to-use iodixanol liposome suspension with a total iodine concentration of 200 mg/mL of which 80 mg were encapsulated inside the liposomes (see Table 18.3). In this single-blind, single-dose efficacy (CT imaging), safety, and pharmacokinetic trial in healthy volunteers ($n = 47$), doses of 10, 30, 70, and 100 mg encapsulated iodine/kg b.w. were studied. Initially, the liposomes were administered using a power injector at a rate of 2 ml/s. Subsequently, the 70 mg dose was administered over 30 minutes with an infusion pump.

At 70 and 100 mg iodine/kg, high attenuation in the aorta was seen 45 seconds after start of the injection amounting to approximately 70–95 ΔHU dropping to approximately 25 ΔHU within 12 minutes. Liver enhancement increased with dose amounting to a maximum of 20, 39, and 45 ΔHU for 30, 70, and 100 mg encapsulated iodine/kg b.w., respectively (within 12.5 minutes at 2 ml/s). Liver attenuation was sustained with less than a 15% decrease over 3 hours giving a long imaging window. Spleen attenuation increased with dose amounting to 49, 126, and 236 ΔHU. Relative liver uptake (percent of dose) decreased with increasing doses indicating a partial saturation of the Kuppfer cells in the liver, whereas it increased in the spleen. Liver and spleen half-lives as measured by CT attenuation were 7 and 2–3 days, respectively. After 28 days, no increase in organ opacification could be observed for the 70 and 100 mg dose.

At doses above 30 mg, encapsulated iodine/kg pronounced depression of leukocytes counts was observed during the first 2 hours. C-reactive protein (CRP) was raised in a dose-dependent manner indicating cytokine-mediated reactions in volunteers. Both parameters normalized spontaneously after 24 hours. At higher doses, there was a prominent increase in neutrophils with rod-shaped nuclei, indicating increased release from the bone marrow. At doses $\geq$ 70 mg iodine/kg b.w., all volunteers reported subjective adverse events (AEs) like chills, back pain, flu-like symptoms, and nausea with vomiting. According to the authors, these AEs showed similarities to the human response of phagocytosing bacteria or viruses which have also been described for other lipoid contrast agents and were thus not classified as unexpected. All AEs were classified as mild or moderate and resolved spontaneously within 3 hours. The type and frequency of the AEs could not be reduced by increasing the administration duration to 30 minutes (studied at 70 mg iodine/kg).

The authors concluded that the tested formulation shows favorable hepato-splenic enhancement in humans, which should allow better delineation of focal lesions (e.g., metastases, hepatocellular carcinomas, and hemangiomas). However, the product elicits acute subjective AEs which are too pronounced for being clinically acceptable.

Spinazzi et al. (2000) studied the safety and pharmacokinetics of iomeprol liposomes (BR21) in non-patient volunteers. The employed sterile, pyrogen-free liposome suspension (BR21) contains 320 mg iodine/mL of which approximately 40 mg were entrapped inside the liposomes (Table 18.3). In this single-blind placebo-controlled study, 30 adult male patients received a single intravenous bolus (2 mL/s via power injector) of either BR21 (0.5, 1.0, 1.5, 2.0, and 2.5 mL/kg, $n = 4$ per dose level) or matched volumes of 0.9% saline ($n = 10$).

Iomeprol content in blood, urine, and feces was detected by high-performance liquid chromatography (HPLC) predose and up to 120 hours after injection. After administration, maximum blood concentrations (Cmax) were obtained 2 minutes after dosing ranging from 1.6 (lowest dose) to 7.4 mg/mL (highest dose). Thereafter, blood levels fell rapidly with a triphasic decline representing (1) a distribution phase ($t_{1/2}$: 0.12–0.21 hours (mean values)), (2) a fast elimination phase ($t_{1/2}$: 1.2–1.5 hours), and (3) a terminal elimination phase ($t_{1/2}$: 3.3–4.5 hours). At the 1.0–2.5 mL/kg doses, blood levels remained above the limit of quantification (3 μg/mL) up to 24 hours after dosing for all subjects.

Iomeprol was mainly excreted in urine (89%–92%) as unchanged drug at all dose levels. During the first 2 hours after administration, approximately 53% of the dose was excreted. By 8 hours 83% and by 24 hours 89–90% of the dose was excreted.

With respect to safety, complete physical examinations, measurement of vital signs, electrocardiographic controls, clinical laboratory (hematology, serum chemistry, and urinalysis), and monitoring of adverse events were performed predose and up to 7 days after application. No serious adverse events were reported in the study. However, starting at the lowest liposome dose (0.5 mL/kg) drug-related adverse events (i.e., asthenia, headache, dizziness, taste perversion, nausea, arthralgia, vasodilatation, back pain, chills, dizziness, tremor, and sweating) were observed in 50% (lowest dose) to 100% (highest dose) of volunteers. Vasodilation was observed only in the highest dose groups (2.0 and 2.5 mL/kg) and was experienced as feeling warm/hot either generally or in specific parts of the body. According to the authors, these events were minor and mild in intensity and rapidly resolved without treatment. No clinically significant changes in vital signs, electrocardiographic components, or laboratory parameters compared to predose were observed during the study.

The authors state that studies are under way to evaluate BR21 in patients with focal liver disease. However, the respective data has not been published to date.

18.6.2. Discussion and Conclusions

During the 1980s, CT contrast-carrying liposomes were produced successfully by various groups (Seltzer, 1988a). However, CT enhancement with these formulations was unsatisfactory in most cases due to the low encapsulation efficiencies of these products. Thus very high iodine doses (i.e., 190–564 mg encapsulated iodine/kg b.w.) and even higher lipid doses (iodine/lipid ratios below 1 in most cases) were required to reach liver enhancement $\geq$ 40 HU (Table 18.2). Hence, there was a need for new production methods which would allow the reproducible large-scale production of CT liposomes with high encapsulation and sufficient in vitro (storage) and in vivo (plasma leakage) stability (Seltzer, 1988b). As was demonstrated by various groups (e.g., Cheng et al., 1987, Henze et al., 1989), preparation of contrast-carrying liposomes by a specific method needs to be optimized for each specific contrast agent (e.g., lipid and iodine concentration, lipid composition and charge).

By the middle of the 1990s, various preparation methods for CT contrast-carrying liposomes were available which allowed to reproducibly produce pharmaceutically acceptable (i.e., sterile and pyrogen-free) liposomes on a scale sufficient to supply larger preclinical, as well as initial clinical (phase I) studies. The respective methods allowed effective iodine encapsulation with encapsulation efficiencies up to 50% and corresponding iodine (encapsulated)-to-lipid ratios (mg/mg) between 0.4 and 4.7 (Table 18.3). Liposomes with narrow size ranges and maximum particle sizes well below 5 μm could be produced. The problem of limited storage stability of liposomal suspensions due to contrast agent leakage could be avoided by not removing the unencapsulated material as first systematically studied by Schneider et al. (1995). Corresponding liposomal formulations showed sufficient plasma stability (e.g., Seltzer et al., 1995), as well as acceptable toxicity in preclinical animal models (e.g., $LD_{50} = 3$ g iodine/kg in mice; Krause et al., 1991).

In various species, satisfactory liver opacification could be demonstrated with density increases at or above 50 ΔHU which have been claimed to be adequate to allow delineation of small focal lesions (Seltzer, 1988a). Indeed, small lesions below 0.5 cm could be detected in rabbits at a liver/tumor density difference of approximately 40 HU (VX-2 carcinoma, Figure 18.4; Sachse et al., 1993) and in rats (chemically induced hepatocellular carcinoma; Fouillet et al., 1995) after administration of iodinated liposomes. Additionally, the usefulness of CT liposomes containing iodixanol for liver abscess imaging could be shown in a rabbit model demonstrating improved overall abscess contrast compared to a conventional contrast agent (iopentol), as well as a prolonged imaging window (Dick et al., 1996).

For all tested formulations, a much higher spleen accumulation (on a per-gram organ weight basis) was found in rats, rabbits, and non-human primates. Higher spleen enhancement is likely due to the greater number of macrophages per gram of tissue in the spleen compared to the liver (Schmiedl et al., 1995). Spleen uptake generally increases with liposome size (Havron et al., 1981).

For the liposomal non-ionic monomeric contrast agents iopromide and iomeprol, rapid elimination could be demonstrated. Thus, in the case of iopromide, 76% of the administered dose were eliminated after 24 hours and complete urinary excretion was found 7 days after injection in rabbits (Sachse et al., 1993). In humans, almost complete urinary excretion of liposomal iomeprol was found after 24 hours amounting to approximately 90% (Spinazzi et al., 2000). In contrast to this, liposomes with the dimeric non-ionic contrast agents iotrolan and iodixanol showed prolonged organ retention. In rats 0.8–2.6 % of dose could still be found after 8 weeks (Seltzer et al., 1995). In humans the liver half-live of iodixanol liposomes was found to be 7 days (Leander et al., 2001). These results suggest that dimeric contrast agents are less effectively cleared from the target cells (e.g., Kuppfer cells in the liver) due to their physico-chemical properties (e.g., molecular size, hydrophilicity).

The only human CT-imaging study published to date demonstrates the potential suitability of liposomes containing free and encapsulated contrast agent (iodixanol) for liver imaging (Leander et al., 2001). Thus, a dose of 100 mg encapsulated iodine/kg b.w. resulted in a maximum liver attenuation of 45 ΔHU which basically remained unchanged over the first hour. Based on the experience in tumor-carrying animals described above, this enhancement should be sufficient to allow improved detection of focal lesions well below 0.5 cm as originally desired.

Unfortunately, dose–dependent, acute, drug-related adverse events (AEs) were consistently reported for all human studies with CT liposomes published thus far (Rozenberg et al., 1993; Spinazzi et al., 2000; Leander et al., 2001). These included chills, back pain, flu-like symptoms, nausea and vomiting, dizziness, taste perversion, and vasodilatation which occurred in 30%–100% of the treated subjects. In one study, a dose-dependent increase in C-reactive protein was observed which indicates cytokine-mediated reactions in the volunteers (Leander et al., 2001). According to the author, such events have also been observed when using other types of particulate contrast agents and are similar to the human response of phagocytizing bacteria and viruses. Based on the fact that the main constituents (phospholipids) of liposomes occur naturally in the human body (Cheng et al., 1987), as well as the earlier experience with therapeutic liposomes, lipid-related toxicity was not expected to be a problem at the dose levels required for acceptable CT organ opacification (Havron et al., 1981). For imaging purposes, ^{99m}Tc-labeled MLVs (DMPC/DMPG 7:3) had already been applied to cancer patients at lipid doses as high as 450 mg/m^2 without observing any adverse side effects (Lopez-Berestein et al., 1984). It is worth noting that infusion-related side effects like dyspnoea, facial flushing, nausea and vomiting, chest pain, facial oedema, and dizziness have also been observed in clinical studies with liposomal doxorubicin (at very low lipid doses), but are usually ascribed to the cytostatic drug rather than the liposomal carrier (e.g., Goebel et al., 1996).

Various factors like lipid dose, lipid composition, or liposome charge may elicit AEs. Thus, for example, saturated lipids like DSPC were shown to result in significant reductions in blood pressure and total peripheral resistance, as well as marked ECG changes compared to SPC in a rat model (Muschik et al., 1995). These effects of DSPC could be reduced by addition of CH or DSPG.

In conclusion, the potential of liposomal CT contrast agents for liver imaging could be confirmed in humans. The subjective AEs observed with the tested formulations in humans are not acceptable for a diagnostic drug so that further research is needed to find formulations with improved tolerability and sustained or even further improved efficacy.

18.7. Blood-Pool Imaging

Originally, radiologic application of liposomes was confined to their use in liver and spleen imaging due to the rapid uptake of such particles by the cells of the reticuloendothelial system (RES) after intravenous application (Senior, 1987). In the early 1990s, however, methods for modification of liposome surfaces were introduced, which were shown to result in liposomes that could avoid the MPS for prolonged time periods (Allen, 1994). It has been suggested that contrast-carrying liposomes with extended lifetimes in the circulation might be useful as blood-pool (vascular) imaging agents, especially in MR imaging and scintigraphy (Tilcock, 1994). Blood-pool imaging would be of special interest for the evaluation of the current state of the blood flow and for the discovery of irregularities caused by pathologic changes (Torchilin and Trubetskoy, 1995).

A useful intravascular contrast agent should selectively enhance the blood pool for at least 5 minutes and must be safe and degradable to allow rapid elimination (Katayama et al., 2001). For CT, an intravascular iodine concentration of $\geq$2 mg/ml (50–60 HU) over a period of at least 15 minutes would be desirable to allow efficient blood-pool imaging without the need for reinjection of the contrast agent. This would be useful especially for imaging of cardiovascular disease (e.g., stenosis, ischemia, atherosclerosis), abnormal capillary permeability (e.g., inflammation, cancer), and tumor neovascularity (Bodganov et al., 1995).

CT blood-pool agents would potentially allow easier imaging of both the macrovasculature and the microvasculature by decreasing the diffusion and interstitial loss of contrast and by creating a longer temporal window for scanning (Katayama et al., 2001).

18.7.1. Prolongation of Liposome Circulation in Blood

The half-life of conventional liposomes in the blood stream after intravenous application can be increased by (1) a reduction in liposome size to below 100 nm, (2) an increase in the cholesterol content, (3) the use of saturated (gel-state) phospholipids, (4) the use of uncharged lipids, and (5) the use of higher liposome doses (Senior, 1987).

Alternatively, liposome surfaces can be altered by inclusion of glycolipids (i.e., monosialoganglioside GM_1) or polymers (principally polyethylene glycol (PEG)). This results in sterically stabilized (Stealth®) liposomes with a more hydrophilic surface which is less able to bind plasma opsonins and, thus, results in decreased RES uptake and prolonged circulation half-lives (Allen, 1994). For PEG-coated liposomes, maximum circulation times can be achieved in the range of DSPE-PEG1000–DSPE-PEG2000, at a molar ratio of 5%–10%. Amphiphatic polymers such as poloxamers, poloxamines, or polysorbates proved to be less effective in prolonging liposome circulation compared to PEG.

As shown in Table 18.4, various groups have studied the utility of conventional or surface modified (PEG-coated) CT liposomes for blood-pool imaging. The results of the respective imaging studies are summarized in the following section.

18.7.2. Animal Experience

The first specific study on the use of liposomes as CT blood-pool agent was published by Sachse et al. in 1997. Iopromide-carrying liposomes (SPC/CH/SPG 6:3:1) were prepared by continuous high-pressure extrusion using 0.2 μm filters for the last extrusion step. Subsequent PEG-coating was performed by simple mixing of the preformed liposomes with 5 mol% of either DSPE-PEG2000 or CHHS-PEG 2000. The unmodified or PEG-ylated liposomes were used without prior removal of the unencapsulated contrast agent. The properties of the resulting vesicles are shown in Table 18.4.

The successful membrane incorporation of the PEG-derivatives in the liposomes membrane by simple mixing was confirmed by measuring the membrane zeta potentials which were remarkably lower for the modified liposomes (Table 18.4) which probably is the result of a shielding of SGP's negative charge. The liposomes remained intact during the PEG-coating procedure, displaying mainly unilamellar characteristics (Figure 18.2). The mean diameter of the DSPE-PEG liposomes considerably increased during coating probably due to vesicle aggregation and fusion. The latter liposomes showed the highest stability in human plasma with relative encapsulation values of 84% and 79% after 1 and 2 hours, respectively. The unmodified liposomes showed the

TABLE 18.4. Properties of blood-pool liposomes

Contrast agent	Lipid composition (molar ratio)	Liposome size (mean diameter) (nm)	Zeta potential (mV)	Total iodine concentration (mg/ml)	Encapsulation efficiency (%)	Iodine/lipid ratio	Vascular enhancement (HU/mg encapsulated iodine)	Reference
Iopromide	SPC/CH/SPG 6:3:1	132	−44.5	61.9	43.2	0.2	0.37 (51 HU at 45 min)	Sachse et al., 1997
Iopromide	see above + 5 mol % CHHS-PEG2000	149	−12.2	61.2	43.7	0.2	–	Sachse et al., 1997
Iopromide	see above + 5 mol % DSPE-PEG2000	204	−12.0	59.4	45.0	0.2	0.63 (71 HU at 45 min)	Sachse et al., 1997
Iopromide	SPC/CH/SPG 6:3:1	195	N/A	96	46	N/A	0.49 (67 HU at 40 min)	Schmiedl et al., 1999
Iomeprol BR22	N/A	N/A	N/A	300	N/A	N/A	Cannot be calculated	Rummeny et al., 1999
Iopromide	SPC/CH/SPG 6:3:1	201	N/A	N/A	50	0.3	0.40 (60 HU, 10–60 min)	Leike et al., 2001
Iohexol	DPPC/CH/DSPE-MPEG2000 5.5:4:0.5	100.6	N/A	34.8 (for CT experiment)	100 (free iohexol removed by dialysis)	N/A	0.27 (130 HU for up to 3 h)	Kao et al., 2003

N/A = not available

lowest plasma stability over the 6-hour study period. This is consistent with steric stabilization of the PEG-ylated liposomes which prevents liposome leakage due to reduced interactions with plasma proteins.

In the biodistribution study in rats (250 mg total iodine/kg b.w.), no significant differences in blood concentration were found between the various formulations up to 1-hour post-injection amounting to approximately 1.2 mg iodine/g or 30% of dose. The maximum liver concentration observed for any of the tested formulations during the 24-hour study period amounted to 0.4 mg iodine/g or 7% of the dose. According to the authors, the observed prolongation in blood circulation for the unmodified liposomes is most likely due to a transient saturation of MPS-uptake (opsonisation or phagocytosis), which is known to occur at high lipid doses (Senior, 1987).

In an orientating rabbit CT-imaging study (250 mg total iodine/kg b.w.), the tested liposomes showed prolonged blood enhancement for at least 1 hour which was not due to the unencapsulated iopromide, as this was no longer detectable after approximately 35 minutes (Figure 18.6). Initially, the DSPE-PEG liposomes showed a very similar imaging behavior compared with the unmodified liposomes with some delineation of the liver vasculature already 28 seconds after injection. After 1 minute, the density in the aorta amounted to 125 ΔHU which corresponded with bright opacification of the liver vessels (Figure 18.7). At 45 minutes p.i., the density in the aorta had dropped to 71 ΔHU. The CHHS-PEG liposomes consistently gave the lowest blood enhancement with a 1-hour value of approximately 45 ΔHU. It can be assumed that an intravascular iodine concentration of $\geq$ 2 mg/ml (50–60 HU) over a period of at

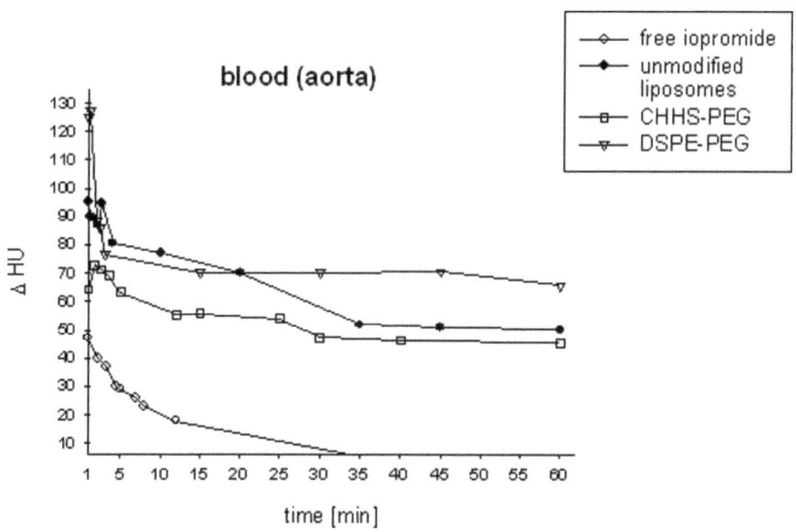

FIGURE 18.6. CT blood densities (aorta) in rabbits at a dose of 250 mg iodine/kg body weight ($n = 1$ each). Reproduced from Sachse et al. (1997), with kind permission of Lippincott Williams & Wilkins.

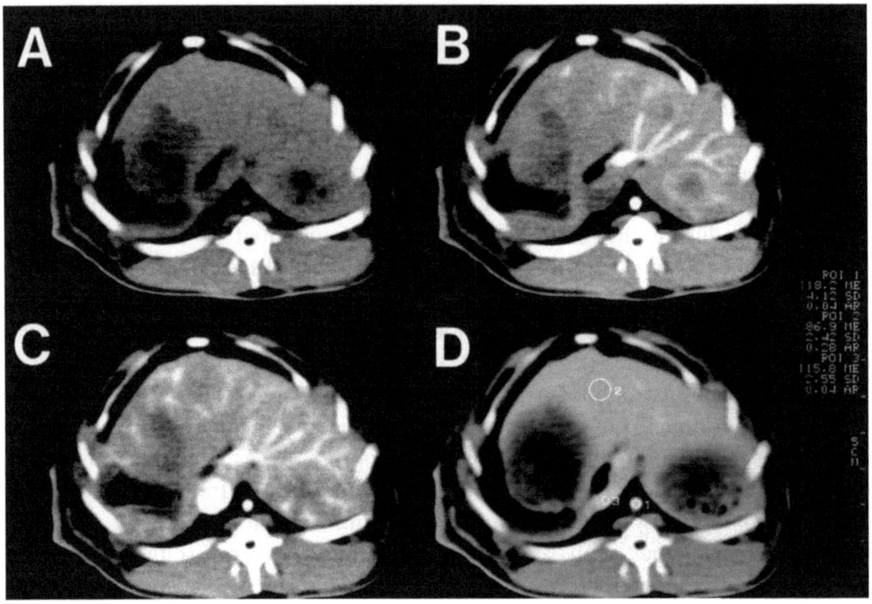

FIGURE 18.7. CT blood-pool imaging in a rabbit with DSPE-PEG liposomes. **A** Precontrast image. **B** 28 seconds p. i. **C** 1 minute p.i. (aorta = circa 125 ΔHU, vena cava = circa 144 ΔHU). **D** 45 minutes p.i. (aorta = circa 71 ΔHU, vena cava = circa 72 ΔHU). Reproduced from Sachse et al. (1997), with kind permission of Lippincott Williams & Wilkins.

least 15 minutes as observed here for the unmodified and DSPE-PEG liposomes (both above 70 ΔHU for up to 20 minutes) should be sufficient for blood-pool imaging without the need for reinjection of the contrast agent. The unmodified liposomes resulted in the highest liver density over the whole observation period reaching a plateau around 40 ΔHU approximately 2 minutes after injection.

Based on these findings, Schmiedl et al. (1999) studied the usefulness of unmodified iopromide liposomes (SPC/CH/SPG 6:3:1) for CT blood-pool imaging in non-human primates (adult baboons). The physico-chemical properties of these liposomes are shown in Table 18.4.

Five normal female adult baboons weighing 15–21 kg received a single dose of 300 mg total iodine/kg b.w. i.v. with a power injector at 1 ml/s. All animals tolerated the injections well and recovered from anesthesia without any adverse reactions. No changes in vital signs, heart rhythm, or ECG were recorded during the course of the injection and the whole observation period.

A biphasic enhancement pattern was found in all organs and vessels except for the bladder. Mean peak enhancement in all vessels was greater than 200% immediately after injection and reached a plateau at approximately 175% between 10 and 20 minutes. This initial peak can be explained by the presence of free iopromide in the liposome suspension which will rapidly extravasate, causing a relative decrease in attenuation. Mean vascular enhancement was 76 ΔHU, 72 ΔHU, and 67 ΔHU at 10, 20, and 40 minutes respectively. During the plateau phase, significant and almost constant parenchymal enhancement was observed in the liver (26 ΔHU), spleen (41 ΔHU), and renal cortex (42 ΔHU, all values at 40 minutes). Attenuation differences between the portal vein and the liver parenchyma were moderate, and the liver vasculature was enhanced relative to parenchyma.

The authors conclude that even though the study showed a certain potential of the tested liposomes for CT blood-pool imaging, it does not answer the question whether the observed enhancement and distribution pattern would have any practical use in either vascular or solid organ imaging.

Another primate study (cynomolgus monkeys) was performed by Rummeny et al. (1999). Liposomal iomeprol (300 mg iodine/ml, BR22, Bracco Research) was injected as a bolus at a dose of 2 ml/kg to seven animals and compared to injection of unencapsulated iomeprol (Imeron®) at the same dose. Details on the formulation were unfortunately not published. However, according to Katayama et al. (2001) the respective formulation consists of small surface-modified (Stealth®) liposomes. Repeated thin section spiral CT scans were performed at 5 seconds, 2 minutes, and 5–20 minutes p.i. in 5 minutes intervals. Peak enhancement in aorta, iliac artery, and femoral artery was not significantly higher for BR22 compared to Imeron®. Time ranges of sufficient enhancement (>60 HU) were significantly longer for BR22 in all arteries lasting up to 20 minutes. At that timepoint, BR22 resulted in an enhancement of 111 ± 9 HU in the femoral artery compared to 66 ± 8 HU for Imeron®. The authors conclude that BR22 is potentially a suitable agent for CT angiography (CTA).

Leike et al. (2001) studied the acute and subchronic toxicity, and pharmacokinetics of iopromide blood-pool liposomes (SPC/CH/SPG, see Table 18.4) in rats and CT-imaging efficiency in rabbits ($n = 5$). The median lethal dose in rats was approximately 4.5 g total iodine/kg b.w. which is remarkably lower than the LD_{50} reported for non-liposomal iopromide in this species (11 g/kg). The reduced tolerance compared to free iopromide is most likely due to the corresponding high lipid dose (initial iodine-to-lipid ratio = 1:1.5) as well as the intracellular uptake of the drug by the MPS. In the subchronic tolerance study in rats, which comprised six doses of 1 g total iodine/kg given twice per week, no adverse effects were observed. Histological examination did not reveal any organ damage; however, in the spleen a discernible foamy cell formation was found.

In the pharmacokinetic (PK) study, with 250+1000 mg iodine/kg b.w., a clear dose dependence of PK parameters was found. Thus, the terminal half-life increased from 4.6 hours for the low dose to 10.7 hours for the high dose, suggesting a longer circulation of intact liposomes at higher doses. Seven days after administration almost complete, elimination of iopromide, mainly via the urine, was found.

In the CT rabbit study, with manually injected i.v. bolus doses of 300 mg total iodine/kg b.w., a biphasic pattern was found in the aorta with peak enhancement

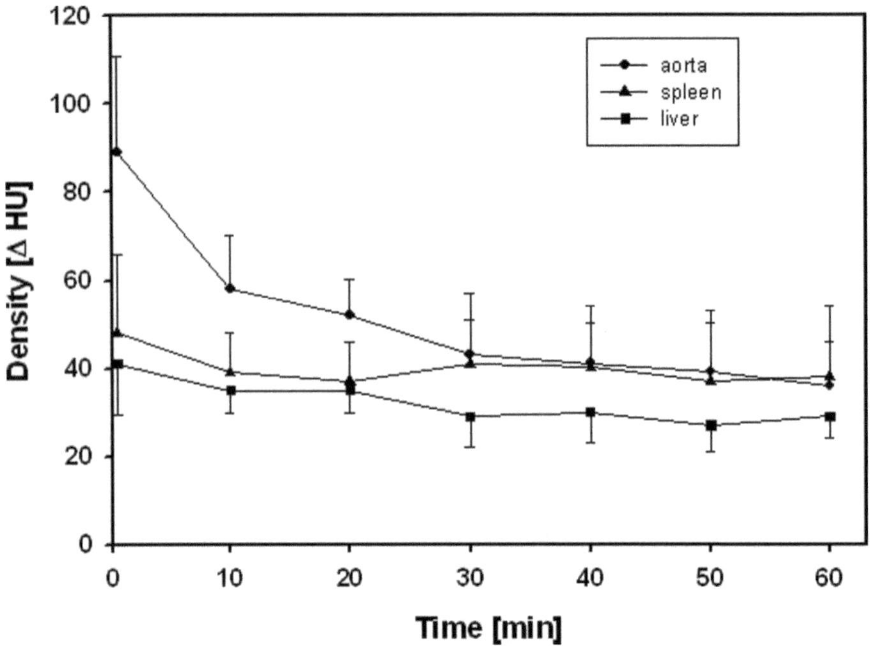

FIGURE 18.8. Time course of CT density differences (ΔHU) in the aorta, liver, and spleen of rabbits after intravenous administration of 300 mg total iodine/kg iopromide liposomes ($n = 5$). Reproduced from Leike et al. (2001), with kind permission of Lippincott Williams & Wilkins.

of approximately 90 ΔHU (Figure 18.8), followed by a plateau phase between 10 and 60 minutes (mean density = 60 ΔHU). At 10 minutes p.i., the liver and spleen densities also reached a plateau around 35 and 40 ΔHU, respectively. According to the authors, the observed vascular enhancement was only moderate, allowing sufficient blood-pool imaging (including vascular reconstructions) only during the first few minutes (Figure 18.9). Based on the good tolerability of the product, they proposed to perform additional studies at higher liposome doses to increase the vascular enhancement and to maximize the vessel-to-background contrast.

The most recent study from Kao et al. (2003) investigated the CT blood-pool–imaging properties of small-sized DPPC/CH/DSPE-PEG (5.5/4/0.5) liposomes containing only encapsulated iohexol in a rabbit. By using smaller liposomes (see Table 18.4), the authors wanted to obtain an agent which remains in the blood for more than 3 hours and is relatively non-toxic to kidneys.

The liposomes were shown to be stable in vitro in phosphate buffer with 7.4 and 7.8% iohexol leakage after 8 hours and 18 days, respectively, and human plasma (2.3% after 8 hours) probably due to stabilization by the presence of PEG. The authors note, however, that leakage stability of the liposomes will have to be improved to allow long-term storage.

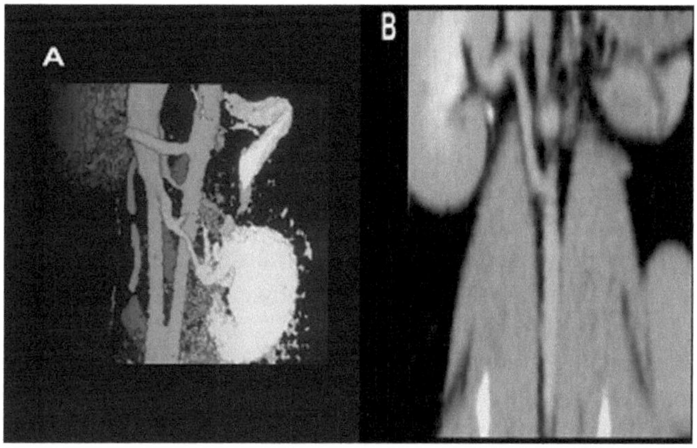

FIGURE 18.9. CT blood-pool imaging in a rabbit after intravenous administration of 300 mg total iodine/kg iopromide liposomes (SPC/CH/SPG 6:3:1). Vascular reconstruction of the abdominal region obtained immediately after administration. **A** Three-dimensional reconstruction. **B** Multiplanar reconstruction. Reproduced from Leike et al. (2001), with kind permission of Lippincott Williams & Wilkins.

One rabbit was dosed with 475 mg iodine/kg (=5 ml liposomal iohexol or 14% of blood plasma volume of animal) by manual i.v. injection. Blood-pool enhancement increased rapidly during the injection remaining almost constant over the observation period (3.5 h). The average attenuation in aorta, pulmonary vein, and liver cortex 3.5 h after injection amounted to 130 ΔHU while that of the kidney cortex amounted to 25 ΔHU. Imaging of the rabbit heart allowed clear demonstration of the anatomy of all the four heart chambers and the associated great vessels. According to the authors, early enhancement in the kidney medulla (peak after 60 minutes at approximately 180 HU) is likely to represent unencapsulated contrast agent, while enhancement in the liver parenchyma (plateau at around 120 HU) suggested iohexol clearance via this route.

The authors concluded that CT liposomes with prolonged blood-pool enhancement as demonstrated in their rabbit model hold great promise for many applications warranting further investigation of this formulation.

18.7.3. Discussion and Conclusions

Starting in the 2nd half of the 1990s, usefulness of CT contrast agent–carrying liposomes for blood-pool imaging was studied by various groups as shown in Table 18.4. In these studies, either conventional liposomes or surface modified (i.e., PEG-ylated) vesicles were employed.

In the study of Sachse et al. (1997), PEG-ylated iopromide liposomes displayed increased plasma stability compared to conventional liposomes, presumably due to reduced interaction with plasma proteins due to steric stabilization of the liposome membrane. Interestingly, at the tested dose levels, no pronounced differences in vascular enhancement was found between unmodified (SPC/CH/SPG 6:3:1) and liposomes additionally containing 5 mol% of DSPE-PEG2000 in rabbits within the first 20 minutes p.i. Thus both formulations resulted in a vascular enhancement (aorta) above 70 ΔHU for up to 20 minutes (Figure 18.6) at a dose of 250 mg total iodine (=500 mg lipid)/kg b.w. which should be sufficient for vascular imaging (e.g., stenosis, ischemia, atherosclerosis) without the need for reinjection. The prolonged circulation time of the unmodified liposomes can be ascribed to a transient saturation of RES-uptake (opsonization or phagocytosis), which is known to occur at high lipid doses (Senior, 1987).

In an attempt to further prolong blood-circulation of PEG-ylated CT liposomes, Kao et al. (2003) prepared iohexol-containing SUVs composed of DPPC/CH/DSPE-PEG2000 which they tested in a rabbit model (Table 18.4). Unencapsulated iohexol was removed from the preparation which resulted in limited storage stability due to drug leakage, as had been observed earlier for liver liposomes (e.g. Henze et al., 1989). At a dose of 475 mg encapsulated iodine/kg (= 15 ml), vascular enhancement (aorta) amounted to 130 ΔHU over 3.5 h. This corresponds to a blood-pool enhancement of only 0.27 HU/mg encapsulated iodine which is dramatically lower than what was found for larger PEG-ylated iopromide liposomes (0.63 HU/mg, Table 18.4). This may either reflect insufficient plasma stability or high organ uptake of the former formulation. Indeed, the authors found marked enhancement in the kidney medulla (max = 180 HU at 60 minutes) as well as pronounced enhancement in the liver parenchyma (plateau around 120 HU).

Due to their larger internal volume-to-surface area ratio, LUVs can potentially encapsulate significantly larger amounts of contrast agent per mol of lipid compared to SUVs. Based on this as well as the fact that PEG-ylated liposomes between 100 (SUVs) and 200 nm (LUVs) have been shown to have very similar circulation times (Allen, 1994), the approach described by Kao et al. should be carefully re-evaluated.

The tolerance, elimination, and imaging properties of unmodified iopromide liposomes (SPC/CH/SPG 6:3:1) were studied by Leike et al. (2001). The median lethal dose LD_{50} in rats was found to be 4.5 g total iodine/kg b.w. in rats which is significantly below the 11 g/kg which have been reported for the unencapsulated contrast agent. This increased toxicity was explained by the high lipid dose (6 × 1.5 g lipid/kg) applied in this study as well as the contribution of intracellular (RES) toxicity of iopromide.

In rats, almost complete urinary excretion was found at doses of 250 and 1000 mg total iodine/kg b.w. within 7 days. The corresponding terminal half-lives were 4.6 hours and 10.7 hours, respectively, reflecting a longer circulation of the liposomes at higher lipid dose.

At a dose of 300 mg iodine/kg in rabbits, a biphasic CT-enhancement pattern was found in the aorta with peak enhancement of approximately 90 ΔHU followed by a plateau phase around 60 ΔHU (10–60 minutes, Figure 18.8). The initial peak reflects the sum of the encapsulated (50%) and unencapsulated material, the latter being rapidly elimininated resulting in a steep decrease in vascular enhancement over the first 10 minutes. Liver enhancement was found to be at or below 40 ΔHU during the first hour indicating effective avoidance of the Kupffer cells in the liver (Figure 18.8).

Schmiedl et al. (1999) studied the CT-imaging behavior of the same type of unmodified liposomes, which displayed a somewhat larger size and higher encapsulation (Table 18.4). At a dose of 300 mg total iodine/kg b.w., mean vascular enhancement in baboons amounted to 76 ΔHU at 10 minutes and 67 ΔHU at 40 minutes. This translates into an enhancement of 0.49 HU/mg encapsulated iodine (40 minutes) which is remarkably higher than what was found for unmodified liposomes in rabbits (0.37 HU/mg at 45 minutes, Table 18.4), but also lower than what was observed for the corresponding surface-modified (DSPE-PEG2000) liposomes (0.63 HU/mg).

Rummeny et al. (1999) demonstrated the suitability of iomeprol-containing Stealth® liposomes (BR22) for blood-pool imaging in cynomolgus monkeys. At a total iodine dose of 600 mg iodine/kg, contrast enhancement above 60 HU was found in all arteries for up to 20 minutes p.i. Unfortunately, no details of the formulation other than the iodine content were disclosed so that a comparison with the other products/studies is not possible.

The long persistence of BR22 in the blood vessels could be demonstrated by three-dimensional reconstruction after CT imaging in monkeys (Katayama et al., 2001). It was found that eventually BR22 is taken up by the RES allowing its use as a liver-specific agent on delayed images, as has also been described by Leike et al. (2001) for conventional blood-pool liposomes.

In conclusion, suitable tolerability, elimination, biodistribution, and imaging properties could be shown for conventional and surface-modified CT liposomes in preclinical animal models including non-human primates. Optimization of formulations should be geared toward increased density increases in blood over even longer timeperiods as well as reduced liposome (i.e., lipid) doses. Eventually, studies are needed to demonstrate acceptable safety and efficacy in relevant animal models of pathologic situations (e.g., cardiovascular disease, abnormal capillary permeability (e.g., inflammation, cancer), and tumor neovascularity) representing potential diagnostic indications in humans.

18.8. Other Recent Developments

The ultimate goal of using liposomes as carriers of therapeutic and diagnostic drugs has been cell-specific targeting to desired tissues and organs (Benita et al., 1984). To this end, targeted liposomes with surface-attached ligands capable of recognizing and binding to cells of interest (i.e., immunoliposomes) have been developed (as reviewed by Torchilin, 2005). Immunoglobulins of the IgG class and their fragments are most widely used for liposomal targeting either by covalent binding to the liposome surface or by hydrophobic insertion of modified monoclonal antibodies. After i.v. application of such immunoliposomes, the majority of the dose still ends up in the RES (mainly liver) as a consequence of insufficient time for their interaction with the target cells. To avoid this problem, attempts have been made to produce long-circulating immunoliposomes by surface modification (i.e., PEG-ylation). Initially, simple co-immobilization of the antibody and PEG on the liposomal surface was performed, although the polymer can create steric hindrance for target recognition and binding. To avoid steric hindrance, it was shown to be advantageous to attach the targeting moiety to the water-exposed terminus of the liposome-grafted PEG, so that the ligand is localized outside of the polymer brush as schematically illustrated elsewhere (Torchilin, 2005).

Recently, Johnston et al. (2003) studied in vitro solid-organ tissue binding of small (approximately 100 nm) unmodified (DSPC/CH/DSPG 2:1:0.1) iohexol-carrying CT immunoliposomes to mouse tissues (liver, spleen, peritoneal fat, quadriceps muscle, and testes). Three different antibodies targeted to the following cell-surface proteins were used: (1) Fas ligand (FasL), expressed on the surface of immune-privileged cells, highly expressed in testicles (anti-FasL IgG antibody), (2) Fc receptor (FcR), expressed by RES cells in liver, spleen, and lymphoid structures (rat anti-mouse FcR IgG antibody), and (3) CD5, expressed on the surface of T-helper and certain B-cell subpopulations (humanized Fab fragments of mouse anti-CD5).

After incubation of tissue samples with 100 μM targeted and non-targeted liposomes, the samples were washed and subsequently scanned on a CT scanner. In a prior dilution experiment with the unconjugated material, the practical limit for liposome detection was found to be 37.5 μg (based on liposomal lipid). Anti-CD5 immunoliposomes resulted in liver, spleen, and muscle enhancement of 27 HU, 48 HU, and 27 HU, respectively. Anti-FcR liposomes enhanced the liver samples by 26 HU and the spleen samples by 67 HU. Finally, testicular enhancement was found for anti-FasL only, amounting to 42 HU.

The authors concluded that tissue-specific enhancement by CT immunoliposomes could successfully be demonstrated. Further improvement of the discriminatory capabilities (i.e., signal amplification) of these liposomes is needed.

Contrast-carrying liposomes can potentially be used for improved detection of tumors as well as inflammations and infections (Torchilin, 2005). Such agents may allow detection of very small foci of metastatic cells, especially in lymph nodes, as well as the in vivo histological characterization of such lesions (Johnston et al., 2003).

The preparation of diagnostic immunoliposomes in general poses formidable challenges which include, but are not limited to (1) avoidance of RES/prolonged circulation, (2) avoidance of steric hindrance, (3) access to extra-vascular targets, (4) selective target binding, (5) sufficient tolerability, and (6) sufficient target/tissue contrast. In addition to this, CT immunoliposomes would require encapsulation of very high concentrations of contrast agent (iodine) in relatively small liposomes in order to allow effective targeting (i.e., long circulation) and tissue enhancement (i.e., density increases $\gg$60 HU). Thus, even though advances in multi-detector row CT have resulted in greatly improved spatial and temporal resolution (Kao et al., 2003), CT still requires ten times higher contrast agent concentrations compared to MR to achieve useful images (1 mg iodine/g tissue = 25-30 HU; Krause et al., 1995). Therefore, while preparation of CT immunoliposomes is clearly scientifically feasible, such preparations are unlikely to confer any advantage over MR agents (Tilcock, 1999).

It remains to be seen whether future advances in the preparation of liposomal contrast agents along with the identification of specific cell-surface proteins which are expressed at high densities in the target tissue will allow clinically meaningful CT-contrast enhancement.

Most recently, Zheng et al. (2006) studied the imaging properties of contrast-carrying liposomes containing the MR agent gadoteridol plus the CT agent iohexol in a rabbit model to examine the feasibility of using such a multi-modal colloidal contrast agent for combined MR + CT imaging. Small (mean diameter approximately 75 nm) surface-modified liposomes composed of DPPC/CH/DSPE-PEG2000 55:40:5 mol% which only contained encapsulated contrast agent were used. In vitro, slightly lower CT attenuation values were observed for the liposomal agent compared to equal concentrations of the unencapsulated agents due to the presence of the liposomal lipids. In line with expectations, liposomal encapsulation of gadoteridol (encapsulation efficiency 18.6%) in the presence of iohexol (encapsulation efficiency 19.6%) also resulted in a decrease in relaxivity. Intravenous injection of 10 ml of the liposomal preparation (36 mg/kg iodine and 12 mg/kg gadolinium) in rabbits resulted in marked signal enhancement (MR + CT) in the major blood vessels. In MR, a signal increase of over 200% was found immediately p.i. which was maintained for 72 hours. For CT, a 60% (approximately 40 ΔHU) increase in opacification was achieved and maintained over 3 hours, thereafter decreasing to 35% after 72 hours.

The authors concluded that the feasibility of engineering a multimodal contrast agent with prolonged blood enhancement could be demonstrated. Such agents may serve as a valuable tool for cardiovascular (potentially cell specific) imaging as well as image registration and guidance applications in radiation therapy.

While scientifically interesting, it is not quite clear whether multimodal liposomal contrast agents would show additional benefits compared to single agent formulations which would outweigh the added technological challenges (e.g., sensitivity differences between MR and CT).

18.9. Summary and Conclusions

Subsequent to the initial studies on the use of liposomal CT contrast agents for liver and spleen imaging published in the early 1980s, significant progress has been made regarding the production and characterization (in vitro and in vivo) of such preparations. By the middle of the 1990s, methods for liposome production were available (mainly filter extrusion) which allowed reproducible large-scale production of CT contrast–carrying liposomes (mainly non-ionic monomers like iomeprol) with adequate storage stability and pharmaceutical quality (i.e., sterility and pyrogenicity) for human use. The lessons from these developments can also be applied to other hydrophilic drugs which have to be used at high doses.

To date, three X-ray contrast-carrying liposome preparations containing different contrast agents have been tested in humans. Due to passive targeting to the Kupffer cells in liver and spleen, favorable hepato-splenic CT enhancement could be observed in one of these studies (up to 45 ΔHU in liver and 236 ΔHU in spleen) which potentially should allow better delineation of focal lesions well below 0.5 cm (e.g., metastases). Unfortunately, dose–dependent, acute drug-related AEs (e.g., chills, back pain, facial flushing) were consistently found in these studies which have also been observed for other particulate contrast agents in humans which render these formulations unacceptable for diagnostic use. Apparently, these side effects could not be predicted based on the preclinical data on these liposomal preparations obtained in the standard tox species (i.e., rodents and non-rodents (e.g., dogs or monkeys)).

With respect to blood-pool imaging, potentially favorable imaging properties, biodistribution, and tolerability could be demonstrated for conventional and surface-modified (i.e., PEG-ylated) CT liposomes in rodents and non-human primates. Most notably no pronounced differences in circulation time were found between unmodified and PEG-ylated liposomes at early timepoints ($\leq$20 minutes), most likely due to a transient saturation of RES-uptake at high lipid doses. To date, studies in relevant animal models of pathologic situations (e.g., cardiovascular disease) representing potential diagnostic indications in humans have not been performed so that the true potential of CT liposomes for blood-pool imaging is yet to be demonstrated.

Most recently, CT blood-pool liposomes which carry an antibody for tissue targeting (i.e., immunoliposomes) have been contemplated. However, while scientifically interesting and potentially feasible, such preparations are unlikely to confer any advantage over MR immunoliposomes which only require 1/10 of the contrast agent concentration to achieve useful images.

Further research should initially focus on elucidating the background of the acute side effects of CT liposomes in humans. Identification of suitable animal models to test for these side effects may potentially allow to identify formulations with remarkably improved tolerability in humans. Subsequently, the liposomal formulations should be further optimized regarding tissue enhancement in liver and blood to fully exploit their diagnostic potential. To this end, it may be possible to develop liposomes which can initially be used for blood-pool imaging during the first 10–20 minutes post-injection while being eventually taken up by the Kupffer cells in the liver and spleen, allowing delineation of small lesions in these organs at later timepoints ($\geq$ 1–2 hours).

Acknowledgment. The help of Jens Leike (DVM, Bayer Schering Pharma, Berlin, Germany) during preparation of this chapter is gratefully acknowledged.

Abbreviations

AEs	Adverse events
CH	Cholesterol
CHHS	Cholesterol hemisuccinate
CT	Computed tomography
DCP	Dicetylphosphate
DPPC	Dipalmitoyl phosphatidylcholine
DSPC	Distearoyl phophatidylcholine
DSPE	Distearoyl phosphatidylethanolamine
EPC	Egg phosphatidylcholine
HU	Hounsfield Units
LUV(s)	Large unilamellar vesicle(s)
MF	Microfluidizer
MLV(s)	Multilamellar vesicle(s)
MPS	Mononuclear phagocytic system
MR	Magnetic resonance imaging
PC	Phosphatidylcholine
PK	Pharmacokinetics
PEG	Polyethylene glycol
RES	Reticuloendothelial system
REV	Reverse-phase evaporation method
SA	Stearylamine
SPC	Soy phosphatidylcholine
SPG	Soy phosphatidylglycerol
SS	Stearic acid
SUV(s)	Small unilamellar vesicle(s)

References

Allen, T.M., 1994. The use of glycolipids and hydrophilic polymers in avoiding rapid uptake of liposomes by the mononuclear phagocyte system. Adv Drug Delivery Rev 13, 285–309.

Benita, S., Poly, P.A., Puisieux, F., Delattre, J., 1984. Radiopaque liposomes: effect of formulation conditions on encapsulation efficiency. J Pharm Sci 73, 1751–1755.

Bodganov, A.A., Weissleder, R., Brady, T.J., 1995. Long-circulating blood pool imaging agents. Adv Drug Delivery Rev 16, 335–348.

Caride, V.J., Sostman, H.D., Twickler, J., et al., 1982. Brominated radiopaque liposomes: contrast agent for computed tomography of liver and spleen: a preliminary report. Invest Radiol 17, 381–385.

Cheng, K.T., Seltzer, S.E., Adams, D.F., Blau, M., 1987. The production and evaluation of contrast-carrying liposomes made with an automatic high-pressure system. Invest Radiol 22, 47–55.

Daeniker, F., 1980. Neues Leberkontrastmittel und Verfahren zu dessen Herstellung. DE 29 35 195 A1.

Desser, T.S., Rubin, D.L., Muller, H., et al., 1999. Blood pool and liver enhancement in CT with liposomal iodixanol: comparison with iohexol. Acad Radiol 6, 176–183.

Dick, A., Adam, G., Tacke, J., et al., 1996. Computed tomography of experimental liver abscesses using a new liposomal contrast agent. Invest Radiol 31, 194–203.

Erdogan, S., Özer, A.Y., Ercan, M.T., et al., 1998. Biodistribution and computed tomography studies on iopromide liposomes. STP Pharma Sciences 8, 133–137.

Fouillet, X., Tournier, H., Khan, H., et al., 1995. Enhancement of computed tomography liver contrast using iomeprol-containing liposomes and detection of small liver tumors in rats. Acad Radiol 2, 576–583.

Goebel, F.D., Goldstein, D., Goos, M., et al., 1996. Efficacy and safety of Stealth liposomal doxorubicin in AIDS-related Kaposi's sarcoma. Br J Cancer 73, 989–994.

Havron, A., Seltzer, S.E., Davis, M.A., Shulkin, P., 1981. Radiopaque Liposomes: a promising new contrast material for computed tomography of the spleen. Radiology 140, 507–511.

Henze, A., Freise, J., Magerstedt, P., Majewski, A., 1989. Radio-opaque liposomes for the improved visualization of focal liver disease by computed tomography. Comput Med Imaging Graph 13, 455–462.

Jendrasiak, G.L., Frey, G.D., Heim, R.C., 1985. Liposomes as carriers of iodolipid radiocontrast agents for CT scanning of the liver. Invest Radiol 20, 995–1002.

Johnston, S.L., Lee, M.W., Hawthorne, M.F., 2003. Development of cell-surface protein targeted CT and MR contrast agents. Suppl Appl Radiol 70–77.

Kao, C., Hoffmann, E.A., Beck, K.C., et al., 2003. Long-residence-time nano-scale liposomal iohexol for x-ray-based blood pool imaging. Acad Radiol 10, 475–483.

Katayama, H., Spinazzi, A., Fouillet, X., et al., 2001. Iomeprol: current and future profile of a radiocontrast agent. Invest Radiol 36, 87–96.

Krause, W., Klopp, R., Leike, J., et al., 1995. Liposomes in diagnostic imaging–comparison of modalities–in-vivo visualization of liposomes. J Liposome Res 5, 1–26.

Krause, W., Leike, J., Sachse, A., Wagner, S., et al., 1991. Preclinical characterization of iopromide-carrying liposomes. Invest Radiol 26, S172–S174.

Lasic, D.D., 1993. Liposomes: from physics to applications, Elsevier Science Publishers B.V., 63 ff.

Leander, P., 1996. A new liposomal contrast medium for CT of the liver–an imaging study in a rabbit tumour model. Acta Radiol 37, 63–68.

Leander, P., Höglund, Kloster, Y., Borseth, A., 1998. New liposomal liver-specific contrast agent for CT: first human phase i clinical trial assessing efficacy and safety. Acad Radiol 5,Suppl 1 S6–S8.

Leander, P., Höglund, P., Borseth, A., et al., 2001. A new liposomal liver-specific contrast agent for CT: first human phase-I clinical trial assessing efficacy and safety. Eur Radiol 11, 698–704.

Leike, J.U., Sachse, A., Rupp, K., 2001. Characterization of continuously extruded iopromide-carrying liposomes for computed tomography blood-pool imaging. Invest Radiol 6, 303–308.

Lopez-Berestein, G., Kasi, L., Rosenblum, M.G., et al., 1984. Clinical Pharmacology of 99m Tc-labeled Liposomes in Patients with Cancer. Cancer Research 44, 375–378.

Martin, F.J., 1990. Pharmaceutical manufacturing of liposomes, in Praveen T. (ed.), Specialized drug delivery systems, Marcel Dekker, Inc., 267–316.

Muschik, P., Sachse, A., Leike, J., et al., 1995. Lipid dependent cardio-haemodynamic tolerability of liposomes in rats. J Liposome Res 5, 933–953.

New RRC, 1990. Preparation of liposomes, in New RRC (ed.), Liposomes a practical approach, Oxford University Press, 33–104.

Olson, F., Hunt, C.A., Szoka, F.C., et al., 1979. Preparation of liposomes of defined size distribution by extrusion through polycarbonate membranes. Biochim Biophys Acta 557, 9–23.

Payne, N.I., Whitehouse, G.H., 1987. Delineation of the spleen by a combination of proliposomes with water-soluble contrast media: an experimental study using computed tomography. Br J Radiol 60, 535–541.

Petersein, J., Franke, B., Fouillet, X., Hamm, B., 1999. Evaluation of liposomal contrast agents for liver CT in healthy rabbits. Invest Radiol 34, 401–409.

Poste, G., 1983. Liposome targeting in vivo: problems and opportunities. Biol Cell 47, 19–38.

Roerdink, F.H., Regts, J., Handel, T., et al., 1989. Effect of cholesterol on the uptake and intracellular degradation of liposomes by liver and spleen; a combined biochemical and γ-ray perturbed angular correlation study. Biochim Biophys Acta 980, 234–240.

Rozenberg, O.A., Loshakova, L.V., Mikhailova, N.Y., et al., 1993. Triombrast in liposomes for liver and spleen imaging: The first phase of the clinical trials. Centr Okt N5, 35–37 (Russian paper).

Rummeny, E.J., Berning, W., Fuest, M., et al., 1999. Evaluation of a new blood pool agent for CT-angiography. Eur Radiol 9, S24.

Rummeny, E.J., Berning, W., Fuest, M., et al., 2002. New RES-specific contrast agents for CT. Acad Radiol 9,Suppl 1, S185–S190.

Ryan, P.J., Davis, M.A., DeGaeta, L.R., et al., 1984. Liposomes loaded with contrast material for image enhancement in computed tomography–work in progress. Radiology 152, 759–762.

Ryan, P.J., Davis, M.A., Melchior, D.L., 1983. The preparation and characterization of liposomes containing X-ray contrast agents. Biochim Biophys Acta 756, 106–110.

Sachse, A., 1998. Large-scale production of liposomes by continuous high pressure extrusion. In: Mueller, R.H., Benita, S., Boehm, B. (eds.), Emulsions and nanosuspensions for the formulation of poorly soluble drugs, medpharm Scientific Publishers, Stuttgart, 257–265.

Sachse, A., Leike, J.U., Rößling, G., et al., 1993. Preparation and evaluation of lyophilized iopromide-carrying liposomes for liver tumor detection. Invest Radiol 28, 838–844.

Sachse, A., Leike, J.U., Schneider, T., et al., 1997. Biodistribution and computed tomography blood-pool imaging properties of polyethylene glycol-coated iopromide-carrying liposomes. Invest Radiol 32, 44–50.

Schmiedl, U.P., Krause, W., Leike, J., et al., 1995. Liver contrast enhancement in primates using iopromide liposomes. Acad Radiol 2, 967–972.

Schmiedl, U.P., Krause, W., Leike, J., Sachse, A., 1999. CT blood pool enhancement in primates with iopromide-carrying liposomes containing soy phosphatidyl glycerol. Acad Radiol 6, 164–169.

Schneider, T., Sachse, A., Rößling, G., Brandl, M., 1995. Generation of contrast-carrying liposomes of defined size with a new continuous high pressure extrusion method. Int Journ Pharm 117, 1–12.

Seltzer, S.E., 1988a. Liposomes in diagnostic imaging. In: Gregoriadis, G. (ed.) Liposomes as drug carriers. Chichester, NY: John Wiley & Sons 509–525.

Seltzer, S.E., 1988b. Contrast-carrying liposomes, current status. Invest Radiol 23,Suppl 1, S122–S125.

Seltzer, S.E., Blau, M., Herman, L.W., et al., 1995. Contrast material-carrying liposomes: biodistribution, clearance, and imaging characteristics. Radiology 194, 775–781.

Seltzer, S.E., Davis, M.A., Adams, D.F., et al., 1984a. Liposomes carrying diatrizoate: characterization of biophysical properties and imaging applications. Invest Radiol 19, 142–151.

Seltzer, S.E., Gregoriadis, G., Dick, R., 1988. Evaluation of the dehydration-rehydration method for production of contrast-carrying liposomes. Invest Radiol 23, 131–138.

Seltzer, S.E., Shulkin, P.M., Adams, D.F., et al., 1984b. Usefulness of liposomes carrying iosefamate for CT opacification of liver and spleen. AJR 143, 575–579.

Senior, J.H., 1987. Fate and behaviour of liposomes in vivo: A review of controlling factors. CRC Crit Rev Therap Drug Carrier Syst 3, 123–193.

Spinazzi, A., Ceriati, S., Pianezzola, P., et al., 2000. Safety and pharmacokinetics of a new liposomal liver-specific contrast agent for CT–results of clinical testing in nonpatient volunteers. Invest Radiol 35 1–7.

Szoka, F., Papahadjopoulos, D., 1978. Procedure for preparation of liposomes with large internal aqueous space and high capture by reverse-phase evaporation. Proc Nat Acad Sci USA 75, 4194–4198.

Tilcock, C., 1994. Liposomal blood pool agents for nuclear medicine and magnetic resonance imaging. J Liposome Res 4, 909–936.

Tilcock, C., 1999. Delivery of contrast agents for magnetic resonance imaging, computed tomography, nuclear medicine and ultrasound. Adv Drug Delivery Rev 37, 33–51.

Torchilin, V.P., 2005. Recent advances with liposomes as pharmaceutical carriers. Nature Reviews 4, 145–160.

Torchilin, V.P., Trubetskoy, V.S., 1995. In vivo visualizing of organs and tissues with liposomes. J Liposome Res 5, 795–812.

Zalutsky, M.R., Noska, M.A., Seltzer, S.E., 1987. Characterization of liposomes containing iodine-125-labeled radiographic contrast agents. Invest Radiol 22, 141–147.

Zerbin, E.A., Davidenkova, E.F., Chanson, K.P., et al., 1985. Ein neues Verfahren zur Kontrastdarstellung von Leber und Milz unter Anwendung wasserlöslicher Röntgenkontrastmittel in Liposomen. Radiol Diagn 26, 285–292. (German paper)

Zheng, J., Perkins, G., Kirilova, A., et al., 2006. Multimodal contrast agent for combined computed tomography and magnetic resonance imaging applications. Invest Radiol 41, 339–348.

Part 5
Quantum Dots: Applications in Optical Imaging

19
Quantum Dots and Targeted Nanoparticle Probes for In Vivo Tumor Imaging

Matthew N. Rhyner, Andrew M. Smith, Xiaohu Gao, Hui Mao, Lily Yang, and Shuming Nie

Abstract: Nanometer-sized particles, such as semiconductor quantum dots and iron oxide nanocrystals, have novel optical, electronic, magnetic, or structural properties that are not available from either individual molecules or bulk solids. When linked with tumor-targeting ligands, such as monoclonal antibodies, peptides, or small molecules, these nanoparticles can be used to target tumor antigens (biomarkers) as well as tumor vasculatures with high affinity and specificity. In the "mesoscopic" size range of 5–100 nm diameter, quantum dots and related nanoparticles also have more surface areas and functional groups that can be linked to multiple diagnostic (e.g., optical, radioisotopic, or magnetic) and therapeutic (e.g., anticancer) agents. Previous research has led to the development of bioaffinity nanoparticle probes for advanced molecular and cellular imaging. In this chapter, we discuss recent advances in the development and applications of bioconjugated quantum dots and multifunctional nanoparticles for in vivo tumor imaging and targeting.

Keywords: quantum dot, nanoparticle, cancer imaging, in vivo, targeting, polymer coating, bimodal.

19.1. Introduction

The development of high-sensitivity and high-specificity probes is of considerable interest in many areas of cancer research, ranging from basic tumor biology to in vivo imaging and early detection. The process to develop new imaging probes, however, remains a slow and expensive undertaking. In this context, the development of quantum dots and biodegradable nanoparticles has attracted considerable attention. Indeed, recent advances have led to the development of functional nanoparticles (electronic, optical, magnetic, or structural) that are covalently linked to biological molecules, such as peptides, proteins, and nucleic acids (Akerman et al., 2002; Alivisatos et al., 2005; Bruchez et al., 1998; Chan et al., 2002; Chan and Nie, 1998; Dahan et al., 2003; Dubertret et al., 2002; Gao et al., 2005; Ishii et al., 2003; Jaiswal et al., 2003; Larson et al., 2003; Lidke et al., 2004; Mattoussi et al., 2000; Medintz et al., 2003; Michalet et al., 2005; Pinaud et al., 2006; Smith et al., 2006, 2004; Stroh et al., 2005; Wu et al., 2003). Due to their size-dependent properties and dimensional similarities to biomacromolecules, these nanobioconjugates are well suited as contrast agents for biomedical imaging (Bulte et al., 2001; Josephson et al., 1999), as controlled carriers for drug delivery, and as structural scaffolds for cellular and tissue engineering (Curtis and Wilkinson, 2001; Gref et al., 1994).

In comparison with traditional contrast agents, such as radioactive small molecules, gadolinium compounds, and labeled antibodies, quantum dots (QDs) and targeted nanoparticles provide several unique features and capabilities. First, their optical and electronic properties are often dependent on size and can be tuned continuously by changing the particle size (Alivisatos, 1996). This "size effect" provides a broad range of nanoparticles for simultaneous detection of multiple cancer biomarkers. Second, nanoparticles have more surface area and functional groups that can be linked with multiple diagnostic (e.g., radioisotopic or magnetic) and therapeutic (e.g., anticancer) agents. This opens the opportunity to design multifunctional or "smart" nanoparticles for multi-modality imaging as well as integrated imaging and therapy. Third, extensive research has shown that nanoparticles in the size range of 10–100 nm are accumulated preferentially at tumor sites through an effect called enhanced permeability and retention (EPR) (Duncan and Discovery, 2003; Jain, 1999, 2001). This effect is believed to arise from two factors: (i) angiogenic tumors produce vascular endothelial growth factors (VEGF) that hyperpermeabilize the tumor-associated neovasculatures and cause the leakage of circulating macromolecules and small particles and (ii) tumors lack an effective lymphatic drainage system, which leads to subsequent macromolecule or nanoparticle accumulation.

These novel properties have opened exciting opportunities in developing new and advanced nanoparticle probes for biomedical imaging, especially for tumor imaging and targeting. In the following, we briefly review the novel features of semiconductor QDs, dual-modality imaging probes, and their applications in tumor imaging.

19.2. Novel Properties of Quantum Dots

QDs are tiny light-emitting particles on the nanometer scale and are emerging as a new class of fluorescent probes for in vivo molecular and cellular imaging. In comparison with organic dyes and fluorescent proteins, QDs have unique optical and electronic properties, such as size-tunable light emission, improved signal brightness, resistance against photobleaching, and simultaneous excitation of multiple fluorescence colors. The following properties are considered most promising for improving the sensitivity of in vivo tumor imaging by at least

10–100 fold. First, QDs have very large molar extinction coefficients in the order of $0.5\text{–}5 \times 10^6 M^{-1}cm^{-1}$ (Leatherdale et al., 2002), which makes them brighter probes under photon-limited in vivo conditions (where light intensities are severely attenuated by scattering and absorption). In theory, the lifetime-limited emission rates for single QDs are 5–10 times lower than those of single organic dyes because of their longer excited state lifetimes (20–50 ns). In practice, however, fluorescence imaging usually operates under absorption-limited conditions, in which the rate of absorption is the main limiting factor of fluorescence emission. Since the molar extinction coefficients of QDs are about 10–50 times larger than that ($5\text{–}10 \times 10^4 M^{-1}cm^{-1}$) of organic dyes, the QD absorption rates will be 10–50 times faster than that of organic dyes at the same excitation photon flux (number of incident photons per unit area). Due to this increased rate of light emission, individual QDs have been found to be 10–20 times brighter than organic dyes (Bruchez et al., 1998; Chan et al., 2002; Chan and Nie, 1998). In addition, QDs are several thousand times more stable against photobleaching than dye molecules and are well suited for continuous in vivo tracking studies (e.g., cell migration, differentiation, and metastasis) over an extended period of time.

Second, the large Stokes shifts of QDs (measured by the separation between the excitation and emission peaks) can be used to further improve detection sensitivity. This factor becomes especially important for in vivo molecular imaging due to the high autofluorescence background often seen in complex biomedical specimens. As reported by Gao et al. (Gao et al., 2004), the Stokes shifts of semiconductor QDs are as large as 300–400 nm, depending on the wavelength of the excitation light. Organic dye signals with a small Stokes shift are often buried by strong tissue autofluorescence, whereas QD signals with a large Stokes shift are clearly recognizable above the background. This "color contrast" is only available to QD probes because the target signal and background can be separated by wavelength-resolved spectral imaging (Gao et al., 2004).

A further advantage is that multicolor QD probes can be used to image and track multiple molecular targets simultaneously. This is an important feature because most complex human diseases, such as cancer, involve a large number of genes and proteins. Tracking a panel of molecular markers at the same time will allow scientists to understand, classify, and differentiate complex human diseases (Gao and Nie, 2003b). Multiple parameter imaging, however, is difficult for MRI, PET, computed x-ray tomography (CT), and related imaging modalities. On the other hand, fluorescence optical imaging provides both signal intensity and wavelength information, and multiple wavelengths (colors) can be resolved and imaged simultaneously. In this regard, QD probes are particularly attractive because their broad absorption profiles allow simultaneous excitation of multiple colors, and their emission wavelengths can be continuously tuned by varying particle size and/or chemical composition. Therefore, different molecular or cellular targets can be tagged with different colored QDs. For organ and vascular imaging in which micrometer-sized particles could be used, optically encoded beads (small beads embedded with multicolor QDs at controlled ratios) could allow multiplexed molecular profiling at high sensitivity (Gao and Nie, 2003a; Gaponik et al., 2002; Han et al., 2001; Xu HX, 2003)

19.3. Advanced Probe Design

Recent work has developed a new class of multifunctional nanoparticle probes based on QDs dots for cancer targeting and imaging in living animals (Gao et al., 2005). The general design involves encapsulating luminescent QDs with a biocompatible copolymer and linking this amphiphilic polymer to tumor-targeting ligands and drug-delivery functionalities (Figure 19.1). In vivo targeting studies of human prostate cancer growing in nude mice indicate that this class of QD probes can be delivered to tumor sites by both the EPR effect and antibody binding to cancer-specific cell surface biomarkers. For instance, using both subcutaneous injection of QD-tagged cancer cells and systemic injection of multifunctional QD probes, Gao et al. (2004) achieved sensitive and multicolor fluorescence imaging of cancer cells under in vivo conditions. A whole-body macro-illumination system was also integrated with wavelength-resolved spectral imaging for efficient background removal and precise delineation of

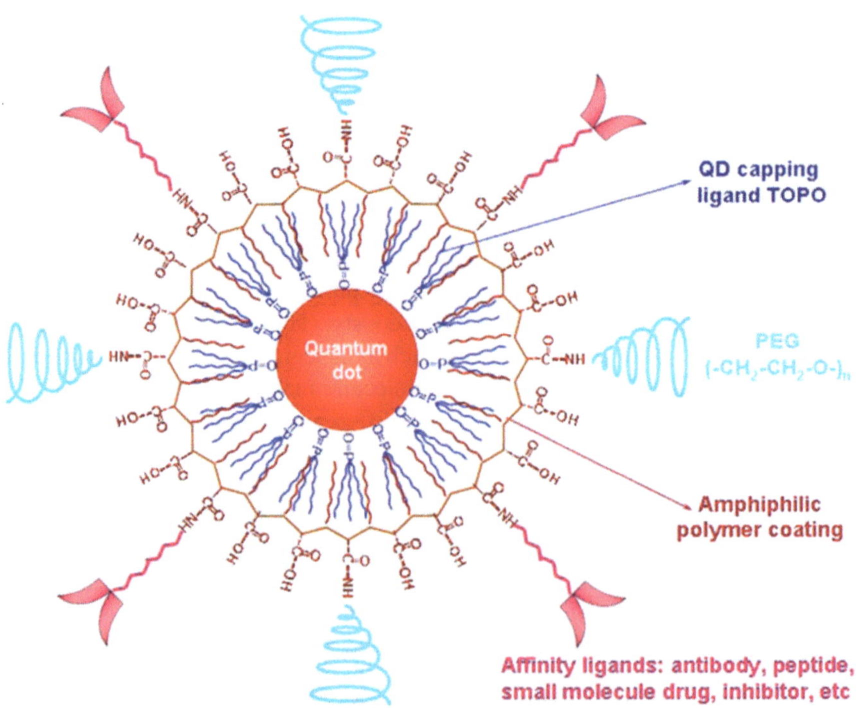

FIGURE 19.1. Schematic diagram of a multifunctional QD probe, showing the capping ligand TOPO, an encapsulating copolymer layer, tumor-targeting ligands (such as peptides, antibodies, or small-molecule inhibitors), and polyethylene glycol (PEG).

weak spectral signatures. With a large number of functional groups on its surface, this encapsulated QD also provides a nanoscale scaffold for linking to additional imaging agents or therapeutics drugs, as discussed in more detail below.

19.3.1. Dual-Modality Nanoparticle Probes

Optical imaging is highly sensitive, but has only limited penetration depth and is poor in providing location information. Although near-infrared wavelengths can be used to improve the penetration depth and 3-D fluorescence tomography can be used to provide depth information (Ntziachristos et al., 2003, 2004), other imaging modalities, such as MRI, are much better for tomography and 3D imaging. Thus, there has been considerable interest in developing dual-modality contrast agents for combined optical and MRI imaging, which has exceptional tissue contrast and spatial resolution and has been widely used in the clinical setting. For example, by reacting superparamagnetic iron oxide nanoparticles with the fluorescent dye Cy5.5, Josephson and coworkers (Kircher et al., 2004) have developed dual magneto-optical probes that are able to bind to apoptotic cells and are detectable by both fluorescence and magnetic resonance imaging. Similarly, dual magnetic and optical imaging probes have been used to yield highly detailed anatomic and molecular information in living organisms (Schellenberger et al., 2004). These probes are prepared by conjugation of peptides to cross-linked iron oxide amine (amino-CLIO), either by a disulfide linkage or a thioether linker, followed by the attachment of the dye Cy5 or Cy7. Fluorescence quenching of the attached fluorochrome occurs by interaction with the iron oxide core, and also by electronic coupling among the dye chromophores (self-quenching). This class of dual-modality probes provides the basis for "smart" nanoparticles, capable of pinpointing their position through their magnetic properties, while providing information on their environment by optical imaging techniques.

Recent research has shown that QDs can be linked with Fe_2O_3 or FePt to generate dual functional nanoparticles (Gu et al., 2004; Patri et al., 2004; Quintana et al., 2002; Wang et al., 2004). Others have entrapped Gd on the QD surface using polymer conjugated lipids to form dual modality probes, but it is not clear whether these types of "hetero" nanostructures would be useful for in vivo medical imaging (Mulder et al., 2006). To explore this area, we have created a new class of dual-modality nanoparticles by attaching a cluster of paramagnetic gadolinium chelates to polymer-coated QDs. Preliminary cellular and in vivo animal studies demonstrated that this class of nanoparticle is biocompatible and detectable by both fluorescence and MRI (Figure 19.2). In comparison with previous work, the polymer-protected QDs offer excellent optical properties (high fluorescence quantum yields, narrow spectral widths, and high photostability), and the attached Gd chelates lead to significant T1 contrast enhancement with a "brightening" effect in MRI, as opposed to the T2 contrast with a "darkening" effect offered by iron oxide-based contrast. By linking to targeting ligands through a biocompatible polyethylene glycol (PEG) spacer, these dual-modality nanoparticle probes are promising for in vivo tumor imaging in animal models.

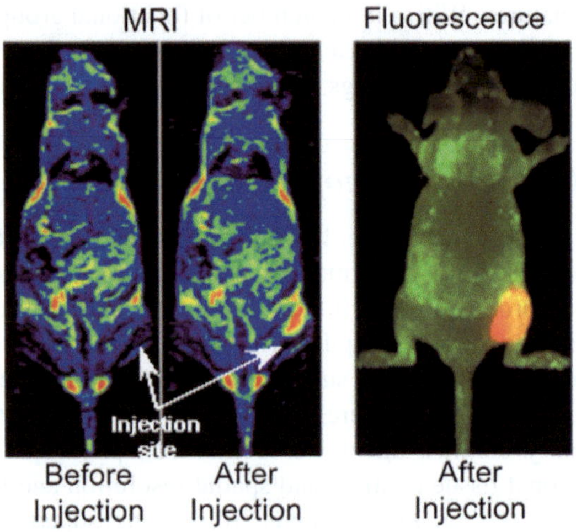

FIGURE 19.2. Correlated MRI (left) and optical (right) imaging of Gd-QD dual-modality probes injected into a live mouse. Note that the T1 weighted spin echo MR images show signal enhancement after probe injection, correlating with the fluorescence signal of optical imaging. Although no active targeting is involved in this study, the correlated images serve to demonstrate feasibility of the proposed dual-modality concept. (Gao, Mao, and Nie, unpublished data).

19.3.2. Multifunctional Nanoparticle Platforms

Another major advantage of nanoparticles is their larger size and surface area, which allow chemical conjugation to multiple diagnostic and therapeutic agents. It is thus possible to design and develop multifunctional nanostructures that could be used for simultaneous tumor imaging and treatment, a major goal in cancer research and development. However, progress has been slow and promising multifunctional platforms, such as dendrimers, liposomes, and PEBBLES (probes encapsulated in biologically localized embedding), have not been able to deliver diagnostic and therapeutic agents to tumors in a selective and efficient manner (Buck et al., 2004; Patri et al., 2004; Quintana et al., 2002; Torchilin et al., 2000, 1994). Most of these studies are still at an early or "proof-of-concept" stage using cultured cancer cells, which are not immediately relevant to in vivo imaging and treatment of solid tumors.

19.4. In Vivo Tumor Imaging and Targeting

19.4.1. Mapping Sentinel Lymph Nodes and Tumor Angiogenesis

In vivo imaging with QDs has been reported for lymph node mapping, blood pool imaging, angiogenic vessels, and cell subtype isolation (Figure 19.3a and b).

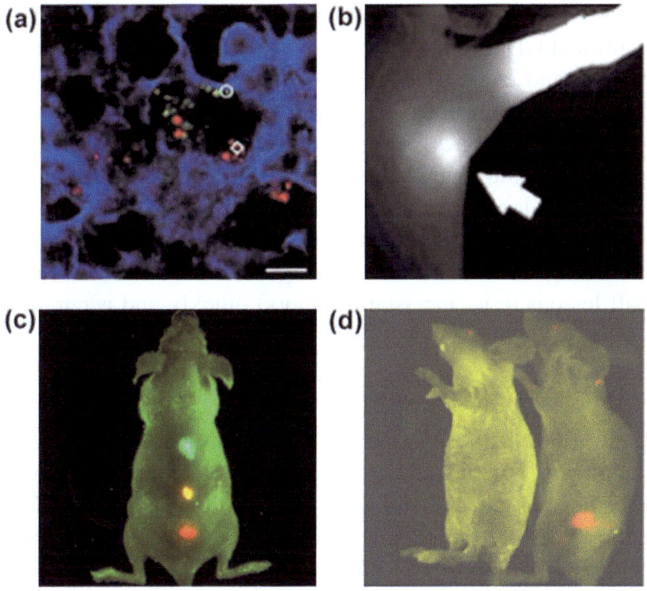

FIGURE 19.3. *In vivo* targeting and imaging with QDs. (a) Ex vivo tissue examination of QD-labeled cancer cells trapped in a mouse lung [51]. (b) Near infrared fluorescence of water-soluble Type II QDs taken up by sentinel lymph nodes [20]. (c) In vivo simultaneous imaging of multicolor QD-encoded microbeads injected into a live mouse [31]. (d) Molecular targeting and in vivo imaging of prostate tumor using QD-antibody conjugate [31].

Ballou et al. and coworkers (2004) injected PEG-coated QDs into the mouse blood stream and studied how the surface coating would affect their circulation lifetime. In contrast to small organic dyes (which are eliminated from circulation within minutes after injection), PEG-coated QDs were found to stay in blood circulation for an extended period of time (half-life more than 3 hours). This long-circulating feature can be explained by the unique structural properties of QD nanoparticles. PEG-coated QDs are in an intermediate size range—they are small and hydrophilic enough to slow down opsonization and reticuloendothelial uptake, while they are large enough to avoid renal filtration. Webb and coworkers took advantage of this property and reported the use of QDs and two-photon excitation to image small blood vessels (Larson et al., 2003). They found that the two-photon absorption cross sections of QDs are 2–3 orders of magnitude larger than that of traditional organic fluorophores. Jain and coworkers have used QDs and QD-doped silica beads for differentiating tumor vessels from perivascular cells and matrix (Stroh et al., 2005). The results demonstrated a much clearer boundary between blood vessels and cells than using traditional high-molecular weight dextran.

For improved tissue penetration, Frangioni and Bawendi prepared a novel core-shell nanostructure called Type II QDs (Kim et al., 2004), with fairly broad emission at 850 nm emission and a moderate quantum yield of ~13 %. In contrast to the conventional QDs (type-I), the shell materials in type-II QDs have valence and conduction band energies both lower than those of the core materials. As a result, the electrons and holes are physically separated and the nanoparticles emit light at reduced energies (longer wavelengths). Their results showed rapid uptake of bare QDs into lymph nodes, and clear imaging and delineation of involved sentinel nodes (which could then be removed). This work points to the possibility that QD probes could be used for real-time intra-operative optical imaging, providing an *in situ* visual guide so that a surgeon could locate and remove small lesions (e.g., metastatic tumors) quickly and accurately.

19.4.2. Tumor Targeting and Imaging

Akerman et al. (2002) first reported the use of QD–peptide conjugates to target tumor vasculatures, but the QD probes were not detected in living animals. Nonetheless, their in vitro histological results revealed that QDs homed to tumor vessels guided by the peptides and were able to escape clearance by the reticuloendothelial system (RES). Gao et al. (2004) used the probes shown in Figure 19.1 for simultaneous targeting and imaging of tumors in live animals. This class of QD conjugates contains an amphiphilic triblock copolymer for in vivo protection, targeting ligands for tumor antigen recognition and multiple PEG molecules for improved biocompatibility and circulation. The use of an ABC triblock copolymer has solved the problems of particle aggregation and fluorescence loss previously encountered for QDs stored in physiological buffer or injected into live animals (Akerman et al., 2002; Gao et al., 2002; Mattoussi et al., 2000). Detailed studies were reported on the in vivo behaviors of QDs probes, including biodistribution, nonspecific uptake, cellular toxicity, and pharmacokinetics. Under in vivo conditions, QD probes are delivered to tumors by both a passive targeting mechanism and an active targeting mechanism. In the passive mode, macromolecules and nanometer-sized particles are accumulated preferentially at tumor sites through the EPR effect (Jain, 1999). For active tumor targeting, Gao et al. (2004) used antibody-conjugated QDs to target a prostate-specific cell surface antigen, PSMA (Figure 19.3d). Previous research has identified PSMA as a cell surface marker for both prostate epithelial cells and neovascular endothelial cells (Chang et al., 2001). PSMA has been selected as an attractive target for both imaging and therapeutic intervention of prostate cancer. Accumulation and retention of PSMA antibody at the site of tumor growth is the basis of radioimmunoscintigraphic scanning (e.g., ProstaScint scan) and targeted therapy for human prostate cancer metastasis (Bander et al., 2003).

By integrating with an x-ray imaging machine, recent work by Nie and workers (S. M. Nie, L. Yang, and A. M. Smith, personal communication) has also achieved optical and structural imaging on the same animal models. Figure 19.4 shows correlated x-ray and fluorescence images of high-quality, deep-red QDs injected into a mouse. The low background and high signal levels create superimposed x-ray and optical images that yield important information about the

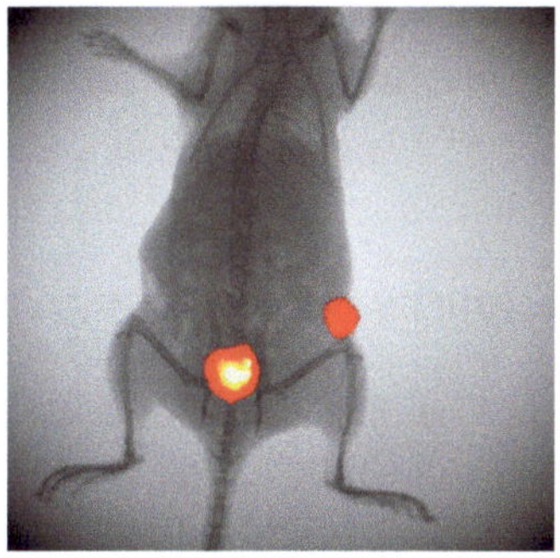

FIGURE 19.4. Combined x-ray and fluorescent imaging of near-infrared-emitting (700 nm) QD probes injected into the peritoneal cavity of a mouse. The near-infrared emission of these QDs is readily detectable above background. Images were captured sequentially with a Kodak in vivo Image Station, with 625 nm excitation for fluorescence.

anatomic locations of small tumors. This type of correlated imaging should provide new insights into cancer development, progression, and metastasis in animal models.

19.5. Prospects for Clinical Human Imaging

These studies using animal models have raised new possibilities for in vivo tumor imaging. However, the current nanoparticle probes have several drawbacks, such as limited tissue penetration, lack of spatial resolution in tumor depth and location, and potential toxicity concerns. Thus, there is an urgent need to develop broadly tunable near-infrared-emitting QDs to improve the tissue penetration depth. For clinical human applications, a major hurdle is likely the potential toxicity of QD probes, which has recently become a topic of considerable discussion and debate. Work by Derfus et al. (2004) indicates that CdSe QDs are highly toxic to cultured cells under UV illumination for extended periods of time. It has also been reported that the polymer-coated QDs could be toxic if significant aggregates are formed on the cell surface (Kirchner et al., 2005). This is not surprising because the energy of UV-irradiation is close to that of a covalent chemical bond and dissolves the semiconductor particles in a process known as photolysis, which releases toxic cadmium ions into the culture medium. In the absence of UV irradiation, QDs with a stable polymer-coating have been found to be essentially nontoxic to cells and animals, for example no effect on cell division or ATP production (D. Stuart, X. Gao, and S. Nie, unpublished data). In vivo studies by Ballou et al. and coworkers (2004) also confirmed the nontoxic nature of stably protected QDs. Still, there is an urgent need to study the cellular toxicity and in vivo degradation mechanisms of QD probes. For polymer-encapsulated QDs, chemical or enzymatic degradations of the semiconductor cores are unlikely to occur. But the polymer-protected QDs might be cleared from the body by slow filtration and excretion out of the body. This and other possible mechanisms must be carefully examined prior to any human applications in tumor or vascular imaging.

19.6. Concluding Remarks

QDs have already fulfilled some of their promises as a new class of molecular probes for cancer research. Through their versatile polymer coatings, QDs have also provided a "building block" to assemble multifunctional nanostructures and nanodevices. In particular, dual-modality imaging probes have been prepared by integrating QDs with paramagnetic or superparamagnetic agents. By correlating the deep imaging capabilities of magnetic resonance imaging (MRI) with ultrasensitive optical imaging, a surgeon could visually identify tiny tumors or other small lesions during an operation and remove the diseased cells and tissue completely. Another desired multifunctional device would be the combination of a QD-imaging agent with a therapeutic agent. Not only would this allow tracking of pharmacokinetics, but diseased tissue could be treated and monitored simultaneously and in real time. Practical applications of these multifunctional nanodevices will not come without careful research, but the multidisciplinary nature of nanotechnology may expedite these goals by combining the great minds of many different fields.

Acknowledgements. The authors are grateful to Amit Agrawal, Gang Ruan, and Leland Chung for insightful discussions. The work was supported in part by NIH grants (P20 GM072069, R01 CA108468-01, and U54CA119338) and the Georgia Cancer Coalition Distinguished Cancer Scholars Program (to S.N.). Andrew M. Smith acknowledges the Whitaker Foundation for generous fellowship support, and Matthew N. Rhyner is grateful to the NSF IGERT Program at Georgia Tech (PI: Dr. Marie Thursby) for financial support.

References

Akerman, M.E., Chan, W.C.W., Laakkonen, P., Bhatia, S.N., Ruoslahti, E., 2002. Nanocrystal targeting in vivo. Proc Natl Acad Sci U S A 99, 12617–12621.

Alivisatos, A.P., 1996. Semiconductor clusters, nanocrystals, and quantum dots. Science 271, 933–937.

Alivisatos, A.P., Gu, W.W., Larabell, C., 2005. Quantum dots as cellular probes. Annu Rev Biomed Eng 7, 55–76.

Ballou, B., Lagerholm, B.C., Ernst, L.A., Bruchez, M.P., Waggoner, A.S., 2004. Noninvasive imaging of quantum dots in mice. Bioconjug Chem 15, 79–86.

Bander, N.H., Trabulsi, E.J., Kostakoglu, L., Yao, D., Vallabhajosula, S., Smith-Jones, P., Joyce, M.A., Milowsky, M., Nanus, D.M., Goldsmith, S.J., 2003. Targeting metastatic prostate cancer with radiolabeled monoclonal antibody J591 to the extracellular domain of prostate specific membrane antigen. J Urol 170, 1717–1721.

Bruchez, M., Moronne, M., Gin, P., Weiss, S., Alivisatos, A.P., 1998. Semiconductor nanocrystals as fluorescent biological labels. Science 281, 2013–2016.

Buck, S.M., Koo, Y.E.L., Park, E., Xu, H., Philbert, M.A., Brasuel, M.A., Kopelman, R., 2004. Optochemical nanosensor PEBBLEs: photonic explorers for bioanalysis with biologically localized embedding. Curr Opin Chem Biol 8, 540–546.

Bulte, J.W.M., Douglas, T., Witwer, B., Zhang, S.C., Strable, E., Lewis, B.K., Zywicke, H., Miller, B., van Gelderen, P., Moskowitz, B.M., Duncan, I.D., Frank, J.A., 2001. Magnetodendrimers allow endosomal magnetic labeling and in vivo tracking of stem cells. Nat Biotechnol 19, 1141–1147.

Chan, W.C.W., Maxwell, D.J., Gao, X.H., Bailey, R.E., Han, M.Y., Nie, S.M., 2002. Luminescent quantum dots for multiplexed biological detection and imaging. Curr Opin Biotechnol 13, 40–46.

Chan, W.C.W., Nie, S.M., 1998. Quantum dot bioconjugates for ultrasensitive nonisotopic detection. Science 281, 2016–2018.

Chang, S.S., Reuter, V.E., Heston, W.D.W., Gaudin, P.B., 2001. Comparison of anti-prostate-specific membrane antigen antibodies and other immunomarkers in metastatic prostate carcinoma. Urology 57, 1179–1183.

Curtis, A., Wilkinson, C., 2001. Nantotechniques and approaches in biotechnology. Trends Biotechnol 19, 97–101.

Dahan, M., Levi, S., Luccardini, C., Rostaing, P., Riveau, B., Triller, A., 2003. Diffusion dynamics of glycine receptors revealed by single-quantum dot tracking. Science 302, 442–445.

Derfus, A.M., Chan, W.C.W., Bhatia, S.N., 2004. Probing the cytotoxicity of semiconductor quantum dots. Nano Letters 4, 11–18.

Dubertret, B., Skourides, P., Norris, D.J., Noireaux, V., Brivanlou, A.H., Libchaber, A., 2002. In vivo imaging of quantum dots encapsulated in phospholipid micelles. Science 298, 1759–1762.

Duncan, R., Discovery, N.R.D., 2003. The dawning era of polymer therapeutics. Nat Rev Drug Discov 2, 347–360.

Gao, X.H., Chan, W.C.W., Nie, S.M., 2002. Quantum-dot nanocrystals for ultrasensitive biological labeling and multicolor optical encoding. J Biomed Opt 7, 532–537.

Gao, X.H., Cui, Y.Y., Levenson, R.M., Chung, L.W.K., Nie, S.M., 2004. In vivo cancer targeting and imaging with semiconductor quantum dots. Nat Biotechnol 22, 969–976.

Gao, X.H., Nie, S.M., 2003a. Doping mesoporous materials with multicolor quantum dots. J Phys Chem B 107, 11575–11578.

Gao, X.H., Nie, S.M., 2003b. Molecular profiling of single cells and tissue specimens with quantum dots. Trends Biotechnol 21, 371–373.

Gao, X.H., Yang, L.L., Petros, J.A., Marshal, F.F., Simons, J.W., Nie, S.M., 2005. In vivo molecular and cellular imaging with quantum dots. Curr Opin Biotechnol 16, 63–72.

Gaponik, N., Radtchenko, I.L., Sukhorukov, G.B., Weller, H., Rogach, A.L., 2002. Toward encoding combinatorial libraries: Charge-driven microencapsulation of semiconductor nanocrystals luminescing in the visible and near IR. Adv Mater 14, 879–882.

Gref, R., Minamitake, Y., Peracchia, M.T., Trubetskoy, V., Torchilin, V., Langer, R., 1994. Biodegradable long-circulating polymeric nanospheres. Science (Washington, D. C., 1883-) 263, 1600–1603.

Gu, H.W., Zheng, R.K., Zhang, X.X., Xu, B., 2004. Facile one-pot synthesis of bifunctional heterodimers of nanoparticles: a conjugate of quantum dot and magnetic nanoparticles. J Am Chem Soc 126, 5664–5665.

Han, M.Y., Gao, X.H., Su, J.Z., Nie, S., 2001. Quantum-dot-tagged microbeads for multiplexed optical coding of biomolecules. Nat Biotechnol 19, 631–635.

Ishii, D., Kinbara, K., Ishida, Y., Ishii, N., Okochi, M., Yohda, M., Aida, T., 2003. Chaperonin-mediated stabilization and ATP-triggered release of semiconductor nanoparticles. Nature 423, 628–632.

Jain, R.K., 1999. Understanding barriers to drug delivery: high resolution in vivo imaging is key. Clin Cancer Res 5, 1605–1606.

Jain, R.K., 2001. Delivery of molecular medicine to solid tumors: lessons from in vivo imaging of gene expression and function. J Control Release 74, 7–25.

Jaiswal, J.K., Mattoussi, H., Mauro, J.M., Simon, S.M., 2003. Long-term multiple color imaging of live cells using quantum dot bioconjugates. Nat Biotechnol 21, 47–51.

Josephson, L., Tung, C.H., Moore, A., Weissleder, R., 1999. High-efficiency intracellular magnetic labeling with novel superparamagnetic-tat peptide conjugates. Bioconjug Chem 10, 186–191.

Kim, S., Lim, Y.T., Soltesz, E.G., De Grand, A.M., Lee, J., Nakayama, A., Parker, J.A., Mihaljevic, T., Laurence, R.G., Dor, D.M., Cohn, L.H., Bawendi, M.G., Frangioni, J.V., 2004. Near-infrared fluorescent type II quantum dots for sentinel lymph node mapping. Nat Biotechnol 22, 93–97.

Kircher, M.F., Weissleder, R., Josephson, L., 2004. A dual fluorochrome probe for imaging proteases. Bioconjug Chem 15, 242–248.

Kirchner, C., Liedl, T., Kudera, S., Pellegrino, T., Javier, A.M., Gaub, H.E., Stolzle, S., Fertig, N., Parak, W.J., 2005. Cytotoxicity of colloidal CdSe and CdSe/ZnS nanoparticles. Nano Letters 5, 331–338.

Larson, D.R., Zipfel, W.R., Williams, R.M., Clark, S.W., Bruchez, M.P., Wise, F.W., Webb, W.W., 2003. Water-soluble quantum dots for multiphoton fluorescence imaging in vivo. Science 300, 1434–1436.

Leatherdale, C.A., Woo, W.K., Mikulec, F.V., Bawendi, M.G., 2002. On the absorption cross section of CdSe nanocrystal quantum dots. J Phys Chem B 106, 7619–7622.

Lidke, D.S., Nagy, P., Heintzmann, R., Arndt-Jovin, D.J., Post, J.N., Grecco, H.E., Jares-Erijman, E.A., Jovin, T.M., 2004. Quantum dot ligands provide new insights into erbB/HER receptor-mediated signal transduction. Nat Biotechnol22, 198–203.

Mattoussi, H., Mauro, J.M., Goldman, E.R., Anderson, G.P., Sundar, V.C., Mikulec, F.V., Bawendi, M.G., 2000. Self-assembly of CdSe-ZnS quantum dot bioconjugates using an engineered recombinant protein. J Am Chem Soc 122, 12142–12150.

Medintz, I.L., Clapp, A.R., Mattoussi, H., Goldman, E.R., Fisher, B., Mauro, J.M., 2003. Self-assembled nanoscale biosensors based on quantum dot FRET donors. Nat Mater 2, 630–638.

Michalet, X., Pinaud, F.F., Bentolila, L.A., Tsay, J.M., Doose, S., Li, J.J., Sundaresan, G., Wu, A.M., Gambhir, S.S., Weiss, S., 2005. Quantum dots for live cells, in vivo imaging, and diagnostics. Science (Washington, D. C., 1883-) 307, 538–544.

Mulder, W.J.M.K., Brandwijk, R. J. Storm, G. Chin, P. T. K. Strijkers, G. J. de Mello Donega, C. Nicolay, K. Griffioen, A. W., 2006. Quantum dots with a paramagnetic coating as a bimodal molecular imaging probe. Nano Letters 6, 1–6.

Ntziachristos, V., Bremer, C., Weissleder, R., 2003. Fluorescence imaging with near-infrared light: new technological advances that enable in vivo molecular imaging. Eur Radiol 13, 195–208.

Ntziachristos, V., Schellenberger, E.A., Ripoll, J., Yessayan, D., Graves, E., Bogdanov, A., Josephson, L., Weissleder, R., 2004. Visualization of antitumor treatment by means of fluorescence molecular tomography with an annexin V-Cy5.5 conjugate. Proc Natl Acad Sci U S A 101, 12294–12299.

Patri, A.K., Myc, A., Beals, J., Thomas, T.P., Bander, N.H., Baker, J.R., 2004. Synthesis and in vitro testing of J591 antibody-dendrimer conjugates for targeted prostate cancer therapy. Bioconjug Chem 15, 1174–1181.

Pinaud, F., et al., 2006. Advances in fluorescence imaging with quantum dot bio-probes. Biomaterials 27, 1679–1687.

Quintana, A., Raczka, E., Piehler, L., Lee, I., Myc, A., Majoros, I., Patri, A.K., Thomas, T., Mule, J., Baker, J.R., 2002. Design and function of a dendrimer-based therapeutic nanodevice targeted to tumor cells through the folate receptor. Pharm Res 19, 1310–1316.

Schellenberger, E.A., Sosnovik, D., Weissleder, R., Josephson, L., 2004. Magneto/optical annexin V, a multimodal protein. Bioconjug Chem 15, 1062–1067.

Smith, A., Ruan, G., Rhyner,M.N., Nie, S.M., 2006. Engineering luminescent quantum dots for in vivo molecular and cellular imaging. Ann Biomed Eng 34, 1–12.

Smith, A.M., Gao, X.H., Nie, S.M., 2004. Quantum dot nanocrystals for in vivo molecular and cellular imaging. Photochem Photobiol 80, 377–385.

Stroh, M., Zimmer, J.P., Duda, D.G., Levchenko, T.S., Cohen, K.S., Brown, E.B., Scadden, D.T., Torchilin, V.P., Bawendi, M.G., Fukumura, D., Jain, R.K., 2005. Quantum dots spectrally distinguish multiple species within the tumor milieu in vivo. Nat Med 11, 678–682.

Torchilin, V., Babich, J., Weissig, V., 2000. Liposomes and micelles to target the blood pool for imaging purposes. J Liposome Res 10, 483–499.

Torchilin, V.P., Trubetskoy, V.S., Milshteyn, A.M., Canillo, J., Wolf, G.L., Papisov, M.I., Bogdanov, A.A., Narula, J., Khaw, B.A., Omelyanenko, V.G., 1994. Targeted delivery of diagnostic agents by surface-modified liposomes. J Control Release 28, 45–58.

Wang, D.S., He, J.B., Rosenzweig, N., Rosenzweig, Z., 2004. Superparamagnetic Fe2O3 Beads-CdSe/ZnS quantum dots core-shell nanocomposite particles for cell separation. Nano Letters 4, 409–413.

Wu, X.Y., Liu, H.J., Liu, J.Q., Haley, K.N., Treadway, J.A., Larson, J.P., Ge, N.F., Peale, F., Bruchez, M.P., 2003. Immunofluorescent labeling of cancer marker Her2 and other cellular targets with semiconductor quantum dots. Nat Biotechnol 21, 41–46.

Xu HX, S.M., Wong, E.Y., et al., 2003. Multiplexed SNP genotyping using the Qbead (TM) system: a quantum dot-encoded microsphere-based assay. Nucleic Acids Res 31, e43.

20
Investigating the Dynamics of Cellular Processes at the Single Molecule Level with Semiconductor Quantum Dots

Maxime Dahan

Abstract: The unique optical properties of semiconductor quantum dots (QDs) make them a powerful tool for ultrasensitive biological detection. Thanks to their large extinction coefficient and photostability, QDs can be detected at the single molecule level with high signal-to-noise ratio. This enables new experiments in which individual QD-labeled biomolecules are tracked with nanometer accuracy as they move in their natural cellular habitats. The ability to observe single biomolecules is essential to account for the stochastic nature of biological processes and to obtain information that remains elusive for biochemical, genetic, and conventional imaging methods. In living cells, single QD imaging has already been used to unravel the diffusion properties of membrane receptors as well as to analyze the motion of intracellular molecular motors. When combined with new optical and biochemical techniques, QDs will contribute to advanced biological imaging at the molecular scale and allow new studies on the dynamics of cellular processes.

20.1. Introduction

Imaging techniques are central in the development of modern cell biology because they provide an invaluable tool to decipher the complex mechanisms controlling the organization of live cells. The ability to observe biological processes in real time is essential for determining their spatial and temporal dynamics. Important mechanisms such as molecular diffusion, gene expression, protein–protein interactions, and cell signaling can now be detected optically (Lippincott-Schwartz and Patterson, 2003; Shav-Tal et al., 2004; Tsien, 2005). The need to quantitatively analyze them has spurred a large effort involving techniques from molecular and cell biology, biochemistry, physics, and computer science.

Over the last decade, progress in biological imaging has stemmed from a combination of factors. The most important is unquestionably the advent of genetically encoded markers, such as green fluorescent protein (GFP) and its variants, which has truly revolutionized the biochemical methods required to selectively label molecules or cellular compartments and record their properties in living systems (Tsien, 2005; Giepmans et al., 2006). A multitude of fluorescent reporters and sensors can now be prepared and inserted in the genome of various organisms. Advanced chemical methods have also been developed to conjugate biomolecules (proteins, nucleic acids, sugars, etc.) to exogenous probes used as contrast agents (Chen and Ting, 2005). In parallel, optical microscopes have significantly gained in sensitivity and dynamics due to the increased quality of their mechanical and opto-electronic components. New techniques (such as two-photon microscopy, evanescent wave microscopy, structured light illumination, second-harmonic generation, fluorescence correlation spectroscopy, etc.) have flourished. These techniques have often become standard tools in research laboratories and are fruitfully complementing more conventional approaches (such as confocal and wide-field epifluorescence microscopy). Finally, the acquisition and processing of large amounts of data have been made possible by elaborate image analysis methods (Meijering et al., 2006) and the ever enhanced capabilities of computers.

The emergence of advanced inorganic nanomaterials adds new elements to the toolbox with which biological systems can be investigated (Alivisatos, 2004). These nanomaterials, with their controlled size, shape, and composition, possess physical properties (optical, electrical, magnetic, etc.) which are often far superior to those of their organic counterparts, opening up fascinating prospects for ultrasensitive sensing and detection in fundamental and applied biomedical research.

20.2. Quantum Dots as Fluorescent Probes for Single Molecule Experiments

Among the new nanomaterials which can be fruitfully employed in biology, semiconductor quantum dots (QDs) are certainly the most prominent example (Michalet et al., 2005). QDs are colloidal nanoparticles with a diameter comprised between 2 and 10 nm. They are composed of different semiconductor materials, the most frequently used being made of a CdSe core capped by a ZnS shell (Figure 20.1). Following their synthesis in organic solvent, they can be chemically modified for water-solubilization and further functionalization. The interest of QDs for biological imaging stems from a combination of unique photophysical properties (Bruchez et al., 1998; Chan and Nie, 1998; Michalet et al., 2005). First, they possess both a large absorption and a narrow emission spectrum whose peak position is determined by the size of the semiconductor core and can be precisely adjusted during their synthesis by inorganic chemistry. As a consequence, QD samples with distinguishable emission wavelength can be excited with a single laser line, which make them ideal for multicolor detection (Wu et al., 2003). Second, their photostability, far superior to that of conventional fluorophores (fluorescent proteins or organic dyes), overcomes the traditional limitation of photodegradation in fluorescent labeling. Finally, the synthesis and

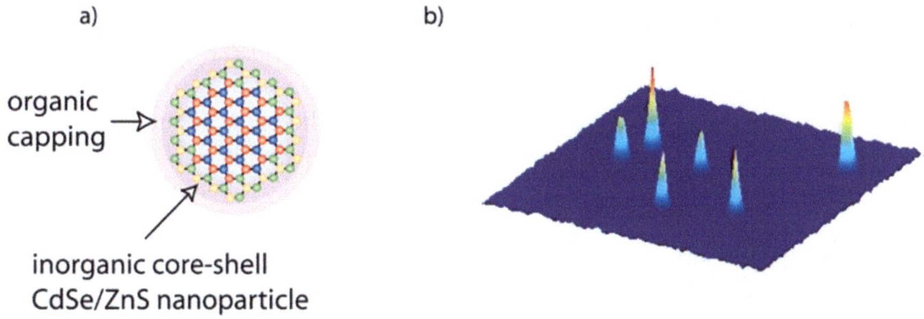

FIGURE 20.1. **a** Structure of a CdSe/ZnS core-shell semiconductor quantum dot surrounded by an organic matrix (in pink) allowing its solubilization and functionalization in aqueous buffers. **b** Fluorescence image of single quantum dots deposited on a glass coverslip and detected in epifluorescence.

engineering of nanoparticles with different semiconductor materials and structures have expanded the range of possible emission wavelengths from the visible, ideally suited for immunofluorescence, to the red and infra-red regions, more appropriate for imaging in tissues and animals (Kim et al., 2004). The potential of QDs as biological probes has now been demonstrated in various contexts and their applications, in vitro either in live cells or in organisms, are rapidly multiplying (Michalet et al., 2005; Medintz et al., 2005).

An attractive feature of QDs is the sensitivity with which individual nanoparticles can be detected (Nirmal et al., 1996) (Figure 20.1). The high signal-to-noise primarily comes from the extinction coefficient ε of the QDs, which is in the order of 10^6 $M^{-1}.l.cm^{-1}$ in the visible, 10 to 100 times higher than that of standard organic fluorophores (dyes or GFP) (Table 20.1). This directly translates into the number of photons *N* detected for a single emitting QD in a given acquisition time *t*:

$$N = \alpha Q(2303\varepsilon/N_A)(I\lambda/hc)\tau$$

TABLE 20.1. Figure of merit for different probes used in single molecule tracking experiments. The photostability corresponds to the typical time during which a probe can be continuously imaged in a single molecule experiment. The extinction coefficient is expressed in $M^{-1}.cm^{-1}.l$. The "stoechiometry" parameter is an indication of how easily a probe can be conjugated to a single biomolecule with a 1:1 ratio.

Probe	Size	Photostability	Extinction Coefficient	Stoechiometry
Organic fluorophores (Cy3, Cy5, Alexi, etc.)	1 nm	~1–10 s	$\sim 10^5$	Good
GFP	2–4 nm	~100 ms	$\sim 10^4$–10^5	Excellent
Quantum dots	~10–20 nm	> 20 mn	$\sim 10^6$	Average
Beads	40 nm–1 μm	∞	Scattered light	Poor

where α is the detection efficiency (∼0.1 % − 5 %), Q the emission quantum yield (c.a. 30 %–80 %), λ the QD emission wavelength, h the Planck constant ($h = 6.6310^{-34}$ J.s), c the speed of light, I the excitation power per surface unit, and N_A the Avogadro number. In other words, for the same input intensity I, one detects 10 to 100 times more fluorescence photons with a QD than with a fluorophore. Alternatively, the time t can be reduced to access faster dynamics and, using state-of-the-art intensified CCD cameras, individual QDs can now be detected with acquisition time as short as 1 ms.

With QDs, high values of N imply that the photon noise $\sigma = \sqrt{N}$ (also known as shot-noise) usually becomes the primary source of noise in the detection system. It dominates other contributions such as σ_b, the background noise (due to scattered light and cellular autofluorescence), and σ_e, the electronic noise (due to read-out and dark current). In this context, the signal-to-noise ratio (SNR) is equal to:

$$\mathrm{SNR} = \frac{N}{\sqrt{\sigma^2+\sigma_b^2+\sigma_e^2}} \approx \frac{N}{\sigma} \approx \sqrt{N}$$

In typical experimental conditions, the detected fluorescence rate can reach 10^5 s^{-1}, such that the SNR is in the order of 30 for $\tau \sim 10$ ms. The photostability of QDs also means that their fluorescence emission can be recorded over long periods of time. For single organic emitters, a useful parameter is the average number N_p of detected photons before photobleaching. N_p is in the order of 10^5 for a fluorescent protein, and increases up to 10^6 for a good dye. In comparison, 10^8–10^9 photons can be routinely detected for a single QD. This number does not necessarily reflect an upper bound imposed by photodestruction but, rather, results from other physical or biological factors limiting the total duration of the experiment.

The ability to detect the fluorescence of single nanoparticles has been first exploited to analyze the photophysical properties of QDs, either at low or at room temperature. In several pioneering experiments, M. Bawendi and collaborators at MIT have amply demonstrated the interest of the single molecule approach (Empedocles et al., 1999; Shimizu et al., 2001). Measurements on individual QD have revealed unexpected phenomena, such as fluorescence intermittency (Nirmal et al., 1996), spectral fluctuations (Empedocles et al., 1999), or photon antibunching (Lounis et al., 2000), which could not have been observed with an ensemble of nanoparticles.

The most striking property is arguably the blinking of individual QDs: under continuous illumination, their emission randomly alternates between bright ('on') and dark ('off') periods (Figure 20.2). Although it is not fully understood, the physical origin of this phenomenon is commonly attributed to transient ionization of the nanoparticule (Shimizu et al., 2001). Upon light excitation, an electron-hole pair is created within the nanocristal. This electron-hole pair

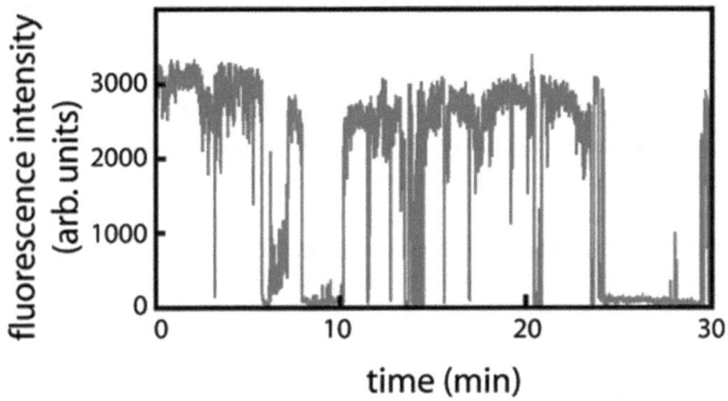

FIGURE 20.2. Intermittency in the fluorescence emission of a single QD on a glass coverslip. The intensity randomly alternates between bright ('on') and dark ('off') periods.

can relax through different competing decay channels. Most frequently, the pair recombines radiatively by emitting a fluorescence photon. However, with a low probability, the electron (or the hole) can tunnel out of the QD toward surface traps. While the QD is in this ionized state, the nonradiative recombination rate is enhanced and the fluorescence emission is dramatically quenched. At a later time, the trapped charge tunnels back into the inorganic core, and the nanoparticule resumes its regular fluorescence cycles. Both the 'on' and 'off' times are distributed according to a power-law (Shimizu et al., 2001; Brokmann et al., 2003), meaning that the mechanisms behind the blinking cannot be simply described in terms of Poisson processes. Blinking is certainly a limitation for the use of QDs as nanoscopic light source. Nevertheless, it is a convenient criterion to evaluate whether individual nanoparticles are observed since it is usually assumed that an intermittent fluorescent spot is indicative of single QD emission.

The value of the SNR has a direct implication on the accuracy with which individual spots can be localized (Bobroff, 1986; Thompson et al., 2002; Ober et al., 2004). Since QDs are much smaller than the excitation wavelength, the shape of the fluorescence spots corresponds to the point spread function (PSF) of the optical system. Its width w is given by the optical resolution of the microscope, i.e. $w \approx \lambda/(2^*NA)$ where NA is the objective numerical aperture. For nanoparticles emitting around 600 nm and an oil-immersion objective with $NA = 1.3$, this corresponds to about 230 nm. However, by fitting the PSF with a two-dimensional Gaussian curve, its center can be pinpointed with a much higher accuracy, on the order of w/SNR when the detection sensitivity is limited by the photon noise (Thompson et al., 2002). For SNR close to 30, as frequently observed with individual QD-tagged membrane receptors, the accuracy can be as low as 5–10 nm. With further optimization of the experimental parameters (excitation intensity, optical filters, acquisition time, etc.), one can reasonably expect to lower the localization accuracy down to 1 nm, comparable to what has recently been achieved in vitro with myosin motors (Yildiz et al., 2003).

20.3. Single Molecule Microscopy and Imaging in Biology

A large effort has been devoted since the late 1980s to the development of optical techniques for the detection of individual fluorophores (Moerner and Orrit, 1999). Compared to conventional methods, measurements at the single molecule level enable the direct determination of molecular properties without averaging over a large number of molecules. Therefore, they permit the identification of distinct molecular subpopulations as well as the analysis of the kinetic parameters controlling their dynamics (Weiss, 1999).

Initially restricted to the study of single emitters at low temperature, these experiments were subsequently extended to measurements at room temperature, opening up exciting prospects for applications to biological systems. This eventually led to the first experiments in which the activity of single enzymes or molecular motors was investigated in vitro by optical means (Lu et al., 1998; Ha et al., 1999; Noji et al., 1997). Soon after, several groups reported the detection of single fluorescently labeled biomolecules in live cells. In most cases, they measured the diffusive movement of individual membrane proteins tagged with either an organic dye or a GFP (Schutz et al., 2000; Harms et al., 2001; Ueda et al., 2001; Iino et al., 2001). In collaboration with A. Triller's laboratory (Biology Department, Ecole normale), our group demonstrated in 2003 that single membrane receptors could be tracked using QDs, expanding the scope of single molecule experiments in live cells (Dahan et al., 2003).

There is now a large choice of probes (fluororophores, GFPs, QDs, beads) to observe single molecule in cells. Selecting the right one depends on a number of physical and biochemical parameters that should be evaluated for each system. Important factors are (i) the photostability, (ii) the extinction coefficient, (iii) the size, (iv) the stoechiometry with which the probe can be bioconjugated to a biomolecule of interest (ideally, it should be achieved in a 1:1 ratio). A figure of merit, based on the value of these parameters for the different probes, is summarized in Table 20.1. For single molecule experiments, CdSe/ZnS colloidal QDs provide a very favorable compromise with their relatively small size (hydrodynamic radius 10–15 nm, including the surface organic groups) and remarkable photostability and brightness in the visible region.

20.4. Tracking Single Membrane Receptors Using QDs

The use of QDs for single molecule tracking in live cells was first demonstrated for glycine receptors diffusing in the membrane of cultured neurons (Dahan et al., 2003). In this experiment, QD labeling was achieved with standard immunochemistry techniques. A primary antibody was used to specifically target the membrane protein and the fluorescent labeling was obtained by means of QDs functionalized with streptavidin or Fab fragments. The density of membrane labeling could be controlled with the incubation time and the reagent concentration. When the density was lowered to the point where the typical distance between individual

QDs exceeded the spatial resolution, individual tagged antigens (corresponding to only a small fraction of the total population of antigens) could be distinguished optically and their positions determined with nanometer accuracy. To evaluate the diffusion dynamics of glycine receptors, sequences of hundreds of images were acquired and the trajectories of single proteins were analyzed over durations as long as 20 minutes.

Tracking single membrane proteins is a powerful tool to understand the mechanisms that control and regulate its lateral dynamics. This has been previously illustrated in experiments, known as single particle tracking (SPT), in which a micron-sized latex bead or a 40-nm gold nanoparticle was specifically coupled to proteins and detected by differential interference contrast microscopy. Using SPT, several questions have been addressed regarding the organization of cell membranes and the diffusion of lipids and membrane proteins (Saxton and Jacobson, 1997).

In practice, each trajectory $(x(t),y(t))$ is analyzed by computing the mean square displacement $\rho(t)$ (MSD) defined as:

$$\rho(\boldsymbol{n}\tau) = \frac{1}{\boldsymbol{N}-\boldsymbol{n}} \sum_{i=1}^{N-n} \left[(\boldsymbol{x}((\boldsymbol{i}+\boldsymbol{n})\tau) - \boldsymbol{x}(\boldsymbol{i}\tau))^2 + (\boldsymbol{y}((\boldsymbol{i}+\boldsymbol{n})\tau) - \boldsymbol{y}(\boldsymbol{i}\tau))^2 \right]$$

where τ is the acquisition time and N the total number of point in the trajectory. Different types of motion can be distinguished based on the temporal dependence of $\ddot{r}\ddot{t}$ (Saxton and Jacobson, 1997). For a simple two-dimensional Brownian motion (with diffusion constant D), $\rho(t)$ varies as $4Dt$. A deviation to this simple linear behavior is an evidence for more complex molecular processes. When $\rho(t)$ grows as $4Dt + v^2t^2$, the dynamics results from a combination of Brownian diffusion and directed motion (with a velocity v). If $\rho(t)$ tends toward a constant value L^2, the motion is confined in a domain with size in the order of L.

For single QD tracking, the intermittency raises several difficulties. Since the nanoparticles can transiently disappear from the fluorescence images, specific algorithms are required to track individual QD-tagged molecules (Bonneau et al., 2004, 2005). In brief, the algorithm can be separated into two parts. First, all the visible spots within a sequence of images are detected using conventional cross-correlation with a Gaussian model for the point spread function (PSF) of the microscope. Second, these spots are associated to reconstruct spatial trajectories of individual proteins. For that purpose, a model of the spatial dynamics (diffusive motion, directed transport, etc.) is used to provide an estimate of the possible location of the molecules, either in consecutive images or after an 'off' period.

In a first approximation, the cell membrane is described as a homogeneous viscous fluid in which lipids and proteins diffuse by Brownian motion. This description is at the core of the fluid-mosaic model introduced by Singer and Nicolson (1972) and has proven extremely rich to understand some of the biological and biophysical properties of the plasma membrane. However, it does not account for the large variety of behavior experimentally measured with membrane molecules and it is now widely acknowledged as too simple.

More elaborate models have thus been proposed and they often imply a compartmentalization of the membrane into micro-domains (Murase et al., 2004). Understanding the nature, the organization, and the dynamics of these micro-domains is one of the major challenges in membrane biophysics and serves as a strong incentive for the development of many optical and biophysical techniques (Marguet et al., 2006).

Compartmentalization of the plasma membrane can also occur on a larger scale. It is for instance the case in neurons in which synapses, either inhibitory or excitatory, can be viewed as specialized membrane subdomains with a typical size of 1 μm. Our experiments on single glycine receptors aimed at understanding how these proteins move in the membrane and reach their synaptic localization (Dahan et al., 2003). Our results, along with other works on AMPA and NMDA receptors (Tardin et al., 2003; Groc et al., 2004) have revealed that, once inserted in the plasma membrane, receptors for neurotransmitters are far more mobile than previously considered. They do not simply stay at synapses where they are known to be stabilized by scaffolding proteins. Synaptic receptors can move and exit the synaptic compartments, upon which they diffuse freely in the extrasynaptic domain (Figure 20.3). Therefore, the spatial distribution of receptors in the neuronal membrane appears to be governed by a dynamic

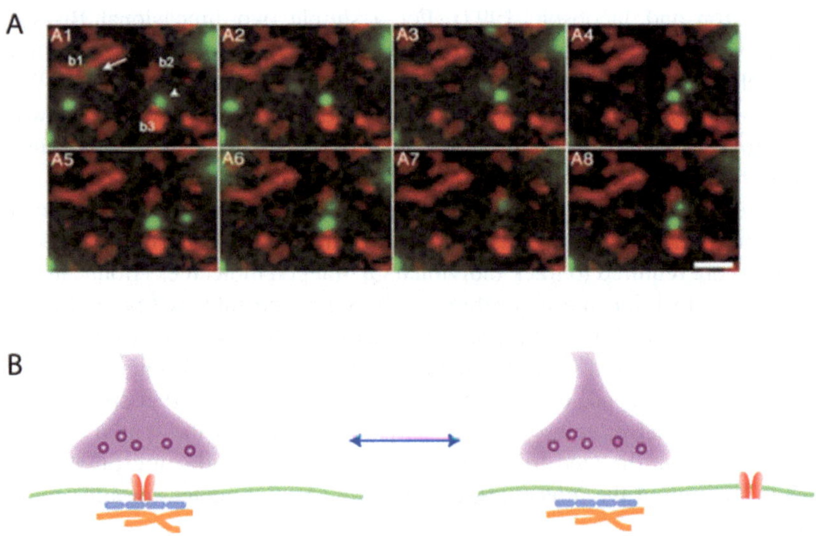

FIGURE 20.3. **A** Motion of individual glycine receptors diffusing in the membrane of live neurons. Reproduced from Dahan et al. (2003). Red spots correspond to synaptic boutons labeled with FM 4-64. One QD (arrow), initially located in bouton b1 diffuses in the extrasynaptic domain and then associates with bouton b2. A second QD (arrowhead) diffuses while remaining associated with bouton b3. Scale bar is 2 μm. **B** Dynamic equilibrium between synaptic and extrasynaptic receptors, as revealed by single molecule experiments.

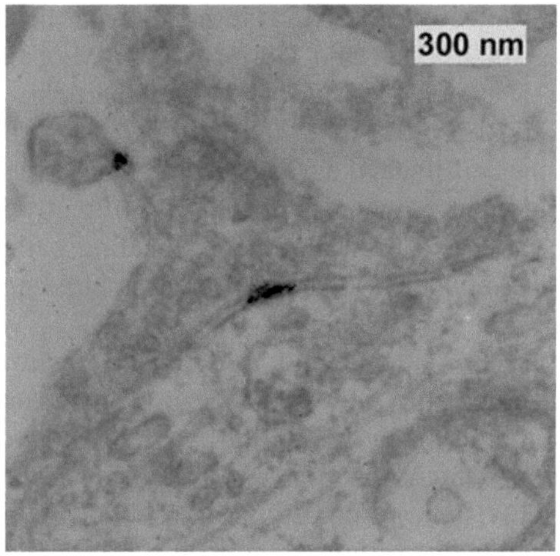

FIGURE 20.4. Detection of QDs by transmission electron microscopy EM detection. Gold toning and silver intensification of QDs on fixed cells following labeling of glycine receptors on live neurons. Reproduced from Dahan et al. (2003). The nanoparticle is located within the synaptic cleft.

equilibrium (Choquet et al., 2003). The biological mechanisms which regulate the properties of this equilibrium are currently the subject of active research.

Experiments on the dynamics of membrane receptors established the potential of QDs as ultrasensitive fluorescent probes for cell biology. In addition, QDs could also be detected in transmission electron microscopy (TEM) after silver amplification (Figure 20.4), thus providing information on the ultrastructural localization of the tagged molecules (Dahan et al., 2003). It was later shown that QDs could be in fact detected directly by TEM, with no need for further amplification (Giepmans et al., 2005).

20.5. Watching Single Intracellular Motors at Work

The high sensitivity achieved in single QD imaging of membrane proteins suggests that QDs could be employed to accomplish a more challenging task: tracking the motion of individual intracellular proteins. Reaching single-molecule sensitivity in the detection of biomolecules within live cells raises several difficulties. First, QDs have to enter the cell cytoplasm (without being trapped in endosomal compartments) and reach their molecular target. Second, the fluorescence of individual QDs has to be detected in a noisy environment due to the autofluorescence of intracellular compartments and organelles. Finally, the motion in the cytosol is likely to be three-dimensional, compared to bi-dimensional diffusion in the membrane. It makes single molecule tracking more complex than in membrane experiments and will likely require new computational methods to extract and analyze trajectories from imaging data.

Several approaches have been tested to insert QDs in the cytoplasm of living cells (Derfus et al., 2004; Chen and Gerion, 2004; Michalet et al., 2005). A major challenge is to avoid sequestering the nanoparticles in the endocytotic pathway. Dubertret et al. (2002) microinjected QDs in Xenopus embryos and successfully traced the cell-lineage during development. Lidke et al. (2004) also used QD-tagged epidermal growth factor (EGF) to directly image the internalization of EGF receptors and the subsequent signaling pathways. However, in these experiments, the number of detected QDs was large and far exceeded the single molecule limit.

In collaboration with G. Cappello and Y. Bellaiche (Institut Curie), our group has recently demonstrated that QDs could be used to measure the activity of single molecular motors in the cytoplasm of live cells (Courty et al., 2006). We studied kinesin motors, which play a central role in many biological processes, such as motility, cell division or intracellular transport. The properties of kinesin molecules have been abundantly investigated by single-molecule in vitro experiments (Howard, 2001). By means of either micromanipulation techniques or fluorescence assays, important quantities such as the velocity along microtubules, the processivity, and the stall force have been determined, thus providing a detailed view of the motor biochemical and mechanical cycles. Relating these in vitro findings to the properties and functional role of individual motors in living cells is an important issue which requires methods to visualize single kinesins in vivo.

Our approach consisted in expressing and purifying biotynilated kinesins in vitro and coupling them to streptavidin-coated QDs. The velocity and processivity of these constructs were first tested by in vitro motility assays to ensure that single kinesins were measured. The constructs were then internalized in the cytoplasm of live Hela cells with a cell-loading technique based on the osmotic lysis of pinocytic vesicles (Courty et al., 2006; Okada and Rechsteiner, 1982). Individual spots displayed phases of directed motion (Figure 20.5) that could be abolished by depolymerizing microtubules. Analysis of these

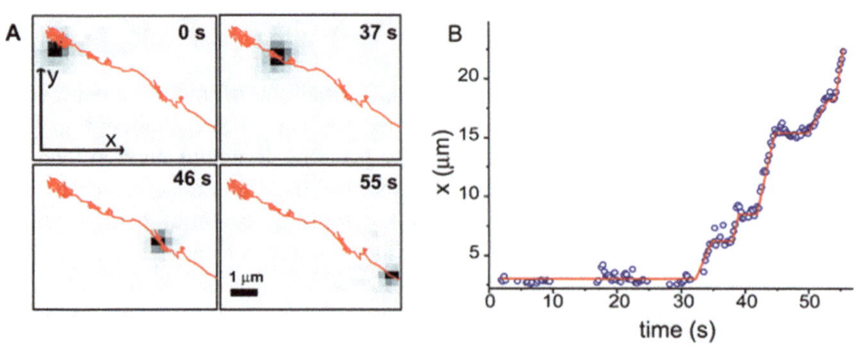

FIGURE 20.5. **A** Motion of an individual kinesin motor labeled with a QD in the cytoplasm of a living Hela cell. Reproduced from Courty et al. (2006). **B** Position along the *x*-axis versus time, indicating a succession of phases of directed and diffusive motion.

movements enabled, for the first time, the characterization of the in vivo motion of individual QD-tagged kinesins. Interestingly, the velocity ($v \sim 0.5\,\mu$m/s) and processivity (2 s) of the constructs was similar to those measured in vitro. This observation indicates that their motility in live cells is not affected by the presence of microtubule-associated proteins and by the traffic of other kinesins or dyneins motors competing for binding sites to microtubules. In addition, we frequently measured that QD-tagged kinesins, after a phase of processive motion, diffused in the close proximity of the microtubules before attaching again and resuming their directed motion. Assuming a pure hydrodynamic regime for the friction, the value D of the diffusion coefficient provided an estimate of the viscosity $h = kT/(6\pi RD) \sim 1.6$ Pas in the local environment of microtubules (with $R \sim 15$nm, the QD hydrodynamic radius). The force necessary to carry a QD was then $6\pi hRv \sim 0.6$ pN, well below the stall force $\sim$ 6 pN of the motor.

These experiments demonstrated that intracellular proteins could be studied at the single molecule level in live cells. They also underlined several aspects that will be critical for future single QD measurements. First, the cytosol is a complex medium in which the mobility of an object critically depends on its size and its specific interactions with the environment (Verkman, 2002). To improve our ability to use QDs as an intracellular probe, it will be thus necessary to account for the colloidal properties of the nanoparticles as well their optical and biochemical properties. In that respect, using the right QD solubilization procedures might be a key element since they lead to nanoparticles with markedly different surface charge and/or hydrodynamic radius (Michalet et al., 2005; Medintz et al., 2005). Second, only $\sim$5 %–10 % of the internalized QD-kinesin constructs displayed a directed motion. This is likely due to the fact that only a small fraction of the kinesins expressed in vitro has a motor activity. In addition, we had to use truncated motors, for which important regulations might not operate. In many situations, it would thus be preferable to label endogeneous molecules possessing an appropriate epitope (e.g., GFP, myc-tag, his-tag, etc.) which can be targeted by functionalized QDs. This approach, although promising, still remains to be validated by experiments.

20.6. Conclusions and Perspectives

QDs are a new class of inorganic fluorophores with unique optical properties. When detected at the single molecule level, they offer exciting possibilities to visualize the motion of individual biomolecules in their cellular environment. Although single QD tracking is still a technique in its infancy and several experimental difficulties remain to be solved, the potential of QDs for ultrasensitive imaging is fast becoming clear. One gets ever closer to a long-term goal, first envisioned when QDs entered the realm of cell biology: being able to observe, in real time, in 3D and at the single molecule level, the motion of different biomolecules labeled with QDs with distinct emission wavelengths (Figure 20.6).

FIGURE 20.6. Principle of a molecular global positioning system. The motions of single biomolecules, corresponding to different molecular species, specifically labeled with QDs having distinct emission colors (here blue, red, green, or brown) and positioned either in the plasma membrane, the cytosol, or the cell nucleus, are recorded in real time and in 3D.

Such a molecular positioning and tracking system will allow detailed investigation of a multitude of biological processes, such as cell migration, cell division, or signal transduction.

This approach will be particularly fruitful for quantitatively analyzing, both in time and in space, the stochasticity of biological processes. The stochasticity, which is not well accounted for by conventional methods and which might be an important operating principle of molecular networks, first results from the nature of the biochemical interactions. Even at equilibrium, biomolecules associate and dissociate randomly and single molecule experiments provide a way to determine *in situ* the kinetic parameters controlling such dynamics. Another stochastic process, Brownian diffusion, decisively contributes to the dynamics of cellular responses (Bray, 1998). Diffusion is the main transport mechanism ensuring the encounters between molecular partners. Controlling the diffusion rates, rather than the binding constants, might be a simple way to regulate the interactions between two biomolecules. Single QD imaging will thus be an important tool for determining how biophysical constraints, such as molecular crowding or diffusion, affect the spatiotemporal organization of molecular networks (Kholodenko, 2006).

Making single QD imaging a standard technique in cell biology will continue to require a multidisciplinary approach. New optical, biochemical, biological, and computer processing methods are needed to functionalize nanoparticles, label biomolecules, and track them in a cellular or multicellular environment. Nevertheless, one can safely expect that these small fluorescent nanocrystals will contribute to enhance our understanding of the cellular organization at a truly molecular scale.

Acknowledgements. I am indebted to Antoine Triller for many valuable discussions and for the support provided by his laboratory. I thank all the members of the "Optics and Biology" group at Ecole normale supérieure, as well as Laurent Cohen, Christophe Tribet, Valérie Marchi-Artzner, Giovanni Cappello, and Yohanns Bellaiche for their contributions and fruitful discussions.

References

Alivisatos, P., 2004. The use of nanocrystals in biological detection. Nat Biotechnol 22, 47–52.

Bobroff, N., 1986 Position measurement with a resolution and noise limited instrument. Rev Sci Instrum 57, 1152–1157.

Bonneau, S., Cohen, L., Dahan, M., 2004. A multiple target approach for single quantum dot tracking, Proceedings of the IEEE International Symposium on Biological Imaging (ISBI 2004), p. 664.

Bonneau, S., Dahan,M., Cohen, L.D., 2005. Single quantum dot tracking based on perceptual grouping using minimal paths in a spatiotemporal volume. IEEE Trans Image Process 14, 1384–1395.

Bray, D., 1998. Signaling complexes: biophysical constraints on intracellular communication. Annu Rev Biophys Biomol Struct 27, 59–75.

Brokmann, X., Hermier, J.P., Messin, G., Desbiolles, P., Bouchaud, J.P., Dahan, M., 2003. Statistical aging and nonergodicity in the fluorescence of single nanocrystals. Phys Rev Lett 90, 120601–120604.

Bruchez, M., Moronne, M., Gin, P., Weiss, S., Alivisatos, A.P., 1998. Semiconductor nanocrystals as fluorescent biological labels. Science 281, 2013–2016.

Chan W.C., Nie, S., 1998. Quantum dot bioconjugates for ultrasensitive nonisotopic detection. Science 281, 2016–2018.

Chen, F., Gerion, D., 2004. Fluorescent CdSe/ZnS nanocrystal-peptide conjugates for long-term, nontoxic imaging and nuclear targeting in living cells. NanoLett 4, 1827–1832.

Chen, I., Ting, A.Y., 2005. Site-specific labeling of proteins with small molecules in live cells. Curr Opin Biotechnol 16, 35–40.

Choquet, D., Triller, A., 2003. The role of receptor diffusion in the organization of the postsynaptic membrane. Nat Rev Neurosci 4, 251–65.

Courty, S., Luccardini, C., Bellaiche, Y., Cappello, G., Dahan, M., 2006. Tracking individual kinesin motors in living cells using single quantum dot imaging. NanoLett 6, 1491–1495.

Dahan, M., Levi, S., Luccardini, C., Rostaing, P., Riveau, B., Triller, A., 2003. Diffusion dynamics of glycine receptors revealed by single-quantum dot tracking. Science 302, 442–445.

Derfus, A.M., Chan, W.C.W., Bhatia, S.N., 2004. Intracellular delivery of quantum dots for live cells experiments and organelles tracking. Adv Mat 16, 961–966.

Dubertret, B., Skourides, P., Norris, D.J., Noireaux, V., Brivanlou, A.H., Libchaber, A., 2002. In vivo imaging of quantum dots encapsulated in phospholipid micelles. Science 298, 1759–1762.

Empedocles, S., Bawendi, M., 1999. Spectroscopy of Single CdSe Nanocrystallites. Acc Chem Res 32, 389–396.

Giepmans, B.N., Adams, S.R., Ellisman, M.H., Tsien, R.Y., 2006. The fluorescent toolbox for assessing protein location and function. Science 312, 217–224.

Giepmans, B.N., Deerinck, T.J., Smarr, B.L., Jones, Y.Z., Ellisman, M.H., 2005. Correlated light and electron microscopic imaging of multiple endogenous proteins using Quantum dots. Nat Methods 2, 743–749.

Groc, L., Heine, M., Cognet, L., Brickley, K., Stephenson, F.A., Lounis, B., Choquet, D., 2004. Differential activity-dependent regulation of the lateral mobilities of AMPA and NMDA receptors. Nat Neurosci 7, 695–696.

Ha, T., Ting, A.Y., Liang, J., Caldwell, W.B., Deniz, A.A., Chemla, D.S., Schultz, P.G., Weiss, S., 1999. Single-molecule fluorescence spectroscopy of enzyme conformational dynamics and cleavage mechanism. Proc Natl Acad Sci USA 96, 893–898.

Harms, G.S., Cognet, L., Lommerse, P.H., Blab, G.A., Kahr, H., Gamsjager, R., Spaink, H.P., Soldatov, N.M., Romanin, C., Schmidt,T., 2001. Single-molecule imaging of l-type Ca(2+) channels in live cells. Biophys J 81, 2639–2646.

Howard, J., 2001. Mechanics of motor proteins and the cytoskeleton, Sinauer.

Iino, R., Koyama, I., Kusumi, A., 2001. Single molecule imaging of green fluorescent proteins in living cells, E-cadherin forms oligomers on the free cell surface. Biophys J 80, 2667–2677

Kim, S., Lim, Y.T., Soltesz, E.G., De Grand, A.M., Lee, J., Nakayama A., Parker, J.A., Mihaljevic, T., Laurence, R.G., Dor, D.M., Cohn, L.H., Bawendi, M.G., Frangioni, J.V., 2004. Near-infrared fluorescent type II quantum dots for sentinel lymph node mapping. Nat Biotechnol 22, 93–97.

Kholodenko, B.N., 2006. Cell-signalling dynamics in time and space. Nat Rev Mol Cell Biol 7, 165–176.

Lidke, D.S., Nagy, P., Heintzmann, R., Arndt-Jovin, D.J., Post, J.N., Grecco, H.E., Jares-Erijman, E.A., Jovin, T.M., 2004. Quantum dot ligands provide new insights into erbB/HER receptor-mediated signal transduction. Nat Biotechnol 22, 198–203.

Lippincott-Schwartz, J., Patterson, G.H., 2003. Development and use of fluorescent protein markers in living cells. Science 300, 87–91.

Lounis, B., Bechtel, H.A., Gerion, D., Alivisatos, P., Moerner, W.E., 2000. Photon antibunching in single quantum dot fluorescence. Chem Phys Lett 329, 399–406.

Lu, H.P., Xun, L., Xie, X.S., 1998. Single-molecule enzymatic dynamics. Science 282, 1877–1882.

Marguet, D., Lenne, P.F., Rigneault, H., Tao, H.E., 2006. Dynamics in the plasma membrane, how to combine fluidity and order. EMBO J 25, 3446–3457.

Medintz, I.L., Uyeda, H.T., Goldman, E.R., Mattoussi, H., 2005. Quantum dot bioconjugates for imaging, labeling and sensing. Nat Materials 4, 435–446.

Meijering, E., Smal, I., Danuser, G., 2006. Tracking in molecular bioimaging. IEEE Signal Processing Magazine 23, 46–53.

Michalet, X., Pinaud, F.F., Bentolila, L.A., Tsay, J.M., Doose, S., Li, J.J., Sundaresan, G., Wu, A.M., Gambhir, S.S., Weiss, S., 2005. Quantum dots for live cells, in vivo imaging, and diagnostics. Science 307, 538–544.

Moerner, W.E., Orrit, M., 1999. Illuminating single molecules in condensed matter. Science 283, 1670–1676.

Murase, K., Fujiwara, T., Umemura, Y., Suzuki, K., Iino, R., Yamashita, H., Saito, M., Murakoshi, H., Ritchie, K., Kusumi, A., 2004. Ultrafine membrane compartments for molecular diffusion as revealed by single molecule techniques. Biophys J. 86, 4075–4093.

Nirmal, M., Dabbousi, B.O., Bawendi, M.G., Macklin, J.J., Trautman, J.K., Harris, T.D., Brus, L.E., 1996. Fluorescence intermittency in single cadmium selenide nanocrystals. Nature 383, 802–804.

Noji, H., Yasuda, R., Yoshida, M., Kinosita, K. Jr., 1997. Direct observation of the rotation of F1-ATPase. Nature 386, 299–302.

Ober, R.J., Ram, S., Ward, E.S., 2004. Localization accuracy in single-molecule microscopy. Biophys J 86, 1185–1200.

Okada, C.Y., Rechsteiner, M., 1982. Introduction of macromolecules into cultured mammalian cells by osmotic lysis of pinocytic vesicles. Cell 29, 33–41.

Saxton, M.J., Jacobson, K., 1997. Single-particle tracking, applications to membrane dynamics. Annu Rev Biophys Biomol Struct 26, 373–399.

Schutz, G.J., Kada, G., Pastushenko, V.P., Schindler, H., 2000. Properties of lipid microdomains in a muscle cell membrane visualized by single molecule microscopy. EMBO J 19, 892–901.

Shav-Tal, Y., Singer, R.H., Darzacq, X., 2004. Imaging gene expression in single living cells. Nat Rev Mol Cell Biol 5, 855–861.

Shimizu, K.T., Neuhauser, R.G., Leatherdale, C.A., Empedocles, S.A., Woo, W.K., Bawendi, M.G., 2001. Blinking statistics in single semiconductor nanocrystal quantum dots. Phys Rev B 63, 205311–205316.

Singer, S.J., Nicolson, G.L., 1972. The fluid mosaic model of the structure of cell membranes. Science 175, 720–731.

Tardin, C., Cognet, L., Bats, C., Lounis, B., Choquet, D., 2003. Direct imaging of lateral movements of AMPA receptors inside synapses. EMBO J. 22, 4656–4665.

Thompson, R.E., Larson, D.R., Webb, W.W., 2002. Precise nanometer localization analysis for individual fluorescent probes. Biophys J 82, 2775–2783.

Tsien, R.Y., 2005. Building and breeding molecules to spy on cells and tumors. FEBS Lett 579, 927–932.

Ueda, M., Sako, Y., Tanaka, T., Devreotes, P., Yanagida, T., 2001. Single-molecule analysis of chemotactic signaling in Dictyostelium cells. Science 294, 864–867.

Yildiz, A., Forkey, J.N., McKinney, S.A., Ha, T., Goldman, Y.E., Selvin, P.R., 2003. Myosin V walks hand-over-hand, single fluorophore imaging with 1.5-nm localization. Science 300, 2061–2065

Verkman, A., 2002. Solute and macromolecule diffusion in cellular aqueous compartments. Trends Biochem Sci 27, 27–33.

Weiss, S., 1999. Fluorescence spectroscopy of single biomolecules. Science 283, 1676–1683.

Wu, X., Liu, H., Liu, J., Haley, K.N., Treadway, J.A., Larson, J.P., Ge, N., Peale, F., Bruchez, M.P., 2003. Immunofluorescent labeling of cancer marker Her2 and other cellular targets with semiconductor quantum dots. Nat Biotechnol 21, 41–46.

21 Targeting Vascular Epitopes Using Quantum Dots

Dardo E. Ferrara, Charles Glaus, and W. Robert Taylor

Abstract: Fluorescent semiconductor quantum dots (Qdots) hold great potential for both ex vivo and in vivo molecular imaging. Traditional imaging with single-photon confocal microscopy and organic fluorophores poses several challenges for the visualization of vascular tissue, including autofluorescence, fluorophore crosstalk, and photobleaching. We will review a recently developed immunohistochemical (IHC) en face method that employs quantum dot bioconjugates and two-photon excitation laser scanning microscopy (TPELSM). These techniques improve contrast resolution and allow detailed cellular structures to be imaged without the common problem of vascular autofluorescence. It is especially useful for multicolor profiling of endothelial structures with only one excitation wavelength. Moreover, this is also a promising technique for mapping the expression of endothelial proteins and their relationship to flow dynamics. In addition, a more reliable fluorescence quantitation of endothelial signals will benefit the ex vivo testing of different interventions. By providing a bridge between in vitro and in vivo studies, this method may facilitate the development of novel drugs and specific multimodal molecular imaging agents.

21.1. Introduction

Imaging of the vascular endothelium has traditionally been limited to either traditional histological and immunostaining approaches or large-scale assessments of overt pathological changes of blood vessels. This has resulted in a void in that it has been extremely challenging to contextually place the dynamic changes occurring in protein expression and activation at the level of the endothelial cell in the perspective of the macroscale of the vasculature.

Perhaps the best example of this limitation is seen in the study of the pathogenesis of atherosclerosis. Atherosclerosis remains a major health problem in the Western societies, with significant morbidity and mortality. This complex disease has been described as a chronic inflammatory process that mainly affects the intima of conduit arteries, with the earliest molecular events occurring at the level of the endothelium. Several imaging techniques are available to assess the atherosclerotic burden of blood vessels. Most of them concentrate more on the anatomical features of the advanced stages of the disease and assess luminal diameter, degree of vascular stenosis, plaque volume, atherosclerotic lesion surface area, thickness of the fibrous cap, or amount of lipid and calcium content. However, none of the available conventional imaging techniques has been able to accurately characterize the dynamic biological properties of diseased vascular walls. This is especially important for evaluation of the vascular wall prior to the development of overt vascular lesions. At the earliest stages of the atherosclerotic disease process, there are dynamic and spatially discrete changes in protein expression and activation that have been difficult to characterize in intact vessels. Thus, the development of enabling imaging technologies to fill this void is an essential element to our further understanding of the disease process.

Novel molecular imaging agents and technologies are rapidly evolving into a very promising field in the basic biological sciences that will help us to better elucidate the fundamental pathogenic mechanisms, develop new pharmacological approaches, and optimize molecular probes (Jaffer and Weissleder, 2004). One of these technologies, biocompatible semiconductor quantum dots (Qdots), has proved to be extremely useful in a broad range of imaging applications.

In this chapter, we will primarily focus on a critical assessment of the utility and applications of Qdots in biological imaging, with a succinct introduction to their optical properties, followed by a discussion of recently developed applications for ex vivo imaging of blood vessels. We will use the example of atherosclerosis as a prototypical situation in which Qdots have proven useful in gaining further insights into the underlying disease process. Finally, we will briefly summarize initial attempts at using these probes for in vivo imaging of small animals.

21.2. The Biology of Atherosclerosis

Atherosclerosis is a multifocal chronic inflammatory disease of medium-sized and large arteries. There has been a tremendous increase in our knowledge of the mechanisms of atherosclerosis during the last two decades and extensive reviews summarizing these data have been recently published (Falk, 2006; Libby, 2002).

The earliest stage of atherosclerosis, the fatty streak, is a pure inflammatory lesion consisting of monocyte-derived macrophages and T lymphocytes. However, it is also known that molecular events at the level of the endothelium contribute significantly to this intimal cellular accumulation. The endothelium is continuously exposed to numerous atherogenic factors, such as hyperlipidemia (Khan et al., 1995), the diabetic milieu, and altered patterns of fluid shear stress (Nerem et al., 1998). The expression of adhesion molecules, such as vascular cell adhesion molecule-1 (VCAM-1) and leukocyte recruitment, play an important role in early atherogenesis at lesion-prone areas of the vascular tree. The initial plaque localization is tightly related to areas exposed to low and oscillatory shear stress where inflamed and dysfunctional endothelial cells (ECs) are present. Understanding these events and how the endothelium responds to different interventions will help to identify new approaches to halt the development of this devastating disease.

Another key factor in early atherosclerosis is nitric oxide (NO), a potent vasodilator and anti-atherogenic agent that is produced by the endothelium. The dysregulation of its enzymatic source, endothelial nitric oxide synthase (eNOS), is thought to play an important role in endothelial dysfunction (Forstermann and Munzel, 2006) and the subsequent development of atherosclerosis. At a later stage, monocyte recruitment is enhanced by soluble mediators (chemokines) secreted by smooth muscle cells, such as monocyte chemoattractant protein-1 (MCP-1). All of these events occur at the cellular and subcellular level but our visualization of these changes on the macroscale has been limited.

21.3. Quantum Dot Nanocrystals as Novel Fluorescent Tags for Two-Photon Microscopy

Quantum dots are tiny light-emitting particles on the nanometer scale (2–10 nm) typically made up of a cadmium selenide semiconductor core surrounded by a zinc sulfide shell. This core-shell complex can be coated with various types of water-soluble polymers and linked to biological molecules such as peptides, proteins, and nucleic acids to make them useful for fluorescence immunohistochemistry (IHC) and in situ hybridization (FISH) (Chan and Nie, 1998; Smith et al., 2006). Because of their size (similar to biological macromolecules) and their unique optical properties, Qdot-bioconjugates are very suitable for molecular imaging applications that require multiplexing and highly sensitive molecular detection and profiling. In contrast to organic fluorophores that suffer from several weaknesses, such as photobleaching, spectral overlap and narrow excitation, Qdots have narrow and symmetric emissions that reduce crosstalk and facilitate multilabeling. Moreover, their broad absorption spectra with large Stokes shifts (usually larger than 100 nm) make possible the use of a single excitation wavelength for multiplexing. Finally, their extremely high resistance to photobleaching makes them ideal probes for fluorescence quantitation (Gao and Nie, 2003; Gao et al., 2005). An in-depth discussion of the synthesis and optical properties of quantum dots can be found in Chapter 22.

Imaging of quantum dots can be achieved using several different modalities including traditional fluorescent microscopy, confocal microscopy, and two-photon excitation laser scanning microscopy (TPELSM). TPELSM offers several advantages over single-photon confocal microscopy (Zipfel et al., 2003). Two optical features improve the signal-to-background ratio and 3D spatial resolution. The first feature, "localization of the excitation volume," is related to the principle of two-photon excitation and has been described elsewhere

(Denk et al., 1990; Rubart, 2004). Briefly, this principle states that two lower-energy or longer wavelength (red-shifted) photons arriving almost simultaneously (femtosecond order) at a fluorophore will induce an electronic transition (excited state) corresponding to twice the energy of each single photon (UV). Because the amount of two-photon excitation depends on the square of the incident illumination intensity, excitation is confined to a small volume around the focal point. Elsewhere along the beam path, out-of-focus fluorescence is not generated, and therefore a pinhole is never necessary. The second useful optical property is related to the fact that intrinsic sources of tissue autofluorescence have low two-photon excitation action cross-sections (a quantitative measure of two-photon absorption) and blue shifts of their spectra (Zipfel et al., 2003). On the other hand, Qdots have two-photon action cross-sections as high as 47,000 GM units, which are two to three orders of magnitude larger than those of conventional fluorescence dyes. The net result is that tissue autofluorescence, which has traditionally been a very significant problem when fluorescently imaging the arterial wall, is dramatically reduced when using Qdots in combination with TPELSM.

Finally, Qdots have a broad two photon absorption between 700 and 1000 nm, offering a significant advantage for simultaneous excitation of multicolored Qdots at a single laser wavelength (Larson et al., 2003). This is particularly useful since the titanium–sapphire (Ti–Sapphire) lasers currently used for TPELSM do not allow for rapid switching of excitation wavelengths. This latter feature is critical for using different size Qdots to label multiple epitopes in the same sample.

21.4. The Missing Target of Conventional Vascular Imaging: The Endothelium

Our appreciation of the importance of a dysfunctional and inflamed endothelium in the pathogenesis of early atherosclerosis and pother vascular disease has grown exponentially during the past 20 years (Hansson, 2005). However, the vast majority of our knowledge is difficult to extrapolate to complex living organisms because it has been generated in static in vitro systems using cultured endothelial cells (Dai et al., 2004) or through the use of histological sections.

Recently, murine models of atherosclerosis have become one of the most valuable tools to study the pathogenic mechanisms of atherosclerosis (Fazio and Linton, 2001). The localization, morphology, size, and composition of atherosclerotic lesions in this animal model are usually analyzed in serial histological sections of the aortic sinus and in flat en face preparations of the descending thoracic aorta (Klinkner et al., 1997; McGillicuddy et al., 2001; Tangirala et al., 1995). Unfortunately, these widespread techniques are not suitable for sensitive visualization of the spatial distribution of endothelial markers of activation. On the other hand, there is a surprising paucity of fluorescence imaging studies of the endothelium using these flat preparations (Hajra et al., 2000; Iiyama et al., 1999). This is likely due to the fact that traditional fluorescence techniques with single-photon confocal microscopy and organic dyes pose several challenges for imaging vascular tissue, including tissue autofluorescence, fluorophore crosstalk, photobleaching, and the lack of capability to place mechanistic changes in protein expression and activation in the context of the macroscale of the disease. As a result, the need for phenotypic information on the complexity of gene expression and the spatial/temporal interactions of endothelial protein expression has recently led to the development of a novel en face technique employing Qdots and TPELSM (Ferrara et al., 2006).

21.5. Imaging System Design: Technical Aspects

We have recently developed a sensitive, yet easy-to-use fluorescence IHC en face method using Qdot-bioconjugates and TPELSM to investigate endothelial changes in relatively thin blood vessels, such as mouse aortas. This method allowed us to obtain accurate multicolored labeling, reliable fluorescence quantitation of low-abundance antigens, and 3D visualization of the vascular endothelium.

A comprehensive description of this methodology can be found elsewhere (Ferrara et al., 2006). In short, this ex vivo two-step immunohistochemical procedure is performed in mouse aortas after pressure-perfusing with cold normal saline solution followed by fixation with 10% formalin. The complete aortas are dissected in situ and then opened longitudinally to make flat preparations. Subsequently, the staining protocol begins with a blocking step followed by incubation with a primary antibody, and finally a reaction with Qdot-bioconjugates at concentrations of 10–20 nM. A permeabilization step with Triton can be done as needed to target intracellular antigens. Nuclear counterstaining is achieved with Hoechst solution due to its convenient two-photon absorption. The specimens are finally mounted between glass slides and cover slips. This procedure takes advantages of the nanocrystals' resistance to photobleaching. Images are collected using a two-photon microscope equipped with a Ti–Sapphire femtosecond laser tuned and mode-locked at 750 nm. A ×40 oil immersion objective connected to an inverted microscope is used with a fully open pinhole. The beam pathway consists of only one beam splitter. Separation of the emission signals is performed by selecting narrow (40 nm) windows, based on Qdots' spectra, using the META detector.

The fluorescence quantitation of the confocal images is carried out using the frequency histogram function of commonly available software (Figure 21.1). Image capture of samples prepared in a standardized way is performed using the same confocal settings during each experiment. Distorted, uneven, or denuded portions of the samples are not considered for quantitation. The positive control sample is imaged first to determine the optimal confocal settings (percent laser transmission and detector gain) to avoid saturation. In this regard, it is advisable to work with 12-bit images, which give a much larger range of pixel intensity. For each mouse sample, a negative control sample is used to establish the mean pixel intensity of the background signal and the threshold needed to subtract

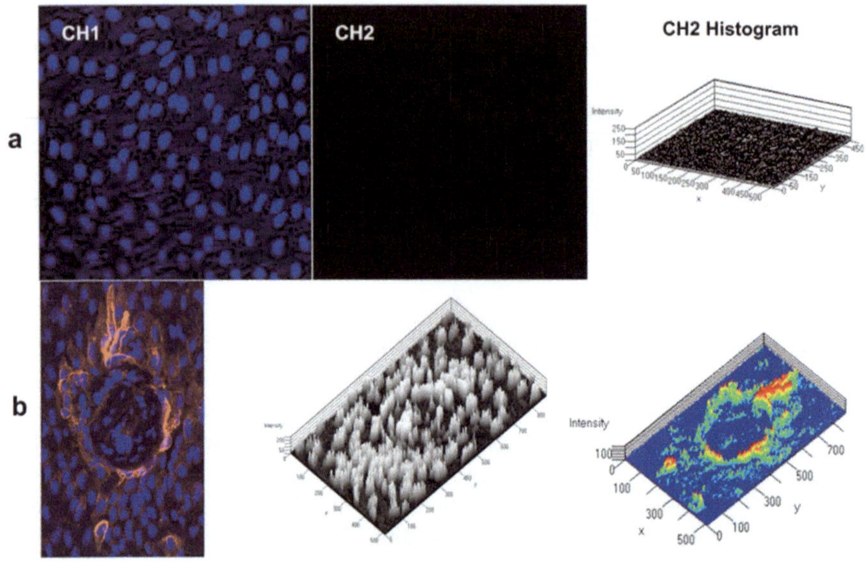

FIGURE 21.1. Fluorescence quantitation frequency histograms. **a** Negative control sample from a non-branching area of a mouse descending aorta. Channel (CH) 1 (400–460 nm) corresponds to the nuclear staining Hoechst (blue). CH2 is for the detection of nanocrystals with peak emission at 585 nm (560–600 nm). CH2 histogram shows very low levels of background fluorescence and no crosstalk from the blue channel that is excited with the same wavelength. **b** Positive sample. On the left is the merged image of an intercostal branch of a normal C57Bl/6 mouse stained for VCAM-1. The images on the center and right side of the panel show the pseudo-3D frequency histograms for both channels (monochrome for Hoechst and rainbow for VCAM-1 stained with Qdot 585 nm streptavidin conjugate).

99% of the background. Next, the mean background intensity and threshold are digitally subtracted by applying the low threshold tool to the images. Finally, the mean fluorescence intensity in arbitrary units and percent surface area are determined for each image (Dobrucki, 2004; Pawley, 2000).

For some experiments, a mouse model of systemic endothelial activation has been very useful as a positive control for quantitation purposes. In this model, diffuse up-regulation of endothelial inflammatory markers can be achieved by intraperitoneal injections of different doses of lipopolysaccharide (LPS) given 5 hours before euthanasia (Henninger et al., 1997).

21.6. Cellular and Molecular Imaging Applications

We have previously shown that our en face technique is very suitable to imaging the endothelial layer of arteries because the combined use of Qdots and TPELSM abolishes vascular autofluorescence. The detection of low levels of antigen expression in vascular tissue is a challenging task due to interference by different native compounds that exhibit fluorescence over a wide spectrum of wavelengths. However, because of the low two-photon excitation (TPE) cross-sections of native organic fluorophores and the broad adsorption of Qdots, the combined use of TPELSM and Qdots has allowed the acquisition of confocal images without the need of subtracting digital algorithms to eliminate the broadband autofluorescence. Indeed, the probability of two-photon excitation of Qdots has been shown to be as much as six orders of magnitude higher than with intrinsic molecules, such as NADH (Larson et al., 2003) (Figure 21.1). In this way, TPE can improve resolution by increasing contrast due to its ability to reduce tissue autofluorescence. Moreover, the high quantum yield, large Stokes shifts, and narrow emission spectra of Qdots further allow easy detection by facilitating spectral separation. In fact, crosstalk removal using the automatic component extraction (ACE) and spectral unmixing functions of new confocal microscopes is straightforward when utilizing nanocrystals. It is also worth noting that novel optical properties can also be used for optimization of signal-to-background ratio (contrast) (Gao et al., 2005). The longer excited-state lifetimes of Qdots provide this new means to separate the Qdot fluorescence from background fluorescence in a technique known as time-gated imaging or Fluorescence Lifetime Imaging (FLIM) (Michalet et al., 2001).

21.7. Multiplexing the Aortic Endothelium with Qdot-Bioconjugates Allows Faster Molecular Profiling

Crosstalk is a common problem in multifluorescence imaging with conventional confocal laser scanning microscopy. Multiple lasers are usually required to excite different colors of organic fluorescent labels. In addition, standard fluorophores have broad emission spectra, which makes their separation by band pass filters imperfect for multilabeling. In contrast, the visualization of multicolored Qdots with only one excitation wavelength is the method of choice due to their broad adsorption and narrow emission spectra.

Figure 21.2 illustrates the ability of this technique to multiplex up to four fluorophores (nuclear staining and three endothelial markers). This can be easily achieved with simple confocal settings (one excitation wavelength and only one dichroic mirror in the beam pathway).

Even though spatial resolution beyond the diffraction limit of light is not necessarily enhanced with two-photon excitation, the effective resolution achieved in confocal microscopy is a function of several factors including contrast. In this regard, Lacoste et al. have demonstrated that ultrahigh resolution multicolor colocalization can be achieved with Qdot nanocrystals using conventional far-field optics (Lacoste et al., 2000). These authors highlight that achieving better spatial resolution between multicolored objects or molecules requires careful a

b

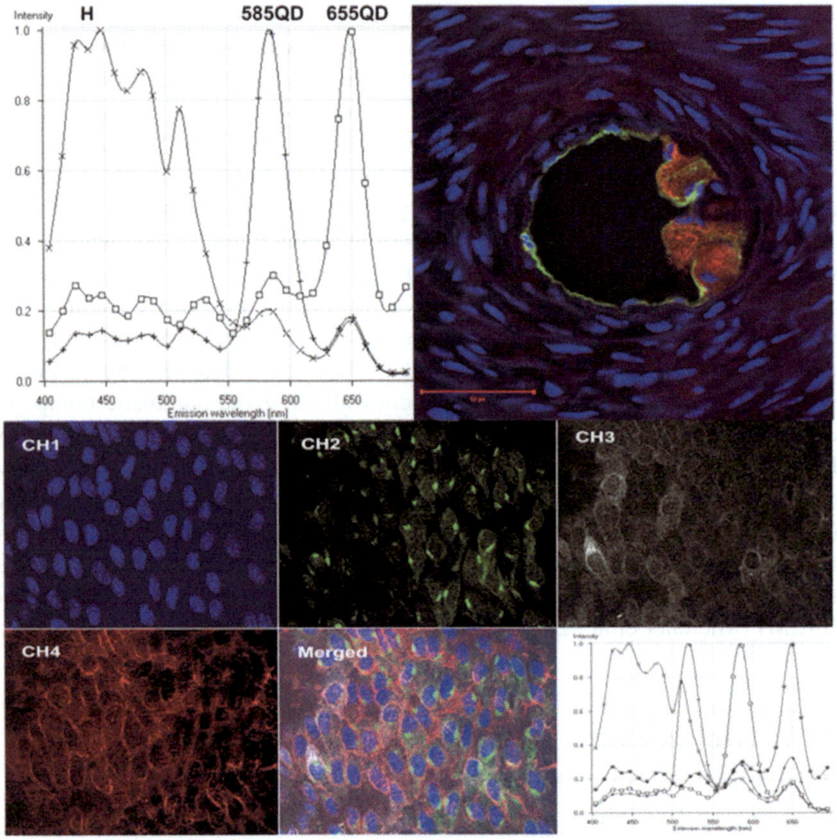

FIGURE 21.2. Multiplexing with one excitation wavelength. **a** Intercostal branch of an apo E KO mouse sectioned at the vascular media. ECs were targeted with a biotinylated anti-VCAM-1 antibody and Qdot–585 nm streptavidin conjugate (pseudo colored green for contrast). Macrophages were stained with a monoclonal rat anti-mouse Ab. (MAC-3) and Qdot–655 nm goat anti-rat IgG conjugate (red). Nuclear counterstaining was done with Hoechst (blue). The elongated nuclei correspond to smooth muscle cells. The emission spectra of the fluorophores are on the left. Only one wavelength (750 nm) was used to excite all three fluorophores. **b** Non-branching area of a normal C57Bl/6 mouse. Endothelial cell nuclei were stained with Hoechst (CH1). eNOS staining is shown in green (CH2). Its perinuclear predominance was demonstrated through the use of Qdot 525 nm bioconjugates. CH3 shows isolated cells that stain positive for VCAM-1 targeted with Qdot 585 nm bioconjugates (in white for contrast). Finally, the red channel (CH4) shows the typical membrane distribution of PECAM-1 marked with Qdot 655 nm IgG conjugate. The lower right panel shows the emission spectra of these four fluorophores.

consideration of chromatic aberration. Therefore, in the case of Qdots that can be excited with one laser wavelength aligned on the optical axis, the effects of chromatic aberration in the excitation pathway can be eliminated (Michalet et al., 2003).

21.8. Quantum-Dot-Based Platform for Quantitative Fluorescence Analysis of Endothelial Markers: A Novel Ex vivo Cyto-ELISA

Our en face technique has many of the ideal features required for image acquisition and quantification: fixed samples stained with a photostable fluorophore that emit with narrow emission spectra, minimal TPE autofluorescence, and optically selected filters using the lambda stack function and META detector (Dobrucki, 2004). Indeed, when using organic fluorophores, photobleaching during sample preparation and image acquisition is almost unavoidable and makes quantitation very unreliable. This results in a high degree of variability which usually prevents the analysis of pooled data and group comparisons (Iiyama et al., 1999). Ex vivo systems are useful tools in translational research because they closely mirror biological phenomena of living organisms. As a result, data obtained with these techniques are physiologically relevant and not as prone to bias as those obtained with in vitro systems (Ferrara and Pierangeli., 2005). The anterior or ventral wall of the descending thoracic aorta provides

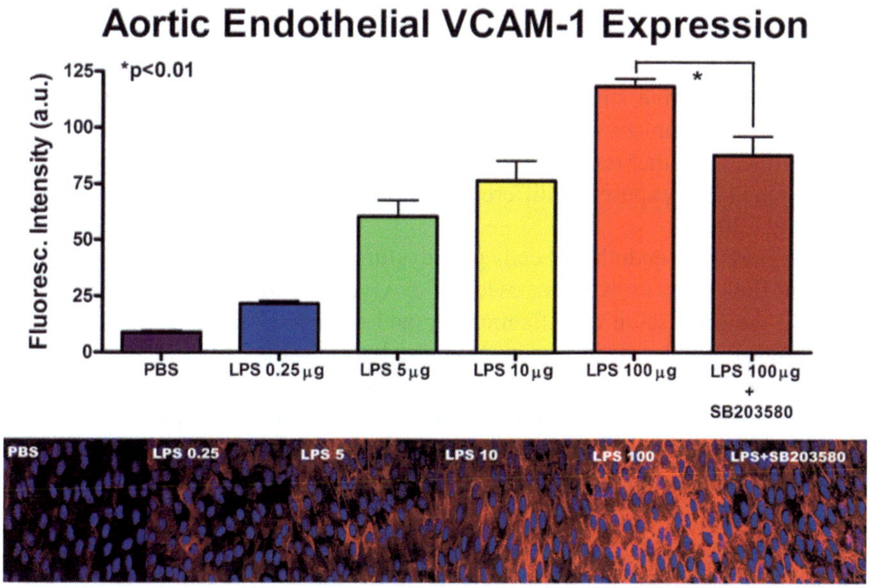

FIGURE 21.3. Fluorescence quantitation. Aortic endothelial VCAM-1 expression induced in wild-type mice by LPS treatment. Approximately 10 areas were imaged per mouse ($n = 2$ mice per group). The dose–response experiment shows significant upregulation of the adhesion molecule by progressively higher doses of the agonist. Significant stimulation is seen after a dose of 100 μg of LPS, with partial inhibition achieved by administration of the MAPK inhibitor 30 minutes before. Values represent relative fluorescence intensity in arbitrary units (a.u.) and are expressed as mean ± SEM.

a large surface for analysis of endothelial cell activation. In these areas, the physiological high wall shear stress exerts atheroprotective effects and keeps the endothelium almost quiescent. Only small areas surrounding the intercostal artery ostia are prone to lesion development and show up-regulation of inflammatory markers even in normal animals (Ferrara et al., 2006). Accordingly, this straight part of the aorta becomes an excellent ex vivo model to test the effects of diverse interventions on the endothelial inflammatory response that can be generated in vivo. Figure 21.3 shows the effects of a dose response experiment treating mice with progressively higher doses of LPS and the inhibitory in vivo effect of the p38 MAPK inhibitor SB 203580 (Lee et al., 1999).

21.9. Mapping the Spatial Distribution of Molecular Markers in Relation to Flow Patterns

Atherosclerotic lesion localization to regions of pronounced curvature and branching points of the arterial tree suggests a key role for local hemodynamic factors (VanderLaan et al., 2004). There has been a recent interest in low and oscillatory wall shear stresses as potential localizing factors in atherogenesis (Zarins et al., 1983) and numerous in vitro studies have been reported on the effects of disturbed flow patterns on several aspects of endothelial cell function. For example, it has been shown that the expression of adhesion molecules and markers of oxidative stress are increased when endothelial cells are exposed to oscillatory shear stress (Chappell et al., 1998; De et al. Keulenaer et al., 1998). However, other researchers have failed to consistently detect low levels of expression in ECs exposed to different flow patterns without cytokine stimulation (Dai et al., 2004).

To explore how endothelial cells gain dysfunctional phenotypes in response to different flow patterns, we have used our ex vivo platform to accurately map and quantify the expression of inflammatory and oxidative stress markers (Ferrara et al., 2006). As shown in Figure 21.4, the lateral walls of the ascending aorta are areas with little adhesion molecule expression, coinciding with a relatively high mean wall shear stress (WSS). Conversely, the inner curvature in the aortic arch is an area of higher VCAM-1/ICAM-1 expression, and the WSS is low and oscillatory. The spatial localization of VCAM-1 is consistent with the distribution of atherosclerotic lesions in humans.

Another interesting example is the membrane translocation of eNOS at areas exposed to disturbed flow. Endothelial nitric oxide synthase (eNOS) is the enzyme responsible for most NO production. Nitric oxide is a key regulator of vascular tone and has numerous antiatherogenic properties including anti-inflammatory, antithrombotic, and antiproliferative effects. Decreased bioavailability of endothelial NO is the main characteristic of endothelial dysfunction, which is involved in the early pathogenesis of hypertension and atherosclerosis. It has been shown that eNOS expression is tightly regulated by shear stress in vitro. However, ex vivo studies of protein expression are scarce and

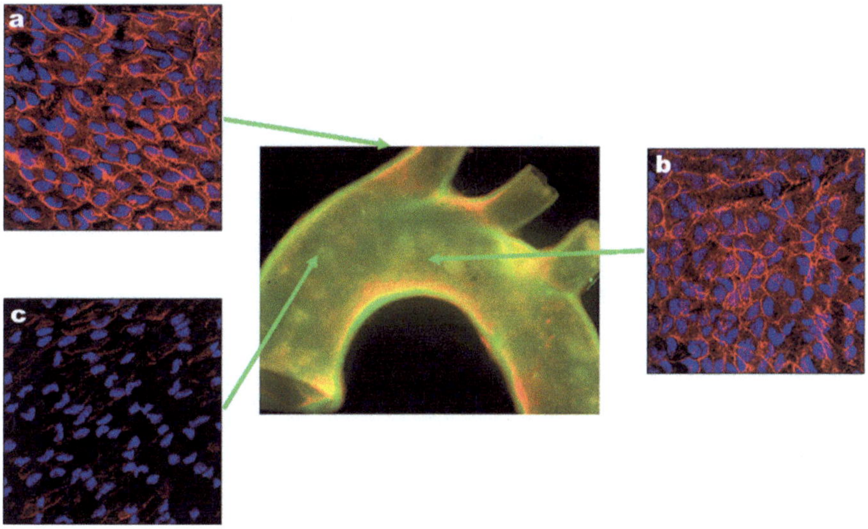

FIGURE 21.4. Spatial distribution of ICAM-1 at different levels of a mouse aorta and its main branches. The differential localization is evident in an epifluorescence image of a whole aortic mount (center). For these images, 655 nm (red) Qdot-bioconjugates were excited at 360 nm and their emission was detected with a 500 nm LP filter. Images were captured with a×4 objective attached to a Zeiss Axioscop 2 microscope. Note the intense distribution of the red fluorophore along the inner curvature of the arch and in the orifices of the arch vessels, areas that correspond to areas exposed to low and oscillatory shear stress. The central image corresponds to a wild-type mouse treated with lipopolysaccharide (LPS) to achieve fluorescence enhancement. However, the peripheral TPELSM images were taken from of a non-LPS treated mouse. **a** Innominate artery. **b** Inner curvature. **c** Lateral wall of ascending aorta.

it remains to be determined whether the mechanisms described are operative in intact vessels. This is very important since both laminar and oscillatory shear can upregulate this enzyme through different pathways. Others and we (Cheng et al., 2005; Ferrara et al., 2006) have shown that eNOS has a differential spatial localization at vascular areas exposed to varying hemodynamics. We found that eNOS had a primarily perinuclear localization at nonbranching areas as opposed to a more diffuse membrane distribution surrounding the intercostal branches (Figure 21.5). This translocation of eNOS to the plasma membrane at areas of disturbed flow is thought to be functionally important (Zhang et al., 2006). In areas of undisturbed flow eNOS expression is quiescent but primed for activation, whereas eNOS is fully functional for NO production when translocated to the membrane at branching areas. In this regard, Cai et al. (2004) have postulated that eNOS upregulation by oscillatory shear stress may represent a compensatory mechanism to increase the bioavailability of NO at areas of high production of reactive oxygen species. Conversely, membrane-bound eNOS at areas of altered shear may become uncoupled and contribute to superoxide and peroxinitrite

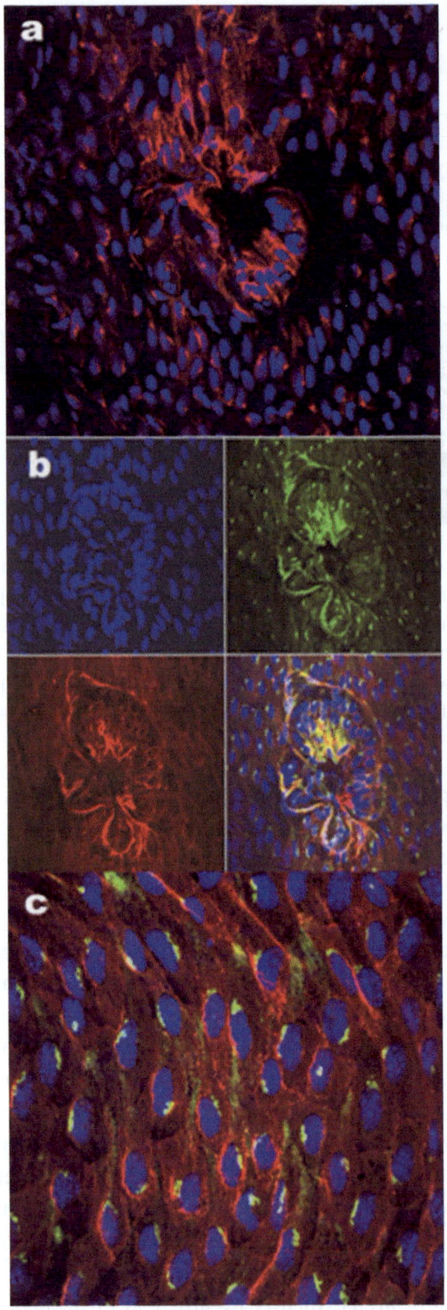

FIGURE 21.5. **a** An intercostal branch of a C57Bl/6 mouse. eNOS was targeted using a polyclonal Rabbit antibody (Santa Cruz Biotech). This antibody has no cross reactivity with either iNOS or nNOS. To assess the binding specificity of this antibody, we also studied eNOS-deficient mice and did competitive studies with a blocking peptide

generation (Forstermann and Munzel, 2006). By addressing these issues, this molecular imaging modality might help elucidate hidden pathogenic mechanisms, and design better therapeutic strategies to prevent endothelial dysfunction and atherosclerotic vascular disease.

The ability to map diverse regions of the vascular tree has important implications for future research. Because a certain intervention may differentially affect the development of atherosclerotic lesions throughout the vasculature, the sites chosen for analysis are critical (Barry-Lane et al., 2001). By accounting for the complex interaction of factors that may operate at these different sites, a better understanding of the prevailing mechanisms that initiate atherosclerotic lesion formation can be achieved (VanderLaan et al., 2004).

21.10. 3D Reconstruction of Atherosclerotic Vascular Walls

Confocal laser scanning microscopy has grown to become the tool of choice for 3D imaging. However, some limitations prevent single-photon confocal microscopy from becoming as popular as epifluorescence microscopy. One of these restrictions, photobleaching, has a negative impact on the collection of high-resolution z-stack images for 3D reconstruction. We have shown that combining low-energy TPE, which provides deep penetration and 3D resolution, with highly photostable Qdots yields a very reliable tool for obtaining large z-stack imaging data files without losing image intensity.

Figure 21.6 depicts a 3D reconstruction of an atherosclerotic lesion that has grown to cover an intercostal branch of a 6-month-old apoE KO mouse. Of note is the striking EC disarray at the dome of the lesion.

21.11. Drawbacks of Using Quantum Dots

The use of Qdot-based imaging is not without limitations. One disadvantage of our technique is the inability to stain deeper structures within the arterial wall. We have been able to stain subendothelial macrophages, but we have been unable to target epitopes in the vascular media. While the plasma membrane is not an obstacle for Qdots by achieving reliable intracellular targeting, penetration of these nanoparticles is poor beyond the internal elastic lamina.

FIGURE 21.5. (continued) (sc-654 P, Santa Cruz Biotech.). 655 nm Qdot-bioconjugates (red) were used and counterstained with Hoechst. Notice the membrane pattern of staining around the intercostal branch and the perinuclear predominance in non-branching or straight areas. **B** and **C** Colocalization studies with PECAM-1 in branching and non-branching areas respectively. eNOS stained with Qdots 525 nm (green) and PECAM-1 with Qdots 655 nm (red). Notice marked colocalization at the ostium of the intercostal artery (yellow).

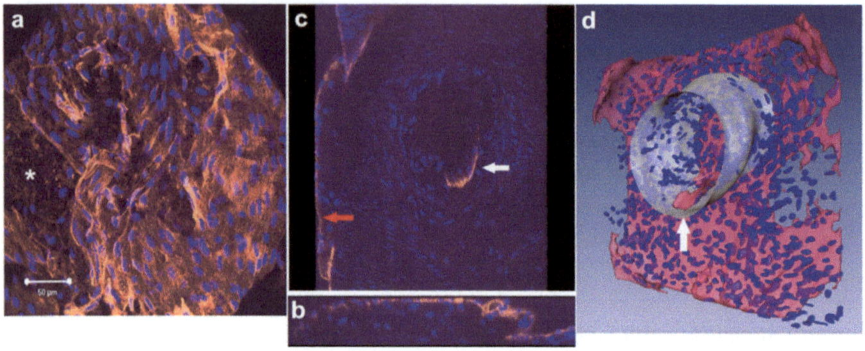

FIGURE 21.6. 3D reconstruction of an intercostal branch occluded by an atherosclerotic lesion. A z-stack of approximately 100 images was collected in 1 hour. Individual components were separately reconstructed using commercially available software (Amira) after preprocessing in Matlab. Positive endothelial VCAM-1 signal (Qdot 585 nm-orange) was used for surface reconstruction of the plaque and the occluded intercostal branch. Cell nuclei were identified using Hoechst counterstaining (blue). Small objects (less than 100 μm^3) were not considered nuclei and were excluded from the reconstruction. **a** Notice the EC disorganization over the plaque with areas of missing endothelium (*). **b** X-Z 90° projection of the plaque. **c** Y-Z 45° pseudo-3D projection of the plaque obtained with the LSM 510 examiner software. The plaque surface (red arrow) and the partially occluded ostium (white arrow) are seen from below. This is the mirror image of **d**. **d** AMIRA 3D reconstruction. The almost completely occluded lumen of the intercostal artery was hand-segmented based on a discernable pattern of VCAM-1 fluorescence (white arrow). This representation of the true occluded lumen was compared with the pattern of circumferentially aligned SMCs (original lumen in gray).

This poor penetration may be due to the size of Qdot-bioconjugates approaching that of a large protein, even when functionalized with $F(ab')_2$ fragments, as is the case with current commercially available conjugates. Functionalization with smaller ligands, such as peptides, has the potential to partially overcome this problem; however, we believe that the difficulty targeting subendothelial structures will persist. This situation is not unique to Qdots, as it is generally appreciated that luminal administration of molecules as small as viral particles does not result in penetration of the particle into the arterial wall beyond the internal elastic lamina.

In addition, concerns have been raised regarding removal of excess probes from the cytoplasm (Michalet et al., 2005). This may become an important consideration in single molecule detection. In this respect, we have seen that prolonged exposure to detergents and long incubation times increase the chances of cytoplasmic entrapment.

One significant additional disadvantage of this technique is that it cannot provide longitudinal analyses at different time points. Because it is an ex vivo study, the number of animals required for evaluating the effects of interventions is larger than with in vivo molecular imaging techniques, which may obtain repeated measurements before and after the application of a particular intervention.

Quantum dots have considerable potential as biologic labels for fluorescence in vivo imaging. Their advantages include an exponentially increasing blue-shifted extinction coefficient, high quantum yield, tunability of emission wavelength, chemical and photostability, and specific targeting. The recent emergence of new technologies, such as NIR fluorescence molecular tomography (Chen et al., 2002), stable NIR Qdots (Kim et al., 2004) along with targeted Qdots functionalized with polyethylene glycol (Gao et al., 2004), have increased expectations for the next generation of quantitative 3D fluorescence imaging in vivo. However, the application of quantum dot-based technologies for in vivo animal experiments has been limited by several problems. First, imaging quantum dots in vivo is challenging owing to the requirement for external illumination, which produces strong background autofluorescence. This still holds true with near-infrared reflectance fluorescence imaging (Cai et al., 2006). In addition, because of absorption and scattering of optical photons in tissues, little light is available for quantum dot excitation deep into the tissues (Frangioni, 2006).

So et al. (2006) have recently reported a new quantum dot technology based on the principle of bioluminescence resonance energy transfer (BRET) that improves imaging in vivo. Using quantum dot–luciferase conjugates, chemical energy from the substrate coelenterazine is converted into photon emission, which excites the quantum dot. No excitation light is required and, because energy transfer is resonant, absorption of excitation photons by other tissue pigments cannot occur.

Finally, the question of potential toxicity remains to be answered because current formulations of Qdots contain heavy metals, such as cadmium. While the absolute amount of cadmium present in Qdots is quite small, localized concentrations are potentially higher at either the site of labeling or during clearance and excretion from the intact animal. Importantly, data on the pharmacokinetics of Qdots in the setting of these applications are lacking (Jiang et al., 2004).

21.12. Conclusions

Molecular imaging technologies have been developing at a rapid pace during the last decade (McVeigh, 2006). The methodologies used in the past have been dominated by the use of in vitro assays. Moreover, the recent introduction of high throughput genomic and proteomic analytical techniques has caused a remarkable need for different approaches to validate their findings. The biochemical analysis of cellular homogenates provides very useful information on the type and relative amount of biomolecules (mRNA, proteins) present under different experimental settings (Lutgens et al., 2005). However, molecular imaging techniques are needed to help link this plethora of information to the behavior of these cells in vivo under normal and pathological conditions. The study of the adaptive responses of transgenic animal models at the molecular imaging level will provide enormous insight into the complex effects of diverse regulatory mechanisms and improve our understanding of cardiovascular physiology in health and disease.

Since the first publications on Qdots as biological labels (Bruchez et al., 1998; Chan and Nie, 1998), there have been important advances in Qdot synthesis, functionalization, and applications for biolabeling in vitro and in vivo. In this chapter, we have outlined the features of a novel molecular imaging method that will help bridge the gap between conventional histology and whole animal in vivo imaging.

We envision that cardiovascular molecular imaging will continue to grow rapidly. In fact, the translation of some of these techniques into the clinical arena has already begun (Jaffer et al., 2006). Nevertheless, the real success of this new applied nanotechnology field rests on the coordinated efforts of multidisciplinary teams of investigators with a wide range of expertise. This will allow the development of specific targeted biological markers and new imaging instruments with improved sensitivity and resolution to answer basic biological and clinical questions.

Acknowledgements. This work was supported by NIH U01 HL080711, NIH P01 HL58000, RO1 HL70531, and R01 DK059499.

References

Barry-Lane, P.A., Patterson, C., van der Merwe, M., Hu, Z., Holland, S.M., Yeh, E.T., Runge, M.S., 2001. p47phox is required for atherosclerotic lesion progression in ApoE(-/-) mice. J Clin Invest 108, 1513–1522.

Bruchez, M., Jr., Moronne, M., Gin, P., Weiss, S., Alivisatos, A.P., 1998. Semiconductor nanocrystals as fluorescent biological labels. Science 281, 2013–2016.

Cai, H., McNally, J.S., Weber, M., Harrison, D.G., 2004. Oscillatory shear stress upregulation of endothelial nitric oxide synthase requires intracellular hydrogen peroxide and CaMKII. J Mol Cell Cardiol 37, 121–125.

Cai, W., Shin, D.W., Chen, K., Gheysens, O., Cao, Q., Wang, S.X., Gambhir, S.S., Chen, X., 2006. Peptide-labeled near-infrared quantum dots for imaging tumor vasculature in living subjects. Nano Lett 6, 669–676.

Chan, W.C., Nie, S., 1998. Quantum dot bioconjugates for ultrasensitive nonisotopic detection. Science 281, 2016–2018.

Chappell, D.C., Varner, S.E., Nerem, R.M., Medford, R.M., Alexander, R.W., 1998. Oscillatory shear stress stimulates adhesion molecule expression in cultured human endothelium. Circ Res 82, 532–539.

Chen, J., Tung, C.H., Mahmood, U., Ntziachristos, V., Gyurko, R., Fishman, M.C., Huang, P.L., Weissleder, R., 2002. In vivo imaging of proteolytic activity in atherosclerosis. Circulation 105, 2766–2771.

Cheng, C., van Haperen, R., de Waard, M., van Damme, L.C., Tempel, D., Hanemaaijer, L., van Cappellen, G.W., Bos, J., Slager, C.J., Duncker, D.J., van der Steen, A.F., de Crom, R., Krams, R., 2005. Shear stress affects the intracellular distribution of eNOS: direct demonstration by a novel in vivo technique. Blood 106, 3691–3698.

Dai, G., Kaazempur-Mofrad, M.R., Natarajan, S., Zhang, Y., Vaughn, S., Blackman, B.R., Kamm, R.D., Garcia-Cardena, G., Gimbrone, M.A., Jr., 2004. Distinct endothelial phenotypes evoked by arterial waveforms derived from atherosclerosis-susceptible and -resistant regions of human vasculature. Proc Natl Acad Sci U S A 101, 14871–14876.

De Keulenaer, G.W., Chappell, D.C., Ishizaka, N., Nerem, R.M., Alexander, R.W., Griendling, K.K., 1998. Oscillatory and steady laminar shear stress differentially affect human endothelial redox state: role of a superoxide-producing NADH oxidase. Circ Res 82, 1094–1101.

Denk, W., Strickler, J.H., Webb, W.W., 1990. Two-photon laser scanning fluorescence microscopy. Science 248, 73–76.

Dobrucki, J.W., 2004. Confocal microscopy: quantitative analytical capabilities. Methods Cell Biol 75, 41–72.

Falk, E., 2006. Pathogenesis of atherosclerosis. J Am Coll Cardiol 47, C7–C12.

Fazio, S., Linton, M.F., 2001. Mouse models of hyperlipidemia and atherosclerosis. Front Biosci 6, D515–525.

Ferrara, D.E., Pierangeli, S.S., 2005. Diverse effects of statins on endothelial cells? Thromb Haemost 93, 186–188.

Ferrara, D.E., Taylor, W.R., 2006. A P47 Phox-Dependent NADPH Oxidase is Not Essential for the in vivo Upregulation of Adhesion Molecules at Aortic Arch Areas Exposed to Altered Patterns of Shear Stress. Arterioscler Thromb Vasc Biol e42.

Ferrara, D.E., Weiss, D., Carnell, P.H., Vito, R.P., Vega, D., Gao, X., Nie, S., Taylor, W.R., 2006. Quantitative 3D fluorescence technique for the analysis of en face preparations of arterial walls using quantum dot nanocrystals and two-photon excitation laser scanning microscopy. Am J Physiol Regul Integr Comp Physiol 290, R114–R123.

Forstermann, U., Munzel, T., 2006. Endothelial nitric oxide synthase in vascular disease: from marvel to menace. Circulation 113, 1708–1714.

Frangioni, J.V., 2006. Self-illuminating quantum dots light the way. Nat Biotechnol 24, 326–328.

Gao, X., Cui, Y., Levenson, R.M., Chung, L.W., Nie, S., 2004. In vivo cancer targeting and imaging with semiconductor quantum dots. Nat Biotechnol 22, 969–976.

Gao, X., Nie, S., 2003. Molecular profiling of single cells and tissue specimens with quantum dots. Trends Biotechnol 21, 371–373.

Gao, X., Yang, L., Petros, J.A., Marshall, F.F., Simons, J.W., Nie, S., 2005. In vivo molecular and cellular imaging with quantum dots. Curr Opin Biotechnol 16, 63–72.

Hajra, L., Evans, A.I., Chen, M., Hyduk, S.J., Collins, T., Cybulsky, M.I., 2000. The NF-kappa B signal transduction pathway in aortic endothelial cells is primed for activation in regions predisposed to atherosclerotic lesion formation. Proc Natl Acad Sci U S A 97, 9052–9057.

Hansson, G.K., 2005. Inflammation, atherosclerosis, and coronary artery disease. N Engl J Med 352, 1685–1695.

Henninger, D.D., Panes, J., Eppihimer, M., Russell, J., Gerritsen, M., Anderson, D.C., Granger, D.N., 1997. Cytokine-induced VCAM-1 and ICAM-1 expression in different organs of the mouse. J Immunol 158, 1825–1832.

Iiyama, K., Hajra, L., Iiyama, M., Li, H., DiChiara, M., Medoff, B.D., Cybulsky, M.I., 1999. Patterns of vascular cell adhesion molecule-1 and intercellular adhesion molecule-1 expression in rabbit and mouse atherosclerotic lesions and at sites predisposed to lesion formation. Circ Res 85, 199–207.

Jaffer, F.A., Libby, P., Weissleder, R., 2006. Molecular and cellular imaging of atherosclerosis: emerging applications. J Am Coll Cardiol 47, 1328–1338.

Jaffer, F.A., Weissleder, R., 2004. Seeing within: molecular imaging of the cardiovascular system. Circ Res 94, 433–445.

Jiang, W., Papa, E., Fischer, H., Mardyani, S., Chan, W.C., 2004. Semiconductor quantum dots as contrast agents for whole animal imaging. Trends Biotechnol 22, 607–609.

Khan, B.V., Parthasarathy, S.S., Alexander, R.W., Medford, R.M., 1995. Modified low density lipoprotein and its constituents augment cytokine-activated vascular cell adhesion molecule-1 gene expression in human vascular endothelial cells. J Clin Invest 95, 1262–1270.

Kim, S., Lim, Y.T., Soltesz, E.G., De Grand, A.M., Lee, J., Nakayama, A., Parker, J.A., Mihaljevic, T., Laurence, R.G., Dor, D.M., Cohn, L.H., Bawendi, M.G., Frangioni, J.V., 2004. Near-infrared fluorescent type II quantum dots for sentinel lymph node mapping. Nat Biotechnol 22, 93–97.

Klinkner, A.M., Bugelski, P.J., Waites, C.R., Louden, C., Hart, T.K., Kerns, W.D., 1997. A novel technique for mapping the lipid composition of atherosclerotic fatty streaks by en face fluorescence microscopy. J Histochem Cytochem 45, 743–753.

Lacoste, T.D., Michalet, X., Pinaud, F., Chemla, D.S., Alivisatos, A.P., Weiss, S., 2000. Ultrahigh-resolution multicolor colocalization of single fluorescent probes. Proc Natl Acad Sci U S A 97, 9461–9466.

Larson, D.R., Zipfel, W.R., Williams, R.M., Clark, S.W., Bruchez, M.P., Wise, F.W., Webb, W.W., 2003. Water-soluble quantum dots for multiphoton fluorescence imaging in vivo. Science 300, 1434–1436.

Lee, J.C., Kassis, S., Kumar, S., Badger, A., Adams, J.L., 1999. p38 mitogen-activated protein kinase inhibitors–mechanisms and therapeutic potentials. Pharmacol Ther 82, 389–397.

Libby, P., 2002. Inflammation in atherosclerosis. Nature 420, 868–874.

Lutgens, E., Faber, B., Schapira, K., Evelo, C.T., van Haaften, R., Heeneman, S., Cleutjens, K.B., Bijnens, A.P., Beckers, L., Porter, J.G., Mackay, C.R., Rennert, P., Bailly, V., Jarpe, M., Dolinski, B., Koteliansky, V., de Fougerolles, T., Daemen, M.J., 2005. Gene profiling in atherosclerosis reveals a key role for small inducible cytokines: validation using a novel monocyte chemoattractant protein monoclonal antibody. Circulation 111, 3443–3452.

McGillicuddy, C.J., Carrier, M.J., Weinberg, P.D., 2001. Distribution of lipid deposits around aortic branches of mice lacking LDL receptors and apolipoprotein E. Arterioscler Thromb Vasc Biol 21, 1220–1225.

McVeigh, E.R., 2006. Emerging imaging techniques. Circ Res 98, 879–886.

Michalet, X., Kapanidis, A.N., Laurence, T., Pinaud, F., Doose, S., Pflughoefft, M., Weiss, S., 2003. The power and prospects of fluorescence microscopies and spectroscopies. Annu Rev Biophys Biomol Struct 32, 161–182.

Michalet, X., Pinaud, F.F., Bentolila, L.A., Tsay, J.M., Doose, S., Li, J.J., Sundaresan, G., Wu, A.M., Gambhir, S.S., Weiss, S., 2005. Quantum dots for live cells, in vivo imaging, and diagnostics. Science 307, 538–544.

Michalet, X., Pinaud, F.F., Lacoste, T.D., Dahan, M., Bruchez, M.P., Alivisatos, A.P., Weiss, S., 2001. Properties of Fluorescent Semiconductor Nanocrystals and their Application to Biological Labeling. Single Mol 2, 261–276.

Nerem, R.M., Alexander, R.W., Chappell, D.C., Medford, R.M., Varner, S.E., Taylor, W.R., 1998. The study of the influence of flow on vascular endothelial biology. Am J Med Sci 316, 169–175.

Pawley, J., 2000. The 39 steps: a cautionary tale of quantitative 3-D fluorescence microscopy. Biotechniques 28, 884–886, 888.

Rubart, M., 2004. Two-photon microscopy of cells and tissue. Circ Res 95, 1154–1166.

Smith, A.M., Ruan, G., Rhyner, M.N., Nie, S., 2006. Engineering luminescent quantum dots for in vivo molecular and cellular imaging. Ann Biomed Eng 34, 3–14.

So, M.K., Xu, C., Loening, A.M., Gambhir, S.S., Rao, J., 2006. Self-illuminating quantum dot conjugates for in vivo imaging. Nat Biotechnol 24, 339–343.

Tangirala, R.K., Rubin, E.M., Palinski, W., 1995. Quantitation of atherosclerosis in murine models: correlation between lesions in the aortic origin and in the entire aorta, and differences in the extent of lesions between sexes in LDL receptor-deficient and apolipoprotein E-deficient mice. J Lipid Res 36, 2320–2328.

VanderLaan, P.A., Reardon, C.A., Getz, G.S., 2004. Site specificity of atherosclerosis: site-selective responses to atherosclerotic modulators. Arterioscler Thromb Vasc Biol 24, 12–22.

Zarins, C.K., Giddens, D.P., Bharadvaj, B.K., Sottiurai, V.S., Mabon, R.F., Glagov, S., 1983. Carotid bifurcation atherosclerosis. Quantitative correlation of plaque localization with flow velocity profiles and wall shear stress. Circ Res 53, 502–514.

Zhang, Q., Church, J.E., Jagnandan, D., Catravas, J.D., Sessa, W.C., Fulton, D., 2006. Functional relevance of Golgi- and plasma membrane-localized endothelial NO synthase in reconstituted endothelial cells. Arterioscler Thromb Vasc Biol 26, 1015–1021.

Zipfel, W.R., Williams, R.M., Webb, W.W., 2003. Nonlinear magic: multiphoton microscopy in the biosciences. Nat Biotechnol 21, 1369–1377.

22
Quantum Dots for Cancer Imaging

Swadeshmukul Santra and Debamitra Dutta

Abstract: During the last century cancer has slowly advanced to become the leading cause of death for patients below the age of 85 in USA in spite of rapid advances in global cancer research toward the understanding of cancer biology in the past several decades. Early cancer diagnosis, in combination with the precise cancer therapies, could eventually save millions of lives. The diagnosis of cancer at the early stage is therefore extremely important and has been an active research area of great interest in current times.

Nanoparticle technology is an interdisciplinary research area of nanoscale science and technology. An emerging branch of nanoparticle technology is the development of optical-based functional nanoparticles, such as luminescent quantum dots (Qdots), having strong potential to revolutionize nanomedicine research, specifically in the area of optical contrast agent development and their applications in various bioimaging techniques including cancer imaging.

This chapter provides a knowledge-based platform to readers who are interested in learning more about the potential of Qdot nanotechnology for diagnostic cancer imaging (Section 22.1). Our goal is to provide readers (i) the basics of luminescent Qdots, (ii) details of Qdot design, synthesis, and bioconjugate chemistries to target cancers (Section 22.2), and (iii) applications of Qdot technology in cancer imaging (Section 22.3) that include a brief description of the benefits, challenges, limitations, and the future scope of Qdot technology in both in vitro and in vivo cancer imaging applications. Finally, Section 22.4 provides concluding remarks and the perspectives of Qdot-based optical imaging of cancers.

22.1. Introduction

The word 'cancer' refers to a group of diseases in which the normal cells undergo a transformation process that triggers rapid and abnormal cell division. These abnormalities in the cell division process often give rise to the formation of lumps called 'tumors'. Tumors can be either benign (non-cancerous) or malignant (cancerous). Malignant tumor tissue grows and invades all adjacent tissues and organs around the primary cancer site, thereby obstructing the normal physiological functions of the body. Gradually, the cancerous cells detach from the primary site and spread through the blood and lymphatic system to various parts of the body, forming new tumors. This process is known as cancer metastasis. Eventually, the malignant tumors cause various organs of the body to start malfunctioning, leading to death. Some of the more common forms of cancer are lung and bronchus, colon and rectum, skin, bladder, and prostate in men and breast, ovarian, and cervical cancer in women.

The American Cancer Society (2005) reported that cancer deaths have surpassed heart diseases to become the leading cause of death in the United States for people under the age of 85. The trend is mainly a reflection of the rapid decrease in heart disease–related deaths as compared to the slow decrease of cancer deaths likely due to the limitations in advanced diagnostic tools. Nearly one-third of the population of this country dies from cancer, making it the most devastating threat to the health of individuals of all ages not only in North America but all around the world.

Early diagnosis of cancer is important for effective treatment and survival of cancer patients. Patients diagnosed early tend to obtain immediate medical care and thus have a better chance of survival. It is therefore extremely important to develop a highly sensitive and robust diagnostic cancer imaging system. In diagnostic cancer imaging, contrast agents play a crucial role in enhancing the image contrast between the normal and the tumor tissues.

Some of the existing technologies for cancer imaging include non-invasive imaging techniques, such as Computed Tomography (CT) Scans, Magnetic Resonance (MR) Imaging Scans, Positron Emission Tomography (PET) Scans, Single Photon emission CT (SPECT), Ultrasound (US) Scans, Bone Scans, and optical imaging, for macroscopically visualizing the tumors. Unfortunately, most of these imaging techniques are somewhat limited to detect abnormalities on the microscopic level. There is a strong possibility of combining the existing optical imaging technology with the sophisticated nanoparticle-based optical contrast agents for high-resolution in vivo cancer imaging. Recently, this strategy has been successfully demonstrated to be useful for detecting tumors (Bremer et al., 2003; Gao et al., 2004; Medintz et al., 2005).

The image quality of tumors (i.e., image contrast) depends on various experimental parameters such as light intensity, wavelength, polarization, coherence, interference, lifetime, and non-linear effects. Of them, the selection of the light wavelength is critical to achieve deep-tissue penetration capability. Optical imaging using near-infrared (NIR) wavelengths (650–900 nm) will provide highest tissue penetration as NIR light has minimal absorption by normal tissue components, such as hemoglobin, lipid, and water. For deep-tissue imaging, the use of NIR light will overcome the limitations of confocal and two-photon microscopy–based imaging techniques (tissue resolution 500–800 μm (Fetcho and O'Malley, 1997; Yager et al., 1997)) that use visible light sources. It is expected that bright optical contrast agents with NIR excitation and emission will provide better tissue contrast. The existing organic NIR cyanine dye–based contrast agents may not be suitable as they rapidly lose their brightness through a process called photobleaching. There is certainly a need to develop better contrast agents for deep-tissue imaging such as tumor imaging. Furthermore, there is a strong necessity to develop optical imaging probes that are highly sensitive to detect the early development of cancers, preferably prior to metastasis. Nanoparticle technology has strong potential for developing such imaging probes.

An emerging branch of nanoparticle technology is the development of optical-based functional nanoparticles (Ben-Ari, 2003; Bremer et al., 2003; Brigger et al., 2002; Feldhein et al., 2001; Licha, 2002; Licha and Olbrich, 2005; Licha et al., 2002; Medintz et al., 2005; Michalet et al., 2005; Morgan et al., 2005; Rosi and Mirkin, 2005; Santra et al., 2004; Sokolov et al., 2003), such as quantum dots (Qdots) (Santra et al., 2005a), having strong potential to revolutionize nanomedicine research, specifically in the area of optical contrast agent development (Licha, 2002; Licha and Olbrich, 2005; Licha et al., 2002; Sokolov et al., 2003) and their applications in bioimaging (Medintz et al., 2005; Michalet et al., 2005; Morgan et al., 2005; Rosi and Mirkin, 2005; Santra et al., 2004). It is believed that the development of robust nanoparticle contrast agents for optical imaging could also lead to the accelerated development of cancer drugs. Our knowledge on the details of the dynamic process of anti-cancer drugs inside cancer cells is limited. This is mainly due to the lack of sophisticated imaging technology. If at all possible, real-time imaging of the therapeutic action of cancer drugs in cells will not only improve our knowledge of the functional activity of drug molecules but will also allow better anti-cancer drug design to combat cancer in the near future.

In this chapter, our goal is to provide readers the basics of luminescent Qdots and their applications for diagnostic cancer imaging, as well as details of Qdot design, synthesis, and bioconjugate chemistries to target cancers. In addition, we will highlight the recent literature on Qdot technology. We will also describe the benefits, challenges, limitations, and the future scope of Qdot technology in both in vitro and in vivo cancer imaging applications.

22.2. Development of Qdots for Bioimaging Applications

Luminescent Qdots are semiconductor-based nanometer-sized crystals (called nanocrystals), typically in the size range of 1–12 nm. They are extremely bright, photostable, and considered to be emerging probes for in vivo molecular and cellular imaging (Akerman et al., 2002; Alivisatos et al., 2005; Bruchez et al., 1998; Chan et al., 2002; Chan and Nie, 1998; Dahan et al., 2003; Dubertret et al., 2002; Gao et al., 2005; Ishii et al., 2003; Jaiswal et al., 2003; Larson et al., 2003; Lidke et al., 2004; Mattoussi et al., 2000; Medintz et al., 2003, 2005; Michalet et al., 2005; Pinaud et al., 2006; Smith et al., 2004, 2006; Stroh et al., 2005; Wu et al., 2003). Their attractive optical properties hold a great promise as a new class of imaging probes for sensitive cancer imaging. It is believed that Qdots could improve tumor imaging sensitivity in vivo by at least 10- to100-fold.

22.2.1. Attractive Features of Qdots for Cancer Imaging

(i) Qdots are several thousand times more photostable than organic dye molecules, allowing continuous imaging of biological samples over an extended period of time (Figure 22.1). For example, continuous tracking of cell migration, differentiation, and metastasis could be possible with Qdots.

(ii) The luminescent excitation and emission peaks in Qdots are largely separated (large Stokes shifts) allowing minimization of high autofluorescence background of tissue samples. As a result the detection sensitivity is further improved and one can achieve quality color-contrast by wavelength-resolved spectral imaging (Gao et al., 2004) where Qdots signals are clearly recognizable above the background. In contrast, the fluorescence signal from organic dyes (small Stokes shifts) is often buried under the strong tissue autofluorescence.

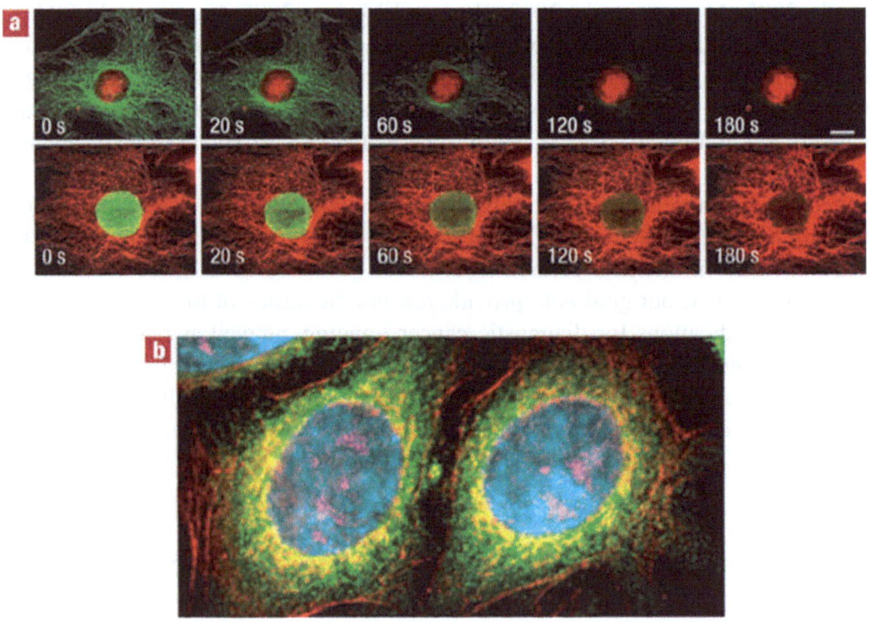

FIGURE 22.1. **a** Photostability of Qdots. *Top row*: Nuclear antigens labeled with QD 630-streptavidin (red) and microtubules labeled with AlexaFluor 488 (green) in a 3T3 cell. *Bottom row*: Microtubules labeled with QD-630 streptavidin (red) and nuclear antigens stained with AlexaFluor 488 (green) (reprinted by permission of the Nature Publishing Group). **b** Pseudocolored image depicting five-color Qdot staining of fixed human epithelial cells. Cyan corresponds to 655-Qdots labeling the nucleus, magenta 605-Qdots labeling the Ki-67 protein, orange 525-Qdots labeling mitochondria, green 565-Qdots labeling the microtubules, and red 705-Qdots labeling the actin filaments (courtesy Invitrogen Corporation). Photobleaching of organic dye, AlexaFluor 488 (green color) is clearly seen over time, demonstrating that Qdots are extremely photostable than organic dyes.

(iii) Qdots have very large molar extinction coefficient values (in the order of $0.5 - 5 \times 10^6 \text{M}^{-1}\text{cm}^{-1}$ (Leatherdale et al., 2002)). As a result, Qdots are much brighter probes (10–20 times) (Bruchez et al., 1998; Chan et al., 2002; Chan and Nie, 1998) than traditional organic dyes even in photon-limited in vivo conditions. This could easily improve the sensitivity of in vivo tumor imaging by at least 10- to 100-fold.

(iv) Because of their longer excited state lifetime, the emission rate of Qdots is 5–10 times slower than organic dyes. It is therefore expected that Qdot emission rates will be lifetime limited. In practice, the rate of absorption is the major limiting factor for emission and hence fluorescence imaging. Because of a much higher molar extinction coefficient of Qdots, the light absorption rate is about 10–50 times faster than that of organic dyes at the same excitation photon flux (number of incident photons per unit area).

(v) Multicolor Qdots can be used to track (Figure 22.1) and image multiple cancer-specific molecular targets using a single excitation wavelength. This is due to the fact that Qdots have size-tunable optical properties. In other words, the Qdot emission wavelengths can be tuned by varying particle size and/or composition. As most complex human diseases such as cancer involves a large number of genes and proteins, simultaneous tracking of a panel of molecular markers using multi-color Qdots will be extremely important for scientists to understand, classify, and differentiate complex human diseases (Gao and Nie, 2003).

22.2.2. Qdot Design and Synthesis

Qdots are made out of hundreds to thousands of atoms that typically belong to group II and VI elements or group III and V elements in the periodic table. For example, cadmium selenium (CdSe), cadmium telluruim (CdTe), and zinc selenium (ZnSe) are group II–VI semiconductor Qdots, whereas indium phosphorus (InP) and indium arsenic (InAs) Qdots are group III–V semiconductors. Almost all bright and photostable Qdots have a core-shell structure (Figure 22.2) protected by a uniform wide band-gap shell material. Upon absorption of light, an electron excites from the valence band to the conduction band of the Qdots creating an electron-hole pair. Qdots emit when the electro-hole pair is recombined. Photophysical studies of Qdots have revealed that strong confinement of excited electrons and holes in these nanocrystals exists. As a result of quantum confinement the electron and hole energy states within the nanocrystals are discrete. However, the electron and the hole energy levels (band-gap) are a function of the Qdot diameter as well as composition. The band-gap of Qdots increases as their size decreases, resulting in shorter emission wavelengths. It is well understood that surface defects in the Qdot crystal structure act as temporary "traps" for the electron or hole, enhancing their non-radiative recombination process. The alteration of trapping and untrapping events causes an intermittent blinking of Qdots at the single particle level, reducing the overall quantum efficiency (the ratio of emitted to absorbed photons). To combat this

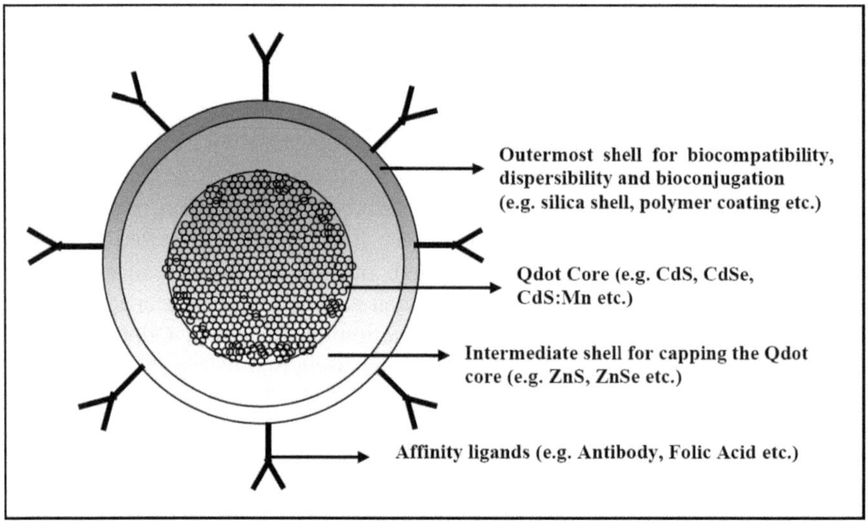

FIGURE 22.2. Schematic representation of core-shell quantum dot (qdot) structure. The qdot core is responsible for optical signaling. The intermediate layer passivates the surface. The outer layer protects the core-shell structure. The surface functional groups (e.g., carboxyls, amines, thiols, etc.) on the outer layer are responsible for attaching the biomolecule/ligand (e.g., antibodies, peptides, proteins, folates, etc.) on the structure.

problem, a wide band-gap surface-passivating shell structure is usually created over the Qdot core. This protective encapsulation protects the core surface atoms from oxidation and other chemical reactions, improving quantum yield close to 90% (Reiss et al., 2002) and photostability by several orders of magnitude relative to conventional organic dyes (Sukhanova et al., 2004).

Hot solution phase–mediated Qdot synthesis is the most popular technique of synthesizing high-quality Qdots. Qdots are synthesized at elevated temperatures in high boiling point non-polar organic solvents. Bawendi group (Murray et al., 1993) reported the synthesis of highly crystalline and monodisperse (size distribution 8%–11%) CdSe Qdots using high-temperature growth solvents/ligands (mixture of trioctyl phosphine/trioctyl phosphine oxide, TOP/TOPO). A combination of tri-n-octylphosphine oxide (TOPO) and hexadecylamine has also been used (Gao et al., 2005). The purpose of using hydrophobic organic molecules as mixed solvents or as a solvent/ligand mixture is 2-fold. The mixture serves as a robust reaction medium and also coordinates with unsaturated metal atoms on the Qdot surface to prevent the formation of bulk semiconductors. The best available colloidal Qdots suitable for biolabeling applications are made of CdSe cores coated with a layer of ZnS (CdSe/ZnS). Colloidal Qdots made of ZnS, CdS, ZnSe, CdTe, and PbSe, emitting from the ultraviolet (UV) to the infrared have been reported in the literature (Artemyev et al., 1995; Bailey and Nie, 2003; Gu et al., 2003, 2004; Hines and Guyot-Sionnest, 1996; Mattoussi et al., 2000; Murray et al., 2000, 1993; Peng et al., 1997; Peng and Peng, 2001;

Pinaud et al., 2004; Suyver et al., 2001; Tsay et al., 2004; Xu et al., 2004); although, fine tuning of the synthesis and surface passivation scheme is required. Following the hot solution phase synthesis strategy, Qu and Peng (2002) have reported the formation of high-quality CdSe nanocrystals having fluorescence quantum yields as high as 85% at room temperature.

The reverse micelle-mediated Qdot synthesis of high-quality CdS:Mn/ZnS core-shell Qdots have been reported (Santra et al., 2005b,c; Yang and Holloway, 2004; Yang et al., 2005,2004a,b). Reverse micelles (also called water-in-oil (W/O) microemulsion system) is an isotopic, thermodynamically stable homogeneous mixture of oil, water, and surfactant molecules, where the surfactant capped (stabilized) water droplets remain uniformly dispersed in the bulk oil phase. The water droplets serve as a tiny reactor (nano-reactor) for the synthesis of Qdots. This is a robust method that allows room temperature synthesis of monodisperse Qdots at normal atmospheric pressure. Yang et al. (2005, 2004a,b) reported the synthesis of manganese-doped cadmium sulfide core and zinc sulfide shell (CdS:Mn/ZnS) Qdots using AOT (dioctylsulfosuccinate sodium salt, a surfactant) with heptane (an oil) using this water reverse micelle system. The bright yellow–emitting CdS:Mn/ZnS Qdots were small (average Qdot size was 3.2 nm) and highly photostable. The maximum reported quantum yield of CdS:Mn/ZnS obtained was about 20% at room temperature.

22.2.3. Qdot Surface Passivation and Aqueous Stabilization

22.2.3.1. Surface Passivation

The effective surface passivation of Qdot nanocrystal core with wide band-gap semiconductor materials (shell) is extremely important (Dabbousi et al., 1997; Hines and Guyot-Sionnest, 1996). For example, in the case of CdSe/ZnS core-shell Qdots, the epitaxially matched ZnS layer effectively passivates the surface defects of the CdSe core (Chen and Gerion, 2004; Dabbousi et al., 1997; Yang et al., 2005), protects the core from oxidation, prevents leaching of highly toxic Cd^{2+} ions, and also drastically improves the quantum yield by reducing surface defects (that act as exciton traps, leading to the non-radiative recombination process). The TOP/TOPO-capped Qdots that are prepared using hot solution phase–mediated synthesis route are hydrophobic having no intrinsic aqueous dispersibility. Therefore, phase-transfer to aqueous solution by surface functionalization with hydrophilic capping ligands is required. This is usually done through either 'cap exchange' or by encapsulating the original nanocrystals in a thick heterofunctional organic coating, driven mainly by hydrophobic absorption onto the TOP/TOPO-capped Qdots. These ligands make Qdots aqueous dispersible and serve as another critical role in insulating, passivating, and protecting the Qdot surface from deterioration in biological media. It can also serve as a point of chemical attachment for biomolecules.

22.2.3.2. Capping Ligands

Synthesis of new capping ligands create an ever-increasing library available for obtaining aqueous Qdot dispersions for various applications (Bruchez et al., 1998; Dubertret et al., 2002; Guo et al., 2003; Hines and Guyot-Sionnest, 1996; Hong et al., 2004; Kim and Bawendi, 2003; Mattoussi et al., 2000; Osaki et al., 2004; Pellegrino et al., 2004; Pinaud et al., 2004; Wu et al., 2003). A representative list of capping ligands and the Qdot-capping strategies is provided in Figure 22.3 as a group of three major routes. The first uses 'cap exchange' which involves the substitution of the native TOP/TOPO with bifunctional ligands. The ligands present a surface-anchoring moiety to bind to the hydrophobic inorganic Qdot surface (e.g., thiol) and an opposing hydrophilic end group (e.g., hydroxyl or carboxyl) to achieve water dispersibility. These include an array of thiol and phosphine mono and multidentate ligands (Figure 22.3a,b) (Chan and Nie, 1998; Goldman et al., 2002c; Mattoussi et al., 2000; Mitchell et al., 1999; Uyeda et al., 2005). The second approach involves formation of polymerized silica shells functionalized with hydrophilic groups (Figure 22.3c) (Bruchez et al., 1998; Gerion et al., 2001). The final method preserves the native TOP/TOPO on the Qdots and uses a variety of amphiphilic 'diblock' and 'triblock' copolymers and phospholipids to tightly bind the alkylphosphine ligands through hydrophobic–hydrophobic attraction, whereas the hydrophilic outer block permits aqueous dispersion and further derivatization to attach desired biomolecules (Figure 22.3c) (Ballou et al., 2004; Dubertret et al., 2002; Gao et al., 2004; Mattheakis et al., 2004; Osaki et al., 2004; Pellegrino et al., 2004; Wu et al., 2003).

22.2.3.3. Stability of Capping Ligands

The compact mono-mercapto ligands (Figure 22.3a), although simple to synthesize, have short shelf lives (less than 1 week) due to dynamic thiol–ZnS interactions (Parak et al., 2003). Substitution from mono to dithiol dihydrolipoic acid ligands improves the long-term stability from 1 week to 1–2 years, suggesting that polydentate-thiolated ligands could be even more effective (Mattoussi et al., 2000; Parak et al., 2003; Uyeda et al., 2005). However, polydentate-thiolated ligands are carboxylic acids which can limit dispersion to basic pH (Parak et al., 2003). Silica shells and polymer/phospholipid encapsulation provides stability over a broader pH range, but result in substantially larger hydrophilic Qdots. For example, phospholipid and block copolymer coatings tend to increase the diameter of CdSe/ZnS Qdots from 4 to 8 nm before encapsulation to 20–30 nm. Larger size Qdots (less than 50 nm), although smaller than most mammalian cells, can still limit intracellular mobility and may preclude fluorescence resonance energy transfer (FRET)-based investigations (Parak et al., 2003). Yang et al. (2004b) reported the water in oil (W/O) microemulsion–mediated synthesis of silica-overcoated CdS:Mn/ZnS Qdots where the silica shell thickness was approximately 2–3 nm.

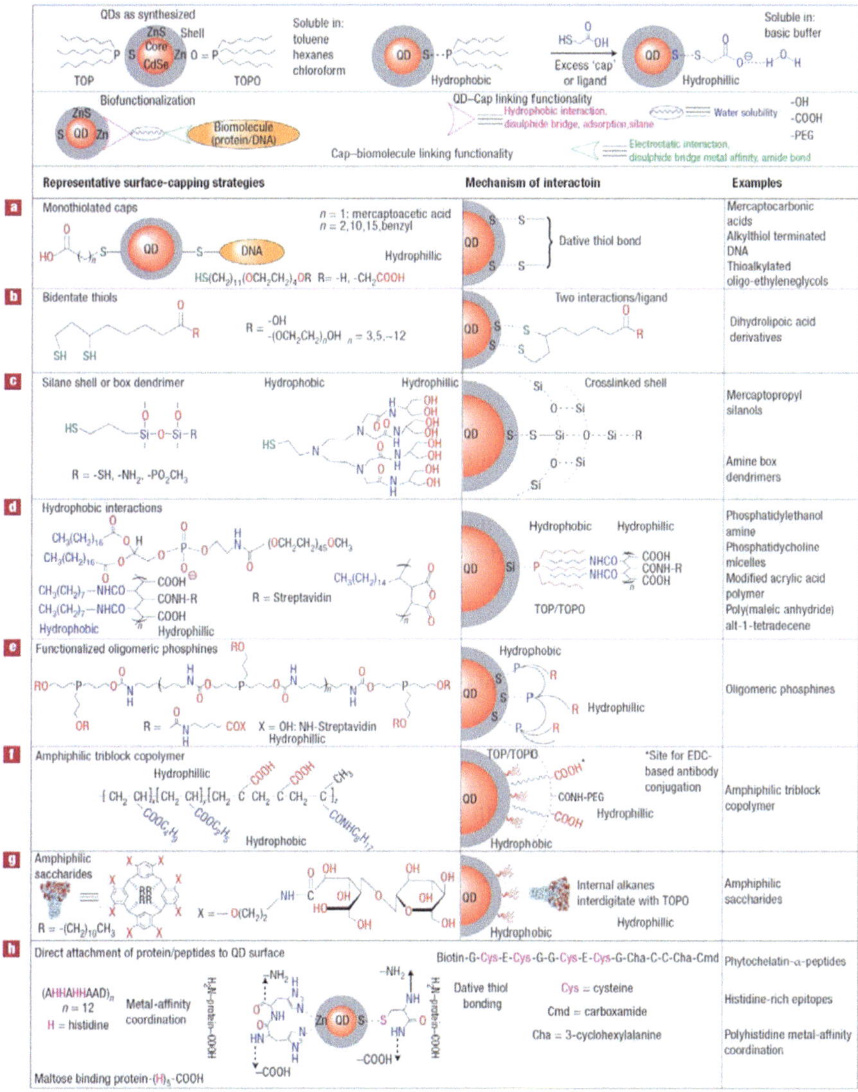

FIGURE 22.3. Schematic of generic Qdot solubilization and biofunctionalization. Biofunctionalization (second panel from *top*) uses caps/ligands to provide three functions: linkage to the Qdot (pink), water solubility (blue), and a biomolecule linking functionality (green). Examples of surface-capping strategies and the mechanism of interaction with the Qdot and the aqueous environment are also shown. For the cap exchange (*top right*), excess thiolated cap displaces the original TOP/TOPO organic coating by binding the ZnS surface with the thiol group and imparting hydrophilicity with the charged carboxyl (or other functionalities) yielding water-soluble colloidal Qdot dispersions. Reprinted by permission from the author.

22.2.4. Qdot Bioconjugation

Water-dispersible Qdots are essentially made of a hybrid structure of inorganic Qdot core and an organic shell. For biological applications, it is important to conjugate such Qdots to affinity ligands that are compatible to both Qdots and living biological systems. The common affinity ligands are antibodies, peptides, small-molecule drugs, inhibitors, etc. Affinity ligands are specifically designed to bind to the target, such as cells, intracellular components, and tissues, thus behaving as a target-specific "bio-glue." A single Qdot can be attached to several affinity molecules simultaneously. This is possible because the size of Qdots is comparable or slightly larger than that of many proteins. The conjugation of Qdots to biological affinity molecules (biomolecules such as proteins or antibodies) is known as Qdot bioconjugation.

Various techniques (Figure 22.4) have been explored for Qdot bioconjugation including passive adsorption (e.g., proteins that are non-specifically adsorbed to Qdots (Hanaki et al., 2003)), multivalent chelation, and covalent coupling. Two popular cross-linking techniques (Hanaki et al., 2003) are carbodiimide-mediated amide bond formation and active ester maleimide-mediated amine and sulfhydryl coupling. In carbodiimide-mediated coupling reaction, carboxyl-modified Qdots are reacted to the amine groups of antibodies (proteins that contain both carboxyl and amine functional groups) to form stable amide bonds using ethyl-3-dimethyl amino propyl carbodiimide (EDAC) as a catalyst. Conjugation of antibody fragments that contains free sulfhydryl groups to amine functionalized Qdots are done using succinimidyl-4-*N*-maleimidomethyl-cyclohexane carboxylate (SMCC) sulfhydryl-amine coupling agents. Manipulating the molecular orientation of affinity ligands as well as their molar ratios with respect to Qdots in using both EDAC and SMCC coupling techniques is a challenging task. Engineering proteins to express positively charged domains allows them to self-assemble onto the surface of negatively charged Qdots through electrostatic assembly (Mattoussi et al., 2000). This approach has been proven to be useful for attaching a variety of engineered proteins to Qdots including maltose binding protein (MBP) (Goldman et al., 2002a). Goldman et al. (2002b) first reported the use of a fusion protein as an adaptor for immunoglobulin G (IgG) antibody coupling. The adaptor protein has two domains, a positively charged leucine zipper domain that binds electrostatically to Qdots' surface and a protein G domain that binds to the antibody Fc region.

→

FIGURE 22.4. Schematic representation of various bioconjugation techniques to attach quantum dots (qdot) to biomolecules. **a** Carboxyl-modified qdots attached to amine-conjugated antibodies by EDAC coupling agent. **b** Carboxyl/amine-modified qdots attached to thiol-conjugated antibodies by SMCC coupling agent. **c** Adaptor-protein–conjugated qdots attached to antibodies by EDAC coupling agent. **d** Ni-NT–conjugated carboxyl-modified qdots attached to His-tagged peptide. **e** Avidin-coated qdots attached to biotinalated antibodies.

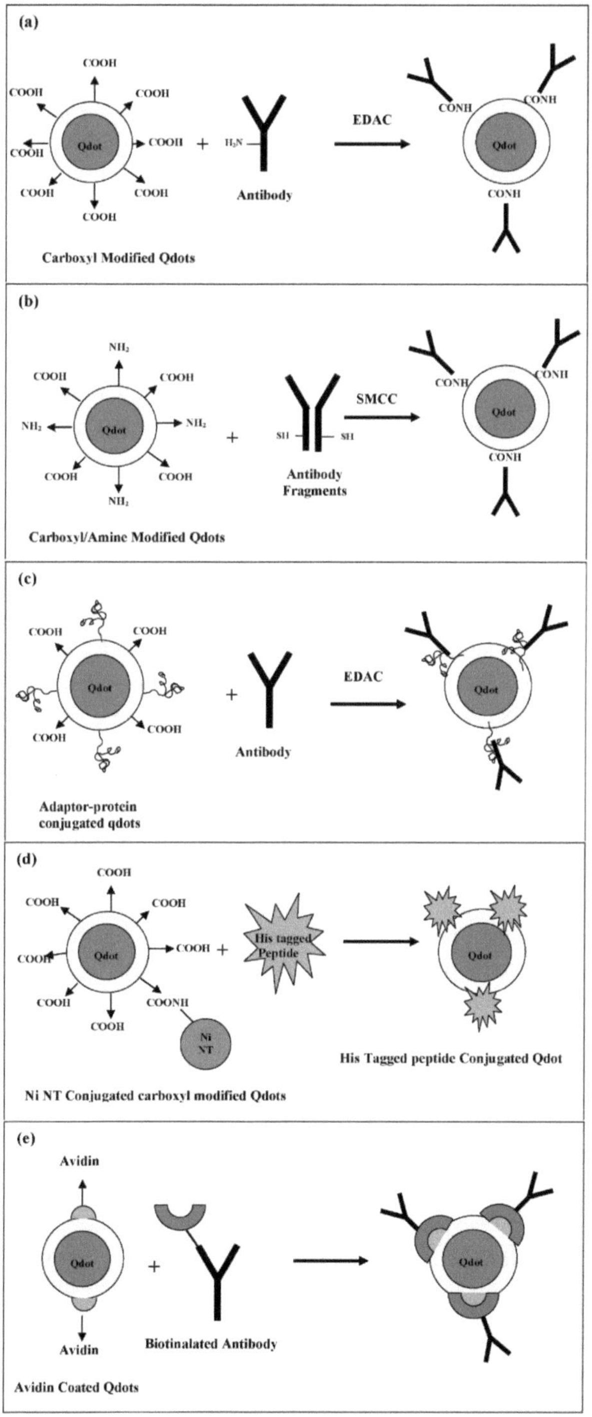
(a)
COOH
Qdot
H2N
Antibody
EDAC
CONH
Carboxyl Modified Qdots
(b)
NH2
COOH
SH
SMCC
Antibody
Fragments
Carboxyl/Amine Modified Qdots
(c)
Adaptor-protein
conjugated qdots
(d)
His tagged
Peptide
COONH
Ni
NT
His Tagged peptide Conjugated Qdot
Ni NT Conjugated carboxyl modified Qdots
(e)
Avidin
Biotinalated Antibody
Avidin Coated Qdots

FIGURE 22.4.

The target-specific $A(ab)_2$ domain remains available on the Qdot surface for target binding. Qdots have also been attached to antibodies via the avidin–biotin coupling method where biotinylated antibodies were linked to avidin-coated Qdots (Goldman et al., 2002b).

Direct attachment of proteins and peptides to the Qdot surface is based on two types of Qdot surface protein interactions: dative thiol-bonding between Qdot surface sulfur atoms and cysteine residues (Akerman et al., 2002; Pinaud et al., 2004) and metal-affinity coordination of histidine (HIS) residues to the Qdot surface Zn atoms (Figure 22.3h) (Hainfeld et al., 1999; Slocik et al., 2002). The Weiss group (Pinaud et al., 2004) demonstrated the former by using phytochelatin-related peptides to cap CdSe/ZnS (core/shell) Qdots, providing not only surface passivation and water solubility, but also a point of biochemical modification (Figure 22.3h). Using peptides for both dispersion and biofunctionalization represents a new class of rationally designed multifunctional biological cap (Mattoussi et al., 2000; Pinaud et al., 2004).

By using metal-affinity coordination, HIS-expressing proteins or peptides can be directly attached to Zn on the Qdot-surface. This strong interaction (Zn^{2+}-HIS) has a dissociation constant, K_D, stronger than most antibody bindings ($10^{-6} - 10^{-9}$) (Hainfeld et al., 1999). Gao et al. (2005) have linked Qdots to a chelating compound (nickel–nitrilotriacetic acid or Ni–NTA) that quantitatively and strongly binds to hexahistidine-tagged biomolecules (K_D for 6-HIS to NTA–Ni^{2+} is 10^{-13}) with controlled molar ratio and molecular orientation. Functional assays with HIS-appended proteins indicate that it can be exerted on the final bioconjugate assembly through the molar ratios of each participant added before self-assembly (Clapp et al., 2004; Medintz et al., 2003, 2004a,b). This strategy allows mixed protein surfaces with HIS-like affinity for other metals (Ni, Cu, Co, Fe, Mn) which is relevant to future materials (Hainfeld et al., 1999).

Regardless of the approach used in forming hybrid conjugates, three important issues remain: (i) the lack of robust bioconjugation strategies, (ii) the lack of homogeneity of attached proteins to Qdot surface, and (iii) the inability to finely control ratios of proteins to Qdots. The lack of homogeneous attachment results in heterogeneous protein orientation on the Qdot surface, which may produce conjugates with surface-attached proteins that are not optimally functional. It is important to understand that the success in bioconjugation directly relies on the development of robust water-dispersible surface-passivating agents.

22.3. Cancer Imaging with Qdots

22.3.1. Strategies of Delivering Qdots into Cells

Qdots have been mostly used in labeling cells. In the last few years, numerous reports (Gao et al., 2004, 2005; Jiang et al., 2004; Medintz et al., 2005; Michalet et al., 2005) have described various strategies of delivering Qdots to cells. Extended visualization of cells with multiple color Qdots have been demonstrated by various researchers (Table 22.1) (Hanaki et al., 2003; Jaiswal et al., 2003; Kaul et al., 2003;

TABLE 22.1. Certain representative cellular components and proteins (along with their functions) that have been labeled with qdots.

Cellular component/ protein	Function
Nucleus/nuclear proteins	Internal organelle containing the genetic information (chromosomal DNA) enclosed by a membrane containing pores that mediate transport in and out
Mitochondria	Organelle providing essential energy-delivery processes, containing its own genome
Microtubules	Cytoskeleton protein that maintains cellular structure and reforms during movement
Actin filaments	Cytoskeleton protein that maintains cellular structure, helps localize other organelles, and reforms during movement
Endocytic compartments	Vescicles that form on cell surfaces and mediate both specific and nonspecific internalization of extra-cellular molecules
Mortalin	Member of Heat Shock 70 Protein Family. Differential staining pattern in normal and precancerous cells
Cytokeratin	Cytoskeleton protein that is over-expressed and differentially stained in many skin cancer cells

Sukhanova et al., 2004; Wu et al., 2003). Qdot labeling of fixed cells (dead cells with chemically cross-linked cellular components to maintain cellular integrity) is comparatively easy as one can harshly treat such fixed cell/samples to promote the delivery of Qdots. Often, Qdots are delivered into cells by chemically creating pores. However, a delicate treatment is required while attempting to label live cells with Qdots. Any adverse labeling conditions can affect cell viability. Therefore, 'live cell' imaging with Qdots is always challenging.

Water-soluble Qdots are relatively large in size to enter into cells by crossing the lipid bilayer of the cellular membrane. Qdots are often endocytosed non-specifically and end up in the endocytic compartments (Table 22.1). Direct microinjection of nanolitre volumes of Qdots is tedious, only a few of them can be labeled. Electroporation technique creates temporary pores in the cell membrane that allows Qdots to enter into cells. The use of lipid-based transfecting agents such as Lipofectamine 2000 greatly facilitates entry of Qdots into cells (Voura et al., 2004) where Qdot aggregates are encapsulated within lipid vesicles. Using cell-penetrating peptides (such as HIV-derived TAT peptides), Qdots have been delivered to specific cellular compartments such as the nucleus (Chen and Gerion, 2004). Specific targeting of cell surface proteins (surface receptors) via antibody-conjugated Qdots followed by monitoring of cellular development for over a week has been successfully demonstrated (Dubertret et al., 2002; Jaiswal et al., 2003; Pellegrino et al., 2003).

22.3.2. In Vitro Cancer Imaging with Qdots

Molecular probes, such as organic fluorescent dyes (e.g., fluorescent proteins), are traditionally used for cancer cell labeling. Being more bright and photostable,

Qdots have demonstrated their superiority in tracking tumor cell migration (Parak et al., 2002), imaging various cancer cells including human breast cancer (MDA-MB-435S (Pellegrino et al., 2003), MCF 7 (Pellegrino et al., 2003) and SK-BR-3 (Wu et al., 2003)), human prostate cancer (Gao et al., 2004), squamous carcinoma (Lidke et al., 2004; Morgan et al., 2005) B16 melanoma (skin cancer) (Voura et al., 2004), human neuro-blastoma (SK-N-SH) (Winter et al., 2001), colon tumor (SW480) (Pellegrino et al., 2003), lung tumor (NCI H1299) (Pellegrino et al., 2003), and bone tumor (Saos-2) (Pellegrino et al., 2003) cells.

Cancer metastasis and the formation of secondary tumors are closely related to cancer cell motility and migration. Parak et al. (2002) used silica-coated CdSe/ZnS Qdots (Gerion et al., 2001) of two different sizes (2.8 nm and 4.1 nm CdSe core sizes with corresponding emission at 554 nm and 626 nm, respectively) for imaging 'phagokinetic tracks'. A phagokinetic track is generated when a cell passes over a layer of markers and ingest them, leaving behind a track. Human mammary epithelial tumor cells (MDA-MB-231), known to be very invasive and migrator, voraciously engulfed Qdots non-specifically as they migrated over the substrate in 3 hours after cell plating. Photostability of Qdot allowed monitoring of these cells continuously over a period of 16 minutes without any noticeable decrease in the intensity, confirming the benefit of using Qdot-based labels.

Tumor and normal blood vessels can be distinguished by their specific molecular markers. Peptides that can be used for targeting these molecular markers are known as homing peptides. Conjugating Qdots to tumor vasculature–specific homing peptides could light-up tumors by the bright and stable Qdot emission. In an attempt to demonstrate this idea, Akerman et al. (2002) have conjugated surface-modified CdSe/ZnS Qdots to F3 (KDEPQRRSARLSAK-PAPPKPEPKPKKAPAKK) and LyP-1 (CGNKRTRGC) peptides that preferentially bind to tumor cells. Studies with human breast carcinoma (MDA-MD-435) cells clearly showed the practical approach both in vitro and in MDA-MB-435 xenograft tumor system.

Human SK-BR-3 breast cancer cells overexpress a cancer marker (Her2) on the cell surface. Anti-Her2 antibody can specifically recognize these Her2 surface receptors, providing a way of targeting SK-BR-3 cells. Wu et al. (2003) used IgG and streptavidin-linked Qdots of two different colors (emission at 560 nm and 608 nm) to detect Her2 cancer markers. Experiments performed on both live and fixed human breast cancer cells (SK-BR-3) demonstrated robustness in terms of brightness and photostability of Qdot-based labels over the traditional organic dyes.

To image cancer tissues that are deeply seeded, it is necessary to employ NIR fluorescent probes. This is due to the fact that most tissue chromophores, including oxyhemoglobin, deoxyhemoglobin, and melanin, absorb comparatively weakly in the NIR spectral range (Ntziachristos et al., 2003). Morgan et al. (2005) reported the successful use of bovine serum albumin (BSA)–coated CdMnTe/Hg Qdots with a broad fluorescence peak in the NIR at 770 nm for imaging squamous carcinoma cells in C3H mice.

Effective tracking of cancer metastasis could lead to better understanding of the metastasis. Voura et al. (2004) used water-soluble CdSe/ZnS Qdots capped with negatively charged dihydroxylipoic acid (DHLA) for imaging B16 melanoma (skin cancer) cells. DHLA-capped Qdots were treated with a cationic lipid–based transfecting agent (Lipofectamine 2000) prior to Qdot loading to cells. With this technique, about 85%–95% cells were consistently loaded with Qdots. These Qdot-loaded cells were then injected into mice to track tumor cell extravasation.

All the above-mentioned in vitro studies have clearly demonstrated the usefulness of Qdots for sensitive cancer imaging including tracking of cancer metastasis. It is expected that the use of Qdot labels for cancer research will continue to grow.

22.3.3. In Vivo Cancer Imaging with Qdots

The application of Qdots for in vivo cancer imaging started in the last few years. Appropriate Qdot surface modification and selection of appropriate affinity ligands play an important role in obtaining success in cancer imaging in vivo. Akerman et al. (2002) reported the use of certain peptides that could specifically target the tumor vasculature and escape clearance by the reticulo-endothelial system. In an MDA-MB-435 xenograft tumor system, they examined the homing of peptide-coated CdSe/ZnS Qdots to tumor vasculature. Intravenously injected F3-coated and LyP-1-coated Qdots were accumulated in tumors, demonstrating the efficacy of Qdot labels. They also found that co-coating of polyethylene glycol (PEG) with peptide on the Qdot surface reduced the accumulation in the liver and spleen by about 95%.

Qdots can be delivered to tumors by both passive and active targeting mechanisms under in vivo conditions. Gao et al. (2004) developed amphiphilic ABC triblock copolymer–coated CdSe/ZnS Qdots. Surface modification of Qdots was performed in two different ways. For passive targeting, Qdots were coated with PEG and for active targeting Qdots were co-coated with both PEG and prostate-specific membrane antigen (PSMA) antibody. In vivo targeting studies (Figure 22.5) of human prostate cancer growing in nude mice indicated that Qdots were accumulated at tumor site by the enhanced permeability and retention (passive targeting) and by PSMA antibody binding to cancer-specific cell surface receptors. The use of an ABC triblock copolymer solved the problems of particle aggregation and fluorescence loss previously encountered for Qdots stored in physiological buffer or injected into live animals (Akerman et al., 2002; Gao et al., 2002; Mattoussi et al., 2000).

Qdots allow tracking of cancer metastasis in vivo by virtue of their photostability, brightness, and multicolor imaging capabilities. Using water soluble CdSe/ZnS Qdots and emission spectrum scanning multiphoton microscopy, Voura et al. (2004) were able to follow cancer metastasis processes in vivo. In their study, Qdot-labeled B16F10 melanoma tumor cells were intravenously injected into C57BL/6 mice. The loading of Qdots into tumor cells were

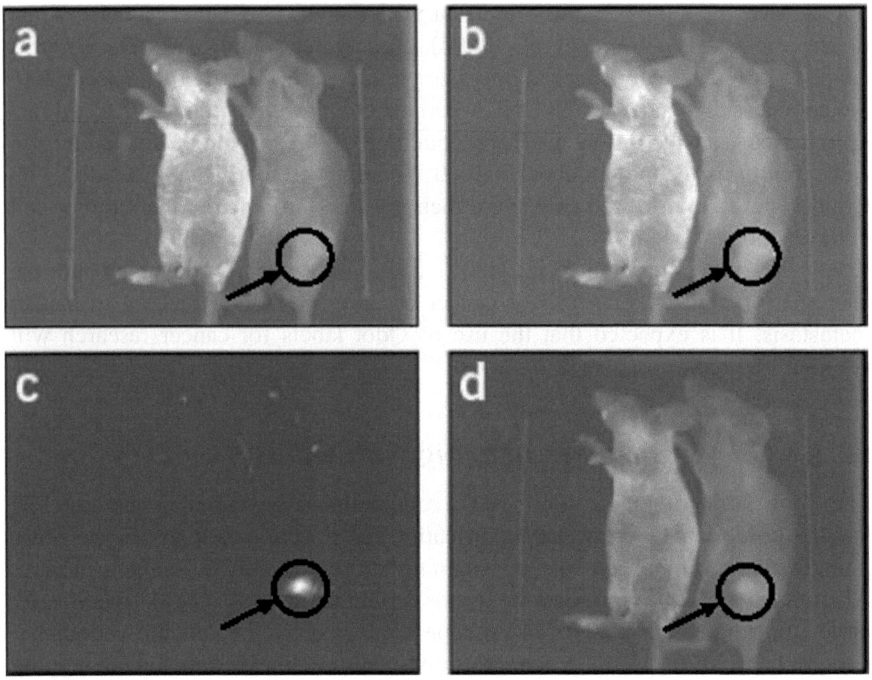

FIGURE 22.5. In vivo imaging using QD-PSMA antibody conjugate in mouse harboring C4-2 tumor xenografts. Arrows indicate the position of a prostate tumor growing in the live mouse (*right*). Control studies using a healthy mouse (no tumor) with same amount of Qdot showed no localized fluorescence signals. **a** Original image. **b** Unmixed autofluorescence image. **c** Unmixed Qdot image. **d** Superimposed image. Reprinted by permission of the author.

performed by means of a lipid-based transfecting agent, lipofectamine 2000. Using a multiphoton microscopy setup, it was possible to follow tumor cells that were extravasated into lung tissue. However, it was not possible to follow tumor cells without Qdot labeling.

For improved tissue penetration, Kim et al. (2004) reported the application of a novel CdTe/CdSe core-shell Qdots (called Type II Qdots), with fairly broad emission at 850 nm emission and a moderate quantum yield of $\sim$13% for cancer surgery. These Qdots allowed in vivo sentinel lymph node mapping in the mouse and pig. Using only 5 mW cm^{-2} excitation, they imaged lymph nodes 1 cm deep in tissue. This work clearly demonstrated the strong possibility of using NIR Qdots for real-time intra-operative surgical guidance to accurately locate and remove sentinel nodes or metastatic tumors. They have also shown that conventional NIR fluorophores, such as IRDye78-CA, photobleached rapidly in a similar excitation condition.

While the future looks quite promising, there remain several challenges in the development of Qdot-based cancer imaging probes.

Ideally, Qdots should have the following features:

a) Bright, photostable, and biocompatible without any issues related to Qdot cytotoxicity.
b) Long blood retention time.
c) Ability to label the target specifically (in case of active tumor-targeting applications).
d) Means to escape from the reticulo-endothelial system (biological particulate filter system such as spleen, liver, etc.).
e) Filtration out of the in vivo system after the completion of imaging studies.
f) Excitation and emission in the NIR spectral range so that deeply seeded tumors can be imaged.

22.4. Conclusions and Perspectives

With the help of early diagnosis and enhanced diagnostic cancer imaging, the future holds promises of prevention and possible cures for cancer patients. Recent developments in nanotechnology including Qdots, with their superior brightness and photostability, have a strong potential for monitoring cancer development at the early stage. Yet, there are still several challenges to overcome before Qdots can be used as diagnostic cancer imaging probes. The first critical step is to use Qdot surface modification chemistry to protect the core from an adverse biological milieu by improving dispersibility in biological media and obtaining appropriate surface functionality for bioconjugation. Next, attachment of appropriate affinity ligands to Qdots is needed for specific targeting of cancers which could reduce nanotoxicity-related side effects. For deep tissue cancer imaging, the development of nanoparticles with NIR excitation and emission is a necessity. The development of multifunctional Qdots (such as fluorescent and paramagnetic or fluorescent and X-ray opaque) have the advantage of pre-operative tumor assessment and intra-operative surgical guidance for the tumor tissue resection. Finally, cellular studies have followed Qdot exposure over time (Dubertret et al., 2002; Hoshino et al., 2004; Jaiswal et al., 2003; Voura et al., 2004; Wu et al., 2003), to date there have been no long-term animal studies assessing Qdot toxicology. Thus, cytotoxicity studies of Qdots in live cell or animal experiments must be completed.

In conclusion, studies using animal models clearly demonstrated new possibilities for in vivo tumor imaging using Qdots. The development of Qdots with broadly tunable NIR emission will overcome the lack of spatial resolution in tumor depth and location. For clinical human applications, a major concern is likely to be the potential toxicity of Qdots. Nevertheless, the introduction of non-toxic, bright NIR Qdots with specific tumor targeting capabilities will allow the precise diagnosis and accurate surgical resection of tumors.

Acknowledgment. The author acknowledges the following people for their aid in editing this chapter: Olajompa Moloye, Cynthia Klein, and Subir Kumar Sabui.

References

Akerman, M.E., Chan, W.C.W., Laakkonen, P., Bhatia, S.N., Ruoslahti, E., 2002. Nanocrystal targeting in vivo. Proc Natl Acad Sci U S A 99, 12617–12621.

Alivisatos, A.P., Gu, W.W., Larabell, C., 2005. Quantum dots as cellular probes. Annu Rev Biomed Eng 7, 55–76.

American Cancer Society, 2005. Cancer Facts and Figures 2005. American Cancer Society, and Atlanta, p. 1.

Artemyev, M.V., Gaponenko, S.V., Germanenko, I.N., Kapitonov, A.M., 1995. Irreversible photochemical spectral hole-burning in quantum-sized CdS nanocrystals embedded in a polymeric film. Chem Phys Lett 243, 450–455.

Bailey, R.E., Nie, S.M., 2003. Alloyed semiconductor quantum dots: tuning the optical properties without changing the particle size. J Am Chem Soc 125, 7100–7106.

Ballou, B., Lagerholm, B.C., Ernst, L.A., Bruchez, M.P., Waggoner, A.S., 2004. Noninvasive imaging of quantum dots in mice. Bioconjug Chem 15, 79–86.

Ben-Ari, E.T., 2003. Nanoscale quantum dots hold promise for cancer applications. J Natl Cancer Inst 95, 502–504.

Bremer, C., Ntziachristos, V., Weissleder, R., 2003. Optical-based molecular imaging: contrast agents and potential medical applications. Eur Radiol 13, 231–243.

Brigger, I., Dubernet, C., Couvreur, P., 2002. Nanoparticles in cancer therapy and diagnosis. Adv Drug Deliv Rev 54, 631–651.

Bruchez, M., Moronne, M., Gin, P., Weiss, S., Alivisatos, A.P., 1998. Semiconductor nanocrystals as fluorescent biological labels. Science 281, 2013–2016.

Chan, W.C.W., Maxwell, D.J., Gao, X.H., Bailey, R.E., Han, M.Y., Nie, S.M., 2002. Luminescent quantum dots for multiplexed biological detection and imaging. Curr Opin Biotechnol 13, 40–46.

Chan, W.C.W., Nie, S.M., 1998. Quantum dot bioconjugates for ultrasensitive nonisotopic detection. Science 281, 2016–2018.

Chen, F.Q., Gerion, D., 2004. Fluorescent CdSe/ZnS nanocrystal-peptide conjugates for long-term, nontoxic imaging and nuclear targeting in living cells. Nano Letters 4, 1827–1832.

Clapp, A.R., Medintz, I.L., Mauro, J.M., Fisher, B.R., Bawendi, M.G., Mattoussi, H., 2004. Fluorescence resonance energy transfer between quantum dot donors and dye-labeled protein acceptors. J Am Chem Soc 126, 301–310.

Dabbousi, B.O., RodriguezViejo, J., Mikulec, F.V., Heine, J.R., Mattoussi, H., Ober, R., Jensen, K.F., Bawendi, M.G., 1997. (CdSe)ZnS core-shell quantum dots: Synthesis and characterization of a size series of highly luminescent nanocrystallites. J Phys Chem B 101, 9463–9475.

Dahan, M., Levi, S., Luccardini, C., Rostaing, P., Riveau, B., Triller, A., 2003. Diffusion dynamics of glycine receptors revealed by single-quantum dot tracking. Science 302, 442–445.

Dubertret, B., Skourides, P., Norris, D.J., Noireaux, V., Brivanlou, A.H., Libchaber, A., 2002. In vivo imaging of quantum dots encapsulated in phospholipid micelles. Science 298, 1759–1762.

Feldhein, D.L., Foss, C.A., (Eds.), 2001. Metal nanoparticles: synthesis, characterization & applications, 1st ed. Marcel Dekker.

Fetcho, J.R., O'Malley, D.M., 1997. Imaging neuronal networks in behaving animals. Curr Opin Neurobiol 7, 832–838.

Gao, X.H., Chan, W.C.W., Nie, S.M., 2002. Quantum-dot nanocrystals for ultrasensitive biological labeling and multicolor optical encoding. J Biomed Opt 7, 532–537.

Gao, X.H., Cui, Y.Y., Levenson, R.M., Chung, L.W.K., Nie, S.M., 2004. In vivo cancer targeting and imaging with semiconductor quantum dots. Nat Biotechnol 22, 969–976.

Gao, X.H., Nie, S.M., 2003. Molecular profiling of single cells and tissue specimens with quantum dots. Trends Biotechnol 21, 371–373.

Gao, X.H., Yang, L.L., Petros, J.A., Marshal, F.F., Simons, J.W., Nie, S.M., 2005. In vivo molecular and cellular imaging with quantum dots. Curr Opin Biotechnol 16, 63–72.

Gerion, D., Pinaud, F., Williams, S.C., Parak, W.J., Zanchet, D., Weiss, S., Alivisatos, A.P., 2001. Synthesis and properties of biocompatible water-soluble silica-coated CdSe/ZnS semiconductor quantum dots. J Phys Chem B 105, 8861–8871.

Goldman, E.R., Anderson, G.P., Tran, P.T., Mattoussi, H., Charles, P.T., Mauro, J.M., 2002a. Conjugation of luminescent quantum dots with antibodies using an engineered adaptor protein to provide new reagents for fluoroimmunoassays. Anal Chem 74, 841–847.

Goldman, E.R., Balighian, E.D., Kuno, M.K., Labrenz, S., Tran, P.T., Anderson, G.P., Mauro, J.M., Mattoussi, H., 2002b. Luminescent quantum dot-adaptor protein-antibody conjugates for use in fluoroimmunoassays. Physica Status Solidi B-Basic Research 229, 407–414.

Goldman, E.R., Balighian, E.D., Mattoussi, H., Kuno, M.K., Mauro, J.M., Tran, P.T., Anderson, G.P., 2002c. Avidin: A natural bridge for quantum dot-antibody conjugates. J Am Chem Soc 124, 6378–6382.

Gu, H.W., Ho, P.L., Tsang, K.W.T., Wang, L., Xu, B., 2003. Using biofunctional magnetic nanoparticles to capture vancomycin-resistant enterococci and other gram-positive bacteria at ultralow concentration. J Am Chem Soc 125, 15702–15703.

Gu, H.W., Zheng, R.K., Zhang, X.X., Xu, B., 2004. Facile one-pot synthesis of bifunctional heterodimers of nanoparticles: a conjugate of quantum dot and magnetic nanoparticles. J Am Chem Soc 126, 5664–5665.

Guo, W.Z., Li, J.J., Wang, Y.A., Peng, X.G., 2003. Conjugation chemistry and bioapplications of semiconductor box nanocrystals prepared via dendrimer bridging. Chem Mater 15, 3125–3133.

Hainfeld, J.F., Liu, W.Q., Halsey, C.M.R., Freimuth, P., Powell, R.D., 1999. Ni-NTA-gold clusters target his-tagged proteins. J Struct Biol 127, 185–198.

Hanaki, K., Momo, A., Oku, T., Komoto, A., Maenosono, S., Yamaguchi, Y., Yamamoto, K., 2003. Semiconductor quantum dot/albumin complex is a long-life and highly photostable endosome marker. Biochem Biophys Res Commun 302, 496–501.

Hines, M.A., Guyot-Sionnest, P., 1996. Synthesis and characterization of strongly luminescing ZnS-Capped CdSe nanocrystals. J Phys Chem 100, 468–471.

Hong, R., Fischer, N.O., Verma, A., Goodman, C.M., Emrick, T., Rotello, V.M., 2004. Control of protein structure and function through surface recognition by tailored nanoparticle scaffolds. J Am Chem Soc 126, 739–743.

Hoshino, A., Hanaki, K., Suzuki, K., Yamamoto, K., 2004. Applications of T-lymphoma labeled with fluorescent quantum dots to cell tracing markers in mouse body. Biochem Biophys Res Commun 314, 46–53.

Ishii, D., Kinbara, K., Ishida, Y., Ishii, N., Okochi, M., Yohda, M., Aida, T., 2003. Chaperonin-mediated stabilization and ATP-triggered release of semiconductor nanoparticles. Nature 423, 628–632.

Jaiswal, J.K., Mattoussi, H., Mauro, J.M., Simon, S.M., 2003. Long-term multiple color imaging of live cells using quantum dot bioconjugates. Nat Biotechnol 21, 47–51.

Jiang, W., Papa, E., Fischer, H., Mardyani, S., Chan, W.C.W., 2004. Semiconductor quantum dots as contrast agents for whole animal imaging. Trends Biotechnol 22, 607–609.

Kaul, Z., Yaguchi, T., Kaul, S.C., Hirano, T., Wadhwa, R., Taira, K., 2003. Mortalin imaging in normal and cancer cells with quantum dot immuno-conjugates. Cell Res 13, 503–507.

Kim, S., Bawendi, M.G., 2003. Oligomeric Ligands for luminescent and stable nanocrystal quantum dots. J Am Chem Soc 125, 14652–14653.

Kim, S., Lim, Y.T., Soltesz, E.G., De Grand, A.M., Lee, J., Nakayama, A., Parker, J.A., Mihaljevic, T., Laurence, R.G., Dor, D.M., Cohn, L.H., Bawendi, M.G., Frangioni, J.V., 2004. Near-infrared fluorescent type II quantum dots for sentinel lymph node mapping. Nat Biotechnol 22, 93–97.

Larson, D.R., Zipfel, W.R., Williams, R.M., Clark, S.W., Bruchez, M.P., Wise, F.W., Webb, W.W., 2003. Water-soluble quantum dots for multiphoton fluorescence imaging in vivo. Science 300, 1434–1436.

Leatherdale, C.A., Woo, W.-K., Mikulec, F.V., Bawendi, M.G., 2002. On the absorption cross section of CdSe nanocrystal quantum dots. J Phys Chem B 106, 7619–7622.

Licha, K., 2002. Contrast agents for optical imaging. Contrast Agents Ii, pp. 1–29.

Licha, K., Olbrich, C., 2005. Optical imaging in drug discovery and diagnostic applications. Adv Drug Deliv Rev 57, 1087–1108.

Licha, K., Riefke, B., Ebert, B., Grotzinger, C., 2002. Cyanine dyes as contrast agents in biomedical optical imaging. Acad Radiol 9, S320–S322.

Lidke, D.S., Nagy, P., Heintzmann, R., Arndt-Jovin, D.J., Post, J.N., Grecco, H.E., Jares-Erijman, E.A., Jovin, T.M., 2004. Quantum dot ligands provide new insights into erbB/HER receptor-mediated signal transduction. Nat Biotechnol 22, 198–203.

Mattheakis, L.C., Dias, J.M., Choi, Y.J., Gong, J., Bruchez, M.P., Liu, J.Q., Wang, E., 2004. Optical coding of mammalian cells using semiconductor quantum dots. Anal Biochem 327, 200–208.

Mattoussi, H., Mauro, J.M., Goldman, E.R., Anderson, G.P., Sundar, V.C., Mikulec, F.V., Bawendi, M.G., 2000. Self-assembly of CdSe-ZnS quantum dot bioconjugates using an engineered recombinant protein. J Am Chem Soc 122, 12142–12150.

Medintz, I.L., Clapp, A.R., Mattoussi, H., Goldman, E.R., Fisher, B., Mauro, J.M., 2003. Self-assembled nanoscale biosensors based on quantum dot FRET donors. Nat Mater 2, 630–638.

Medintz, I.L., Konnert, J.H., Clapp, A.R., Stanish, I., Twigg, M.E., Mattoussi, H., Mauro, J.M., Deschamps, J.R., 2004a. A fluorescence resonance energy transfer-derived structure of a quantum dot-protein bioconjugate nanoassembly. Proc Natl Acad Sci U S A 101, 9612–9617.

Medintz, I.L., Trammell, S.A., Mattoussi, H., Mauro, J.M., 2004b. Reversible modulation of quantum dot photoluminescence using a protein-bound photochromic fluorescence resonance energy transfer acceptor. J Am Chem Soc 126, 30–31.

Medintz, I.L., Uyeda, H.T., Goldman, E.R., Mattoussi, H., 2005. Quantum dot bioconjugates for imaging, labelling and sensing. Nat Mater 4, 435–446.

Michalet, X., Pinaud, F.F., Bentolila, L.A., Tsay, J.M., Doose, S., Li, J.J., Sundaresan, G., Wu, A.M., Gambhir, S.S., Weiss, S., 2005. Quantum dots for live cells, in vivo imaging, and diagnostics. Science 307, 538–544.

Mitchell, G.P., Mirkin, C.A., Letsinger, R.L., 1999. Programmed assembly of DNA functionalized quantum dots. J Am Chem Soc 121, 8122–8123.

Morgan, N.Y., English, S., Chen, W., Chernomordik, V., Russo, A., Smith, P.D., Gandjbakhche, A., 2005. Real time in vivo non-invasive optical imaging using near-infrared fluorescent quantum dots. Acad Radiol 12, 313–323.

Murray, C.B., Kagan, C.R., Bawendi, M.G., 2000. Synthesis and characterization of monodisperse nanocrystals and close-packed nanocrystal assemblies. Annu Rev Mater Sci 30, 545–610.

Murray, C.B., Norris, D.J., Bawendi, M.G., 1993. Synthesis and Characterization of Nearly Monodisperse CdE (E = S, Se, Te) Semiconductor Nanocrystallites. J Am Chem Soc 115, 8706–8715.

Ntziachristos, V., Bremer, C., Weissleder, R., 2003. Fluorescence imaging with near-infrared light: new technological advances that enable in vivo molecular imaging. Eur Radiol 13, 195–208.

Osaki, F., Kanamori, T., Sando, S., Sera, T., Aoyama, Y., 2004. A quantum dot conjugated sugar ball and its cellular uptake on the size effects of endocytosis in the subviral region. J Am Chem Soc 126, 6520–6521.

Parak, W.J., Boudreau, R., Le Gros, M., Gerion, D., Zanchet, D., Micheel, C.M., Williams, S.C., Alivisatos, A.P., Larabell, C., 2002. Cell motility and metastatic potential studies based on quantum dot imaging of phagokinetic tracks. Adv Mater 14, 882–885.

Parak, W.J., Gerion, D., Pellegrino, T., Zanchet, D., Micheel, C., Williams, S.C., Boudreau, R., Le Gros, M.A., Larabell, C.A., Alivisatos, A.P., 2003. Biological applications of colloidal nanocrystals. Nanotechnology 14, R15–R27.

Pellegrino, T., Manna, L., Kudera, S., Liedl, T., Koktysh, D., Rogach, A.L., Keller, S., Radler, J., Natile, G., Parak, W.J., 2004. Hydrophobic nanocrystals coated with an amphiphilic polymer shell: A general route to water soluble nanocrystals. Nano Letters 4, 703–707.

Pellegrino, T., Parak, W.J., Boudreau, R., Le Gros, M.A., Gerion, D., Alivisatos, A.P., Larabell, C.A., 2003. Quantum dot-based cell motility assay. Differentiation 71, 542–548.

Peng, X.G., Schlamp, M.C., Kadavanich, A.V., Alivisatos, A.P., 1997. Epitaxial growth of highly luminescent CdSe/CdS core/shell nanocrystals with photostability and electronic accessibility. J Am Chem Soc 119, 7019–7029.

Peng, Z.A., Peng, X.G., 2001. Formation of high-quality CdTe, CdSe, and CdS nanocrystals using CdO as precursor. J Am Chem Soc 123, 183–184.

Pinaud, F., King, D., Moore, H.P., Weiss, S., 2004. Bioactivation and cell targeting of semiconductor CdSe/ZnS nanocrystals with phytochelatin-related peptides. J Am Chem Soc 126, 6115–6123.

Pinaud, F., Michalet, X., Bentolila, L.A., Tsay, J.M., Doose, S., Li, J.J., Iyer, G., Weiss, S., 2006. Advances in fluorescence imaging with quantum dot bio-probes. Biomaterials 27, 1679–1687.

Qu, L., Peng, X., 2002. Control of photoluminescence properties of CdSe nanocrystals in growth. J Am Chem Soc 124, 2049–2055.

Reiss, P., Bleuse, J., Pron, A., 2002. Highly luminescent CdSe/ZnSe core/shell nanocrystals of low size dispersion. Nano Letters 2, 781–784.

Rosi, N.L., Mirkin, C.A., 2005. Nanostructures in biodiagnostics. Chem Rev 105, 1547–1562.

Santra, S., Dutta, D., Walter, G.A., Moudgil, B.M., 2005a. Fluorescent nanoparticle probes for cancer imaging. Technology in Cancer Research and Treatment 4, 593–602.

Santra, S., Xu, J.S., Wang, K.M., Tan, W.H., 2004. Luminescent nanoparticle probes for bioimaging. J Nanosci and Nanotechnol 4, 590–599.

Santra, S., Yang, H., Stanley, J.T., Holloway, P.H., Moudgil, B.M., Walter, G., Mericle, R.A., 2005b. Rapid and effective labeling of brain tissue using TAT-conjugated CdS: Mn/ZnS quantum dots. Chem Commun, 3144–3146.

Santra, S., Yang, H.S., Holloway, P.H., Stanley, J.T., Mericle, R.A., 2005c. Synthesis of water-dispersible fluorescent, radio-opaque, and paramagnetic CdS : Mn/ZnS quantum dots: A multifunctional probe for bioimaging. J Am Chem Soc 127, 1656–1657.

Slocik, J.M., Moore, J.T., Wright, D.W., 2002. Monoclonal antibody recognition of histidine-rich peptide encapsulated nanoclusters. Nano Letters 2, 169–173.

Smith, A.M., Gao, X.H., Nie, S.M., 2004. Quantum dot nanocrystals for in vivo molecular and cellular imaging. Photochem Photobiol 80, 377–385.

Smith, A.M., Ruan, G., Rhyner, M.N., Nie, S.M., 2006. Engineering luminescent quantum dots for In vivo molecular and cellular imaging. Ann Biomed Eng 34, 3–14.

Sokolov, K., Aaron, J., Hsu, B., Nida, D., Gillenwater, A., Follen, M., MacAulay, C., Adler-Storthz, K., Korgel, B., Descour, M., Pasqualini, R., Arap, W., Lam, W., Richards-Kortum, R., 2003. Optical systems for In vivo molecular imaging of cancer. Technology in Cancer Research * Treatment 2, 491–504.

Stroh, M., Zimmer, J.P., Duda, D.G., Levchenko, T.S., Cohen, K.S., Brown, E.B., Scadden, D.T., Torchilin, V.P., Bawendi, M.G., Fukumura, D., Jain, R.K., 2005. Quantum dots spectrally distinguish multiple species within the tumor milieu in vivo. Nat Med 11, 678–682.

Sukhanova, A., Devy, M., Venteo, L., Kaplan, H., Artemyev, M., Oleinikov, V., Klinov, D., Pluot, M., Cohen, J.H.M., Nabiev, I., 2004. Biocompatible fluorescent nanocrystals for immunolabeling of membrane proteins and cells. Anal Biochem 324, 60–67.

Suyver, J.F., Wuister, S.F., Kelly, J.J., Meijerink, A., 2001. Synthesis and photoluminescence of nanocrystalline ZnS : Mn2+. Nano Letters 1, 429–433.

Tsay, J.M., Pflughoefft, M., Bentolila, L.A., Weiss, S., 2004. Hybrid approach to the synthesis of highly luminescent CdTe/ZnS and CdHgTe/ZnS nanocrystals. J Am Chem Soc 126, 1926–1927.

Uyeda, H.T., Medintz, I.L., Jaiswal, J.K., Simon, S.M., Mattoussi, H., 2005. Synthesis of compact multidentate ligands to prepare stable hydrophilic quantum dot fluorophores. J Am Chem Soc 127, 3870–3878.

Voura, E.B., Jaiswal, J.K., Mattoussi, H., Simon, S.M., 2004. Tracking metastatic tumor cell extravasation with quantum dot nanocrystals and fluorescence emission-scanning microscopy. Nat Med 10, 993–998.

Winter, J.O., Liu, T.Y., Korgel, B.A., Schmidt, C.E., 2001. Recognition molecule directed interfacing between semiconductor quantum dots and nerve cells. Adv Mater 13, 1673–1677.

Wu, X.Y., Liu, H.J., Liu, J.Q., Haley, K.N., Treadway, J.A., Larson, J.P., Ge, N.F., Peale, F., Bruchez, M.P., 2003. Immunofluorescent labeling of cancer marker Her2 and other cellular targets with semiconductor quantum dots. Nat Biotechnol 21, 41–46.

Xu, C.J., Xu, K.M., Gu, H.W., Zhong, X.F., Guo, Z.H., Zheng, R.K., Zhang, X.X., Xu, B., 2004. Nitrilotriacetic acid-modified magnetic nanoparticles as a general agent to bind histidine-tagged proteins. J Am Chem Soc 126, 3392–3393.

Yager, T.D., Ikegami, R., Rivera-Bennetts, A.K., Zhao, C., Brooker, D., 1997. High-resolution imaging at the cellular and subcellular revels in flattened whole mounts of early zebrafish embryos. Biochemistry and Cell Biology-Biochimie Et Biologie Cellulaire 75, 535–550.

Yang, H., Holloway, P.H., 2004. Efficient and photostable ZnS-Passivated CdS : Mn luminescent nanocrystals. Adv Funct Mater 14, 152–156.

Yang, H., Santra, S., Holloway, P.H., 2005. Synthesis and Application of Mn-Doped II-VI Semiconductor Nanocrystals. J Nanosci Nanotechnol 5, 1364–1375.

Yang, H.S., Holloway, P.H., Cunningham, G., Schanze, K.S., 2004a. CdS : Mn nanocrystals passivated by ZnS: Synthesis and luminescent properties. J Chem Phys 121, 10233–10240.

Yang, H.S., Holloway, P.H., Santra, S., 2004b. Water-soluble silica-overcoated CdS: Mn/ZnS semiconductor quantum dots. J Chem Phys 121, 7421–7426.

23
Bimodal Liposomes and Paramagnetic QD-Micelles for Multimodality Molecular Imaging of Tumor Angiogenesis

Willem J.M. Mulder, Gustav J. Strijkers, Rolf Koole, Celso De Mello Donega, Gert Storm, Arjan W. Griffioen, and Klaas Nicolay

Abstract: Among many pathological processes, angiogenesis, the formation of new blood vessels from pre-existing blood vessels, has received much attention during the last few decades in the field of oncology, since it was postulated that tumor growth is angiogenesis dependent. Therefore, therapies that are aimed to inhibit angiogenesis are promising interventions for cancer. Imaging methods to evaluate angiogenesis are becoming increasingly important and especially the area of molecular imaging highly depends on suitable and specific contrast agents.

Nanotechnology offers the unique possibility to create nanoscale devices that can have multiple properties integrated.

In this chapter the synthesis and application of two types of nanoparticles with fluorescent and magnetic properties for multimodality imaging, i.e. magnetic resonance imaging (MRI) and optical techniques, will be discussed. The first part deals with a nanoparticle based on liposomes which was used to visualize and quantify tumor angiogenesis in vivo with MRI, after applying angiostatic therapy. The second part of this chapter focuses on a nanoparticle that is based on fluorescent quantum dots with a paramagnetic micellar coating. We demonstrate the applicability of this contrast agent for parallel intravital microscopy and MRI in vivo.

Both nanoparticles were shown to be very useful for molecular imaging of tumor angiogenesis.

Keywords: Tumor angiogenesis, molecular imaging, magnetic resonance imaging, intravital microscopy, nanoparticles, liposomes, micelles, quantum dots.

23.1. Introduction

23.1.1. Angiogenesis

Among many physiological and pathological processes, angiogenesis, the formation of new blood vessels from pre-existing blood vessels, has received much attention during the last few decades (Griffioen and Molema, 2000). In normal tissue growth, such as in embryonic development, wound healing, and the menstrual cycle, new vessel formation for the supply of oxygen and nutrients as well as removal of waste products is a normal phenomenon. Importantly, many pathological disorders are also accompanied by angiogenesis. In tissue damage after reperfusion of ischemic tissue, for example, the formation of new blood vessels is required to improve tissue condition. On the other hand, in diseases, such as cancer, excessive new vessel formation is part of the pathology. Many other pathologies are also accompanied by excessive angiogenesis including cardiovascular diseases (atherosclerosis), chronic inflammation, and diabetes.

In 1971, it was postulated by Judah Folkman that tumor growth is angiogenesis dependent and therefore the inhibition of angiogenesis would lead to the inhibition of tumor growth (Folkman, 1971). Since then, an era of new research in the field of tumor angiogenesis started which aimed to investigate the angiogenesis processes in all its aspects, ranging from a fundamental molecular biological approach to the application of angiostatic therapy to inhibit tumor growth. Today, angiogenesis inhibitors have received FDA approval for many types of cancer (Folkman, 2003).

Since angiogenesis is a process of the blood vessels, the endothelium, the cellular barrier between the blood vessel lumen and the blood vessel wall, is of prime importance. Numerous molecular targets expressed on angiogenically activated endothelial cells have been identified. These may be used not only to identify angiogenic blood vessels, but also to serve as targets for therapeutic intervention. In the latter application, the advantage is that angiogenesis inhibitors target cells that support tumor growth, i.e. endothelial cells, which are genetically stable and therefore drug-induced resistance is less likely to develop. In addition, endothelial cells are directly accessible from the blood stream. Important angiogenic markers that have been identified include the $\alpha_v\beta_3$, $\alpha_v\beta_5$, and $\alpha_5\beta_1$ integrins, CD105, VEGF receptor, and CD36 (van Beijnum and Griffioen, 2005). In Figure 23.1, a cartoon of the angiogenesis cascade is given. Next to targets expressed on endothelial cells, circulating angiogenic factors, such as VEGF or bFGF, released by tumor cells are abundant in the blood and therefore can also serve for the identification of angiogenesis.

23.1.2. Imaging Angiogenesis

Imaging methods to depict angiogenesis have been proven to be very valuable at various levels (McDonald and Choyke, 2003). They may be helpful for the identification of angiogenic sites and to visualize abnormal morphology of angiogenic blood vessels. Furthermore, the severity of ongoing angiogenesis can be

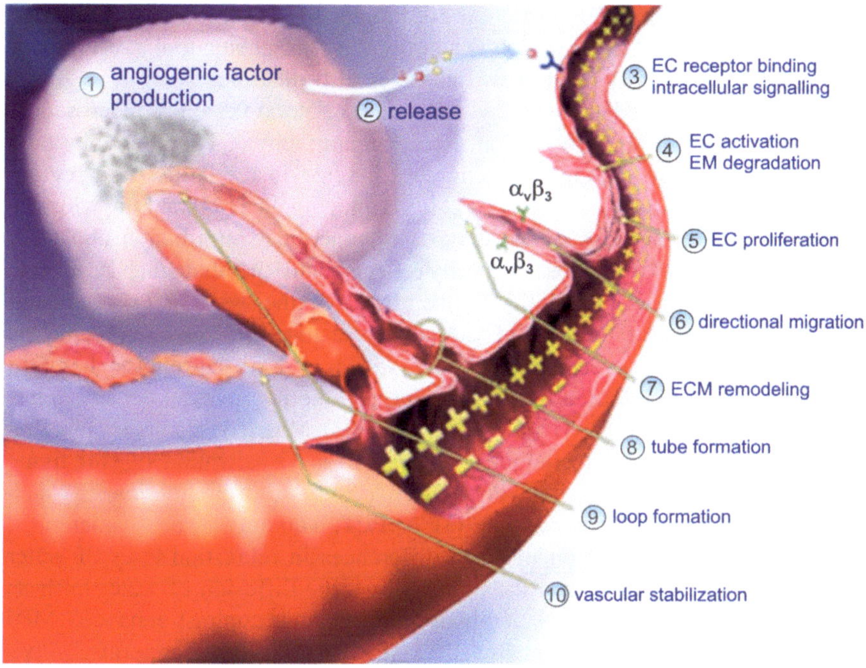

FIGURE 23.1. The angiogenesis cascade. After the release of angiogenic factors, endothelial cells become activated and start proliferating. Next, they migrate out of the existing vessel toward the tumor. This is associated with the expression of adhesion molecules $\alpha_v\beta_3$- and $\alpha_v\beta_5$-intergins that help pull the sprouting new blood vessel sprout forward. The migrating endothelial cells form a tube and maturate to become new blood vessels. (This figure is adapted from the figure "Angiogenesis: Cascade of Events," with permission from the Angiogenesis Foundation.)

visualized and the effect of angiostatic therapy may be evaluated with imaging techniques. Other parameters, like the number and the density of blood vessels, vascular permeability, and blood flow, can be quantified.

Visualization of (angiogenic) blood vessels can be realized in many ways. At the microscopic level, blood vessels can be depicted using techniques like electron microscopy or fluorescence microscopy. Detection of molecular targets expressed at the endothelial cells of such blood vessels is realized with immunohistochemical stainings. CD31 and CD34, for example, are commonly used for the identification of blood vessels in histological tissue sections. Microscopic techniques that can be applied in vivo, such as intravital microscopy (Jain et al., 2002) or two photon scanning laser microscopy (Stroh et al., 2005), are rapidly being developed and allow the visualization of blood vessels and the tumor microenvironment in live animals. This requires the application of sophisticated fluorescent probes, such as quantum dots (Michalet et al., 2005), targeted to activated blood vessels. Transgene expression of fluorescent proteins by the endothelial cells also allows the visualization of blood vessels with intravital microscopy. Furthermore, microscopic techniques can be used to determine the microvessel density (MVD), which is an important prognostic tool as well as an indicator for the severity of the new blood vessel formation.

Angiographic methods allow the visualization of blood vessels at macroscopic level, be it with a limited resolution. X-ray angiography, CT angiography, and MR angiography are among the techniques applied in the clinic on a regular basis (McDonald and Choyke, 2003). PET imaging in combination with the appropriate agent provides different types of functional information about the tumor. [^{18}F]fluorodeoxyglucose primarily reflects tumor metabolism, [^{11}C]carbon monoxide can be used to quantify tumor blood volume, while [^{15}O]water scan provides a quantitative measure of vascular perfusion (McDonald and Choyke, 2003). Furthermore, several studies have shown a good correlation between methionine uptake measured by PET in vivo and the microvessel density determined on tissue sections ex vivo (Kracht et al., 2003). Magnetic resonance imaging (MRI) is a non-invasive technique that allows the visualization of soft tissue with a relatively high resolution without the exposure to harmful radiation, which does occur in imaging methods like CT or PET. Tumors can be characterized not only with MRI based on intrinsic differences in relaxation rates of water between the normal tissue and the tumor tissue, the diffusion of water in these different tissues, but also with methods like dynamic contrast-enhanced MRI (de Lussanet et al., 2005), which allow the investigation of vascular permeability.

23.1.2.1. Molecular Imaging

A relatively new discipline in the field of imaging is molecular imaging, in which the aim is to visualize (pathophysiological) processes and transgene expression at the cellular and molecular level (Weissleder, 2002; Weissleder and Mahmood, 2001). This discipline originally was the domain of nuclear imaging methods like PET and SPECT. MRI was believed to be less suitable for the visualization of biological markers, because of the low inherent sensitivity of the technique. The development of sophisticated and powerful contrast agents during the past decade has resulted in a strongly increased interest in the use of MRI for molecular imaging purposes (Aime et al., 2002). For cellular MR imaging studies, cells of interest, for example stem cells or leukocytes, need to be labeled with magnetic agents. Different strategies to achieve a high labeling efficiency have been developed. Transfection agents, originally used for enhanced uptake of nucleotides/RNA/DNA, have been employed successfully to mediate the uptake of contrast materials by cells. Another approach for efficient labeling is the conjugation of the HIV-derived TAT-peptides to the surface of nanoparticles (Lewin et al., 2000).

For molecular imaging studies, aimed to visualize the expression of cell surface receptors, the previous strategy is also employed. Most of the MRI contrast agents used for molecular and cellular imaging are nanoparticulate probes that allow the incorporation of a high payload of contrast-generating materials. Nanoparticles are conjugated with multiple targeting ligands to introduce specificity for the receptor of interest (Mulder et al., 2006b). The ligands employed include antibodies (Abs), antibody fragments (Fabs), proteins, peptides, peptidomimetics, or small molecules with target specificity (Mulder et al., 2006b). In the case of angiogenesis, different receptors expressed at the endothelium can be used for molecular imaging, since many cell surface receptors are strongly expressed on activated endothelial cells of angiogenic vessels, as compared to resting endothelial cells of blood vessels in non-diseased tissue.

23.1.2.2. The $\alpha_v\beta_3$-Integrin and Molecular Imaging of Angiogenesis

Particularly, the $\alpha_v\beta_3$-integrin has been shown to be very useful as a target for therapies and molecular imaging contrast agents. This integrin mediates endothelial cell adhesion and migration on a variety of extracellular matrix proteins (Giancotti and Ruoslahti, 1999). Integrins are accessible to and readily inhibited by antibodies, peptides, and peptidomimetics, making them excellent targets. Different ligands are available. Next to $\alpha_v\beta_3$-specific antibodies (LM609), the $\alpha_v\beta_3$-specific RGD peptide and peptidomimetics have been used in several studies. The expression of $\alpha_v\beta_3$ has been non-invasively detected in tumor-bearing mice with a combination of radio-labeled RGD and positron emission tomography (Haubner et al., 2001). Furthermore, gene delivery and drug targeting with $\alpha_v\beta_3$-targeted lipidic nanoparticles have been successfully employed in tumor-bearing mice to inhibit tumor growth (Hood et al., 2002).

The first successful attempt to detect angiogenesis with MRI making use of the $\alpha_v\beta_3$-integrin was done with polymerized liposomes conjugated with $\alpha_v\beta_3$-specific LM609 antibodies in tumor-bearing rabbits (Sipkins et al., 1998). The polymerized liposomes used in this study were 300 to 350 nm in size and contained 30% of a Gd-lipid. This means that these liposomes have an extremely high payload of gadolinium per particle, which makes them a very potent MRI contrast agent. The liposomes were applied intravenously in the tumor-bearing rabbits, which resulted in a statistically significant signal intensity enhancement 24 hours after injection. Importantly, the signal enhancement correlated with the immunohistochemical determination of $\alpha_v y\beta_3$-integrin distribution. Perfluorcarbon nanoparticles of 250 nm, containing 90,000 Gd-DTPA-BOA, targeted to the $\alpha_v\beta_3$-integrin with RGD peptidomimetics have also been used for detecting angiogenesis in tumors implanted in rabbits (Winter et al., 2003).

In the studies described in this chapter, a cyclic variant of the RGD-peptidic sequence, synthesized with a protective acetyl group, was used (Figure 23.2).

23.1.3. Multimodality Imaging

All previous examples have focused on the visualization of angiogenesis with a single imaging modality. The drawback of this approach is that the interpretation of the results is usually difficult and the mechanism of contrast change due to the accumulation of contrast material cannot be unraveled. Furthermore, tomographic imaging techniques, like PET, SPECT, and MRI, have a limited resolution which does not permit the visualization of angiogenesis with the subcellular resolution.

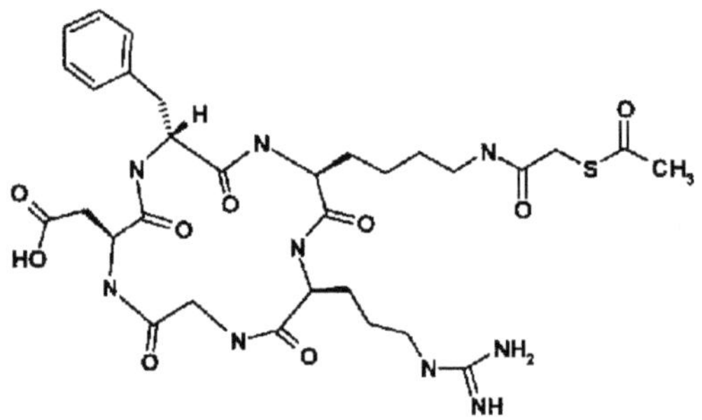

FIGURE 23.2. The cyclic pentapeptide c(RGDfV). Cyclic peptides containing a Arg-Gly-Asp (RGD) sequence have specificity for the $\alpha_v\beta_3$- and $\alpha_v\beta_5$-intergins. The above cyclic pentapeptide c(RGDfV) consists of a head-to-tail cyclic ring containing one D-amino acid and has favorable affinity for $\alpha_v\beta_3$-intergin.

Optical techniques, like fluorescence microscopy, are highly complementary to MRI and have the capability to depict tissue at the subcellular resolution, be it with a limited depth penetration and a small scanning window. Furthermore, in most cases, fluorescence microscopy is used to examine excised and fixed tissue, although in vivo fluorescence microscopy is also being employed and becoming more important. Furthermore, optical imaging is becoming increasingly important in recent years and a rapid progress is seen in tomographic methods to allow three-dimensional optical imaging.

To achieve the above bimodal imaging approach, MRI probes that also contain fluorescent reporter groups were soon recognized to be useful. One of the first examples of such a bimodal probe was reported by Huber et al. (1998). They describe bimodal contrast-enhancing agents for optical and magnetic resonance imaging. The main advantage of this combination is that it integrates optical properties for high-resolution fluorescence microscopy and imaging and magnetic properties for in vivo MRI visualization of intact and opaque organisms.

For the assessment of increased vascular permeability that is typically seen in tissues with angiogenic blood vessels, albumin labeled with a fluorophore and Gd-DTPA has been developed (Dafni et al., 2002). The paramagnetic and fluorescent properties of this interstitial macromolecular agent allow the investigation of angiogenesis-associated vascular hyperpermeability with MRI. Active clearance of this contrast material, which was also biotinylated, from the circulation was achieved with an avidin-chase. This allowed early differentiation between vascular leak and lymphatic drain. Furthermore, with histology and confocal microscopy this contrast agent could also be visualized in tumor and tissue sections.

This chapter deals with the development and application of two novel fluorescent and paramagnetic nanoparticles that have been employed to visualize tumor angiogenesis with both MRI and optical techniques. The nanoparticles are conjugated with multiple cyclic RGD-peptides (Figure 23.2) and therefore have a high affinity for the $\alpha_v\beta_3$-integrin. The first nanoparticle is based on pegylated liposomes of 150 nm in size that carry approximately 80,000 amphiphilic Gd chelates for MRI and rhodamine-based phospholipids for fluorescence microscopy. The second nanoparticle is based on quantum dots capped within the core of pegylated and paramagnetic micelles. In this agent the quantum dots exhibit fluorescent properties, while the micellular coating generates MRI contrast. Both contrast agents were fully characterized and their specificity was assessed in vitro on human endothelial cells (HUVEC). In tumor-bearing mice the agents were successfully applied for the visualization of newly formed tumor vessels with MRI and fluorescence microscopy.

23.2. Bimodal Liposomes

23.2.1. Introduction in Liposomes

Liposomes were discovered in the early 1960s by Alec D. Bangham, who found that egg lecithin phospholipids combined with water self-organized immediately into spheres because of the amphiphilic character of the lipids (Bangham et al., 1965). Liposomes can be defined as spherical, self-closed structures formed by one or several concentric lipid bilayers with an aqueous phase inside and between the lipid bilayers (Torchilin, 2005). Liposomes have been used extensively as a model to study the properties of biological membranes. Liposomes can vary in size and lamellarity and are therefore subdivided into multilamellar vesicles (MLV, consisting of several concentric bilayers), large unilamellar vesicles (LUV, in a size range of 200–800 nm), and small unilamellar vesicles (SUV, in a size range of 50–150 nm). Soon after the discovery of liposomes, they were suggested for use as a drug carrier vehicle, because of their striking biological properties. First, liposomes are composed of naturally occurring lipids, or derived synthetic lipids, and therefore are biocompatible. Liposomes can carry water-soluble pharmaceutical agents in the aqueous interior and amphiphilic or hydrophobic agents in the lipid bilayer. Furthermore, liposomal pharmaceutical agents are protected from interactions with plasma proteins, deactivation, and exhibit enhanced circulation times and biodistribution properties. Altering their surface properties makes improved delivery of liposomes to diseased tissue and into cells possible. These properties also make liposomes excellent candidates to carry or deliver contrast agents for MRI. In the 1980s, the first studies on the use of liposomes as a carrier of MR contrast agents appeared in the literature. For a recent review see Mulder et al. (2006b).

23.2.2. Synthesis of Bimodal RGD-Liposomes

Several methods have been developed for the preparation of liposomes. Most commonly, amphiphilic lipids are dissolved in organic solvents to realize a proper mixing of the different components. Upon evaporation of the solvents, a dry lipid film is formed, which is subsequently hydrated. This results in the formation of the so-called "multilamellar vesicles". MLVs are sized to LUVs by means of sonication or extrusion. The fluorescent and paramagnetic liposomes discussed in this chapter were developed recently (Mulder et al., 2004, 2005) and have shown their potential for parallel detection of cell surface receptors with fluorescence microscopy and MRI, both in vitro and in vivo. The liposomal contrast agent is composed of DSPC (1,2-distearoyl-sn-glycero-3-phosphocholine) and cholesterol, PEG-DSPE (1,2-distearoyl-sn-glycero-3-phosphoethanolamine-N-[methoxy(polye thylene glycol)-2000]) Mal-PEG-DSPE (1,2-distearoyl-sn-glycero-3-phosphoethanolamine-N-[maleimide(polyethylene glycol)2000]), rhodamine-PE (1,2-dipalmitoyl-sn-glycero-3-phosphoethanolamine-N-(lissamine rhodamine B sulfonyl)), and Gd-DTPA-BSA (Gd-DTPA-bis(stearylamide)). The main component DSPC (37%) functions as a carrier lipid, while cholesterol (33%) and PEG-DSPE (2.5%) function to stabilize the liposome. Mal-PEG-DSPE (2.5%) is used for functionalization, since maleimide can be used to conjugate ligands via a sulfhydryl–maleimide coupling method. Gd-DTPA-BSA is a DTPA bis(amide) with two aliphatic chains of 18 hydrocarbon residues. This latter amphiphile is incorporated in large quantities (25%) to provide the liposomal agent with excellent paramagnetic properties. The fluorescent properties are introduced by the incorporation of a small amount (0.1%) of rhodamine-PE.

The preparation procedure is schematically depicted in Figure 23.3 and starts with dissolving the different components in a mixture of chloroform and methanol (3/1 v/v ratio), which is followed by evaporation of the solvents until a dry mixed lipid film is obtained. This film is hydrated in an appropriate buffer with a slightly acidic pH (between 6.5 and 6.7), necessary for efficient conjugation of the targeting ligand. Upon hydration, multilamellar vesicles are formed which are vesicles composed of several concentric bilayers. Next, the lipid suspension is forced through a polycarbonate filter with a defined pore size to yield unilamellar vesicles with a diameter near the pore size (typically in between 80 and 200 nm) of the filter used. This procedure is called extrusion and is done at 65 °C, which is above the phase transition temperature of the phospholipids, which facilitates the process.

The relaxivities r_1 and r_2, the potency to shorten the T_1 and T_2 relaxation time of water, and thus the potency to generate contrast in MRI, of a MRI contrast agent can be determined from the following equation:

$$\frac{1}{T_{1,2,obs}} = \frac{1}{T_{1,2,dia}} + r_{1,2} \cdot [\mathrm{CA}]$$

which shows that the inverse of the relaxation time T_1 and T_2 (i.e., $1/T_1$ and $1/T_2$) depend linearly on the contrast agent concentration [CA] and that r_1 and r_2 can be obtained by determining the slope of a plot of $1/T_{1,2}$ versus [CA]. $1/T_{1,2,\mathrm{dia}}$ represents the inverse of the intrinsic T_1 or T_2, without contrast agent. The ionic r_1 of the bimodal liposomes at a clinically relevant field strength of 1.5 T is

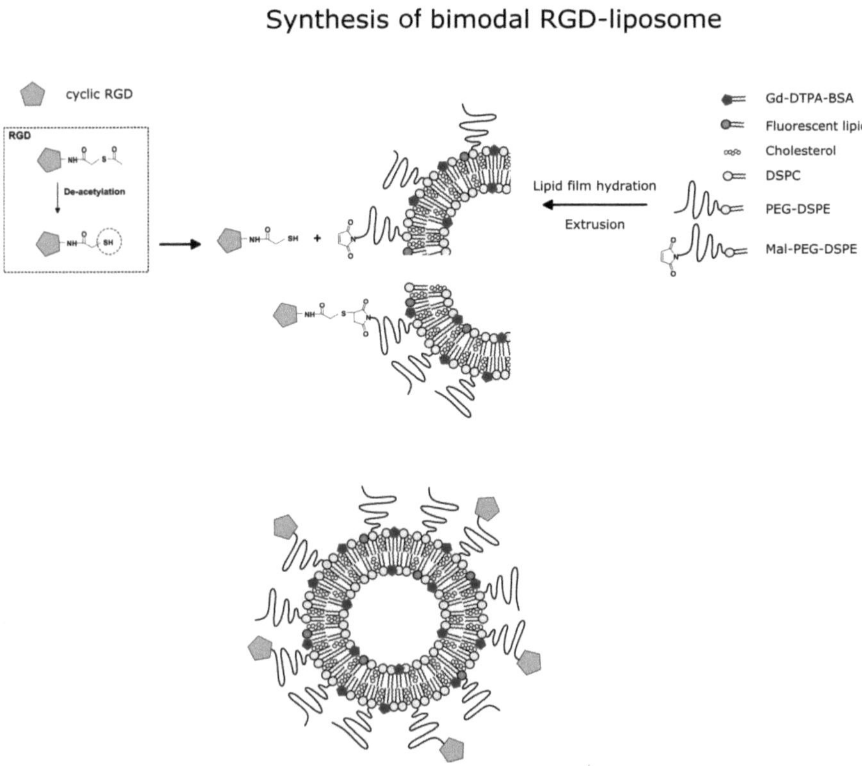

FIGURE 23.3. Schematic representation of the synthesis of bimodal RGD-liposomes. Liposomes composed of the amphiphilic molecules depicted *right* are prepared by lipid film hydration, followed by extrusion. Protected RGD-peptides are activated to obtain a thiol moiety for covalent linkage to maleimide-containing liposomes.

approximately 4 $mM^{-1}s^{-1}$, which is comparable with that of Gd-DTPA at this field strength. Depending on the size of the liposomes, they can carry up to 200 000 Gd-DTPA-BSA molecules, which enormously boosts up the relaxivity per particle, necessary to allow MRI-based molecular imaging of cell surface receptors with a relatively low abundance.

Next, RGD-peptides with a protective group are activated by cleaving of the acetyl moiety to obtain a free thiol for the bond formation via maleimide. This is achieved with hydroxylamine and results in the conversion of the thioacetate into a thiol group. The thiol-ligands react with the maleimide-containing liposomes and form a covalent thioether linkage, like schematically depicted in Figure 23.3. After overnight incubation at 4 °C the preparation is ultracentrifuged to separate uncoupled RGD-peptides from RGD-liposomes and the pH of the buffer is adjusted to a physiological value (pH 7.4).

23.2.3. In Vitro Evaluation of the Specificity

The specificity of this RGD-conjugated contrast agent was tested on cultured human endothelial cells (HUVEC) expressing $\alpha_v\beta_3$ and compared with non-specific RAD-liposomes (Mulder et al., 2005). Massive association with HUVEC was observed in case of the RGD-liposomes, while minimal uptake was observed for endothelial cells incubated with RAD-liposomes as revealed with both MRI and confocal laser scanning microscopy (CSLM). Since CSLM allows 3D scanning, it was revealed that the RGD-liposomes were internalized in the cytosol of the endothelial cells.

23.2.4. Molecular Imaging of Tumor Angiogenesis Using RGD Conjugated Bimodal Liposomes

As an in vivo model for tumor angiogenesis, nude mice were inoculated with human LS174T colon carcinoma cells subcutaneously on the right flank and tumors were grown to a size of 500–1000 mm^3. RGD-liposomes were employed to visualize angiogenically activated blood vessels with MRI, while ex vivo fluorescence microscopy on histological sections was used to reveal the mechanism of contrast agent accumulation.

Tumors were localized and identified with T_2-weighted MRI (Figure 23.4A), which also facilitates the drawing of a region of interest for analysis. Next, T_1-weighted MR scans were collected before liposomes were intravenously injected (Figure 23.4B), directly after liposome injection, and 35 (Figure 23.4C), 70, 105, and 140 minutes after injection. Mice that received RGD-liposomes revealed an increase in MR signal intensity of the tumor within 35 minutes, predominantly at the rim of the tumor (Figure 23.4D). As controls, mice were either injected with non-specific RAD-liposomes or a competition assay was carried out. In the latter experiment, the specificity was evaluated by first injecting mice with

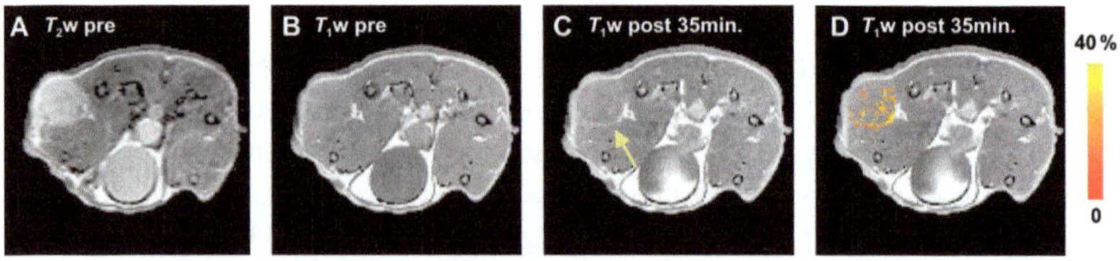

FIGURE 23.4. MRI. MR images of a slice through the tumor of an animal that was injected with paramagnetic RGD-liposomes. (**A**) A T_2-weighted image to visualize the tumor contours. T_1-weighted image (**B**) measured before and 45 minutes after (**C, D**) the injection of the RGD-conjugated liposomes. In (**D**) pixels in the tumor with signal enhancement of at least three times, the noise levels are color coded according to the pseudo-color scale on the *right*. (This figure is adapted from Figure 2 of Mulder et al. (2005), with permission from The Federation of American Societies for Experimental Biology.)

non-paramagnetic RGD-liposomes in order to occupy and block the $\alpha_v\beta_3$-integrin receptors. Thereafter, mice were injected with paramagnetic RGD-liposomes. Signal enhancement was observed in all three cases and appeared comparable for RGD- and RAD-liposomes. The percentages of the tumor volume with signal enhancement were high for RAD-liposomes (22% after 35 minutes) and RGD-liposomes (18% after 35 minutes). In case mice were first injected with non-paramagnetic RGD-liposomes, a fourfold reduction of this number would be observed as compared to mice that were injected with RGD-liposomes only.

Fluorescence microscopy was used to validate the MRI findings. RGD-liposomes were exclusively found to be associated with endothelial cells of tumor blood vessels, indicative for a specific association with $\alpha_v\beta_3$ expressed at the angiogenic endothelium (Figure 23.5A). In contrast, after RAD-liposome administration, a diffuse pattern of fluorescence was found around blood vessels within the tumor tissue, indicative of extravasation of the liposomes (Figure 23.5B). Fluorescence microscopy findings on histological slices of tumors from mice

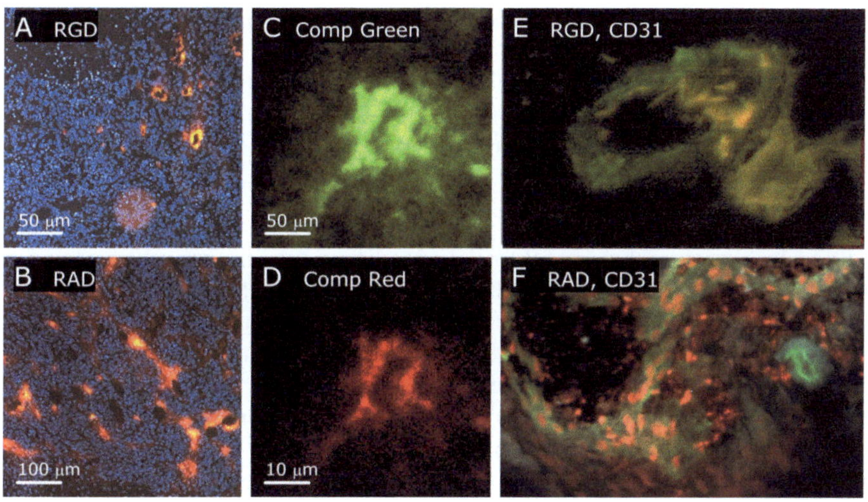

FIGURE 23.5. Fluorescence microscopy of histological sections. Fluorescence microscopy of 4′, 6-diamidino-2-phenylindole (DAPI) colored 10-μm sections from dissected tumors of mice that were injected with RGD-liposomes (**A**) or RAD-liposomes (**B**). Fluorescence microscopy of a section from a tumor of a mouse that was first injected with non-paramagnetic fluorescein RGD-liposomes (**C**), followed by an injection of paramagnetic rhodamine RGD-liposomes (**D**). (**E, F**) Endothelial cell and liposome co-registration. Vessel staining was done with an endothelial cell specific CD31 antibody (green). The red fluorescence represents the RGD-liposomes (**E**). RGD-liposomes were exclusively found within the vessel lumen or associated with vessel endothelial cells, indicative of a specific association with $\alpha v\beta 3$ expressed at the angiogenic endothelium. RAD-liposomes (**F**) were also found outside blood vessels within the tumor, indicative of extravasation of these liposomes. (This figure is adapted from Figure 6 and 7 of Mulder et al. (2005), with permission from The Federation of American Societies for Experimental Biology.)

that were first injected with non-paramagnetic fluorescein RGD-liposomes before injection of the paramagnetic rhodamine RGD-liposomes revealed similar fluorescence distribution patterns, indicating that both liposomes target the vessel wall (Figure 23.5C and D). The difference between RGD- and RAD-liposome localization was further validated by immunohistochemical detection of blood vessels using the endothelial cell–specific CD31 antibody (Figure 23.5E and F). In case of the RGD-liposomes, the liposomes were exclusively associated with endothelial cells, while RAD-liposomes were also found in the extravascular compartment, not associated with endothelial cells

23.2.5. The Evaluation of Angiostatic Therapy Using Molecular MRI

Non-invasive diagnostic imaging methods to establish the efficacy of angiostatic therapies are becoming increasingly important with the first FDA approvals of such agents. The above-mentioned $\alpha_v\beta_3$-targeted bimodal liposomes were applied to quantitate the therapeutic efficacy of the angiogenesis inhibitors anginex and endostatin in a mouse tumor model. To quantitatively assess the angiogenic activity in the tumors, as well as the effects of angiostatic agents, the signal enhancement observed in the MR images needs to be analyzed in an objective way. A custom image analysis procedure was developed in Mathematica (Wolfram Research, IL) for this purpose.

23.2.5.1. Data Analysis

The analysis starts with a careful spatial co-registration of the pre-contrast images with the post-contrast images to allow for a pixel-to-pixel comparison. Since in the studies described here the post-contrast images are made without removing the mice from their animal bed inside the MRI scanner, usually very little change in the position of the animal over time after injection of the contrast agent is observed. In case corrections do prove necessary, a simple two-dimensional translation of one or two pixels suffices. This translation can be done unbiased fully automatically by considering the feature space of the two images that need to be co-registered. The feature space of two images is the combinations of gray values in each of the two images for all the corresponding spatial coordinates, as illustrated in Figure 23.6A. In the ideal situation, when two images are perfectly matched, all the corresponding anatomical structures overlap and the feature space is a diagonal line. In case the images are shifted with respect to each other, the feature space shows a dispersion, which can be quantified by using the Shannon entropy. The transformation that minimizes the feature space entropy results in the best co-registration of the two images. More complex transformations, such as rotation or image warping, which may be needed when pre- and post-contrast images are not made in a single session, can also be applied.

After the co-registration, the tumors are manually segmented by drawing an ROI around the tumor area for all the slices of a T_2-weighted pre-contrast scan

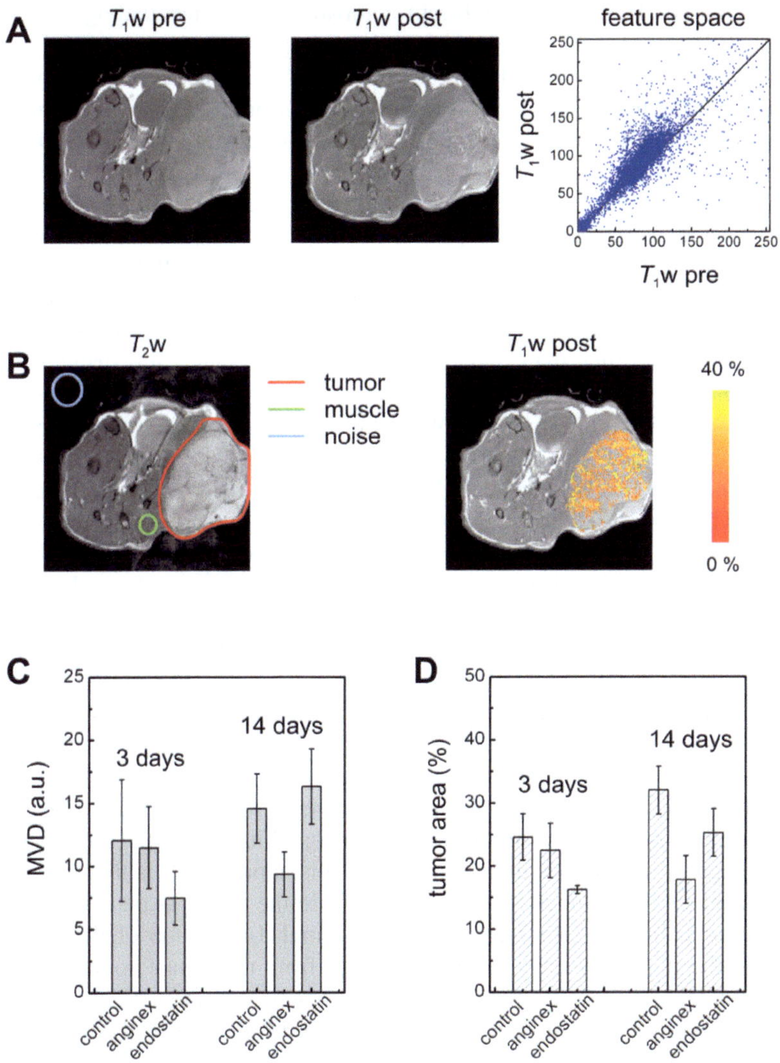

FIGURE 23.6. Data analysis of MR images for the evaluation of angiostatic therapy. (**A**) T_1-weighted MR images of a transversal slice through a mouse with a subcutaneously implanted tumor pre- and 45 minutes post-contrast agent injection. The feature space represents the combinations of gray values in each of the two images for all the corresponding spatial coordinates which can be minimized in order to co-register both images. (**B**) T_2-weighted MR image of the same tumor with three ROIs drawn around the tumor, in the muscle, and in a portion of the image that contains only noise, respectively. The right-hand side shows a T_1-weighted image with significantly enhanced pixels within the tumor ROI, color coded according to the color scale. (**C**) Comparison of histologically determined micro-vessel density (MVD) with (**D**) significantly enhanced tumor area using $\alpha_v\beta_3$-targeted molecular MRI. The mice received angiogenic treatment with anginex and endostatin for 3 or 14 days.

(Figure 23.6B). Tumor tissue is bright compared to the surrounding muscle on T_2-weighted images which makes it easy to recognize the complete tumor. The ROIs around the tumor are drawn somewhat larger than the visible boundary of the tumor to make sure that no tumor tissue is omitted. This is no problem for further analysis since the surrounding muscle does not show any significant increase in intensity post-contrast. The reason for not analyzing the whole slice is that in many of the post-contrast T_1-weighted images bladder and large veins show considerable increase in signal intensity, which should not be analyzed together with the intensity increase in the tumor. A second ROI in each slice of the T_1-weighted images is drawn inside muscle near the tumor for comparison. Finally, a third ROI is placed in a portion of each slice that contains no signal or artifact to determine the standard deviation of the Gaussian-distributed noise level in the images.

Pixel intensities in the tumor ROIs of the T_1-weigthed images after injection of the contrast agent are compared to the baseline pre-contrast scans on a pixel-to-pixel basis. A pixel is considered significantly enhanced when its intensity is increased by at least five times the standard deviation of the noise level, resulting in a confidence of 99.8% that the pixels are statistically significantly enhanced.

The mean intensity increase of the significantly enhanced pixels may be analyzed and compared between different groups of animals or contrast agents used. There are problems with this approach, however. The applied selection criterion that a pixel should be at least five times enhanced compared to the noise level and the fact that the maximum increase in signal intensity of the MRI signal intensity that can be obtained is finite and levels off at high concentrations of the contrast agent result in average enhancements in signal intensities that are very similar between different groups. Moreover, a specific contrast agent may enhance fewer pixels with similar increase in signal intensity than a non-specific contrast agent, which goes undetected using this approach. It is therefore better to also analyze the number of enhanced pixels in the tumor as percentage of the total tumor volume. Furthermore, also the spatial distribution of the enhanced pixels in the tumor, in terms of their distance from the tumor boundaries, may be considered. The results for the different groups can then be averaged. All averages are reported as mean $\pm$ standard error of mean. Statistical comparison of the means is performed using student's t-tests. $P < 0.05$ is considered statistically significant.

23.2.5.2. Quantitation of the Effects of Angiostatic with Molecular MRI and the Correlation with Microvessel Density

Thirty 6-week-old C57BL/6 mice were inoculated with 1×10^5 B16F10 melanoma cells subcutaneously on the right flank. Mice were treated either 3 or 14 days with the angiogenesis inhibitors anginex or endostatin using subcutaneously placed osmotic minipumps. The percentage of the tumor area with significant signal enhancement, upon intravenous injection of the contrast agent, was quantified according to the method described above. After the MRI measurements, the tumors were dissected and frozen. Microvessel density, which is as an ex vivo surrogate marker for angiogenic activity, was determined by counting vessels in five random tumor areas.

The ex vivo determined MVD of the 3 days treatment group and 14 days treatment group (Figure 23.6C) were compared to the in vivo MRI data (Figure 23.6D). Overall, the trend was found that in vivo molecular MRI of tumor angiogenesis reflected closely the treatment effects as deduced from ex vivo MVD determinations.

This methodology allows the rapid evaluation of the efficacy of angiostatic therapy by quantifying angiogenesis with molecular MRI in vivo. Molecular MRI may, thus, serve as a useful surrogate marker of angiogenic activity and as such be used to stage the disease, to facilitate treatment planning, and to predict survival chances.

23.3. Paramagnetic QD-Micelles

23.3.1. Semiconductor Nanocrystals

Semiconductor nanocrystals (Quantum Dots, QDs) have unique optical properties, which make them ideally suited for a number of applications in biomedical imaging, with several important advantages over fluorescent dye-molecules (Medintz et al., 2005). In contrast to fluorescent dyes, the absorption spectrum of a QD is characterized by a very broad band, since any photon with energy equal to or higher than the band-gap is absorbed, while the emission spectrum is rather narrow (full width at half maximum: 25–40 nm). Due to quantum confinement effects, the band-gap of a QD increases upon decreasing its diameter, reaching the bulk band-gap value for sizes larger than the so-called "exciton Bohr diameter" (a typical dimension describing the spatial extension of free charge carriers in semiconductors). The optical properties of QDs can thus be tuned by a judicious control of composition and size, allowing emission wavelengths spanning from the near-UV to near-infrared (NIR) to be obtained, making QDs particularly suitable for multiplexed imaging. For the visible range CdSe, CdTe, or InP can be used, whereas the NIR can be easily covered by PbSe or PbS nanocrystals. QDs are also more stable, brighter, and better suited for time-gated detection than fluorescent dyes. The radiative lifetime of QDs (viz. 20 ns to 1 μs) is longer than that of intrinsic tissue fluorophores (and fluorescent dyes), allowing the auto fluorescence background to be greatly suppressed by using time-gated detection. Moreover, the surface of QDs can be easily modified, so that new functionalities and properties can be conferred. However, QDs are potentially cytotoxic if allowed to interact with the cellular environment and are prone to photochemical degradation, albeit to a lesser extent than dyes. These shortcomings have been (partially) overcome by the use of core/shell QDs and/or suitable coatings, as will be discussed below.

High-quality QDs are usually prepared by a chemical synthesis using suitable precursors (e.g., Trioctylphosphine-Se and $Cd(CH_3)_2$ or cadmium oleate for CdSe QDs) in a high-boiling coordinating solvent (e.g., phosphineoxides, alkylamines, or alternatively a non-coordinating solvent containing a coordinating ligand, such as fatty acids in octadecene). By injecting a cold mixture of the precursors into the hot (180–320 °C) coordinating solvent, nucleation and subsequent growth of the nanocrystals takes place (Donega et al., 2005) (Figure 23.7A). In general, the organically capped QDs, thus, obtained have a poor stability under ambient conditions because the organic ligands (e.g., hexadecylamine, trioctylphosphine, or oleic acid) surrounding the QDs do not prevent photo-oxidation of the surface. The (photo)stability can be greatly improved by over-coating the QDs with a shell of a wider band-gap semiconductor (e.g., CdSe/ZnS), yielding the so-called "core-shell QDs" (Donega et al., 2005; Xie et al., 2005). An outer ZnS shell has also the added benefit of lower cytotoxicity. However, the large lattice mismatch between CdSe and ZnS (ca. 12%) induces

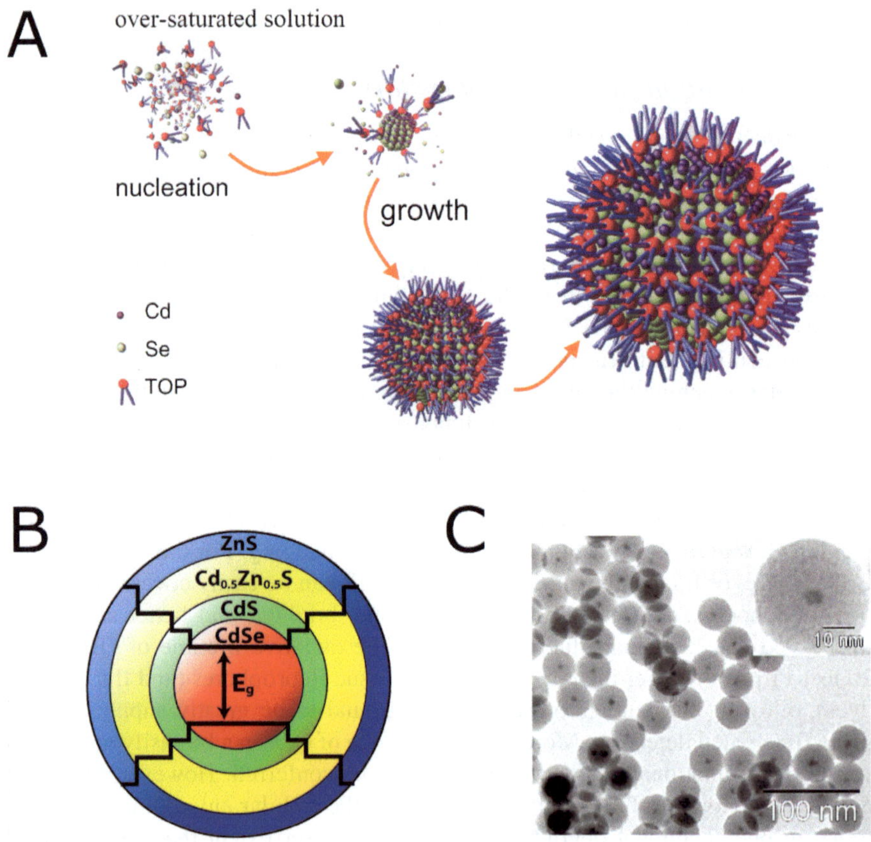

FIGURE 23.7. Quantum dots. **A** Cartoon of the nanocrystal synthesis. **B** Schematic representation of the composition of multi-shell CdSe QDs. The bandgap (Eg) of the different materials is indicated by solid lines and is increasing from inside to outside. **C** TEM-image of silica-coated CdSe QDs with a final size of ∼35 nm. Inset shows a magnification, with the dark dot (diameter ∼5 nm) representing the QD core.

strain at the core/shell interface. To achieve high photostability and negligible toxicity, thick shells are needed (~6 monolayers ZnS), but the accumulated interfacial strain will induce defects, negatively affecting both the quantum efficiency (QE) and the stability of the resulting core/shell QDs (Donega et al., 2005; Xie et al., 2005). Recently, Xie et al. (2005) reported the synthesis of multi-shell QDs (Figure 23.7B) consisting of a CdSe core (3–4 nm) and several successive shells of CdS (3 monolayers), (Cd0.5Zn0.5)S (3.5 monolayers), and ZnS (2 monolayers), using the so-called "SILAR" (successive ion layer adsorption and reaction) method. The multi-shell strategy minimizes interfacial strain in core/shell systems by using intermediate shells with smaller lattice mismatches between the CdSe core and the outermost ZnS shell. The reduced strain between the different materials allows the synthesis of thicker and more robust shells, which more effectively shield the QD core from the environment, thereby inhibiting both the QD photo-oxidation and the interaction with the cellular environment. The resulting multi-shell QDs with a diameter up to 10 nm are extremely air-stable under UV-illumination (QE of 70–5%) and can be made biocompatible by surface functionalization.

A second important improvement in the biocompatibility of QDs is the incorporation of QDs into silica particles. The silica coating of the QDs avoids their exposure to the (biological) environment, making them non-cytotoxic. Moreover, the silica surface is easy to functionalize for biological purposes (Roy et al., 2005) and the particles are readily soluble in water. Figure 23.7C shows the CdSe/multi-shell QDs discussed above incorporated into highly monodisperse (<5% size dispersion) 35 nm silica spheres, following an adaptation of a method developed by Darbandi et al. (2005). The size of the silica spheres can easily be varied from 20 nm up to any desired size, and the number of QDs per silica sphere can be controlled.

A particularly interesting future application is envisaged for (silica coated) PbSe QDs as optical probes for in vivo imaging. The absorption and emission wavelengths of PbSe nanocrystals are tunable in the near-infrared spectral region, for which the transparency of biological tissue is much higher than for visible light. The higher penetration depth of NIR-light emitted by PbSe QDs may facilitate the in vivo investigation of the biological process of interest.

23.3.2. Synthesis of Paramagnetic QD-Micelles

Dubertret et al. (2002) introduced QDs with a micellular coating for in vivo imaging. Water insoluble QDs are usually dissolved in apolar solvents like toluene or chloroform. The hydrophobicity of the QDs is caused by hydrophic capping molecules. A micellar coating can subsequently be applied by simply adding an excess of pegylated lipids to a QD/chloroform solution. Upon evaporation of the solvents a mixed film, containing PEG-lipids and QDs, is formed. Lipid preparations with large quantities of PEG-lipids are known to form micelles in aqueous solutions. Hydrating and heating this film results in the formation of micelles that contain a QD in their core.

Mulder et al. (2006a) extended this procedure by incorporation of a paramagnetic amphiphile in the micellar shell to allow parallel detection of this nanoparticle by both optical methods and MRI. Furthermore, the PEG chains of the micelles can be used for functionalization with targeting ligands. In Figure 23.8 the synthesis of RGD conjugated paramagnetic QD-micelles is schematically depicted. High-quality CdSe/ZnS core/shell QDs are synthesized by injection of precursors (Cd(acetate)2 and Se in trioctylphosphine) into a hot coordinating solvent mixture (trioctylphosphineoxide/Hexadecylamine, TOPO/HDA). The micellar and paramagnetic coating is applied to the QDs in order to make them MR detectable, water soluble, and biocompatible. The coating is comprised of 40 mol% PEG-DSPE (1,2-distearoyl-sn-glycero-3-phosphoethanolamine-N-[methoxy(polyethylene glycol)-2000]) and 50 mol% paramagnetic lipid, Gd-DTPA-BSA (Gd-DTPA-bis(stearylamide). For functionalization of the QDs, 10% Mal-PEG-DSPE (1,2-distearoyl-sn-glycero-3-phosphoethanolamine-N-[maleimide(polyethylene glycol)2000]) is added as well. The lipid mixture is added to the purified QDs in chloroform and the solvents are evaporated gently until a dry film of lipids and QDs is obtained. Thereafter, the lipid film is heated to 70 °C and hydrated with a HEPES buffer (pH 6.7) of the same temperature. This suspension is heated and vigorously stirred until a clear suspension is obtained. Empty micelles are separated from the micelles containing QDs (pQDs) by ultracentrifugation.

Next, the pQDs are conjugated to cyclic RGD to make them specific for activated and angiogenic vascular endothelium. The cyclic 5-mer RGD, synthesized with a thioacetyl group at the lysine residue for coupling, is used. After deacetylation of the peptide the thiol group forms a thioether bond between maleimide-functionalized PEG-lipids, present in the micellular coat of the QDs, and cRGD (Figure 23.8).

The ionic r_1 of the pQDs at a clinically relevant field strength of 1.41 T was more than 12 $mM^{-1}s^{-1}$, which is three times higher than that of Gd-DTPA at this field strength. Since the QD-micelles contain approximately 300 lipids, half of which is Gd-DTPA-BSA, the relaxivity per mM pQD is estimated to be circa 2000 $mM^{-1}s^{-1}$. This high relaxivity makes the pQD contrast agent an attractive candidate for molecular MRI purposes.

23.3.3. *In Vitro Evaluation of Specificity*

To assess the biological specificity of the paramagnetic RGD-conjugated QD-micelles, human endothelial cells (HUVEC) were used as an in vitro test system. Proliferating HUVEC express cell surface receptors, including the $\alpha_\nu\beta_3$-integrin, which are also expressed at angiogenic blood vessels. HUVEC were incubated with the $\alpha_\nu\beta_3$-targeted QD-micelles (RGD-QD-micelles) at 37 °C. As a control, QD-micelles that were not conjugated with RGD were used. In Figure 23.9A, a fluorescence microscopy image of HUVEC that had been incubated with green-emitting RGD-pQDs is depicted. The RGD-pQDs were clearly associated with the cells and were found internalized, at a perinuclear location (inset).

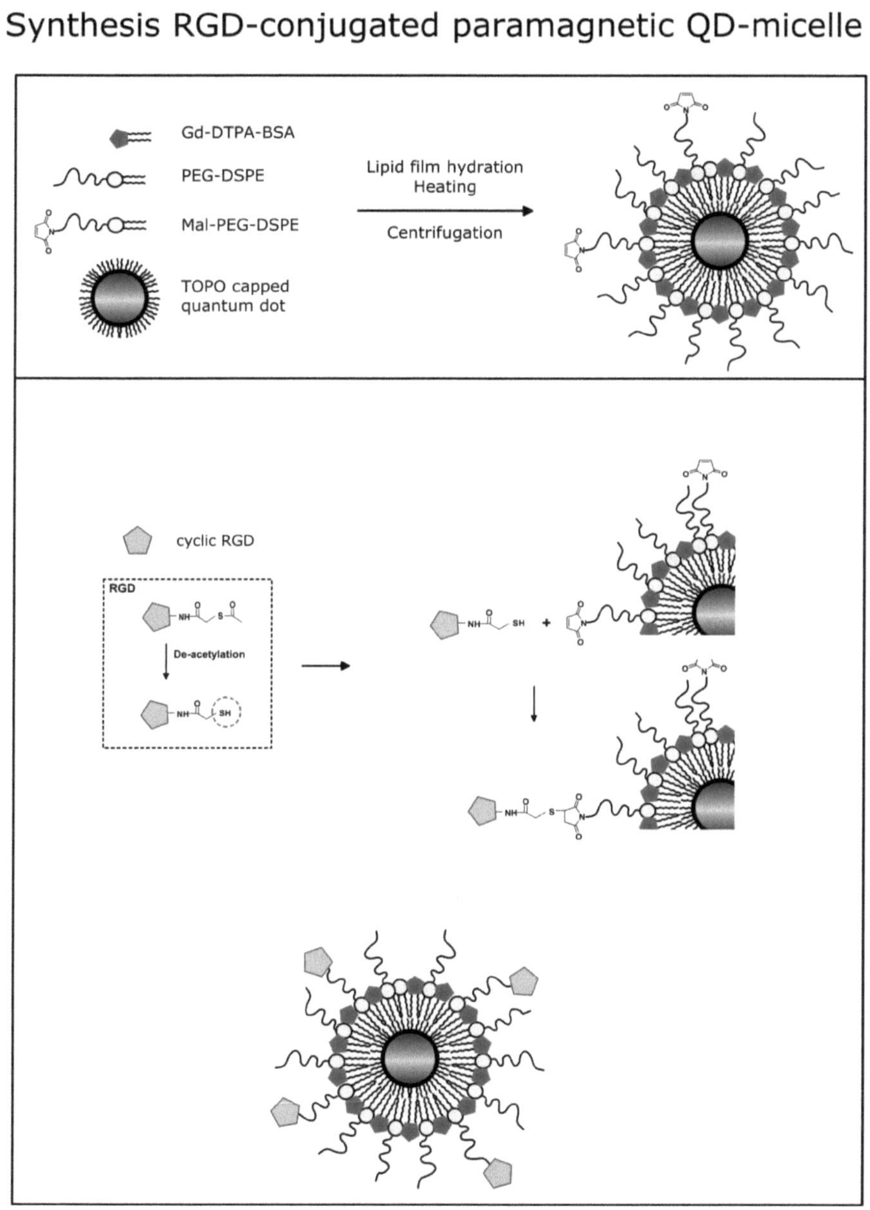

FIGURE 23.8. Synthesis of RGD-conjugated paramagnetic QD-micelles. Schematic representation of the preparation of QDs with a paramagnetic micellular coating (*top*) and the coupling of cyclic RGD to maleimide incorporated in the micellular coating after activation of the peptide (*bottom*).

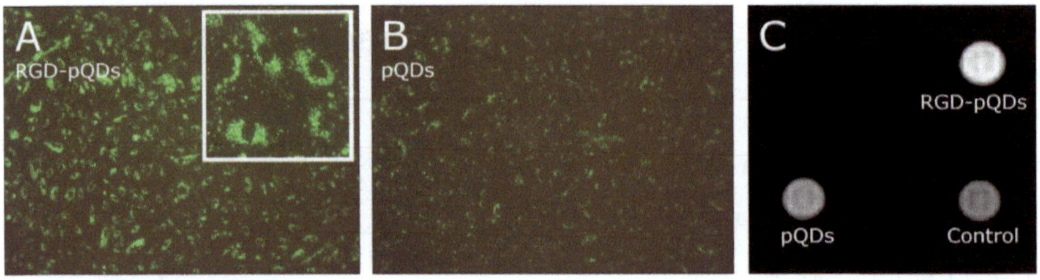

FIGURE 23.9. In vitro targeting and imaging of QD-micelles. Fluorescence microscopy of HUVEC incubated with (**A**) RGD-pQDs and (**B**) bare pQDs. (**C**) T_1-weighted image of cells that were incubated with RGD-pQDs, pQDS, or without contrast agent (control). (Adapted from Figures 5 and 6 of Mulder et al. (2006a), with permission from the American Chemical Society.)

Fluorescence images of HUVEC incubated with bare pQDs showed much less green fluorescence (Figure 23.9B). In addition, MR imaging was performed on preparations of 1.5 million packed cells. In a T_1-weighted image, cells with a high level of association of Gd-containing paramagnetic QDs appear brighter than cells with a low content of Gd or no Gd. The MR image of a cell pellet that was incubated with RGD-conjugated QD-micelles was much brighter than those that were incubated with non-targeted QD-micelles and that were not incubated with contrast agent (Figure 23.9C).

23.3.4. Multimodality Molecular Imaging of Tumor Angiogenesis

The QD-based contrast agent was used to demonstrate molecular imaging of tumor angiogenesis at the cellular and anatomical level. Mice were inoculated with B16 melanoma cells and had a tumor grown to a size of approximately 500 mm^3. The animals were intravenously injected with the contrast agent and were studied with either IVM or MRI in vivo.

23.3.4.1. Intravital Microscopy

Intravital microscopy was used for the real-time monitoring of the fate of injected QDs. At several time points, post-injection fluorescence microscopy was performed at different locations in and around the tumor. Within 5 to 10 minutes, endothelial cells of the tumor blood vessels were labeled with the RGD-pQDS. In Figure 23.10A, a typical bright field image of blood vessels of a tumor is depicted 30 minutes after the injection of contrast agent. The corresponding fluorescence image (Figure 23.10B) revealed labeling of the tumor endothelial cells as indicated by the arrows. Labeling of activated blood vessels by RGD-pQDs was found at a length of 1 cm from the tumor periphery, indicative for a widespread activation of the blood vessels at this site. No endothelial cell labeling of blood vessels in the hind limb of the animals and of blood vessels

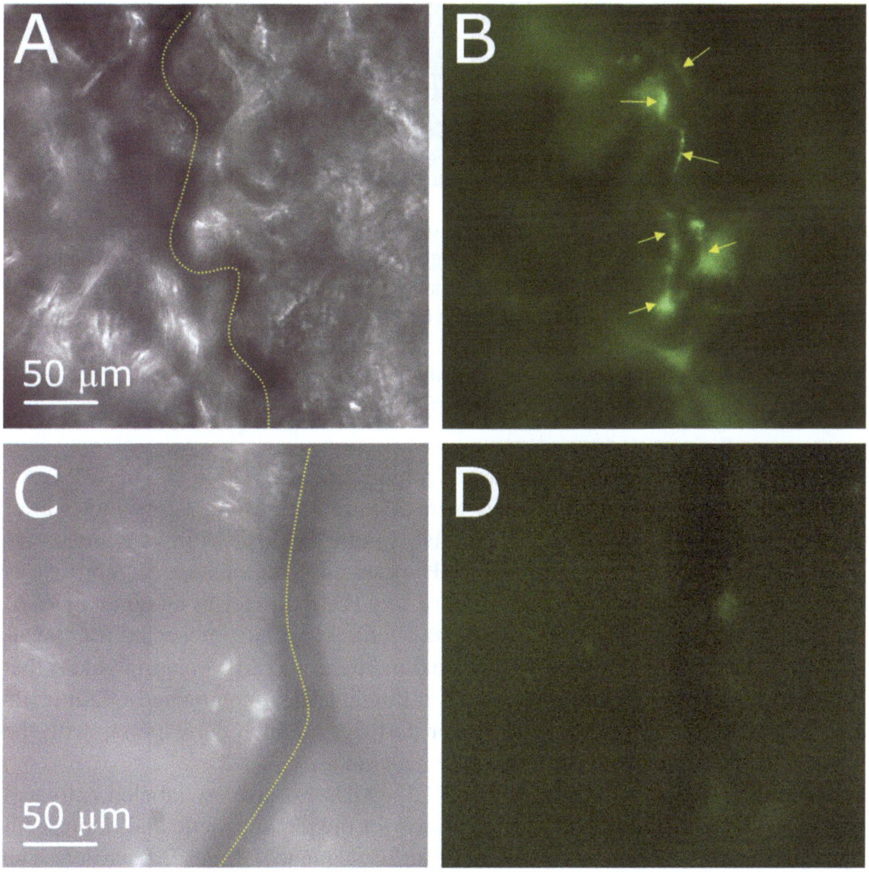

FIGURE 23.10. Intravital microscopy upon injection of RGD-conjugated QD-micelles. Intravital microscopy of microvessels inside (**A**, **B**) and outside (**C**, **D**) a tumor in a mouse after intravenous injection of RGD-pQDs. Brightfield microscopy (**A**, **C**) was used to select blood vessels (indicated by the yellow lines). Fluorescence microscopy (**B**) revealed labeling of endothelial cells in tumor blood vessels of mice that were injected with RGD-pQDs, as indicated by the yellow arrows. Endothelial cells in blood vessels of normal tissues (skin of the ear) in the same mouse showed no accumulation of contrast agent (**C**, **D**).

in the ears of the animals was observed (Figure 23.10C and D). Endothelial cell labeling was not observed for blood vessels in the tumor upon injection of control pQDs that were not conjugated to RGD-peptides (not shown). These data demonstrate the specificity of this contrast agent for the angiogenically activated endothelium.

23.3.4.2. MRI

MRI was also performed on animals that were intravenously injected with the QD-based contrast agent. First, the tumor was localized with T_2-weighted

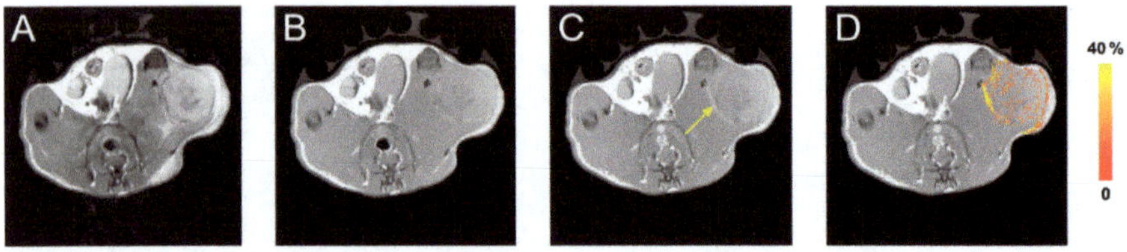

FIGURE 23.11. Molecular MR imaging of ongoing angiogenesis as detected by RGD-conjugated QD-micelles. T_2-weighted image (**A**), collected before the contrast agent was injected, shows the contour of the tumor on the flank. T_1-weighted images (**B–D**) measured before (**B**) and 45 minutes after (**C**, **D**) the injection of the RGD-pQDs. The arrow in (**C**) indicates a bright region at the periphery of the tumor. In (**D**) pixels in the tumor with signal enhancement of at least three times, the noise levels are color coded according to the pseudo-color scale on the *right*.

imaging, in which the tumor appears as a hyper intense region (Figure 23.11A). High-resolution T_1-weighted images were acquired prior to and 30 minutes after contrast agent injection (Figure 23.11B). Figure 23.11C shows an image 45 minutes after contrast agent injection. Hyperintense regions were identified as indicated by the yellow arrows. In Figure 23.11D, the pixel-by-pixel analysis of the images before and after injection of the contrast agent is presented for visualization of the statistically significant signal increase. The MR imaging voxels that showed significant signal enhancement following RGD-conjugated QD-micelle injection was mainly found at the tumor periphery, which corresponds with the regions of the tumor with highest angiogenic activity.

The above study demonstrates that RGD-pQDs exclusively labeled activated blood vessels as has also been demonstrated for the paramagnetic and fluorescent liposomes conjugated to the same peptidic sequence discussed earlier. The extraordinary optical properties of the quantum dot–based contrast agent allowed continuous exposure to excitation, without bleaching. Furthermore, the sensitivity of the technique allowed the survey of different sites in the animal for angiogenic activity. In vivo MRI was demonstrated to be useful for the localization of sites with high angiogenic activity throughout the whole tumor, which was mainly found at the rim of the tumor. In conclusion, the unique possibilities of the paramagnetic quantum dot preparation for parallel optical and MR imaging studies was demonstrated. In addition, the complementarity of in vivo MRI and IVM used, allowed to study the kinetics of the contrast agent targeting, the specificity of the labeling of activated endothelium and the identification of angiogenic sites at the macroscopic level.

23.4. Conclusions and Limitations

The nanoparticles described in this chapter have shown great potential and applicability for studying tumor angiogenesis. Nevertheless, several improvements should be realized, especially for translation of the technology to the clinical setting. The amphiphilic Gd chelate, Gd-DTPA-BSA, incorporated in both nanoparticles has some drawbacks that need to be considered. Firstly, Gd3+ is a toxic ion that needs to be chelated when applied in vivo. The DTPA bisamide lipid has three instead of five carboxylic acids available for complexation with Gd^{3+}. The thermodynamic stability constants of Gd-DTPA and, for example, a DTPA-bisamide–like Gd-DTPA-BMA are $10^{22}$and 10^{17}, respectively. The lower stability of the Gd-DTPA-BSA complex compared to Gd-DTPA might facilitate the release of Gd^{3+} and, thus, increases the toxicity. Furthermore, the exchange of bulk water with paramagnetic gadolinium, which is at the basis of a T_1 shortening effect, is somewhat limited because of its position at the membrane interphase. Improvement of both aspects may be realized by linking DTPA to DSPE using isocyanate-functionalized DTPA building blocks instead of the cyclic dianhydride of DTPA. This results in an improved binding of Gd^{3+} and a better exposure of Gd-DTPA, which most likely will lead to a higher relaxivity of constructs composed of this molecule. The quantum dots used consisted of colloidal nanocrystals composed of CdSe covered by a protective layer of ZnS. Accumulation and degradation of such particles will lead to the release of cadmium-ions and may result in acute and long-term toxicity. Novel and less-toxic QD formulations are under development, which may eventually enable the use of QDs in humans. Another improvement is more stable packing of the QDs in the micellular shell. This may be realized by using amphiphiles that can be polymerized, for example under UV illumination. As discussed previously, another way to reduce toxicity may be achieved by coating QDs with silica.

A major drawback of targeting $\alpha_\nu\beta_3$-integrin with therapeutic or diagnostic nanoparticles is the massive uptake by the spleen and liver. In case of therapeutic nanoparticles, like doxorubicin liposomes for the treatment of cancer, this will result in loading these organs with drugs. Evidently, this may result in unwanted and severe side effects. Although Gd-DTPA is FDA approved and widely used clinically, active targeting of this contrast agent and long residence times may also result in adverse health effects. Especially, nanoparticles with a high payload of Gd are actively targeted and taken up by vital organs like the liver. The ongoing search for novel targets expressed at the angiogenic endothelium may lead to the identification of receptors that can be used to more specifically target angiogenesis.

The application of nanoparticles for therapeutic purposes has been studied extensively. During the past three decades, they have been employed as carriers to deliver drugs, and more recently their application has expanded to gene delivery. Nanoparticle-based molecular imaging may benefit greatly from the drug-targeting field. In addition, the combination of both technologies (therapy and diagnostics) opens exciting opportunities for image-guided delivery of therapeutics, which may be relatively easily realized by combining existing drug-delivery platforms with diagnostic imaging procedures, including MRI (Mulder et al., 2006b).

In conclusion, this chapter described the development and application of two novel nanoparticles, i.e. bimodal liposomes and paramagnetic QD-micelles, for

MR- and optical-based molecular imaging. The ease of preparation, the flexibility, and the biocompatibility of these nanoparticles make them fruitful tools for biomedical imaging purposes. Their application may be extended to other imaging modalities, to other disease processes than angiogenesis, and for drug-targeting purposes. Furthermore, they can serve as an excellent template for the development of any new (molecular) MRI nanoparticle.

References

Aime, S., Dastru, W., Crich, S.G., Gianolio, E., Mainero, V., 2002. Innovative magnetic resonance imaging diagnostic agents based on paramagnetic Gd(III) complexes, Biopolymers 66,6, 419–428.

Bangham, A.D., Standish, M.M.,Watkins, J.C., 1965. Diffusion of univalent ions across the lamellae of swollen phospholipids J Mol Biol 13, 1, 238–252.

Dafni, H., Israely, T., Bhujwalla, Z.M., Benjamin, L.E., Neeman, M., 2002. Overexpression of vascular endothelial growth factor 165 drives peritumor interstitial convection and induces lymphatic drain: magnetic resonance imaging, confocal microscopy, and histological tracking of triple-labeled albumin, Cancer Res 62, 22, 6731–6739.

Darbandi, M., Thomann, R., Nann, T., 2005. Single quantum dots in silica spheres by microemulsion synthesis, Chemistry of Materials 17, 23, 5720–5725.

de Lussanet, Q.G., Langereis, S., Beets-Tan, R.G., van Genderen, M.H., Griffioen, A.W., van Engelshoven, J.M., and Backes, W.H., 2005. Dynamic contrast-enhanced MR imaging kinetic parameters and molecular weight of dendritic contrast agents in tumor angiogenesis in mice, Radiology 235, 1, 65–72.

Donega, C.D., Liljeroth, P., Vanmaekelbergh, D., 2005. Physicochemical evaluation of the hot-injection method, a synthesis route for monodisperse nanocrystals, Small, 1,12, 1152–1162.

Dubertret, B., Skourides, P., Norris, D.J., Noireaux, V., Brivanlou, A.H., Libchaber, A., 2002. In vivo imaging of quantum dots encapsulated in phospholipid micelles, Science, 298,5599, 1759–1762.

Folkman, J., 1971. Tumor angiogenesis: therapeutic implications, N Engl J Med 285,21, 1182–1186.

Folkman, J., 2003. Angiogenesis inhibitors: a new class of drugs, Cancer Biol Ther 2,4 Suppl 1, S127–S133.

Giancotti, F.G., Ruoslahti, E., 1999. Integrin signaling, Science285,5430, 1028–1032.

Griffioen, A.W., Molema, G., 2000. Angiogenesis: potentials for pharmacologic intervention in the treatment of cancer, cardiovascular diseases, and chronic inflammation, Pharmacol Rev 52,2, 237–268.

Haubner, R., Wester, H.J., Weber, W.A., Mang, C., Ziegler, S.I., Goodman, S.L., Senekowitsch-Schmidtke, R., Kessler, H., Schwaiger, M., 2001. Noninvasive imaging of alpha(v)beta3 integrin expression using 18F-labeled RGD-containing glycopeptide and positron emission tomography, Cancer Res 61,5, 1781–1785.

Hood, J.D., Bednarski, M., Frausto, R., Guccione, S., Reisfeld, R.A., Xiang, R., Cheresh, D.A., 2002. Tumor regression by targeted gene delivery to the neovasculature, Science 296,5577, 2404–2407.

Huber, M.M., Staubli, A.B., Kustedjo, K., Gray, M.H., Shih, J., Fraser, S.E., Jacobs, R.E., Meade, T.J., 1998. Fluorescently detectable magnetic resonance imaging agents, Bioconjug Chem 9,2, 242–249.

Jain, R.K., Munn, L.L., Fukumura, D., 2002. Dissecting tumour pathophysiology using intravital microscopy, Nat Rev Cancer 2, 4, 266–276.
Kracht, L.W., Friese, M., Herholz, K., Schroeder, R., Bauer, B., Jacobs, A., Heiss, W.D., 2003. Methyl-[11C]- l-methionine uptake as measured by positron emission tomography correlates to microvessel density in patients with glioma, Eur J Nucl Med Mol Imaging 30,6, 868–873.
Lewin, M., Carlesso, N., Tung, C.H., Tang, X.W., Cory, D., Scadden, D.T., Weissleder, R., 2000. Tat peptide-derivatized magnetic nanoparticles allow in vivo tracking and recovery of progenitor cells, Nat Biotechnol 18,4, 410–414.
McDonald, D.M., Choyke, P.L., 2003. Imaging of angiogenesis: from microscope to clinic, Nat Med, 9, 6, 713–725.
Medintz, I.L., Uyeda, H.T., Goldman, E.R., Mattoussi, H., 2005. Quantum dot bioconjugates for imaging, labelling and sensing, Nat Mater 4,6, 435–446.
Michalet, X., Pinaud, F.F., Bentolila, L.A., Tsay, J.M., Doose, S., Li, J.J., Sundaresan, G., Wu, A.M., Gambhir, S.S., Weiss, S., 2005. Quantum dots for live cells, in vivo imaging, and diagnostics, Science 307, 5709, 538–544.
Mulder, W.J., Koole, R., Brandwijk, R.J., Storm, G., Chin, P.T., Strijkers, G.J., de Mello, D.C., Nicolay, K., Griffioen, A.W., 2006a. Quantum dots with a paramagnetic coating as a bimodal molecular imaging probe, Nano Lett, 6, 1, 1–6.
Mulder, W.J., Strijkers, G.J., Griffioen, A.W., van Bloois, L., Molema, G., Storm, G., Koning, G.A., Nicolay, K., 2004. A liposomal system for contrast-enhanced magnetic resonance imaging of molecular targets, Bioconjug Chem, 15, 4, 799–806.
Mulder, W.J., Strijkers, G.J., Habets, J.W., Bleeker, E.J., van der Schaft, D.W., Storm, G., Koning, G.A., Griffioen, A.W., Nicolay, K., 2005. MR molecular imaging and fluorescence microscopy for identification of activated tumor endothelium using a bimodal lipidic nanoparticle, FASEB J, 19,14, 2008–2010.
Mulder, W.J., Strijkers, G.J., van Tilborg, G.A., Griffioen, A.W., Nicolay, K., 2006b. Lipid-based nanoparticles for contrast-enhanced MRI and molecular imaging, NMR Biomed, 19,1, 142–164.
Roy, I., Ohulchanskyy, T.Y., Bharali, D.J., Pudavar, H.E., Mistretta, R.A., Kaur, N., Prasad, P.N., 2005. Optical tracking of organically modified silica nanoparticles as DNA carriers: a nonviral, nanomedicine approach for gene delivery, Proc Natl Acad Sci U S A., 102,2, 279–284.
Sipkins, D.A., Cheresh, D.A., Kazemi, M.R., Nevin, L.M., Bednarski, M.D., Li, K.C., 1998. Detection of tumor angiogenesis in vivo by alphaVbeta3-targeted magnetic resonance imaging, Nat Med, 4,5, 623–626.
Stroh, M., Zimmer, J.P., Duda, D.G., Levchenko, T.S., Cohen, K.S., Brown, E.B., Scadden, D.T., Torchilin, V.P., Bawendi, M.G., Fukumura, D., Jain, R.K., 2005. Quantum dots spectrally distinguish multiple species within the tumor milieu in vivo, Nat Med, 11,6, 678–682.
Torchilin, V.P., 2005. Recent advances with liposomes as pharmaceutical carriers, Nat Rev Drug Discov, 4,2, 145–160.
van Beijnum, J.R., Griffioen, A.W., 2005. In silico analysis of angiogenesis associated gene expression identifies angiogenic stage related profiles, Biochim Biophys Acta, 1755,2, 121–134.
Weissleder, R., 2002. Scaling down imaging: molecular mapping of cancer in mice, Nat Rev Cancer, 2,1, 11–18.
Weissleder, R., Mahmood, U., 2001. Molecular imaging, Radiology, 219,2, 316–333.

Winter, P.M., Caruthers, S.D., Kassner, A., Harris, T.D., Chinen, L.K., Allen, J.S., Lacy, E.K., Zhang, H., Robertson, J.D., Wickline, S.A., Lanza, G.M., 2003. Molecular imaging of angiogenesis in nascent Vx2 rabbit tumors using a novel alpha(nu)beta3-targeted nanoparticle and 1.5 tesla magnetic resonance imaging, Cancer Res, 63,18, 5838–5843.

Xie, R., Kolb, U., Li, J., Basche, T., Mews, A., 2005. Synthesis and characterization of highly luminescent CdSe-core CdS/Zn0.5Cd0.5S/ZnS multishell nanocrystals, J Am Chem Soc, 127, 20, 7480–7488.

Index

MIX
Papier aus verantwortungsvollen Quellen
Paper from responsible sources
FSC® C105338

If you have any concerns about our products,
you can contact us on
ProductSafety@springernature.com

In case Publisher is established outside the EU,
the EU authorized representative is:
Springer Nature Customer Service Center GmbH
Europaplatz 3, 69115 Heidelberg, Germany

Printed by Libri Plureos GmbH
in Hamburg, Germany